FOREWORD

The National Constitution of the Independent State of Papua New Guinea states:

"We declare our fourth goal to be for Papua New Guinea's natural resources and environment to be conserved and used for the collective benefit of us all, and be replenished for the benefit of future generations."

"WE ACCORDINGLY CALL FOR-

1. wise use to be made of our natural resources and the environment in and on the land or seabed, in the sea, under the land, and in the air, in the interests of our development and in trust for future generations; and

2. the conservation and replenishment, for the benefit of ourselves and posterity, of the environment and its sacred, scenic, and historical qualities; and

3. all the necessary steps to be taken to give adequate protection to our valued birds, animals, fish, insects, plants and trees."

To be able to realise this declaration it is necessary to produce guides for the identification of the Flora of Papua New Guinea. This is to ensure that adequate steps can be taken for its protection. The publication of The Flowering Plants of Papuasia, a guide to the Dicotyledons of Papua New Guinea, Irian Jaya and the Solomon Islands, will assist in these aims. It also reaffirms the commitment of the Papua New Guinea University of Technology to produce publications relevant to the Melanesian region.

A. Tololo

Sir Alkan Tololo KBE, CBE.
Hon. LLD U.P.N.G. Hon D Tech. P.N.G.U.T.

INTRODUCTION

The Flowering Plants of Papuasia is designed as a guide to the diverse flora of the largest tropical island and one of the richest archipelagos in the Tropics. It is intended for use by botanists and other scientists, students requiring an introduction to the flora, and knowledgeable amateurs. The Flora, to be published in nine parts, will include all the dicotyledons known to be indigenous or introduced in Papuasia. It should complement the Keys to the Genera of New Guinea Plants by R. Barker which is at present being edited for publication at the University of Papua New Guinea by Dr. M. Hopkins. The revised publication of the Common Forest Trees of Papua New Guinea by R. J. Johns is also being rewritten to follow the system of classification proposed by A. J. Cronquist, as adopted for the Flowering Plants of Papuasia. Cross referencing with the revised volume on the forest trees will enable the practical forester to gain a better understanding of the diversity of the tree genera in Papuasia.

The Flowering Plants of Papuasia was originally written as a basis for teaching second year Botany in the Forestry Degree course at the Papua New Guinea University of Technology in Lae. Publication of these notes (over 1000 pages of manuscript) was suggested because of the lack of any comprehensive guide to the Flora. Flora Malesiana, the authoritative publication on the Malaysian Flora, provides the base for floristic studies in the region. Unfortunately it is not readily available and also it excludes the Solomon Islands, and initially excluded the Bismarck Archipelago. The geographical area covered in the Flowering Plants includes areas in three countries: Irian Jaya a Province of Indonesia and the Independent Nations of Papua New Guinea and of the Solomon Islands. Together they comprise a distinct floristic region which is called Papuasia. Within Papuasia many areas are still poorly collected and it is expected that in future years many species will be collected which are at present unknown to science, and that the distribution records of many genera and species will change significantly.

Papuasia is divided into six areas. Within the mainland of New Guinea the divisions follow the political regions: these follow the demarkations laid down in 1884 when New Guinea was divided by the Colonial powers into Dutch, British and German New Guinea. The Papuasian region does not include the Santa Cruz Islands which are floristically related to Vanuatu. For the convenience of scientists in the Solomon Islands, notes are included where distinct species are recorded from Santa Cruz.

The distribution pattern of each family and genus within Papuasia is shown by using six small circles, each refering to one of the regions (I - VI) of Papuasia. When the circles are blacked in, the family or genus occurs in that region of Papuasia. Consequently it should be easy for research workers, in (for example) the Solomon Islands, to mark the families occuring

THE FLOWERING PLANTS OF PAPUASIA

DICOTYLEDONS

PART 1 MAGNOLIIDAE

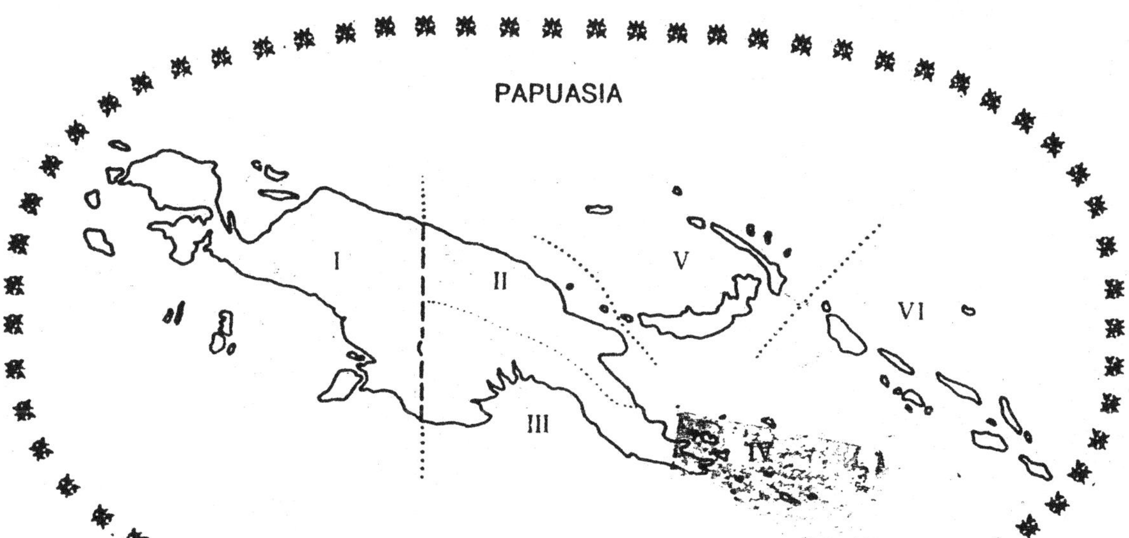

R. J. Johns
FORESTRY DEPARTMENT

1987

P.N.G. UNIVERSITY OF TECHNOLOGY
LAE
PAPUA NEW GUINEA

REGIONS OF PAPUASIA

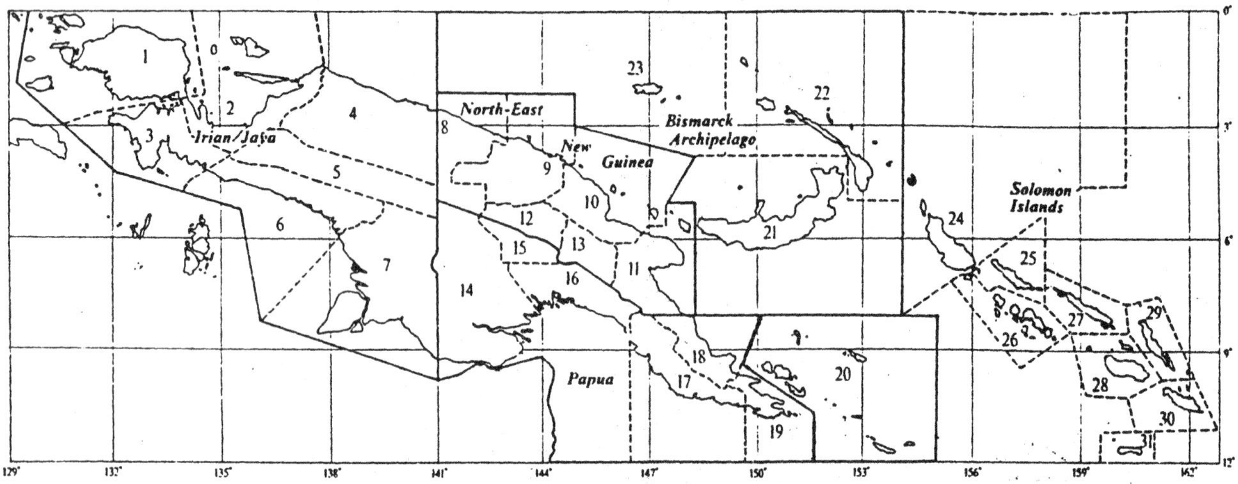

O O O O O O
I II III IV V VI

I. Irian Jaya

1. Vogelkop
2. Geelvink Bay
3. Fakfak
4. Jayapura
5. Snow Mountains
6. Mimika
7. Digul

II. North-east New Guinea

8. West Sepik
9. East Sepik
10. Madang
11. Morobe
12. Western Highlands
13. Eastern Highlands

III. Papua

14. Western
15. Southern Highlands
16. Gulf
17. Central
18. Northern
19. Milne Bay

IV. Papuan Islands

20. Papuan Islands

V. Bismarck Archipelago

21. New Britain
22. New Ireland
23. Manus

VI. Solomon Islands

24. Bougainville
25. Choiseul
26. New Georgia
27. Santa Isabel
28. Guadalcanal
29. Malaita
30. San Cristobal
31. Rennell Group

in that region. As an example the distribution of the genus *Knema* (Myristicaceae), a genus known only from Irian Jaya, is shown by:

● ● ● ● ● ●

In an attempt to reduce printing space the authorities for the species were not included in the draft of Part 1. Following discussions with Dr. H. Hopkins it has been decided to include these in future parts, and in the revised edition of Part 1. In cases where only a *List of Species in Papuasia* is given, the authority, date and page of publication will be listed. Where the publication is not in the *References* the entry will be sufficient for locating the description of the species. Synonyms and basonyms are listed in this manner:

Endiandra brassii Kostermans
(1980: 276)
Endiandra pachycaulis (Synonym)

ACKNOWLEDGEMENTS

During the preparation of these parts many people have provided assistance and encouragement. I wish to express my sincere thanks to them. Mrs M Popeu has typed the keys to *Myristica*, *Horsfieldia*, and *Drimys*. The remainder of the typing of the manuscript, and the layout was done by the author. The numerous typing errors and errors of omission were carefully corrected by Mr A. E. K. Tisseverasinghe and Mrs R. Payet. I am grateful to them for their assistance but accept all responsibility for any errors in this draft. I also wish to thank Mr G. J. Mills who designed the front cover and prepared the base map which is used to show the distribution patterns for each family of plants. The National Herbarium at Lae kindly allowed access to their index cards, illustrations and herbarium records.

The present volume includes 21 families, 105 genera and some 728 species. Several families are in urgent need of revision and the production of keys to the species. These include the Annonaceae, the Lauraceae and the Peperomiaceae. I am grateful to Dr. M. Jebb for giving permission to publish his keys to *Piper* and *Nepenthes*. I would be most grateful to receive any assistance in the form of advice, reprints, keys or maps and illustrations, which could be included in this or subsequent parts. These should be sent to:

R. J. Johns,
Department of Forestry,
P. N. G. University of Technology,
P.O. Private Bag,
LAE,
PAPUA NEW GUINEA.

KEY TO THE SUBCLASSES OF THE DICOTYLEDONS

(Based on Cronquist 1983)

1. Plants relatively archaic, the flowers typically apocarpous, always polypetalous or apetalous, generally with an evident perianth, usually with numerous stamens; seeds very often with a tiny embryo and copious endosperm, but sometimes with a larger embryo and reduced or no endosperm; cotyledons occasionally more than 2. .... MAGNOLIIDAE PART 1

1. Plants more advanced in one or more respects than the Magnoliidae; not more than 2 cotyledons; stamens generally with well defined filament and anther. 2

2. Flowers more or less strongly reduced and often unisexual, the perianth poorly developed or wanting, the flowers often borne in catkins, never with numerous seeds on parietal placentas. .... HAMAMELIDAE PART 2

2. Flowers usually more or less well developed and with an evident perianth, but if not so, then usually either grouped into bisexual pseudanthia or with numerous seeds on parietal placentas (rarely with all of the characters of the Hamamelidae as listed above).... 3

3. Flowers sympetalous (rarely polypetalous or apetalous); stamens generally the same number as the corolla-lobes or fewer, never opposite the lobes; carpels most commonly 2, occasionally 3-5 or more; plants only seldom with tannins.. ASTERIDAE PARTS 8-9

3. Flowers polypetalous or less often apetalous or sympetalous, if sympetalous then usually either with more stamens than corolla-lobes, or with the stamens opposite the corolla-lobes, carpels distinct or more often united to form a compound pistil; plants often with tannins mustard oils............... 4

4. Stamens, when numerous usually initiated in centripetal (seldom centrifugal) sequence; flowers seldom with parietal palcentation (notable exception in Saxifragaceae) and also seldom (except in parasitic species) with free-central or basal palcentation in a unilocular, compound ovary. Very often (especially in species with few stamens) with 2-several locules that have only 1 or 2 ovules in each; flowers polypetalous or less often apetalous, only rarely sympetalous; plants often with tannins... ROSIDAE PARTS 5-7

4. Species with few stamens and axile placentation usually either bearing several or many ovules per locule, or with a sympetalous corolla, or both. Stamens, when numerous, usually initiated in centrifugal sequence; placentation various, but often parietal or free-central or basal, but usually axile. 5

5. Plants with free-central to basal placentation (in a compound ovary), tanniferous only in the two smaller orders; plants most commonly herbaceous or nearly so, the woody species usually with anomalous secondary growth or otherwise anomalous stem-structure; petals distinct or wanting, except in the Plumbaginales; the seeds usually with perisperm instead of endosperm. .... CARYOPHYLLIDAE PART 3

5. Plants placentation only rarely (except in the Primulales) free-central or basal; plants variously woody or herbaceous, many species ordinary trees; petals distinct or less often connate to form a sympetalous corolla, seldom wanting; seeds only seldom with perisperm.... DILLENIIDAE PART 4

The above key is included to assist in identification but characters that are difficult to observe are frequently used and the numerous exceptions are necessarily minimized or ignored.

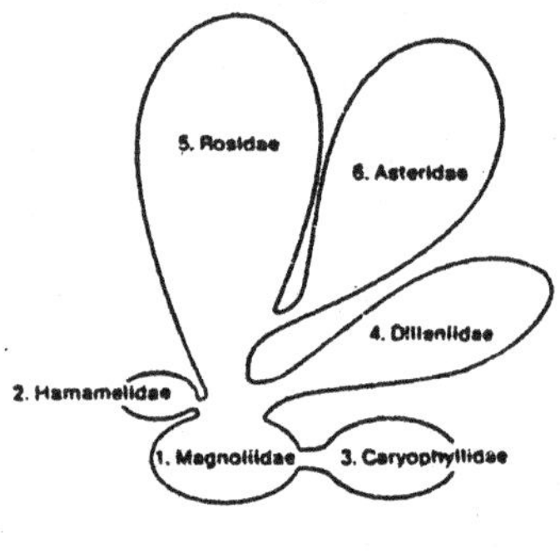

Putative evolutionary relationships among the subclasses of dicotyledons. The size of the balloons is proportional to the number of species in each group. (Source: Cronquist (1981)

SUBCLASS: MAGNOLIIDAE

The Magnoliidae consists of 8 orders, 39 families and some 12,000 species. In Papuasia it is represented by 7 orders (the Papaverales is introduced), 21 families, 105 genera and at least 728 species. The largest orders are the Magnoliales, the Laurales and the Ranunculales.

Most members are woody or herbaceous dicotyledons that have retained one or several primitive features. The most obvious feature, the retentention of numerous stamens, is found in more than half the species. Some species show the dispersal of vascular bundles in the stem, considered characteristic of the monocotyledons.

The largest families in Papuasia are:

	Genera	Species
Annonaceae	32	134
Myristicaceae	5	70
Monimiaceae	9	82
Lauraceae	12	274
Ranunculaceae	3	38
Menispermaceae	19	33

Most other families are represented by only a few genera and species.

Within the Magnoliidae the Magnoliales are clearly the most archaic order. Cronquist (1981) in his book An integrated system of classification of flowering plants gives the following diagram to show the putative evolutionary relationships amongst the orders of the Magnoliidae.

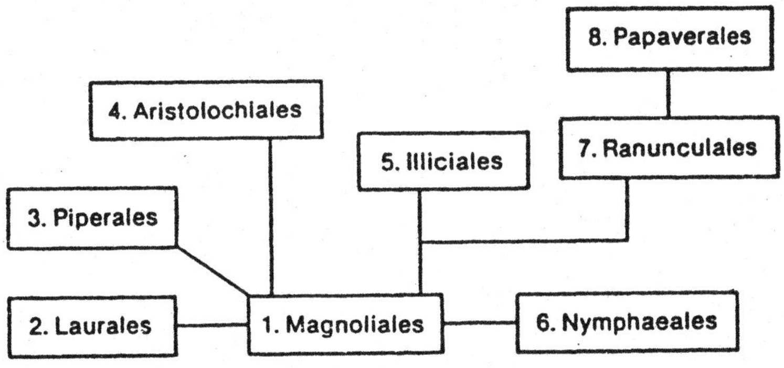

Putative evolutionary relationships among the orders of the Magnoliidae.

LIST OF ORDERS AND FAMILIES IN MAGNOLIIDAE

MAGNOLIALES

Magnoliaceae
Winteraceae
Himantandraceae
Austrobaileyaceae
Canellaceae
Annonaceae
Myristicaceae
Degeneriaceae

ILLICIALES

Illiciales
Schisandraceae

LAURALES

Eupomatiaceae
Lactoridaceae
Gomortegaceae
Trimeniaceae
Spenostemonaceae
Monimiaceae
Calycanthaceae
Chloranthaceae
Lauraceae
Hernandiaceae

PIPERALES

Piperaceae
Peperomiaceae
Saururaceae

ARISTOLOCHIALES

Aristolochiaceae
Nepenthaceae

NYMPHAEALES

Ceratophyllaceae
Nymphaceae

RANUNCULALES

Berberidaceae
Ranunculaceae
Lardizabalaceae
Menispermaceae

PAPAVERALES

Papaveraceae
Fumariaceae

SARRACENIALES

Sarraceniaceae

Annonaceae
Aristolochiaceae
Ceratophyllaceae
Chloranthaceae
Eupomatiaceae
Hernandiaceae
Himantandraceae

Lauraceae
Magnoliaceae
Menispermaceae
Monimiaceae
Myristicaceae
Nepenthaceae
Nymphaceae

Papaveraceae
Piperaceae
Peperomiaceae
Ranunculaceae
Sphenostemonaceae
Trimeniaceae
Winteraceae

The orders and families that are underlined occur in Papuasia.

MAGNOLIALES

Early classification systems for the flowering plants were based primarily on the gross morphology of the plants, the emphasis being placed on their reproductive structures: that is the flowers and fruits. Recent research work on wood anatomy, pollen morphology, chromosome analysis, floral anatomy, comparative biochemistry, seedling development and embryology have given a more complete picture of the relationships of the families. Based on these features most students would consider that the Magnoliales are the most primitive group of the flowering plants.

The families occuring in the Magnoliales are the Magnoliaceae, Winteraceae, Himantandraceae, Canellaceae (not in Papuasia), Annonaceae, Myristicaceae and Degeneriaceae (a monotypic family found only in Fiji). The classification system adopted here follows that used by Heywood (1978) which is based on the system developed by Cronquist (1968). Other authorities have their own systems of classification. Several families such as the Eupomatiaceae (here placed in the Laurales) are placed by some authors in the Magnoliales.

MAGNOLIACEAE

● ● ● ● ● ●

The family consists of some 220 species in 12 genera. Eighty percent occur in the region from S E Asia to Papuasia: the other 20 percent occur in America. The family is easily recognised in Papuasia because of the prominent stipule scar on the twigs.

Diagnostic features

Leaves: alternate, simple, petiolate, with large stipules that leave a characteristic scar around the node.

Flowers: bisexual, usually large and showy, either terminal (at the ends of the branches) or axillary. The peduncle bears bracts which enclose the young flower; these fall when it expands. The perianth is composed of 2 or 3 whorls of free tepals which are petaloid. The stamens are numerous with stout filaments; free and spirally arranged. The anthers have two locules opening by longitudinal slits. Carpels are numerous to few, spirally arranged, free or partly fused.

Seeds: large, with copious endosperm and a minute embryo.

Notes:

1. This is clearly one of the oldest families of the angiosperms. Like other members of the Magnoliales they have their own combination of primitive and advanced characters and cannot be considered directly ancestral to any other family.

MAGNOLIACEAE

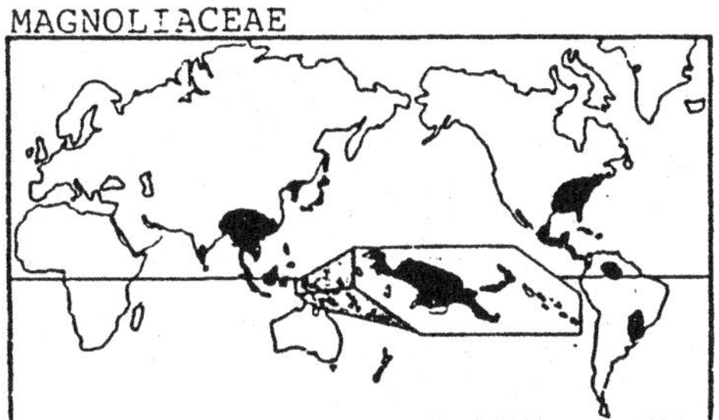

	WORLD	PAPUASIA
Number of genera:	12	2 (1)
Number of species:	220	2 (1)

Distribution: Centered in T and temperate S.E. Asia.

Economic uses: Ornamentals (Magnolia) and timbers Wau beech.

2. Keng (1978) states that Magnolia should be treated as a broader genus to include Talauma, Manglietia and Aromadendron.

References

Croft, J.R. (1978). Handbooks of the Flora of Papua New Guinea. 1: 129 - 134.

Dandy, J.E. (1927). The genera of the Magnolieae. Kew Bull. Art.7: 257 - 264.

Dandy, J.E. (1928). Malayan Magnolieae. Kew Bull. Art.29: 183 - 193.

Hutchinson, J. (1964). The Genera of Flowering Plants. 1: 50 - 57.

Johns, R.J. (1982). Common Forest Trees of Papua New Guinea. 2: 48 - 85.

Keng, H. (1978). The delimitation of the genus Magnolia. Gard. Bull. Sing. 31: 127 - 131.

Nooteboom, H.P. (1985). Notes on Magnoliaceae with a revision of Pachylarnax and Elmerrillia and the Malesian species of Maglietia and Michelia. Blumea 31: 65 - 121.

Royen, P. van (1965). Manual of Forest Trees of Papua New Guinea. Part 6: 7 - 19.

Family in Papuasia

In Papuasia the family is represented by 2 genera, each with a single species. In addition the introduced genus Michellia is potentially widely grown as an avenue and ornamental plant.

KEY TO GENERA IN PAPUASIA

(Source: Dandy 1928; Croft 1978).

1. Inflorescence terminal. Stipules adnate to petioles. Petiole flat with stipule scar extending to base of lamina. 2. Magnolia

1. Inflorescence axillary. Stipules free from petioles. Petiole terete in section, finely grooved above. 2

2. Ovaries stipitate, anthers latrorsely dehiscent. Cultivated. .... 3. Michelia

2. Ovaries sessile, anthers introrsely dehiscent. Common tree in lower montane forest. .... 1. Elmerrillia

1. ELMERRILLIA

A genus of 4 species, throughout Malesia (Nooteboom 1985). One widespread species, E. tsiampacca (L) Dandy occurs from Sumatra and Borneo to mainland New Guinea and New Britain. In New Guinea this species is usually referred to as E. papuana: a synonym.

Large trees without exudate. Twigs terete, densely hirsute to glabrous. Leaves spirally arranged, oblong to elliptic, secondary nerves anastomosing at margin. Stipules large, enveloping buds, usually densely hirsute, free from petioles. Flowers solitary, rarely in pairs, axillary; perianth segments 9 - 17, reflexed in flower. Fruit a syncarp of 1-seeded follicles.

Key to Species in Papuasia

1 widespread species:

Elmerrillia tsiampacca

Notes:

1. Nooteboom (1985) recognised 2 varieties in subspecies tsiampacca from Papuasia:

var. tsiampacca
var. glaberrima

2. Restricted to Papuasia and Moluccas. Not in Bismark Arch. or Solomon Islands.

3. It is locally common in the lowlands and the lower montane zone from sea level to 1800m, where it forms a tree to 60m tall. The timber is marketed as Wau Beech.

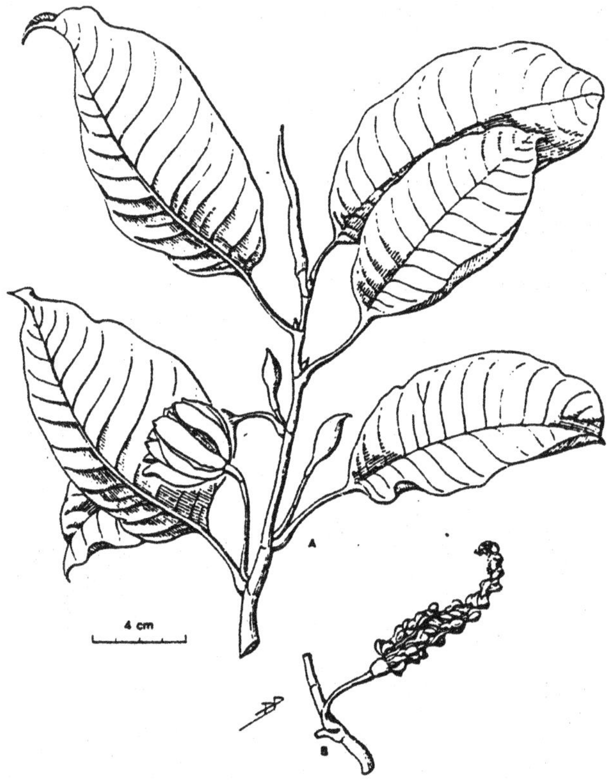

Elmerrillia tsiampaca: (A) flowering leafy shoot; -(B) fruiting twig.
(Source: National Herbarium, Lae)

2. MAGNOLIA

A tropical genus of some 120 species. The New Guinea species Magnolia creadum occurs in the Highlands at altitudes from 1200 to 2700m. In cloud covered areas it can occur as low as 450m. The species occurs from the Moluccas to Aseki in the Morobe Province. Nooteboom (1985) places the Papuasian species in Section Blumiana of the subgenus Talauma which includes some 40 tropical and subtropical species distributed from S E Asia to New Guinea. Croft (1978) places the New Guinea material in the genus Talauma. Keng (1978) proposes that members of the genus Talauma should be treated as species of Magnolia.

Trees or shrubs. Twigs with annular stipular scars. Leaves alternate and spirally arranged, entire, apex and base acute; stipules caducous, adnate to petiole. Flowers solitary, terminal; peduncle with spathulate bracts; perianth segments 9, in threes, outer 3 larger, broadly attached at base; stamens with acuminate connectives; ovary superior. Fruit a syncarp.

Key to Species in Papuasia

1 highlands species:

Magnolia oreadum

Notes:

1. This species is often found in Castanopsis or mixed Lauraceous / Fagaceous forests, swamp forests (1900m) and secondary forests particularly in the Aseki area. At Lake Kopiago it has been collected growing along forest margins.

2. I cannot locate a publication with the recombination from Talauma oreadum Diels placing this species in the genus Magnolia.

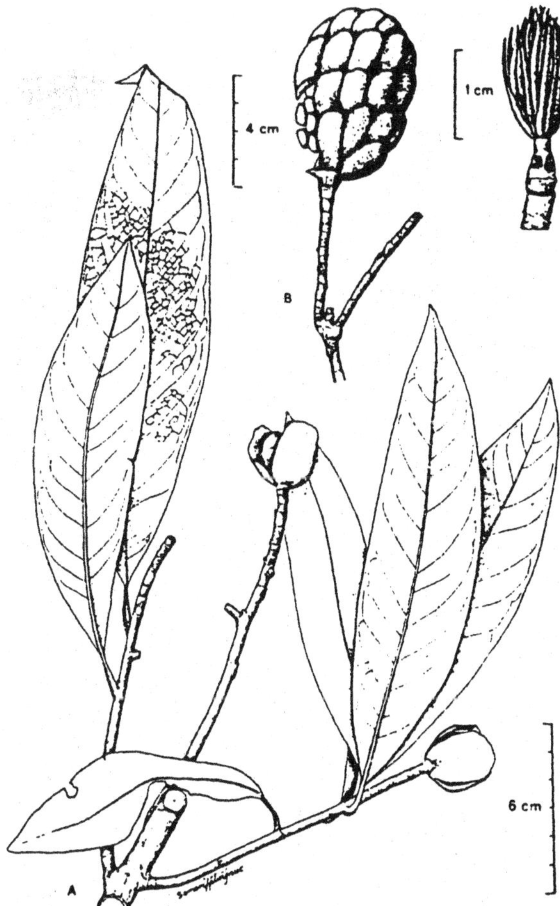

Magnolia oreadum: (A) flowering leafy shoot; (B) ripe fruit; (C) flowers with petals removed.
(Sources: National Herbarium, Lae)

3. MICHELIA

A genus of some 30 species in T Asia and China to Indonesia. It does not occur in the Celebes or Papuasia. M. champaca L. is widely cultivated throughout the tropics for its perfumed flowers. Plants are grown in the National Botanic Gardens, Lae and it has potential as an avenue tree. Other species yield commercial timbers but none are cultivated in Papuasia.

Key to Species in Papuasia

1 introduced species:

Michelia champaca

WINTERACEAE

The small family Winteraceae is of considerable interest to botanists because the wood lacks vessels, a characteristic of the gymnosperms. Representatives in the family are probably the least specialised of all living flowering plants.

Most of the estimated 120 species in the family occur in the wet tropical montane rainforests and cool temperate rain forests fringing the Pacific basin in the southern hemisphere. The family is easily recognised by the glaucous appearance of the underside of the leaves, due to the waxy deposits plugging the stomata.

WINTERACEAE

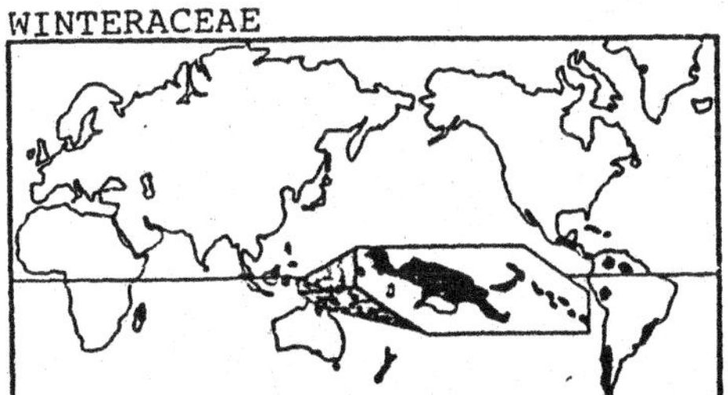

	WORLD	PAPUASIA
Number of genera:	5.7	2
Number of species:	60	C.10

Distribution: Montane to cool temperate forests in areas bordering South Pacific.
Economic uses: Some horticultural and medicinal uses.

Diagnostic features

Leaves: leathery, margins entire, simple, exstipulate.

Flowers: bisexual, or functionally unisexual, with parts mostly separate, spirally arranged and indefinite in number. Stamens are short and broad.

Fruits: indehiscent berries; seeds have a rudimentary embryo and abundant endosperm.

References

Burtt, B.L. (1939). The taxonomic position of *Tetrathalamus*. *Kew Bull.* Art. 62: 458 - 461.

Royen, P. van (1982). *The Alpine Flora of New Guinea*. 2: 1233 - 1249.

Smith, A.C. (1942). Taxonomic notes on Old World Species of Winteraceae. *Journ. Arn. Arb.* 23: 438 - 440.

Smith, A.C. (1943). Taxonomic notes on Old World Species of Winteraceae. *Journ. Arn. Arb.* 24: 119 - 164.

Smith, A.C. (1969). A reconsideration of the genus *Tasmannia*. *Taxon* 18: 286 - 290.

Vink, W. (1970). The Winteraceae of the Old World. I. *Pseudowintera* and *Drimys* - morphology and taxonomy. *Blumea* 23: 225 - 354.

Vink, W. (1977). The Winteraceae of the Old World. II. Zygogynum-Morphology and Taxonomy. *Blumea* 23: 219 - 250.

Vink, W. (1985). The Winteraceae of the Old World. V. *Exospermum* links *Bubbia* to *Zygogynum*. *Blumea* 31: 39 - 55.

Family in Papuasia

In Papuasia the family is represented by 2 genera. *Drimys* is locally common and very diverse. Vink (1970) recognised only 1 species, *Drimys piperita*, in Papuasia: this he subdivided into 39 entities.

Species described in the genus *Belliolum* are treated by Vink (1985) as *Zygogynum*. The genus *Belliolum* was restricted to New Caledonia and to the Solomon Islands, including Bougainville, where it is represented by 4 species. Vink (1985) listed new combinations in *Zygogynum* for species previously described in *Bubbia* and *Belliolum*: but 'for only those names that are expected to survive the current revision'. The genus is represented by some 20 species in Papuasia.

Lauterbach in 1905, described *Tetrathalamus* for the species *T. montanus*. This species was placed in *Bubbia* by Smith and more recently in *Zygogynum* by Vink (1985).

KEY TO GENERA IN PAPUASIA

1. Petals in bud protruding past the the sepals, sepals regular ± cup-like or short and dentate. .... 2. *Zygogynum* (including *Belliolum* and *Bubbia*)

1. Petals in bud covered by sepals, sepals irregularly separating. .... 1. *Drimys* (including *Tasmannia*)

1. DRIMYS

• • • • • •

A single highly variable species *Drimys piperita* Hooker in which Vink (1970) combined the 39 species previously described into a single polymorphic species due to the number of overlapping specimens from different mountains. On an individual mountain the 'entities' are often distinct. van Royen (1982) treated this genus as *Tasmannia* in *The Mountain Flora of New Guinea*. Material of *Drimys* is treated under the variable species *Tasmannia piperita* (Hooker) Miers.

Key to the Entities in Papuasia

(Source: Vink (1970)

1. Flowers always solitary 2

1. Flowers in at least one axil per inflorescence in groups of three or more 27

2. Apex of leaves acuminate........ 3

2. Apex of leaves acute to rounded. 6

3. Shrublet 0.15-0.3 m high. .... 12. *nettoti*

3. Shrub over 1 m high 4

4. Occurring in West New Guinea. .... 35. *oligandra*

4. Occurring in East New Guinea.... 5

5. Petals (0-) 2 (-6). .... 25. *heteromera*

5. Petals 4-11. .... 29. *macrantha*

6. Lower side of leaves densely papillate...... 7. *papillata*

6. Lower side of leaves smooth to tub- 7

7. Petals 4-11.................... 8

7. Petals (0-) 2 (-5)............. 16

8. Leaf blades 5.5-10 x 2-4 cm (Waigeo Island)... 20. *buffelhoorn*

8. Leaf blades 0.5-5 x 0.2-2.3 cm (New Guinea mainland)............... 9

9. Leaves recurvate. .... 13. *fistulosa*

9. Leaves that or margins recurved. 10

10. Petals 12-23 x 5-11 mm. .... 17. *elongata*

10. Petals 2.5-11 x 0.6-5.5 mm...... 11

11. Petals 2-5.5 mm wide. .... 16. *myrtoides*

11. Petals 0.6-2.5 mm wide.......... 12

12. Leaf blades 0.5-1.0 x 0.2-0.5 cm. .... 14. *microphylla*

12. Leaf blades 1.2-5 x 0.5-1.5 cm.. 13

13. Petals 4.5-8 mm long............ 14

13. Petals 2.5-4 mm long............ 15

14. Shrub c. 2m high; leaves coriaceous .... 11. *lamii*

14. Shrub 0.2-0.3 m high; leaves (stiffly) chartaceous. .... 12. *nettoti*

15. Leaf blades 1.2-2.5 cm long, coriaceous; pedicels 8-14 mm... .... 3. *buxifolia*

15. Leaf blades 2.5 - 5 cm long, chartaceous to subcoriaceous; pedicels 15-12 mm. .... 35. *oligandra*

16. Leaf blades 0.15 - 0.2 cm wide. .... 15. *vacciniodes*

16. Leaf blades 0.4 - 4.2 cm wide... 17

17. Bases of leaves all abruptly narrowly rounded. .... 5. *pittosporoides*

17. Bases of leaves at least partly cuneate to obtuse............... 18

18. Seeds orange-brown.............. 19

18. Seeds black..................... 20

19. Petals 5-8 mm long. .... 12. *nettoti*

19. Petals 2.5-4 mm long. .... 35. *oligandra*

20. *In sicco* midrib grooved above; *in vivo* lower side of mature leaves distinctly glaucous. .... 10. *giluwe*

20. *In sicco* midrib flat to slightly crested above; *in vivo* lower side of mature leaves not glaucous... 21

21. Petals 0.7 - 1.3 mm wide. .... 35. *oligandra*

21. Petals 1.5 - 5 mm wide 22

22. Leaf blades 0.6 - 2.1 cm long. .... 4. *versteegii*

22. Leaf blades 1.0-6.5 cm long..... 23

23. Occurring on Mt Wilhelm in open subalpine vegetation. .... 2. *subalpina*

23. Occurring outside Mt Wilhelm, or in shrubberies or forest (margins). 24

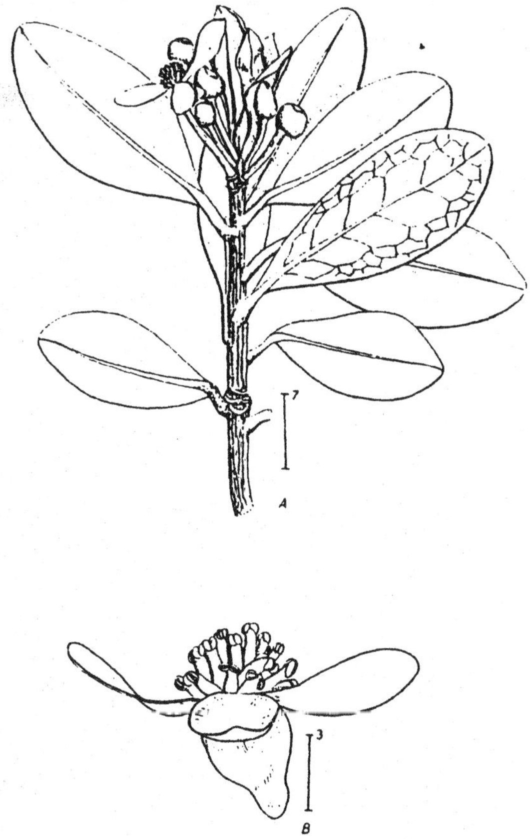

Drimys piperita entity *montis-wilhelmi*: (A) branch with male flowers; (B) male flower.
(Source: van Royen (1982)

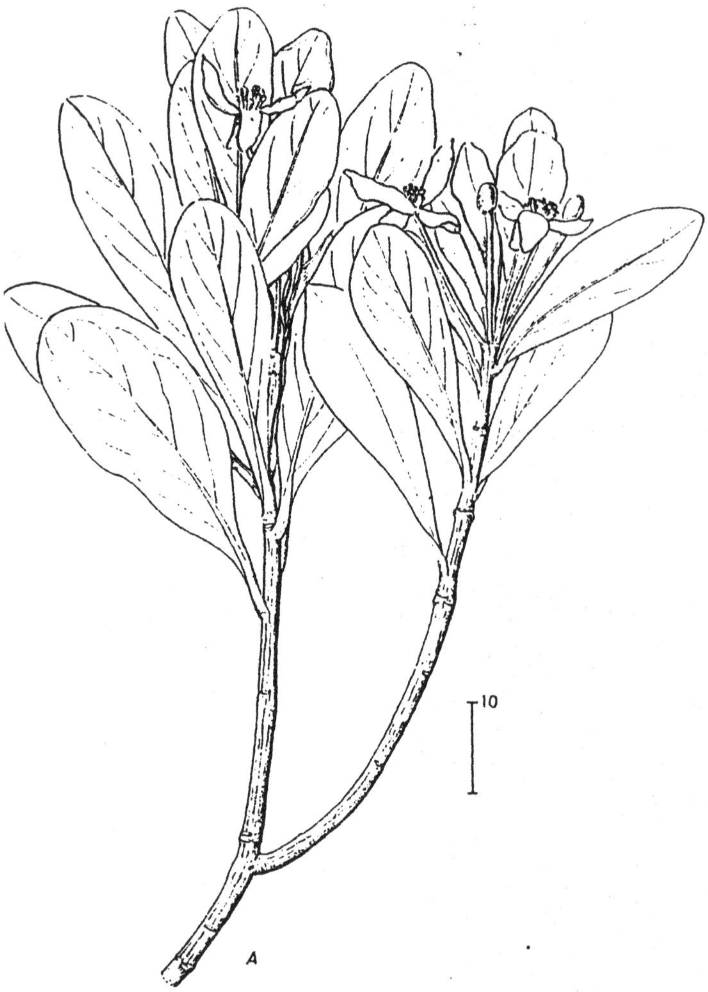

<u>Drimys</u> <u>piperita</u> entity <u>subalpina</u>: (A) branch with male flowers.
(Source: van Royen (1982)

24. Shrub 0.1-1(-3) m high; leaf blades 1.0-3.5 x 0.4-1.2 cm. .... 1. <u>reducta</u>

24. Shrub 1 - 6 m high; leaf blades 1.0 - 6.5 x 0.5 - 4.2 cm........... 25

25. Occurring in Irian Jaya. .... 6. <u>subpittosporoides</u>

25. Occuring in Papua New Guinea ... 26

26. Twigs erect, little branched. .... 8. <u>crassipes</u>

26. Twigs not erect, repeatedly branched......... 9. <u>'montis-wilhelmi</u>

27. Petals (0 -) 2 (- 6) 33

27. Petals (3 or) 4 - 9 28

28. Leaves scattered 31

28. Leaves pseudoverticillate....... 29

29. Leaves 3 - 7 mm petioled; leaf base cuneate to acute. .... 3. <u>verticillata</u>

29. Leaves (sub)sessile; leaf base rounded to auriculate............ 30

30. Leaf base narrowly rounded to minutely auriculate Irian Jaya. .... 31. <u>rosea</u>

30. Leaf base cordate to auriculate. Papua New Guinea. .... 32. <u>cordate</u>

31. Leave (sub)sessile; leaf base narrwly rounded to minutely auriculate .... 31. <u>rosea</u>

31. Leave 2 - 12 mm petioled; leaf base cuneate to obtuse............... 32

32. Leaf blades 1.5 - 4 x 1 - 2.5 cm; occurring on Mt Doorman, Irian Jaya .... 34. <u>pachyphylla</u>

32. Leaf blades 3-10 x 1-3.5 cm; throughout, not known from Mt Doorman. .... 26. <u>polymera</u>

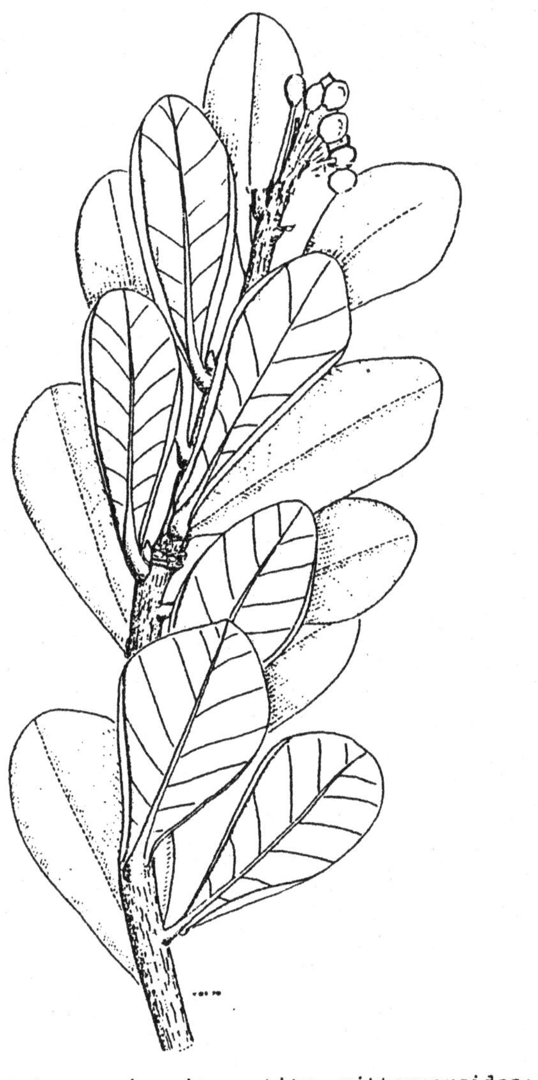

<u>Drimys</u> <u>piperita</u> entity <u>pittosporoides</u>: (A) branch with flowerbuds.
(Source: van Royen (1982)

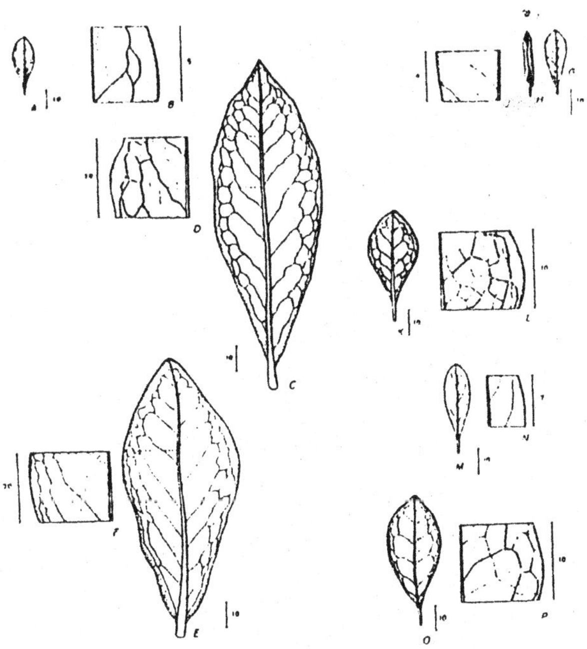

<u>Drimys piperita</u>: (A-B) entity <u>buxifolia</u> (A) leaf; (B) part of leaf; (C-D) entity <u>coriacea</u> (C) leaf; (D) part of leaf; (E-F) entity <u>densifolia</u> (E) leaf; (F) part of leaf; (G-J) entity <u>fistulosa</u> (G) leaf expanded; (H) leaf rolled up; (I) cross section rolled up leaf; (J) part of leaf; (K-L) entity <u>pachyphylla</u> (K) leaf; (L) part of leaf; (M-N) entity <u>reducta</u> (M) leaf; (N) part of leaf; (O-P) entity <u>rubiginsa</u> (O) leaf; (P) part of leaf enlarged.
(Source: van Royen (1982)

33. Leaf base abruptly and narrowly rounded to minutely auriculate. .... 30. <u>densifolia</u>

33. Leaf base cuneate to obtuse, some leaves of a specimen sometimes rounded........................ 34

34. Branchlets stout to very stout; leaves firmly coriaceous, blades 8 - 20 x 2.5 - 6.5 cm; Arfak Mts. .... 21. <u>robusta</u>

34. Branchlets slender to stout; leaves chartaceous to coriaceous, blades 1.5 - 23.5 x 0.5 - 11 cm; throughout New Guinea................. 35

35. Leaf blades 1.5 - 5 cm long, obtuse to rounded. .... 36. <u>rubiginosa</u>

35. Leaf blades 3.5 - 23 cm long; acute to rounded or acuminate to cuspidate........................... 36

36. Occurring in Papua New Guinea. .... 37c, 37e, 37g, 37h.

36. Occurring in Irian Jaya. .... 37a, 37b, 37c, 37d, 37e, 37f, 37g.

37a. Leaf blades 3.5 - 9.0 x 0.9 - 2.6 cm, apex acute to cuspidate; midrib often impressed above; flower buds often acuminate. .... 18. <u>acutifolia</u>

37b. Leaf blades 3.5 - 14 x 1 - 3.5 cm, apex obtuse to acuminate; midrib flat to impressed above; flower buds rounded to acute. .... 28. <u>hatamensis</u>

<u>Drimys piperita</u> entity <u>versteegii</u>: (A) fruiting branch.
(Source: van Royen (1982)

37c. Leaf blades 4 - 10 x 1.5 - 3 cm, apex acute to acuminate; midrib flat to impressed above; flower buds rounded to acute; venation more densely reticulate than in (b) .... 27. reticulata

37d. Leaf blades 4 - 12 x 1 - 4 cm, apex rounded to acuminate; midrib flat to crested above; flower buds rounded....... 19. beccariana

37e. Leaf blades 5.5 - 23.5 x 2 - 11 cm, apex acute to acuminate; midrib flat to crested above; flower buds rounded. .... 22. coriacea

37f. Leaf blades 7 - 16 x 1.5 - 6 cm, apex rounded to acuminate; midrib usually flat above; flower buds usually rounded. .... 23. obovata

37g. Leaf blades 8 - 19 x 2 - 6 cm, apex acuminate; midrib usually flat above; flower buds usually rounded .... 24. chartacea

37h. Leaf blades 3.5 - 18 x 0.5 - 6.5 cm, apex (obtuse to) acuminate; midrib often impressed aobve; flower buds rounded; veryvariable. .... 25. heteromera

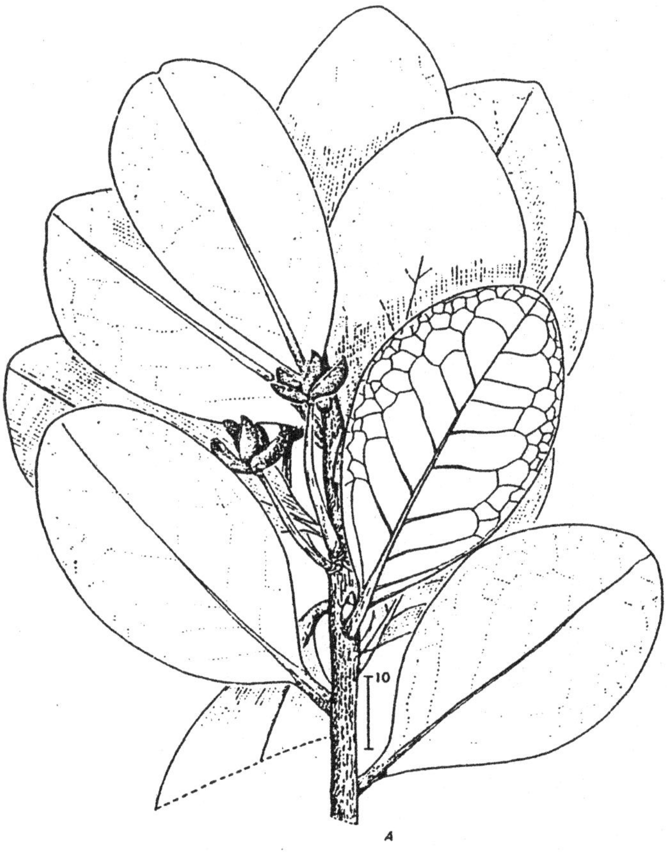

Drimys piperita entity crassipes: (A) fruiting branch.
(Source: van Royen (1982)

2. ZYGOGYNUM
(including BELLIOLUM and BUBBIA)

A genus of 32 species, New Guinea has 19 species 5 of which are at high altitudes in Papuasia (van Royen 1982). The genus is characterised by terminal inflorescences of few to many bisexual flowers, each lateral flower with a single bracteole. The calyx is persistent in the fruit.

Four species were described from the Solomon Islands in the genus Belliolum. Vink (1985) does not list any of these so, presumably, they will not be retained in his current revision. The species listed were: Belliolum burttianum, B. gracile, B. haplopus and B. kajewski.

HIMANTANDRACEAE

● ● ● ● ● ●

The family consists of a single genus, with 1 (or according to some authors 3) species found in N E Australia, New Guinea and the Moluccas. All are aromatic trees.

Diagnostic features

Leaves: covered with numerous minute scales each attached by a central stalk. Leaves simple, alternate and without stipules; densely covered with copper coloured scales.

Flowers: regular, bisexual and solitary. The flower has two, unusually leathery sepals, which form a cap around the bud. Perianth is described as salmon-pink in Galbulimima. Stamens (to 40) are very much like the petals (petaloid) in shape and texture, with the anthers at the edge of the flat blades (golden yellow in colour). The 7-10 superior carpels fuse to form a globose fruit.

Fruit: small, brown, succulent with a solitary flattened seed in each locule.

NOTE: Hutchinson (1970: 163) remarks that the family is 'a small and curious relic, with petaloid stamens' and that it is 'remarkable in having a peltate, scaly indusium'.

Authors classify the family either in the Magnoliales or the Laurales. It is an isolated end line, not closely linked to any other family.

References

Bailey, I.W. Nast, C.G. and Smith, A.C. (1943). The Family Himantandraceae. J. Arn. Arb. 24: 190 - 206.

Bullock, A.A. (1957). Galbulimima versus Himantandra. Kew Bull. : 409.

Croft, J.R. (1978). Handbooks to the Flora of Papua New Guinea. 1: 126 - 128.

Royen, P. van (1959). Himantandraceae. Nova Guinea, Bot. NS. 10: 127 - 135.

Sauer, W. & Ehrendorfer, F. (1970). Chromosomen, verwandtschaft und evolution tropischer Holzpflanzen. II Himantandraceae. Osterr. Bot. Z. 118: 38 - 54.

Smith, A.C. (1942). A nomencultural note on the Himantandraceae. J. Arn. Arb. 23: 366 - 368.

HIMANTANDRACEAE

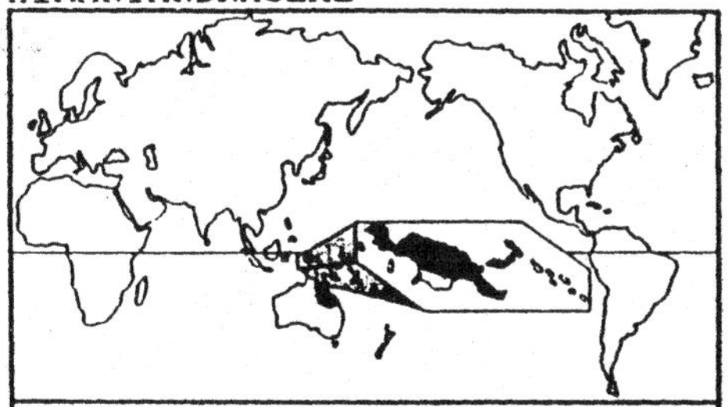

	WORLD	PAPUASIA
Number of genera:	1	1
Number of species:	2-3	1

Distribution: N.E. Australia, New Guinea and Moluccas.

Economic uses: Limited use for timber.

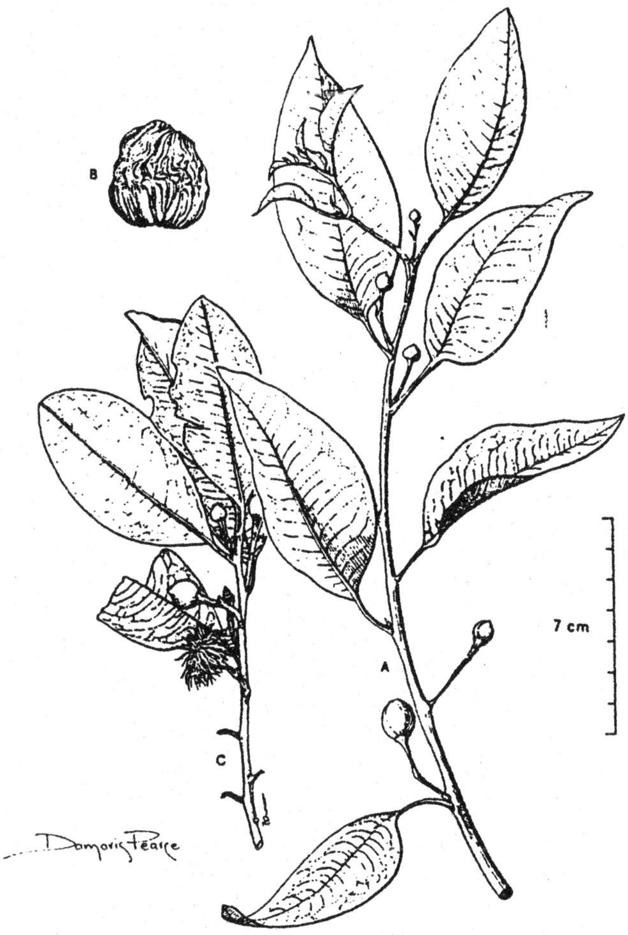

Galbulimima belgraveana: (A) twig with flower buds; (B) fruit; (C) twig with flower and buds.
(Source: National Herbarium, Lae)

Family in Papuasia

In New Guinea a single species Galbulimia belgraveana (F. Muell.) Sprague. It is used as a timber tree where available. Not in Solomon Islands.

KEY TO GENERA IN PAPUASIA

1 genus: 1. Galbulimima

1. GALBULIMIMA

A monotypic genus occuring from Moluccas into Papuasia and N E Australia. Tree to 35m tall with straight bole especially common in Nothofagus forest between the altitudes of 1200 - 2700 m. It however does occur down to sealevel. The bark, which is rich in alkaloids, is chewed to induce hallucinations.

Key to Species in Papuasia

1 widespread species:

Galbulimima belgraveana

(F. Muell.) Sprague

Notes:

1. Galbulimima is widespread throughout the Highlands from 1850 to 2000m altitude. It occurs in Lithocarpus and Castanopsis forest and also in Nothofagus forests.

ANNONACEAE

The family consists of some 130 genera and about 2,300 species of mainly tropical trees and shrubs. About 50 genera occur in Asia (including Oceania), ca. 45 in Africa and c. 35 in America (Okada and Ueda, 1984). Few genera are in common between three or even two of the areas. Most species occur in the Old World Tropics, mostly confined to the moist tropical lowland rain forest, though a few are in montane forests. The family is allied to the Magnoliaceae but differs in having ruminate seeds and having no stipules (exstipulate). The primitive floral morphology contrasts with the advanced wood anatomy in the family.

ANNONACEAE

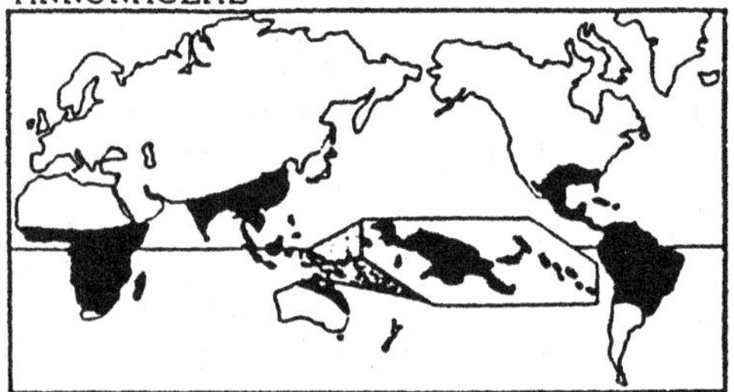

	WORLD	PAPUASIA
Number of genera:	120	32
Number of species:	2100	C.95

Distribution: Pantopical, centered in Old World Tropics.

Economic uses: Cultivated for fruits (soursop) and aromatic oils. Several timber trees.

The family is well defined, but a satisfactory taxonomic division has not yet been established for the Annonaceae, and generic delimitations are still uncertain. The family is notoriously difficult to divide into natural groupings of species. The basic division of the family is into 2 subfamilies: the Uvarieae with imbricate petals and the Unoneae with valvate petals. Walker (1971) has more recently made a study of the pollen and found it is highly distinctive for some genera.

The family is one of the primitive families of the flowering plants, often with an indefinite number of free floral parts and spirally arranged stamens.

Two South American species are widely cultivated in the tropics for their fruits: soursop Annona muricata and custard apple Annona reticulata. The flowers of Canaga odorata contain aromatic oils which are extracted and used as a perfume. Some are logged in New Guinea for timber: particularly Cananga and Polyalthia.

The division into climbing and non-climbing genera is a useful systematic character. The following genera are climbers: 3: Artabotrys; 6: Cyathostemma; 8: Fissistigma; 9: Friesodielsia; 14: Mitrella; and 26: Rauwenhoffia. The two genera 7: Desmos and 12: Melodorum are climbing or straggling shrubs. The remainder of the species are shrubs or trees some of which grow to 20m tall. Also important in classification are the petals. Plants with imbricate petals occur in the Uvarieae; plants with valvate petals are in other tribes.

Diagnostic features

Habit: trees, shrubs or climbers.

Leaves: simple, alternate, 2-ranked, entire, exstipulate.

Flowers: often solitary, actinomorphic, bisexual (more rarely unisexual), sometimes formed on old wood. Perianth parts 9 in 3 rows of 3. Outer row (3) sepaloid; inner two rows (3+3) petaloid. Stamens: many, spirally arranged on receptacle above perianth. Carpels numerous, above stamens on receptacle; ovules numerous.

Fruits: aggregate of berries, in Annona coalesce with an edible, fleshy receptacle. Seeds have ruminate endosperm.

References

Backer, C.A. and Bakhuizen van den Brink, R.C. (1963). Flora of Java. 1: 100 - 116.

Diels, L. (1931). Annonaceae novae. Notizblatt des Botanisches Garten und Museum. Nr. 102: 73 - 86.

Hutchinson, J. (1923). XXI. Contributions towards a phylogenetic classification of flowering plants. II. The genera of the Annonaceae. Kew Bull. No.7: 241 - 261.

Kostermans, A.J.G.H. (1952). Notes on New Guinea plants. IV. Saccopetalum koolsii Kosterman. Reinwadtia 1: 459 - 461.

Sinclair, J. (1955). A revision of Malayan Annonaceae. Gard. Bull. Sing. 14: 149 - 516.

Sinclair, J. (1956). Notes on New Guinea Annonaceae. Pt. I. Gard. Bull. Sing. 15: 4 - 13.

Steenis, C. G. G. J. van (1964). An account of the genera Richella and Oxymitra. Blumea 12: 353 - 361.

Walker, J.W. (1971). Pollen Morphology, phytogeography and phyllogeny of the Annonaceae. Contr. Gray Herb. 202: 1 - 131.

KEY TO GENERA IN PAPUASIA
(Source: Sinclair 1955)

1. Carpels free, or if slightly joined, then always free in fruit. 2

1. Carpels united into a fleshy mass, especially in fruit. Fruit surface usually with knobs or bulges. Fruit edible, cultivated. .... 2. Annona

2. Petals 6. 3

2. Petals 3 (see also 3b). .... 20. Petalophus

3. Petals in 2 series, one or both series imbricate in bud. Sepals imbricate or valvate. Stamens many, closely packed, their connectives concealing top of anther cells. Leaf indumentum stellate or lepidote. TRIBE UVARIEAE................ 4

3. Petals all valvate, usually in 2 distinct series, more rarely 1 series, usually 6m rarely 4, and very rarely 3. Sepals valvate.. 7

4. Sepals imbricate. Trees. .... 29. Stelechocarpus

4. Sepals valvate or connate. Climbers or scandent or sarmentose shrubs......................... 5

5. Inner petal with abroad base. 6

5. Inner petal with narrow claws. .... 27. Rauwenhoffia

6. Petals 1-4cm long, flower opening wide when mature. .... 30. Uvaria

6. Petals usually not more than 3mm long, incurved at apex, not expanding much in mature flower. Woody climbers. 6. Cyathostemma

List of Genera in Papuasia

Alphonsea*	Friesodielsia*	Orophea	Pseudovaria
Annona	Goniothalamus	Papualthia	Rauwenhoffia*
Artabotrys*	Meiogyne	Petalophus	Schefferomitra
Cananga	Miliusa	Phaeanthus*	Stelechocarpus
Cyathocalyx*	Mitrella	Platymitra	Uvaria
Cyarhostemma	Mitrephora*	Polyalthia	Xylopia
Desmos*	Oncodostigma	Polyaulax	
Fissistigma*	Oreomitra	Popowia*	

Genera marked with an * have extra-axillary or terminal flowers. Other genera have flowers axillary or from axils of fallen leaves.

7. Anther cells not concealed by flat-topped, rounded or pointed connectives. Stamens few to many. Outer petals usually smaller then inner, free. Larger inner petals bag-shaped, free or adhering by margins. TRIBE MILIUSEAE.... 8

7. Anther cells not concealed by connectives. 12

8. Inner petals forming a mitre over sexual organs such that 3 species are present at base of the petals. 9

8. Inner petals not forming a mitre. 10

9. Claw long and narrow, distinct from blade. Carpels 3-5, walls 1-2mm thick. Stamens 3 - 12. .... 18. Orophea

9. Claw not distinct from blade which is narrowed at base. Carpels 1-2, walls 5mm thick. Stamens many. .... 22. Platymitra

10. Anther connective truncate at apex. Carpels 1-seeded. .... 21. Pheanthus

10. Anther connective projecting, pointed or rounded. Carpels 1-many seeded. 11

11. Outer petals and sepals ± equal; inner petals larger, often bag-shaped at base. .... 13. Miliusia (including Saccopetalum)

11. Outer petals much larger than sepals; inner and outer petals +- equal, bag-shaped at base and apex. .... 1. Alphonsea

12. Inner petals adhering to form a cap or dome over the sexual organs, or if not cohering, their edges at first united for a short time. Petals clawed, claws usually long and narrow, sometimes vestigial or absent. TRIBE MITREPHOREAE. 13

12. Inner petals not cohering to form a cap, though the base of the petals may press closely over sexual organs......................... 19

13. Petals +- equal. Erect or climbing shrubs with rusty tomentum...... 14

13. Petals unequal.................. 15

14. Carpels numerous. .... 28. Schefferomitra

14. Carpels 3-6. .... 17. Oreomitra

15. Inner petals shorter than outer. 16

15. Inner petals longer than outer.. 18

16. Trees. Inner petals clawed; claws long, lower half narrower. Flower buds not conical. ripe carpels with usually more than 2 seeds. .... 15. Mitrephora

16. Climbers, shrubs or trees. Inner petals sessile or with short, broad claw. Inner petals very much shorter than outer ones. Flower buds conical. Ripe carpels 1-2 seeded.......................... 17

17. Erect shrubs or trees. Ripe carpels bright red. .... 10. Goniothalamus

17. Climbers. Inner petals not clawed. Flowers extra-axillary. .... 9. Friesodielsia

18. Flowers bisexual. Inner petals with a short or rudimentary claw, edges adhering at first, later free but remaining incurved. .... 25. Popowia

18. Flowers unisexual. Inner petals with 7 long narrow claws, mitriform. Leaves membranous or thinly leathery, with straight +- parallel veins. .... 26. Pseuduvaria

19. Inner petals usually shorter than outer petals, nearly always 3-angled in cross-section. Inner petals often remaining erect and touching each other; outer petals spreading, lanceolate. TRIBE XYLOPIEAE............. 20

19. Inner and outer petals ± equal, ± similar, spreading or apices free. TRIBE UNONEAE............... 23

20. Climbers....................... 21

20. Trees or shrubs................. 22

21. Flowers terminal, cymes or branched panicles. Leaves usually hairy, without secondary veins between parallel lateral veins. Pedicels, flowers and carpels, densely brown, hairy. .... 8. Fissistigma

21. Flowers axillary, solitary or 2-3 in a fascicle. Leaves glabrous above, minutely pubescent below. Pedicels and flowers minutely pubescent, carpels glabrous. .... 14. Mitrella

22. Inner petals deeply grooved at base. Anthers not transversely septate. Erect shrub. .... 24. Polyaulax

Variation in Flower Structure of Selected Genera in the Annonaceae.

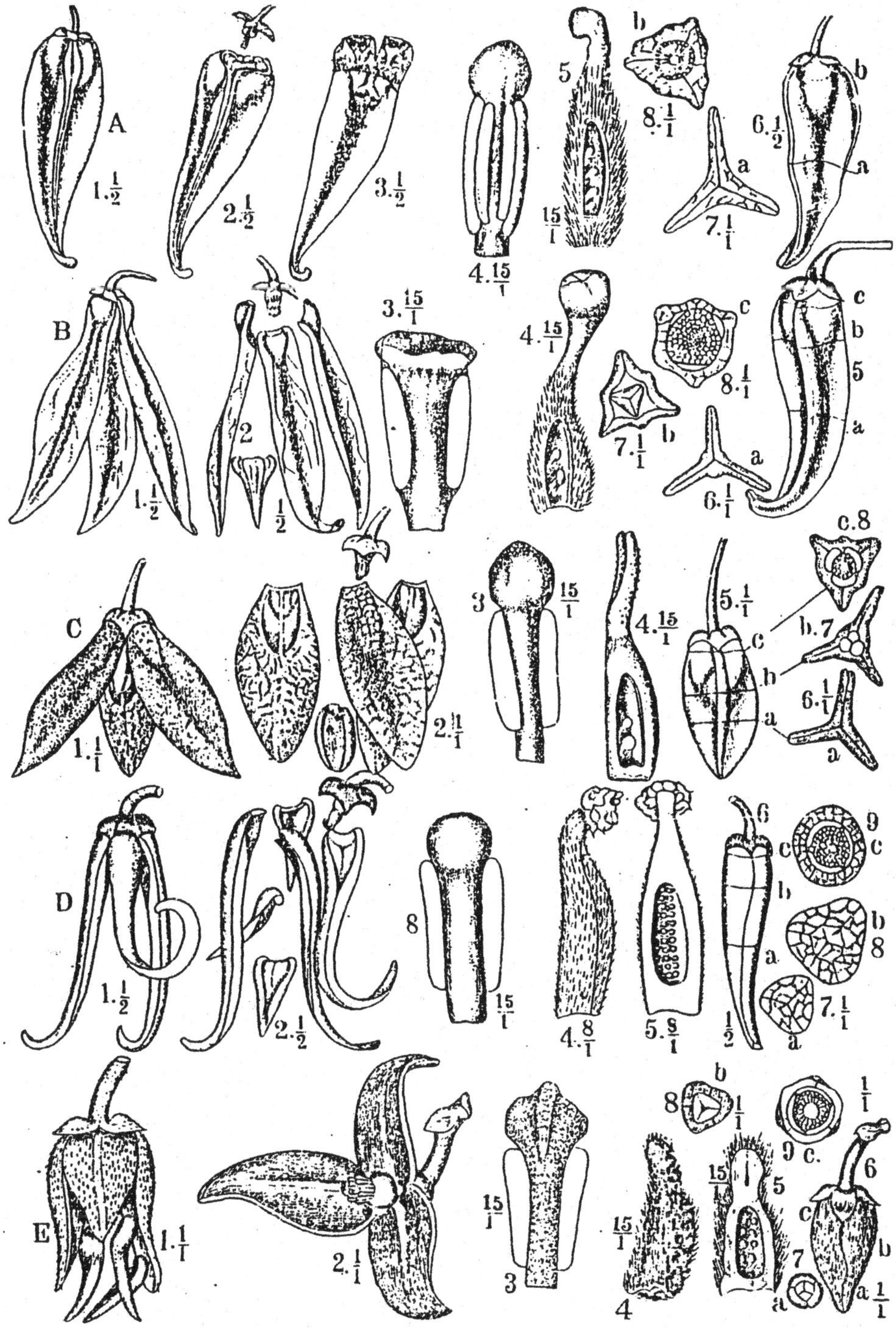

List of Genera: A. Unona sect. Dasymaschala
B. Oxymitra
C. Mitrella
D. Pyramidanthe
E. Melodorum
(Source: Icones Bogoriensis II. Tab. XLIII)

22. Inner petals not deeply grooved at base. Anthers transversely septate, pollen grains visible with naked eye........ 31. Xylopia

23. Petals spreading above, but constrictive at junction of blade and claw. Claw orbicular, concave, conniving and closely pressed over sexual organs. Blade linear or filliform, oblong, flat or terete. The three inner petals separate at base but falling off in one piece. 24

23. Petals flat, ± spreading from base, not constricted above the claw and not closely pressed over the sexual organs. The three inner petals not falling off in one piece.......................... 25

24. Trees. Carpels 1-6, ovules more than 2..... 5. Cyathocalyx

24. Climbers with woody peduncles, hooked as a means of climbing. Carpels few to many, ovules 2. .... 3. Artabotrys

25. Fruit moniliform with several seeds in a single row. (Petals sometimes reduced to 2 or 3 in Desmos Sect. Dasymascholon). .... 7. Desmos

25. Fruits not moniliform........... 26

26. Climbing shrubs. Young parts with stellate hairs. Inner petals much narrower than outer petals. .... 19. Papualthia

26. Trees or shrubs. Indumentum mostly absent, if present consisting of simple hairs.................... 27

27. Carpels hard and thick-walled (2-7mm). Sessile or subsessile.... 28

27. Carpels thin-walled (1mm thick). Mostly stalked, rarely sessile. Seeds often with characteristic longitudinal circumferential groove 29

28. Veins of leaves faint, not prominent, not straight, secondary veins present. Seeds 3-5. Petals with thick edges. Trees. .... 16. Oncodostigma

28. Veins of leaf prominent, straight, oblique, without secondary veins. Seeds 8-10. Petals not thick at edges. Trees or shrubs. .... 11. Meiogyne

29. Stamen connectives flat topped or convex. Flowers axillary or tubercled on woody outgrowths of trunk, scented or not. Seeds with 1-5 longitudinal grooves about the circumference. .... 23. Polyalthia

29. Stamen connectives apiculate, erect. Flowers axillary, strongly scented. Seeds 2-12, in 2 rows, pitted. .... 4. Cananga

Introductory Notes

The family Annonaceae is poorly known in Papuasia due to the lack of recent research. The last comprehensive attempt to treat the family in Papuasia is the work of Diels published in 1913 (Bot. Jahrb 49: 113 - 167). Translations of the descriptions of endemic genera have been based on this work. Descriptions of more widely dispersed Malesian genera have been taken from the studies of Sinclair (1955).

Many presumably undescribed species occur in the National Herbarium at Lae. No attempt has been made to include the diverse records although these include at least one undescribed genus from the Milne Bay Region (Croft pers. comm.).

It is important to collect both flowers and fruit from the same plant. DO NOT MIX COLLECTIONS.

1. ALPHONSEA

Tall trees or shrubs of very local distribution and quite rare (poorly represented in Herbaria). A genus of some 30 species from India and Sri Lanka to China and Papuasia. Related to Miltiusa, Orophea and Platymitra. Most closely related to Milusa which has only the inner sepals saccate at the base, while Alphonsea has both sets of sepals saccate at base.

No specimens are identified as Alphonsea in the National Herbarium at Lae.

Tall trees or shrubs. Leaves usually coriaceous, shining, mostly glabrous, margins slightly revolute. Peduncles 1-several, fasiculate, extra-axillary, bearing several pedicels. Flower buds conical. Sepals 3, valvate, several times smaller than petals. Petals 6, valvate in 2 series, bases saccate and apices reflexed, inner set about the same length as outer but slightly narrower. Torus cylindric or hemispheric. Stamens indefinite, miliusoid, the connectives apiculate, produced, not concealing the dorsal anther-cells. Ovaries few, 1-8, stigma sessile, U- or horseshoe-shaped with a slit down the adaxial side; ovules 4-10 in 2 rows. carpels subsessile or stalked, with thick walls, glabrous, tomentose or verrucose (Sinclair 1955).

List of Species in Papuasia

Alphonsea ovata
    Orophea ovata
Alphonsea papuasica
Alphonsea stenogyra
    Orophea stenogyra

2. ANNONA

A genus of over 100 species mostly distributed in America. The genus includes several very important tropical fruits, most native in tropical America. The genus is sometimes spelt Anona. The species of Annona found in Papuasia are not native but cultivated, although some have become naturalised.

Flowers extra-axillary, sometimes cauliflorous, never axillary. sepals 3, valvate. Petals 6, or inner whorl very small or absent, coriaceous with thick edges and concave at base inside, valvate, inner petals sometimes imbricate. stamens with convex or apiculate connectives, filaments short. Style club -shaped, stigma muriculate. Ovaries with 1 basal ovule, free at first or united from the beginning into a fleshy syncarp, the surface covered with bulges, spines or less often smooth.

List of Introduced Species in Papuasia

Annona cherimola L.
Annona muricata L. (Soursop)
Annona reticulata L.
Annona squamosa L.

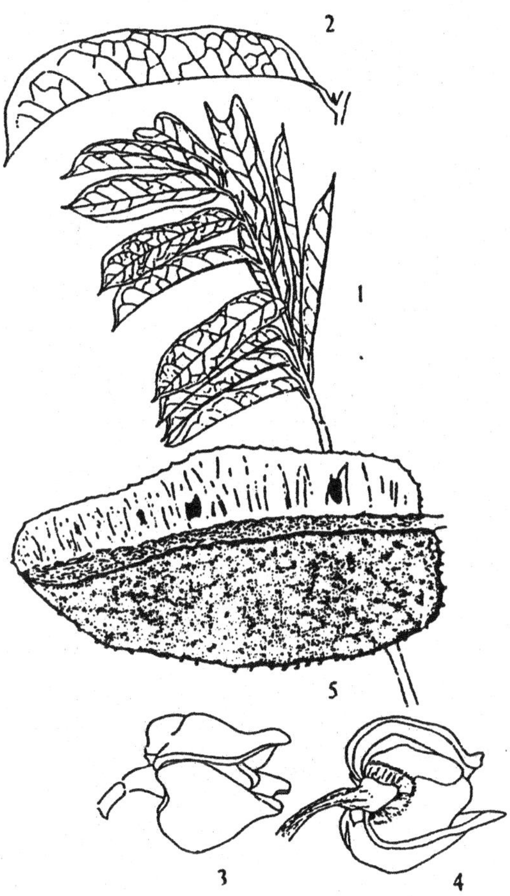

Annona muricata (A) habit; (B) leaf; (C) (D) flowers; (E) fruit. (Source: Purseglove (1977))

Key to Common Cultivated Species
(Source: Sinclair (1955)

1. Fruit oblong, dull green, covered with prickles; petals 6. .... A. muricata

1. Fruit round, not prickly; inner petals very small or absent.... 2

2. Fruit nearly smooth, reddish brown; leaves sharply acute or acuminate. .... A. reticulata

2. Fruit with knobby, separable bulges, light green; leaves rather blunt..... A. squamosa

Notes:

1. A. cherimola is cultivated in the Kainantu area. It is commonly called the cherimoyer and originates from the Andes.

2. A. reticulata is naturalised on Yule Is., Brown River, Buka and Bougainville. Common in gully forest in Brown River. Custard Apple or Bullock's Heart. Warburg in 1891 (Bot. Jahrb 13: 307) recorded this species from Port Dore in Irian Jaya.

3. A. squamosa, the Sweet Sop, Sugar Apple or Wild Custard Apple, grows as a small tree to 5m tall in deciduous woodland around Port Moresby.

4. A. muricata forms a shrub 3-5m tall, it is more common than the other species. Fruit of the Soursop is often used to flavour milk.

3. ARTABOTRYS

A genus of over 100 species distributed from Tropical Africa and Asia to Papuasia. The genus is easily recognisedd by its hooked, flattened peduncular branches by which the plant climbs. 4 species are described from Papuasia. Some species have narrow juvenile leaves.

Climbers or scandent shrubs. Midrib raised on upper surface or flush with it. Flowers hermaphrodite, solitary or fascicled, white or yellow, often fragrant. Peduncles extra-axillary, opposite the leaves, reflexed, flattened and hooked, bearing the pedicels. Bracts small, often caducous. Sepals 3, valvate. Petals 6, in 2 series, valvate, the blade flat, oblong, ovate-lanceolate or linear and terete; claws distinct, orbicular, often with a projecting rim on the inside where the blade joins it, connivent, closely adpressed over the sexual organs as in Cyathocalyx; the inner 3 petals remaining attached at their bases when they fall. Stamens

Annona squamosa (A) flowering shoot; (B) fruit.
(Source: Purseglove (1977)

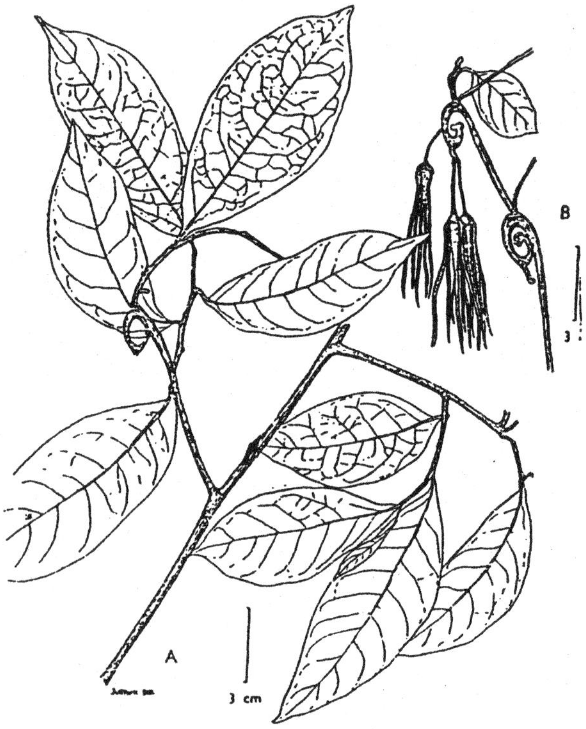

Artabotrys arachnoides (A) leafy; (B) flowers. (Source: Sinclair (1956)

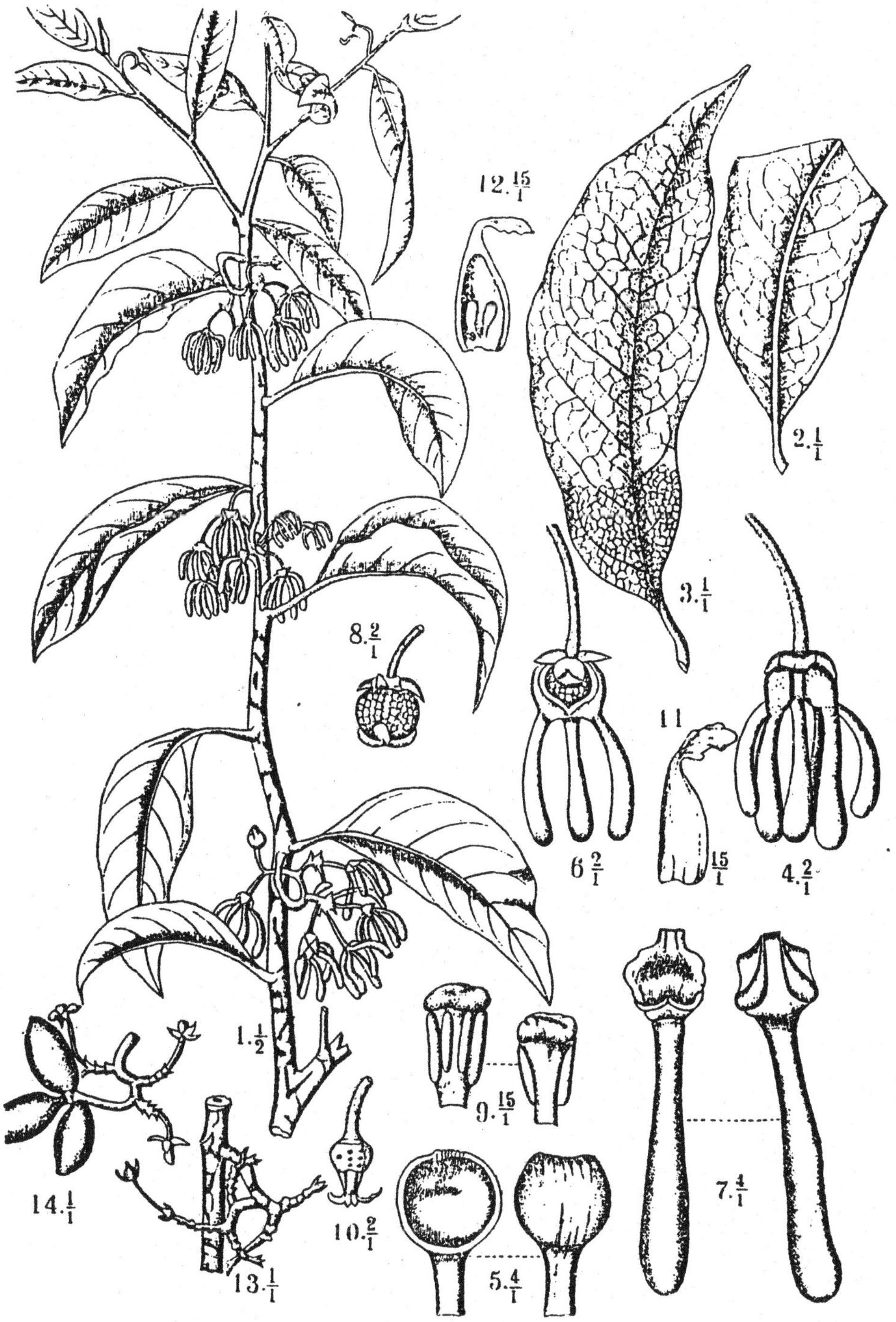

<u>Artabotrys</u> <u>suaveolens</u> var. <u>parviflorus</u>
(Source: <u>Icon</u>. <u>Bog</u>. Tab. 39)

with flat-topped or convex connectives concealing the dorsal anthers. Torus flat or concave. Ovaries few to many; stigma elongated, rod-like or leaf-like in which case it may be lobed, the lobe often bent, a slight groove running down tits adaxial side to base of ovary, sometimes seen as a faint line in the mature fruit. Seeds 2, erect, collateral.

Key to Species in Papuasia

1 Petals narrow, linear or cylindrical, especially the inner petals. 2

1. Both sets petals broad, flat, lanceolate.... A. camptopetala

2. Petals long 5 - 6cm, narrow 2 - 3mm (spider-like). .... A. arachnoides

2. Petals short 0.7 - 1cm long, 3 - 4 mm wide.. A. suaveolens
A. inodorus

Notes:

1. Diels in 1913 (Bot. Jahrb. 53: 135) lists A. suaveolens Scheffer non Blume (Burck Nova Guinea 8 (1911) 428.) as a synonym of A. inodorus Zippel.

4. CANANGA

A genus of 2(?) species distributed from T Asia to N Australia. Both species trees with large, greenish-yellow, fragrant flowers. In Papuasia 1 species distributed from India to N Australia and Papuasia. The second species is confined to Burma and N Malaya. The genus is allied to Poyalthia and Desmos as shown by the similarity in the strap-shaped petals.

Trees. Flowers large, greenish-yellow, fragrant, on short or very short axillary peduncles. Sepals 3, ovate or triangular, valvate. Petals 6, valvate, flat, strap-shaped, several-veined, spreading, the inner series sub-equal or slightly smaller. Stamens linear, about 3 mm long, the connective produced to about 1/3 the length of the stamen and apiculate; anthers lateral or sub-introrse. Ovaries many, oblong; style slender, stigmas agglutinated, forming a pileate, convex, lamellate disc, each separate stigma club-shaped, with a U-shaped groove continued down the inner side of the ovary; ovules indefinite. Ripe carpels oblong, pulpy stalked. Seeds several in 2 rows, pitted, sending spinous processes into the albumen.

Key to Species in Papuasia

1 widespread species:

Cananga odorata
Uvaria odorata

Notes:

1. Corner states that two forms exist in Malaya: wild and cultivated.

BECCARIODENDRON

This monotypic genus was described by Warburg (Pl. Pap. 452). The single species, B. grandiflorum, has been transferred to the genus Goniothalamus (Sch. & Laut. Fl. Deut. Schutz. Sudsee 2 (1901) 321).

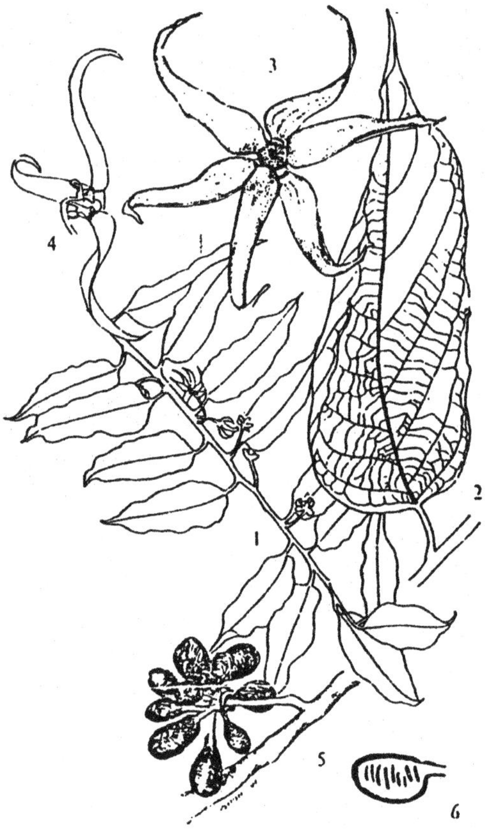

Cananga odorata (A) shoot; (B) leaf; (C) (D) flower; (E) fruit; (F) carpel. (Source: Purseglove (1977)

5. CYATHOCALYX (Including DREPANANTHUS)

• • • • • •

The genus extends from India to S E Asia, throughout Malaysia to Papuasia. It includes some 38 species in 2 sections. 8-9 species are described from Papuasia. Drepananthus has been segregated as a genus but this was not followed by Sinclair (1955). Sinclair (loc. sit.) concluded that the differences, as listed in the key, do not appear sufficient for separation into two distinct genera. Aratabotrys is very similar to Eucyathocalyx except that the former are climbers, not trees. Cyathocalyx is a genus of monopodial trees with mast-like trunks and a crown of branches near the top.

Monopodial trees with mast-like trunks and a crown of branches near the top. Leaves 2-ranked. Flowers extra-axillary, usually opposite the leaves, fragrant. Peduncles decurrent. Sepals valvate, free or united into a 3-lobed cup. Petals 6, valvate in 2 series, sub-equal, the claws orbicular, concave, conniving and closely pressed over the sexual organs, the inner three petals not separating when they fall, slightly constricted or bent at the junction of the blade; blades linear, filiform, oblong, flat ot terete. Stamens indefinite, the connectives flat-topped, concealing the anther cells, filament short. Ovaries 1-several, oblong or cylindric, stigmas discoid, peltate, pileate and lamellate, convex, often agglutinated; ovules several in two rows. Ripe carpels large or small, sessile or shortly stalked.

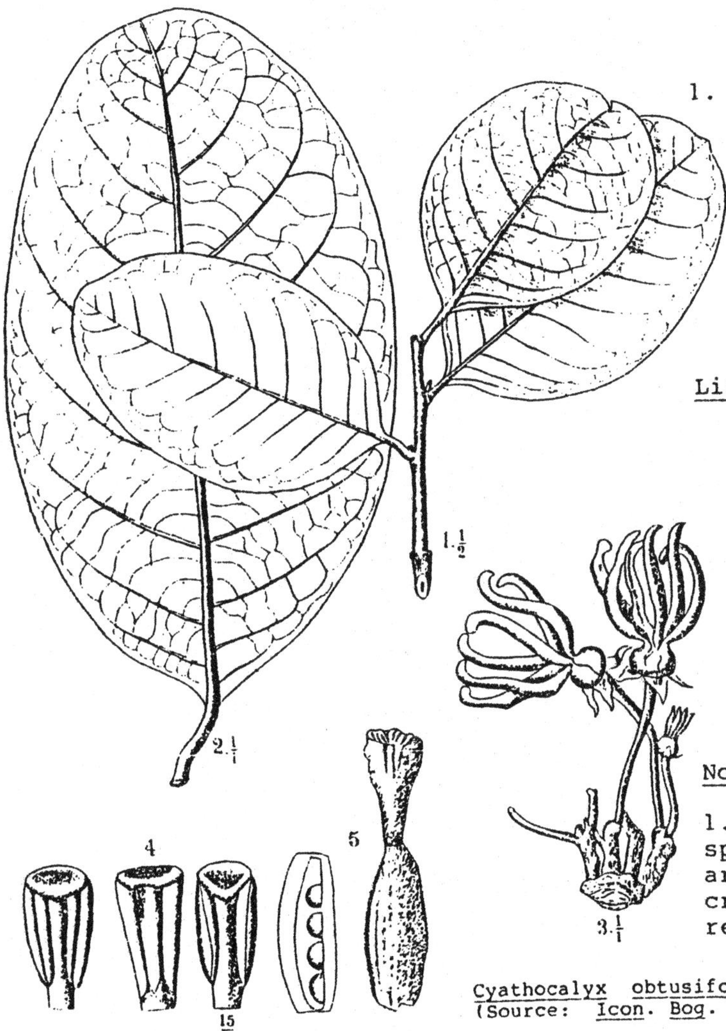

Cyathocalyx obtusifolius
(Source: Icon. Bog. II. Tab. 35)

Key to Sections in Cyathocalyx

1. Midrib and nerves raised or flush with upper surface; midrib broad at base. Ovary 1, glabrous.
.... Sect. Eucyathocalyx

1. Midrib and nerves on upper surface of leaves always sunk. Ovary 2 to several, pubescent with sericeous hairs.
.... Sect. Drepananthus

List of Species in Papuasia

Cyathocalyx cauliflorus
Cyathocalyx lucidus
Cyathocalyx novoguineensis
Cyathocalyx obtusifolius
Cyathocalyx obtusifolius
var. angustipetalus
C. limus
Cyathocalyx osmanthus
Cyathocalyx aff. osmanthus
Cyathocalyx papuanus
Cyathocalyx petiolatus
Cyathocalyx polycarpum

Notes:

1. Unidentified collections of several species are lodged in the National Herbarium at Lae. Description of any undescribed species must await a proper revision.

6. CYATHOSTEMMA

A genus of some 8 species: 3 in Papuasia. The genus is closely allied to Uvaria but differs in the globose flowers with petals which do not expand and are imbricate only at the tips. The reticulations of the leaves form a very close network.

Woody climbers mostly with simple hairs, sometimes stellate. Flowers globose in pendulous or clustered cymes from old wood or extra-axillary. Sepals valvate, petals valvate at base, imbricate at tips, not expanding, opening only to 3 - 5mm.

Key to Species in Papuasia

1. Flower colour not yellow, greenish or white.. C. grandifolium

1. Flowers yellow, greenish or white. 2

2. Leaves glabrous on lower surface. Ripe carpels glabrous.......... 3

2. Leaves not so.

3. Ripe carpels large (+ 1.5cm long), thick-walled, with stalks more than 2mm thick. C. hookeri

3. Ripe carpels small (less than 1.5cm long), thin-walled, with slender stalks less than 2mm thick. .... C. micranthum
   Uvaria micrantha
   Guatteria micrantha

Notes:

1. The collection NGF 9796 has been identified as C. excelsa. No other records of this species in Papuasia have been seen.

2. Diels (Bot. Jahrb. 49 (1913) 124) lists Uvaria glabra and U. parviflora as synonyms of C. hookeri.

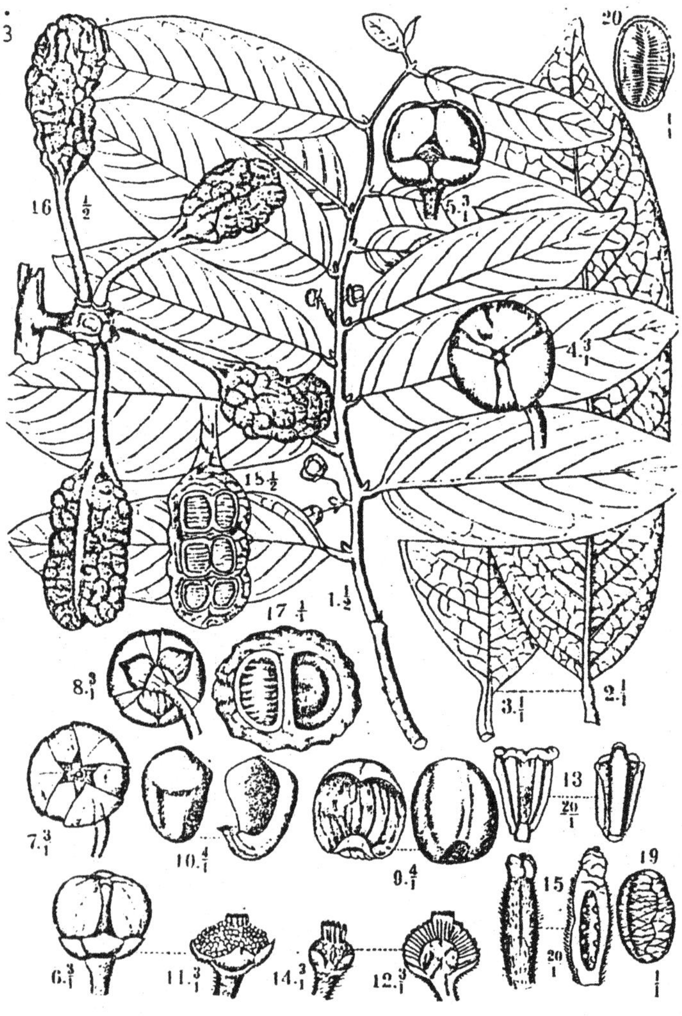

Cyathostemma hookeri
(Source: Icon. Bog. II. Tab. 42)

7. DESMOS (including DASYMASCHALON)

• • • • • •

A genus extending from India to Papuasia with c. 25 or more species. Some authors divide Desmos into 2 sections. Taxa in section Dasymascholon are sometimes treated as a distinct genus.

Shrubs or scandent shrubs. Leaves sometimes glaucous beneath. Flowers often fragrant, extra-axillary, opposite the leaves or axillary. Sepals 3, valvate. Petals valvate in 2 series or reduced to 1 series, clawed at the base. Stamens numerous, connectives convex, concelaing the extrorse anther cells. Ovaries numerous; s_igma oblong, ovoid or clavate, bent with a U shaped opening and a groove continued down the adaxial side of the ovary. Torus small. Ripe carpels monoliform with 1 - 8 segments. Seeds 1-seriate, a seed in each segment.

Key to Sections in Desmos

1. Petals 6, in 2 series, spreading, the apices free. Inflorescence extra-axillary or axillary. .... Sect. Eudesmos

1. Petals 4 - 3 or 2, inner series absent. Edges of petals cohering, sometimes free later. Inflorescence axillary. .... Sect. Dasymascholon

Notes:

1. Two specimens of a single species collected at 900m on Mt Susu, Watut Divide, both have yellow/red flowers. (Collections NGF 26345, NGF 30743).

2. Uvaria micrantha could belong in this genus.

8. FISSISTIGMA

• • • • • •

A genus of 50 or more species from Africa to N E Australia and Papuasia. The genus shows an affinity to Xylopia but is also closely related to Mitrella and Pyramidanthe. It is not related to Melodorum as often stated (Sinclair 1955).

Climbers. Leaves, pedicels, flowers and carpels usually pubescent or tomentose, sometimes copiously so. secondary veins of leaves absent; reticulations when visible scalariform, at right angles to the nerves. Flowers terminal in cymes or in branched panicles opposite the leaves but sometimes a part of the inflorescence appearing axillary due to branching low down between the main axis and the leaf. Sepals 3, valvate, united at the very base. Petals valvate in 2-series, coriaceous and with thickened edges, usually elongated, lanceolate but sometimes ovate or oblong; the outer erect or spreading, flat; the inner slightly shorter, triquetrous and concave at base inside. Stamens numerous, connectives slightly produced, oblique, obtuse or slightly apiculate. Ovaries pubescent, slightly curved , grooved on inner side; style present, sometimes narrowly clavate at the stigmatic tip but usually not enlarging, the stigmatic portion being very short and not differentiated clearly from the style. Carpels usually large, 2 cm or more in diameter; spherical, ovoid or oblong, thick walled; stalked. several seeds in 2 rows, smooth, shining.

List of Species in Papuasia

No recorded species.

Notes:

1. The following collections in the National herbarium represent unnamed species:

Sp. 1 Climber in Nothofagus forest, 1200 - 1500m altitude. Fruit orange. Tomentose. Kalkman 5321, NGF 39974, NGF 41021, NGF 42013. Common Telefomin and Kopiago.

Sp. 2 Scandent shrub with tomentose fruits. NGF 37475.

9. FRIESODIELSIA (including OXYMITRA)

A genus of over 60 species distributed over the range: Africa to Moluccas and Papuasia (van Steenis 1964). In Papuasia it is represented by F. biglandulosa (Fries 1959: 171). The nearest genus to Friesodielsia is Goniothalamus: the inner sepals are not distinctly clawed as they are in Goniothalamus. Also the flowers are extra-axillary (not axillary) and most species are climbers.

Mostly woody climbers (some trees). Leaves with oblique, rather straight veins which do not anastomose at the edge in distinct loops but form broken or indistinct ones; reticulations scalariform. Flowers extra-axillary. Sepals 3, valvate. Petals 6, valvate; the outer large, long, flat or trigetrous and narrow, more or less spreading in mature flowers; the inner much smaller, ovate-lanceolate, oblong or narrow, conniving over the stamens and ovaries, their edges adhering, the base not clawed or occasionally forming a rudimentary claw. Stamens numerous, with oblique, flat-topped or slightly convex connectives. Ovaries oblong or cylindric, pubescent; style short, stigma sub-capitate or compressed-capitate, slightly falcate, usually pubescent and split on the top and on the inner side. Ripe carpels ovoid or oblong, thin-walled, usually stalked. Seeds 1, pale brown, rarely 2, sub-basal.

List of Species in Papuasia

Friesodielsia bigladulosa
- Gualteria biglandulosa
- Oxymitra biglandulosa

Notes:

1. Material of 2 unidentified species is in the National herbarium at Lae:
Sp. 1 Climber- perianth orange. 60m. Hoogland 4924. Gogol near Mawan.

Sp. 2 Vine. Busu River NGF 41827.

2. F. bigladulosa is a Climber to 15-35m high. Undersurface of leaf with characteristic golden colour (not always in dried specimens).

3. The proposal to treat Oxymitra as a nomen cons. was rejected (Taxon 1954: 115) but the generic name was still used in the treatment by Sinclair (1955).

10. GONIOTHALAMUS

A large genus of up to 114 species distributed from India and Sri Lanka to Papuasia. Shrubs or small trees with axillary flowers, sometimes flowers cauliflorous or terminal. Goniothalamus is closely related to Friesodielsia which can be distinguished by the extra-axillary flowers and the lack of clawed inner petals. The genus can be grouped into two unnatural sections: species with apiculate stamen connectives and species with convex stamen connectives.

Shrubs or small trees. Leaves coriaceous or membranous; nerves prominent, oblique, straight and parallel with scalariform reticulations or very fine, sometimes scarcely distinct, not straight or parallel but with a lax network of reticulations, not scalariform. Flowers usuallu axillary, someimes terminal and axillary or cauliflorous. Pedicels with several minute, lanceolate, distichous bracts at the base. Sepals valvate, large or small, usually membranous, several-veined and reticulate, free or forming a cup, often persistent in fruit. Petals valvate, coriaceous; outer flat or nearly so, one-veined, slightly clawed or truncate at base; inner smaller, half as long or les, clawed, cohering above to form a vaulted cap over the stamens. Stamens many, linear-oblong, pollen grains large, visible under hand lens, connectives apiculate or convex. Ovaries numerous, cylindrical, pubescent or glabrous; style linear, grooved on the anterior side; stigma funnel-shaped or narrowly so, split down the inner side, often two lobbed, rarely cylindrical and truncate. Carpels stalked or sessile; 1-2 seeded.

List of Species in Papuasia

Goniothalamus aruensis
Goniothalamus caloneurus
Goniothalamus cauliflorus
Goniothalamus cleistogamus
Goniothalamus coriaceus
Goniothalamus grandiflorus
- Beccariodendron grandiflorum
- Mitrephora grandiflora
- Oxymitra macrantha

Goniothalamus imbricatus
Goniothalamus inaequilatera
Goniothalamus longirostris
Goniothalamus myrmeciocarpa
Goniothalamus rhynchocarpus
Goniothalamus uniovulatus
Goniothalamus viridiflorus

11. MEIOGYNE

A genus of some 7 species from India to Papuasia where it is represented by a single species. Both Sinclair (1955) and Merrill (1915) follow Boerlarge who in 1899 maintained the distinction based on the following characters, from Cyathocalyx (in brackets). The flowers are axillary (extra-axillary or opposite leaves); petals diverging from a broad base (clawed and constricted); base of petals not closely adpressed over stamens (closely adpressed).

Trees or shrubs with pale straw coloured twigs. Leaves membranous with prominent, often steeply ascending veins on lower surface. Flowers axillary medium to large. Sepals 3, valvate, connate at base. Petals 6 (3 + 3), valvate, diverging, tapering gradually from base upwards and diverging, densely tomentose or sericeous-tomentose; the inner slightly shorter in length with a warted patch at base inside. Stamens numerous, with flat-topped, somewhat oblique connectives concealing the anther cells when viewed from above. Torus covex. Ovaries 2-5 with several ovulea in 2 rows; stigma discoid, sessile. Carpels thick-walled, sessile or sub-sessile.

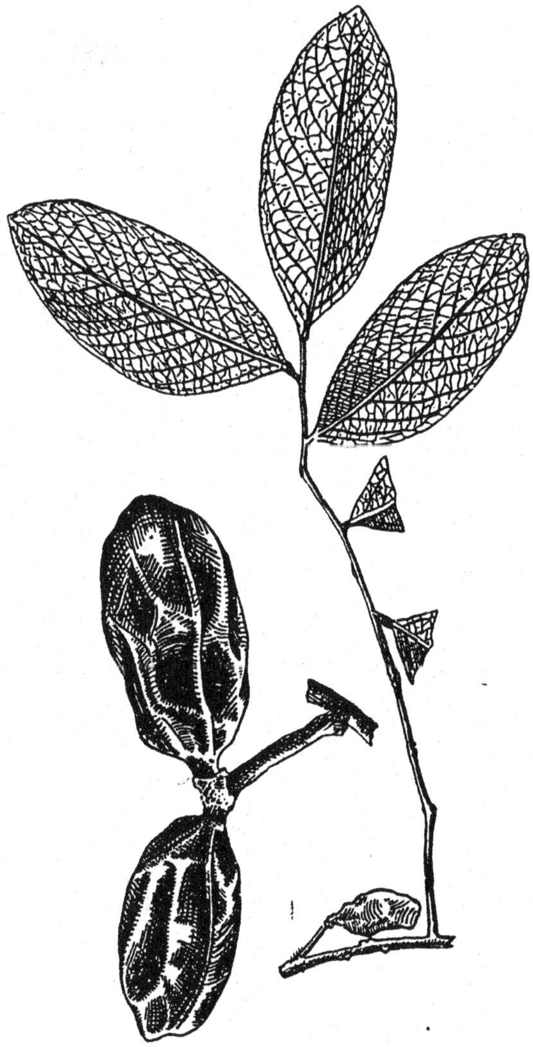

Meiogyne macrocarpa
(Source: Nova Guinea 8: Tab. 71)

List of Species in Papuasia

1 species:

Meiogyne macrocarpa

12. MELODORUM

A small genus of 2 - 3 species from Burma through Indo-China to Malaya and possibly Papuasia from which are described three species. The genus resembles Mitrella in having secondary veins, axillary flowers, concave petals and a similar style (Sinclair 1955). The connate sepals persistent in the fruit are another important similarity.

Shrubs or climbing shrubs. Leaves glabrous. Flowers solitary, axillary or terminal. Sepals 3, valvate, connate, persistent in fruit. Petals 6, valvate, concave inside, outer petals spreading, inner petals slightly smaller, closed over sexual organs (eventually opening). Anthers linear, connectives flat-topped, filaments short. Style grooved, not expanding at stigmatic tip.

List of Species in Papuasia

Melodorum beccarii
Mitrella beccarii
var. lanceolata
Melodorum micranthum

Notes:

1. Sinclair (1955) notes that most species described this century are not congeneric with Loureiro's Melodorum and many are now put in Fissistigma.

2. Diels (1913) placed Melodorum micranthum as a synonym of Papualthia longirostis. He also listed the following synonyms: Goniothalamus inaequilaterus and Trivalvaria longrostris. The basionym is Goniothalamus longirostris Scheffer.

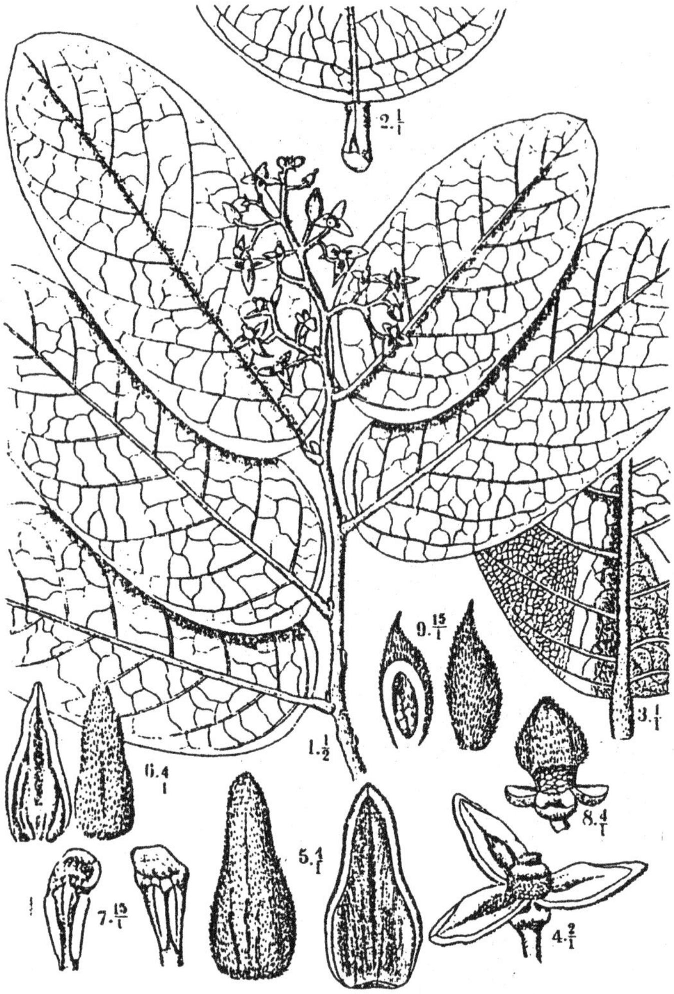

Melodorum parviflorum
(Source: Icon. Bog. Tab. 45)

13. MILIUSA (including SACCOPETALUM)

● ● ● ● ● ●

A genus of about 40 species from India through S E Asia to N Australia and Papuasia. The genus resembles Pheanthus with outer petals and sepals alike and much larger inner petals but differs in that the stamen connectives are apiculate or rounded not truncate.

Small trees or shrubs. Flowers usually hermaphrodite, occasionally dioecious or polygamous. Solitary or in 2's, 3's or cymose fascicles. Pedicels usually slender. Axillary (when in axils of fallen leaves appear extra-axillary). Sepals and outer petals, similar, equal, valvate. Inner petals valvate, larger, thickened, fleshy or saccate to various degrees at the base, edges cohering in younger flowers.

Key to Species in Papuasia

1 species:

Miliusa koolsii

14. MITRELLA

● ● ● ● ● ●

A Malesian genus with 5 species. It is represented in Papuasia by 3 species. The genus is related to Fissistigma and Pyramidanthe (Sinclair 1955). from which it differs in the shape of the petals, the glabrous ovary, the non-capitate stigma and the pitted seeds.

Climbers with leaves with feint veins. Flowers axillary: sepals 3, valvate persisting in the fruit. Petals 6, in 2 series: outer series ovate or oblong-ovate with thickened edges and concave base: the inner nmuch shorter, concave at base. Stamens numerous, filaments half as long as anthers; connectives orbicular, obtuse. Ovary glabrous. Seeds black, pitted.

List of Species in Papuasia

Mitrella ledermannii
Mitrella schlechteri
Xylopia micrantha
Mitrella silvatica

15. MITREPHORA

● ● ● ● ● ●

A large genus distributed from India and Sri Lanka to Papuasia. The genus is distinct from Pseuduvaria (in brackets) which several researchers have included in Mitrephora. The major difference is that the flowers are extra-axillary (axillary). The leaves are also distinct.

Trees or shrubs, leaves with nearly parallel nerves and fine reticulations. Flowers hermaphrodite, extra-axillary, opposite the leaves. Sepals valvate, petals valvate, outer petals spreading, narrowed slightly to the base; inner petals slightly shorter, cohering by their edges, free below with very narrow claws. stamens numerous, connectives flat topped. Stigmas sub-capitate. Torus hairy. Seeds 4-8 in two rows.

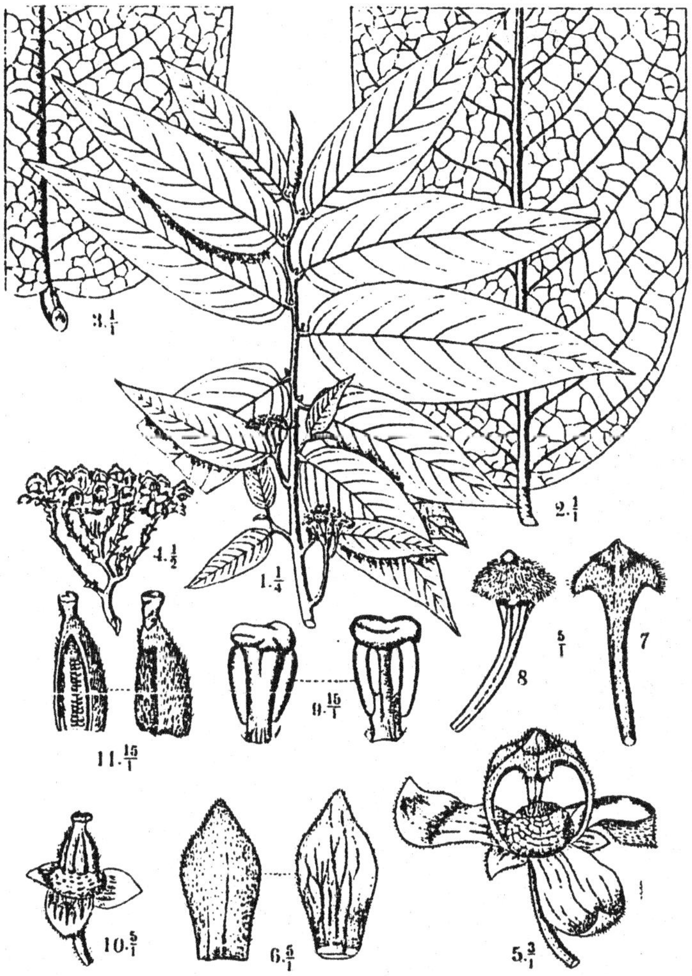

Mitrephora celebica
(Source: Icon. Bog. Tab. 46)

	Mitrephora	Pseuduvaria	Orophea
Flowers			
1. Position	Extra-axillary	Axillary	Supra-axillary or axillary
2. Size	Large, average 2–4 cm. diam.	Small, average under 2 cm.	Small, average under 2 cm.
3. Sex	Hermaphrodite	Unisexual	Hermaphrodite
4. Proportion of length of outer petals to inner	Outer petals longer than inner	Outer petals shorter than inner	Outer petals shorter than inne
Stamens	Uvarioid, numerous	Uvarioid, numerous	Miliusoid, few
Carpels	Globose or oblong, never narrow or elongate	Globose	Globose or narro and elongate
Stigma	Sub-capitate, discoid	Sessile, sub-capitate, fleshy, discoid	Sessile, slightly elongate or rod-shaped, bent not erect

List of Species in Papuasia

Mitrephora ochracea
M. celebica
Uvaria ochiracea
Mitrephora subaequalis
var. macrocalyx
Mitrephora versteegii
M. reticulata
Pseduvaria reticulata

16. ONCODOSTIGMA

A genus of 3 species distributed from the Malay Penin. through Sumatra, Borneo to Papuasia. Closely related to Meiogyne . This genus however lacks the characteristic texture or the straight prominent veins of Meiogyne (Sinclair 1955).

Trees with papery or slightly coriaceous leaves: primary nerves slender. Flowers axillary, sub-sessile or shortly stalked. Petals 6, valvate, often very thick with flattened edges, inner petals with base: warted, concave,

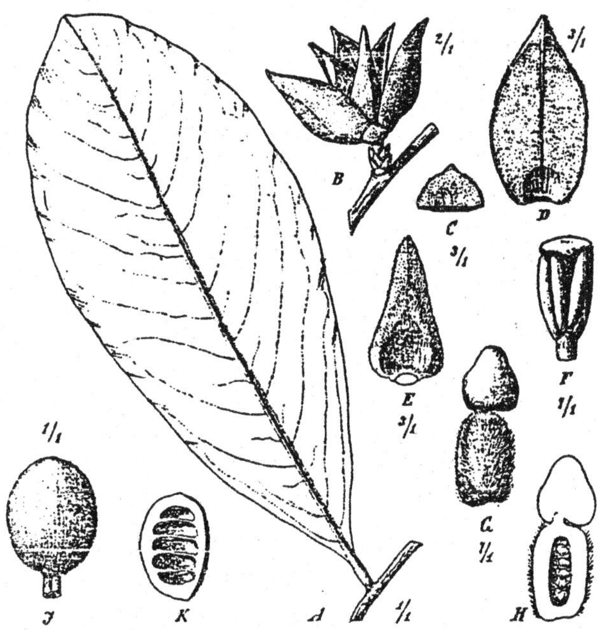

Oncodostigma leptoneura (A) leaf; (B) flower; (C) sepal; (D) outer petal; (E) inner petal; (F) stamen; (G) (H) fruit; (J) (K) fruit.
(Source: Diels (1931)

glabrous. Stamens numerous, connectives flat-topped. Ripe carpels very shortly stalked, tomentose or becoming glabrous. Seeds 3-5.

Key to Species in Papuasia

Oncodostigma leptoneura

Notes:

1. Diels (1913) described the genus as monotypic. The type was collected by Schlechter (No. 17657) at 800m altitude in the Kani Ranges. The genus is distributed in the Malay Peninsula, Borneo, Sumatra and New Guinea (Sinclair 1955: 274).

17. OREOMITRA

A monotypic genus described by Diels in 1913 (Bot. Jahrb. 49: 151) from Papuasia. The genus is similar in its flower construction to Orophea but in its fruits is related to Popowia. The surfaces of the leaves, at least when dried, have a distinct appearance (Diels 1913).

Stems densely pilose, hairs ferrugineous. Leaves subsessile, lanceolate to ovate-lanceolate, apex acuminate; to 5.5-7 cm long by 2-3cm broad. Flowers axillary, solitary, sparsely on branches, pendent, peduncles from 15-25mm long. Sepals 3, to 2.5mm diameter. Petals 6, in 2 whorls of 3: outer petals 5-6 mm long, 4-5 mm wide; inner petals 5mm long by 6-7 mm wide. Carpels 3 - 6, pubescent, style elongate to 1.5mm long.

List of Species in Papuasia

1 endemic species:

Oreomitra bullata

Notes:

1. The type collection was made by Schlechter (No. 17655) at an altitude of 600m in the Kani Ranges.

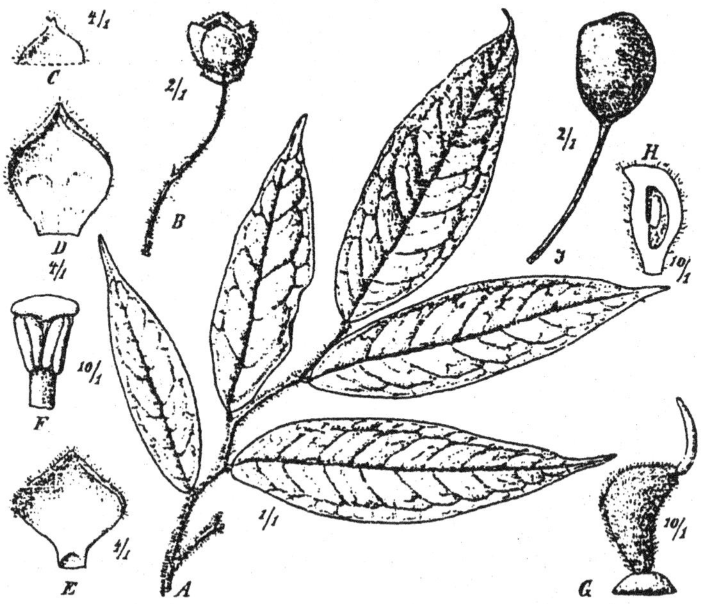

Oreomitra bullata (A) habit; (B) flower; (C) sepal; (D) outer petal; (E) inner petal; (F) stamen; (G) ovary; (H) ovay; (I) fruit. (Source: Diels (1931)

18. OROPHEA

A genus of some 60 species distributed from E Asia to Papuasia. Orophea is related to Mitrephora and Pseduvaria but can easily be distinguished by reference to the following table reprinted from Sinclair (1955). It is interesting to note that in the following list of species in Papuasia many have subsequently been placed in the genus Pseuduvaria.

Shrubs or small trees. Flowers small, 0.4-2.0cm diameter, hermaphrodite supra-axillary or axillary. Peduncles slender bearing 1-several pedicels. Sepals 3, valvate smaller than outer petals. Petals 6, valvate: the outer ovate shorter than the inner; inner petals with narrow claws, usually cohering by margins into a cap, or not. Stamens 6-12, connectives never truncate. Staminodes 0 or 3-6. Stigma sessile, bent to one side. Carpels 1-several seeded, globose or elongate.

List of Species in Papuasia

Orophea aurantiaca
Orophea filipes
Orophea ovata
    Alphonsea ovata
Orophea rhytidophylla
    Orophea rhytidocarpa
Orophea setigera

Notes:

1. Diels (1913) included in the genus Orophea, O. ovata and O. stenogyra. These species cannot be placed in Orophea because the petals are saccate at the base, the inner ones are not mitriform, and the stamens have apiculate, produced connectives. These are characters of Alphonsea and Sinclair (1956) placed these two species in that genus.

19. PAPUALTHIA

The endemic genus Papualthia can be distinguished by the petals being connate at the base into a distinct tube, a character it apparently shares with several African (Asteranthe and Hexalobus) and C and S American (Cardiopetalum and Rollinia) genera (Sinclair 1955). The genus consists of some 8 species with possibly several undescribed. The last critical review was that of Diels in 1913. With the exception of Pseuduvaria mollis transferred from Papualthia by Sinclair (1956) all the genera described remain as treated by Diels (1913).

Trees. Foliage assymetrical. Peduncles short to subsessile. Stamens infinite. Carpels infinite to 3, ovules 2.

List of Species in Papuasia

Papualthia auriculata
- Goniothalamus auriculatus

Papualthia bracteata

Papualthia grandifolia
- Cyathostemma grandifolium

Papualthia longirostris
- Goniothalamus longirostris
- Goniothalamus inaequilaterus
- Trivalvaria longirostris
- Melodorum micranthum

Papualthia micrantha

Papualthia pilosa

Papualthia roemeri

Papualthia rudolphi

Key to Species in Papuasia
(Source: Diels (1913)

1. Flowers dioecious or polyganous. Leaves barely assymetrical. .... P. bracteata

1. Flowers hermaphroditic......... 2

2. Petals in free. Leaf almost completely bare, leafless...... 3

2. Petals stunted................ 4

3. Peduncle 10-15mm long. .... P. auriculata

3. Peduncle 1-2mm long. .... P. grandifolia

4. Ovaries over 10. Flowers short stemmed. Leaves naked. .... P. rudolphi

4. Ovaries less than 10. Flowers sessile or almost sessile...... 5

5. Undersurface of flowers, particularly the central axis, densely covered with hairs. .... P. pilosa

5.

.... P. longirostris

.... P. roemeri

Notes:

1. Most species are distributed throughout mainland New Guinea, particulary along the northern regions.

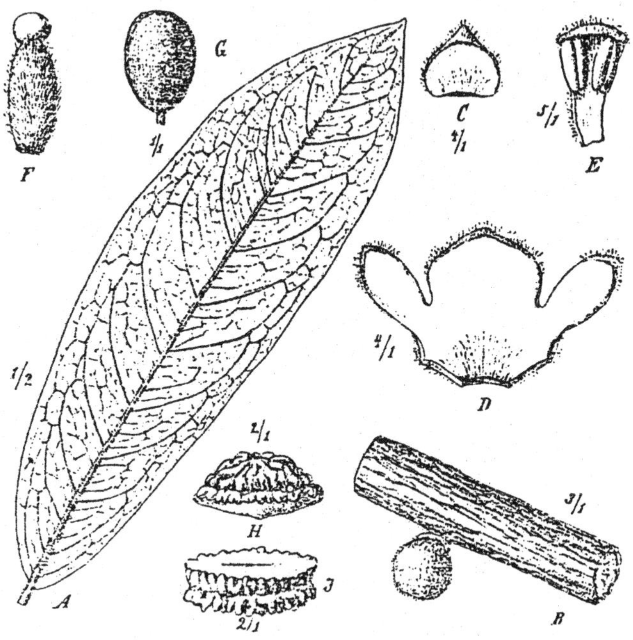

Papualthia pilosa (A) leaf; (B) flower; (C) sepal; (D) petals; (E) stamen; (F) ovary. P. longirostris (G) fruit; (H) seed. (Source: Diels (1931)

20. PETALOPHUS

The genus Petalophus was described by K. Schum in Lauterbach and K. Kchumann (Nachtr. Fl. D. Schutzgeb. Suds (1905) 265). The family is monotypic, the species Petalophus megalopus K. Schumman was collected from the Sattelberg area.

List of Species in Papuasia

1 species:

Petalophus megalopus

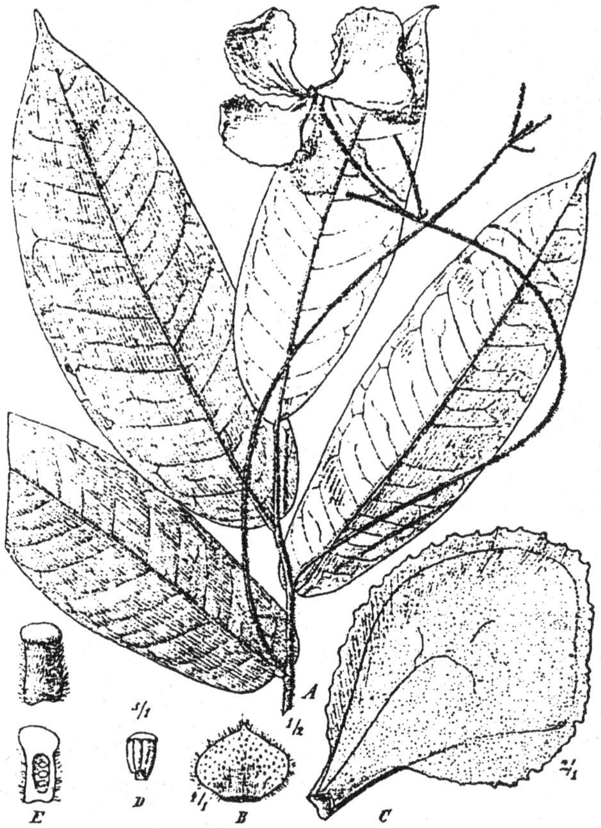

Petalophus megalopus (A) twig with flower; (B) outer petal; (C) inner petal; (D) stamen; (E) ovary. (Source: Diels (1931)

21. PHEANTHUS

A genus of some 20 species distributed from India and Burma to Papuasia. The stamens are definite in number (to 20) a character which is termed miliusoid. The most closely related genus is probably Marsypopetalum (Siclair 1955).

Small trees or shrubs. Leaves with rather straight, obliquely ascending nerves; lacking secondary nerves. Flowers hermaphroditic, extra-axillary, solitary or 2-3 on a very short, bracteolate peduncle which bears the pedicels. Sepals and outer petals similar; much smaller than inner petals. Stamens several, connectives flat-topped, filaments about same length as anthers. Ripe carpels several, stalked, thin-walled.

List of Species in Papuasia

Phaeanthus macropodus
    P. nutans
    Guatteria macropoda
Phaeanthus macropodus
    var. mollifolius
Phaeanthus schefferi
    P. crassipetala var. papuana
    P. natans

Notes:

1. Diels (1913) lists Pheanthus schefferi Boerl. as a synonym of P. macropodus.

22. PLATYMITRA

• • • • • •

A small genus from Burma to Java, records from Papuasia are doubtful. The genus is close to *Orophea* and *Pseuduvaria*. It resembles *Orophea* in the miliusoid stamens, both genera in the mitriform inner petals and *Pseuduvaria* in the horse-shoe shaped, sessile stigma (Sinclair 1955).

Tall trees. Leaves glabrous with feint nerves. Flowers small, fascicled in the axils of fallen leaves. Sepals 3, valvate, connate at base forming a 3-lobed cup. Petals 6, valvate, outer peatsl sessile, ovate, spreading as flower matures; inner same size, narrow and diverging at base but not clawed, cohering in a mitriform fashion over the stamens and carpels. Stamens indefinite, 20 or more, miliusoid. Ovaries 2-3, ovules 10 in 2 rows; stigma sessile, horse-shoe shaped (slightly bilobed). Ripe carpels large, globose or ovoid, solitary or 2, sessile with a hard, thick wall, many-seeded.

List of Species in Papuasia

The record from Papuasia is doubtfull.

23. POLYALTHIA

• • • • • •

A large, widely dispersed genus which is very uniform. It includes over 100 species throughout the tropics of the Old World. Both *Desmos* and *Cananga* are offshoots of *Polyalthia* (Sinclair 1955).

Trees or shrubs. Leaves glabrous or with simple hairs. Flowers hermaphrodite, axillary or extra-axillary (opposite the leaves or below them on woody outgrowths on the older branches). Sepals 3, valvate. Petals 6 (3+3), valvate. Petals large and showy, coloured red, orange, yellow, greenish or white, spreading flat, variously shaped. Stamens usually numerous with flat-topped or slightly convex connectives. Style nearly always absent, stigma rectangular with a faint groove on inner side. Seeds 1 - 5 with a longitudinal circumferential groove.

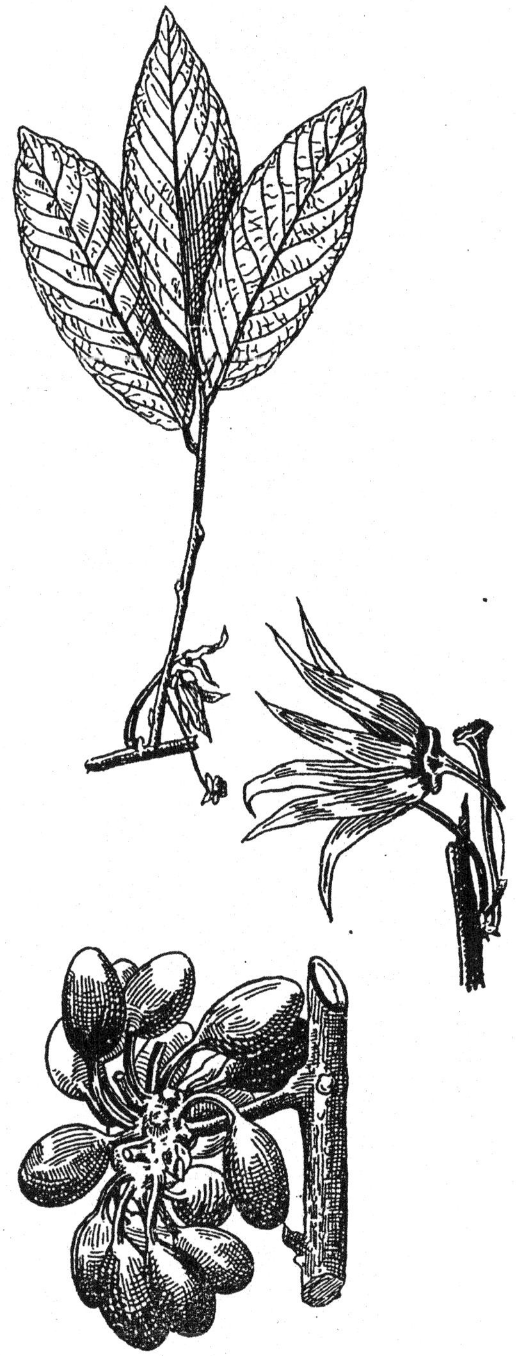

Polyalthia polycarpa (A) stem with flowers; (B) flowers; (C) fruit. (Source: Diels (1931))

List of Species in Papuasia

Polyalthia chlorantha
Polyalthia chloroxantha
Monoon chloroxantha
Polyalthia discolor
Polyalthia elegans
Polyalthia forbesii
Polyalthia glauca
Monoon glaucum
Polyalthia gracilis
Polyalthia hirta
Monoon hirtuum
Polyalthia leptopoda
Polyalthia macropoda
Monoon macropodum
Polyalthia multinervis
Polyalthia nitidissima
Polyalthia oblongifolia
Polyalthia polycarpa
Polyalthia papuana
Polyalthia rumphii
Polyalthia trichoneura

24. POLYAULAX

A monotypic genus described by Backer (Blumea 5 (1942) 492 - 493) for the species previously described as Mitrephora cilindrocarpa by Burck (Nova Guinea 8 (1911) 473) from S New Guinea. Only fruiting specimens were known. The genus resembles some species of Polyalthia but Polyaulax has 6-7-ovulate ovaries while Polyalthia has only 1-2-ovulate ovaries. Both Diels (1912) and Sinclair (1956) noted that this species should not be placed in Mitrephora. Backer (1942) noted that the genus is most closely related to Ararocarpus? a genus described by Scheffer from Java, not collected after 1885.

Small tree to 1-3m tall. Leaves ovate to oblong or lanceolate; base cuneate to rotund, apex acuminate to avute or subobtuse; adult leaves glabrous, 5-10cm long, 1-4cm wide; petiole to 3-5mm long. Flowers solitary, axillary; pedicel 2-3mm long. Sepals 3, base connate, 2-3mm long. Petals 6, valvate, in 2 series; outer petals 10-15mm long, 4-7mm wide; inner petals 10-13mm long, 4-5mm wide. Stamens numerous, to 1.75mm long. Ovary subsessile.

Key to Species in Papuasia

1 rare species:

Polyaulax cylindrocarpus
Mitrephora cilindrocarpa

Notes:

1. Backer collected 2 flowering specimens of this species from Madoera in Java.

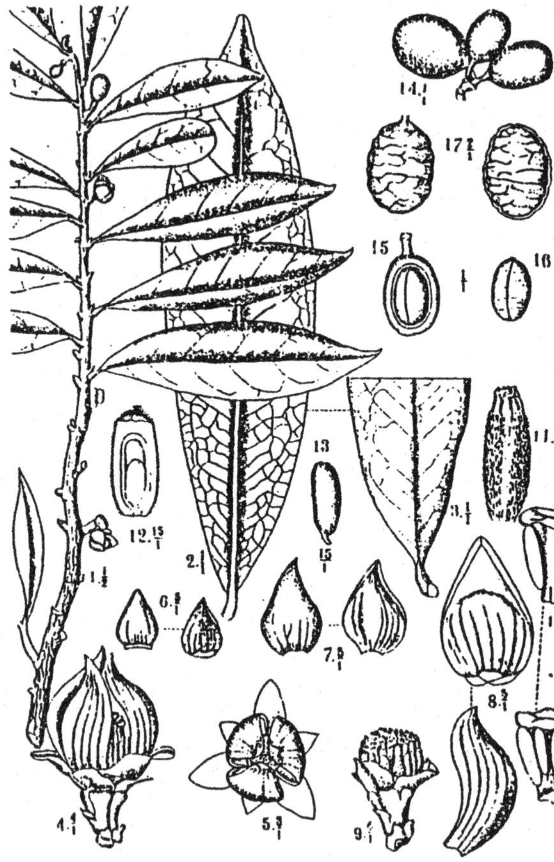

Polyalthia littoralis
(Source: Icones Bog.)

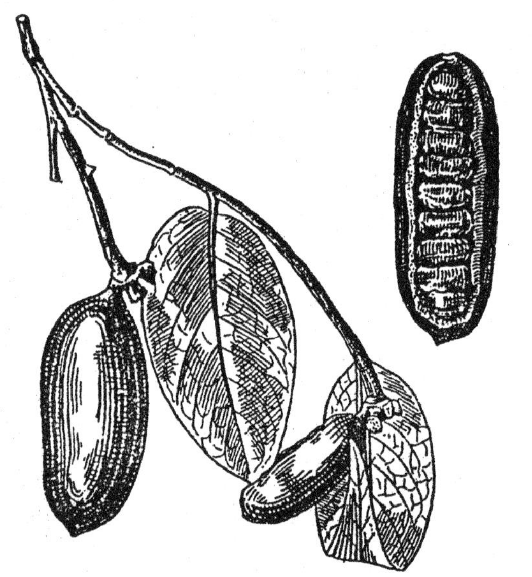

Polyaulax cylindrocarpus (A) stem with fruit; (B) LS fruit.
(Source: Nova Guinea 8: Tab. 71)

25. POPOWIA

A genus including some 90 species, distributed from T Africa (center of genus) to India Malaya, Papuasia and N Australia. It is probably most closely related to Pseuduvaria, Mitrephora and Neouvaria (Sinclair 1955).

Shrubs or small trees. Leaves granular in texture, glabrous, pubescent or tomentose. Flowers small, single or in fascicles opposite the leaves. Sepals 3, valvate. Petals 6, valvate; outer petalssmall, spreading, slightly larger than sepals; inner petals larger than outer, concave inside, edges at first adhering, later free, apices often incurved. Style very short: stigma grooved at top and down the inner side. Seeds 1 - 4, rugose or pitted and with circumferential ridge.

List of Species in Papuasia

Popowia beccarii
Popowia cynocarpa
P. piscocarpa
Popowia cynocarpa
var. vestita
Popowia lavata
Popowia filipes
Popowia novo-guineensis
Popowia pachypetala
Popowia papuana
Popowia parviflora
Popowia platyphylla
Popowia pisocarpa
Popowia polytricha
Popowia schefferiana

Notes:

1. The genus Popowia in Papuasia is difficult due to variability of some species.

26. PSEUDUVARIA

A genus of some 17 species distributed from Burma to Papuasia. 13 species are recorded from Papuasia and presumably Sinclair did not include this in the above estimates for the genus of 17 species. Neither did he include New Guinea within the distributional range of the species even though he included the new combination of P. dielsiana in his paper published in 1955.

Airy -Shaw (1939) and Merrill (1915) both retain it as separate from Mitrephora with which, as noted by Sinclair (1955) it has many similarities. The presence of unisexual flowers in

Pseuduvaria sessilifolia (A) flowering and fruiting twig.
(Source: Sinclair (1956)

Pseuduvaria distinguishes this genus from both Orophea and Mitrephora; however often staminodes are present and thus similarities can be seen to Orophea in which some species have 6 stamens and 6 staminodes. The main differences are summarised in the Table under Orophea.

Shrubs or trees. Leaves with well-marked, straight and nearly parallel veins. Flowers small, unisexual, solitary or fascicled in the axils of leaves or fallen leaves. Sepals 3, valvate, membranous. Petals valvate; outer 3 like the sepals but slightly larger; inner larger with narrower, linear claws, the blades usually hastate or diamone-shaped, united at top forming a mitriform hood, galnds sometimes present on the inner surface of the blades. Stamens numerous, very small, cuneate with flat-topped connectives which project over the anther cells at the top. Ovaries in male flowers absent. female flowers often with a ring of staminodes; sometimes 1-2 fertile or reduced stamens (ie. with 2 pollen sacs only) may be present. Ovaries tomentose, slightly elongated and incurved. Stigma sessile, sub-capitate with a U-shaped split at top and also a groove on the inner side. Ripe carpels densely and minutely tomentose, stalked or sessile, smooth or warted, 1-several seeded.

Pseuduvaria lignocarpa (A) flowering twig; (B) flower; (C) inner petals; (D) (E) stamens; (F) (G) fruit.
(Souce: Sinclair (1956)

List of Species in Papuasia

Pseuduvaria beccarii
Orophea beccarii
Pseuduvaria costata
Orophea costata
Pseuduvaria dielsiana
Orophea dielsiana
Goniothalamus dielsiana
Pseuduvaria dolichonema
Orophea dolichonema
Pseuduvaria filipes
Orophea filipes
Pseuduvaria grandifolia
Stelechocarpus grandifolia
Mitrephora grandifolia
Pseuduvaria lignocarpa
Pseuduvaria mollis
Goniothalamus mollis
Mitrephora mollis
Papualthia mollis
Pseuduvaria nova-guineensis
Pseuduvaria pulchella
Orophea pulchella
Pseuduvaria rhytidophylla
Orophea rhytidophylla
Pseuduvaria sessilifolia
Pseuduvaria silvestris
Orophea silvestris

27. RAUWENHOFFIA

● ● ● ● ● ●

A genus of 5 species distributed from Burma to New Guinea. The genus is very close to Uvaria the only difference being that the inner petals have narrow claws, a character absent in Uvaria (Sinclair 1955). Sinclair loc. cit. (229), suggests that the genus Rauwenhoffia should be combined with Uvaria.

Scandent shrubs. Leaves with stellate hairs. Flowers extra-axillary, opposite the leaves. Sepals 3, valvate, united at base. Petals 6, imbricate, the outer flat, the inner slightly smaller with a narrow claw. Stamens with convex connectives. Stigma funnel-shaped with opening on inner side, continuing as a groove down the ovary. Seeds several in 2 rows.

List of Species in Papuasia

Rauwenhoffia forbesii
Rauwenhoffia leichhardtii
    Melodorum leichhardtii
    Unona leichhardtii
    Uvaria lutescens
    Rauwenhoffia uvarioides
Rauwenhoffia oligocarpa
Rauwenhoffia papuasica

RICHELLA

A genus of 3 species in Fiji, New Caledonia and in Borneo. It could be collected in Papuasia.

SACCOPETALUM

Saccopetalum koolsii was described by Kostermans (Reinwardtia 1:3 (1952) 459 - 461. It is a species of Milusia.

28. SCHEFFEROMITRA

● ● ● ● ● ●

A monotypic endemic genus described by Diels in 1913 (Bot. Jahrb. 49: 152) based on Mitrephora subaequalis Scheffer (Ann. Jard. Buitenzorg 2 (1885) 20). Scheffer placed these as a monotypic variant of Mitrphora due to the similarities in structure of the flowers. Diels concluded that 'the two genera are not closely related'. Schefferomitra is deceptively similar in vegetative structure and fruits to some species of Oxymitra and Melodorum.

Hermaphroditic. Foliage glaucous. Sepals 3, triangular. Petals valvate; outer petals 3, pilose. Stamens numerous, connectives oblique. Styles terminal, ovule 1. Fruit scandent.

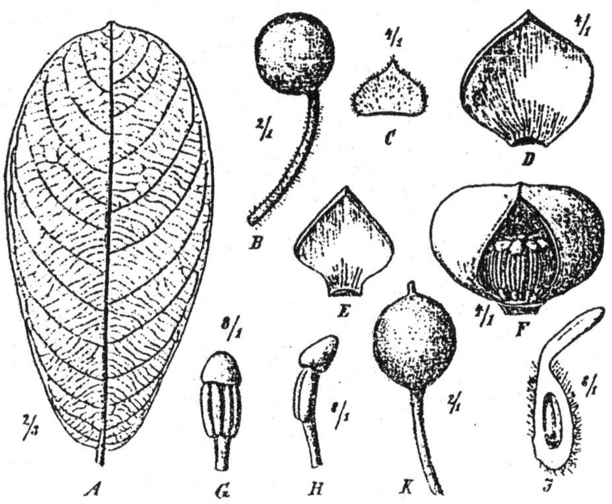

Schefferomitra subaequalis (A) leaf; (B) flower; (C) sepal; (D) outer petal; (E) inner petal; (F) flower; (G) (H) stamens; (J) ovary; (K) fruit.
(Source: Diels (1931)

List of Species in Papuasia

1 endemic species:

Schefferomitra subaequalis
Schefferomitra subaequalis var. *macrocalyx*

Notes

1. The type collection by Beccari (No. 365, 341) was collected at Andai. var. *macrocalyx* was collected by Beccari (No. 378, 368 and 371) at Putat near Arfak; and at Ramoi.

29. STELECHOCARPUS

A small genus of some 5 species distributed from Burma to New Guinea. Closely related to *Sageraea* in flower structure but differs in its cauliflorous habit, in the different stamens with anthers not embedded in connective tissue and the conical or hemispheric torus. The midrib is raised on the upper surface, a character distinguishing it from *Sageraea* (Sinclair 1955).

Trees with coriaceous leaves and raised midrib on upper surface. Flower buds globose. Flowers unisexual, fascicled on woody projections on the trunk or older branches, remote from the leaves. Sepals 3, imbricate, elliptic to orbicular. Petals 6, imbricate, concave, ronded, coriaceous. Stamens in male flowers many in several series, filaments short, connectives slightly lobed, flat-topped. Ovaries numerous, styles absent, stigma orbicular with several minute lobes. Ripe carpels globose, sessile with 4-8 seeds in 2 rows.

List of Species in Papuasia

Stelechocarpus burahol
Uvaria burahol

30. UVARIA

A genus of over 100 species mainly in T Asia but distributed to Africa, N Australia and Papuasia where it is represented by some species. Most species have showy red or brown flowers. Many species have been wrongly described in the genus and need to be removed to other genera.

Scandent or sarmentose shrubs usually with stellate pubescence. Flowers conspicuous, terminal or leaf-opposed, not axillary. Two bracts on the pedicel, one basal the other medial or at base of calyx. Sepals 3, valvate, often connate. Petals 6, imbricate, spreading, sometimes connate at base. Stamens many, the outer sometimes sterile and leaf-like. Style very short or inconspicuous, stigma horse-shoe or funnel shaped with a slit down the inner side. Ripe carpels dry or succulent, often sweet and edible.

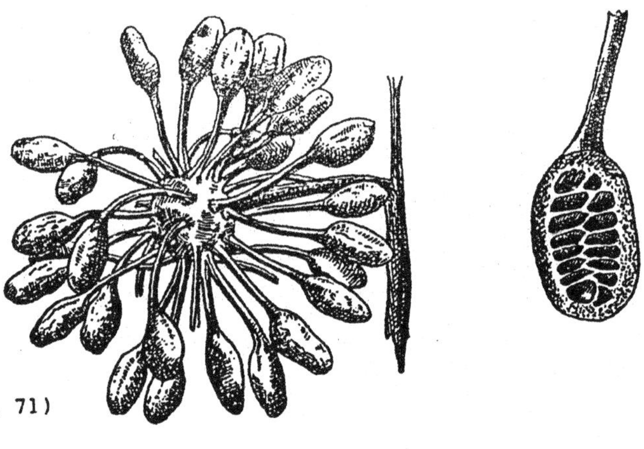

Uvaria branderhorstii
(Sources: *Nova Guinea* 8: Tab. 71)

List of Species in Papuasia

Uvaria branderhorstii
Uvaria cordata
Uvaria lauterbachii
Uvaria ?lutescens
Uvaria macrophylla
Uvaria purpurea
Uvaria purpurea
var. neoguineensis
Uvaria neoguineensis
Uvaria rosenbergiana
Uvaria rufa
Uvaria branderhorstii

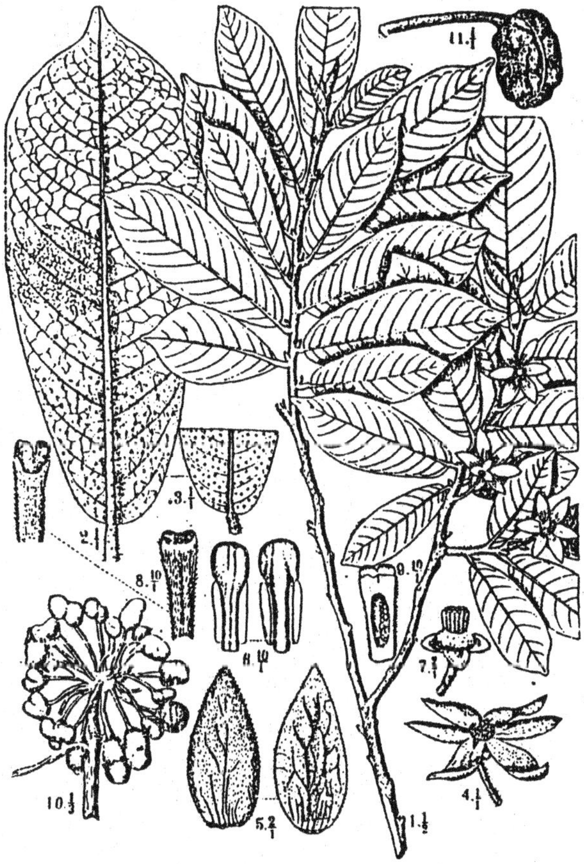

Uvaria javanica

31. XYLOPIA

• • • • • •

A genus of about 100 species in the Old and New World Tropics, numerous in Africa. Remotely related to Fissistigma but there are several differences between them.

Shrubs or trees. Flowers hermaphrodite, axillary, solitary, often fragrant; buds triquetrous or conic. Sepals 3, valvate, connate at base. Petals 6, valvate, linear, often elongated or ovate. When linear, outer petals flat, 1-veined, concave at base. Stamens elongated, the anthers transversely septate with large pollen grains. Ovaries 1 or more, style filiform, stigma clavate narrowly clavate or not broader than style. Ripe carpels elongated, cylindric or ovoid-oblong or obovoid. Seeds several (rarely 2), 1-2 seriate.

List of Species in Papuasia

Xylopia calosericea
Xylopia papuana
Xylopia peekelii

Notes:

1. Probably all Papuasian species have fragrant flowers and odour should be recorded by collectors.

2. Throughout the genus the stamens and ovaries are remarkably uniform and vary little. The septate anthers and filiform styles distinguish this genus.

MYRISTICACEAE

The family includes some 17 genera and 380 species. The largest genus Myristica with 125 species is centred in New Guinea. The trees mainly occur in the lowland tropical rainforest, but in New Guinea some members can be locally common in the montane zone. The family is distinctive and its relationships appear to be with the Monimiaceae and the Annonaceae.

Trees with male and female flowers on separate trees (dioecious) or monoecious in Endocomia. Copious red sap when trunk cut.

Diagnostic features.

Leaves: alternate, entire, often with glandular dots containing aromatic oils. Many species have the branches whorled and leaves almost horizontal.

Flowers: unisexual, small, inconspicuous, borne in inflorescences. Sepals 3, partly united. No petals. Male flowers have 2-45 stamens, filaments fused in a column or disc; anthers shed the pollen through two longitudinal slits. Ovary superior: 1-celled; style absent; stigmas 2. The flowers have a unilocular ovary.

Fruit: fleshy berry which splits on maturity into 2 or 4 valves (a rare example of a dehiscent berry). Seed covered in a brightly coloured (red or orange-red) aril, which can entirely cover the seed or be formed by a network of tissue.

References

Corner, E.J.H. (1983). The myristicaceous seed. Blumea 28: 419 - 420.

Foreman, D.B. (1974). Notes on Myristica in Papuasia. Contr. Herb. Aust. 9: 35 - 44.

Foreman, D.B. (1978). Myristicaceae. Handbook of the Flora of Papua New Guinea. 1: 175 - 215.

Schouten, R.T.A. (1986). Revision of the genus Gymnacranthera. Blumea 30: 173 - 196.

Sinclair, J. (1958). The genus Gymnacranthera in Malaysia. Gardens Bull. Sing. 16: 250 - 252, 435.

Sinclair, J. (1958). The genus Knema in Malaysia and outside Malaysia. Gardens Bull. Sing. 16: 254 - 332.

Sinclair, J. (1961). The genus Knema in Malaysia and outside Malaysia. Gardens Bull. Sing. 18: 102 - 327.

Sinclair, J. (1968). The genus Myristica in Malesia and outside Malesia. Gardens Bull. Sing. 23: 1 - 540.

MYRISTICACEAE

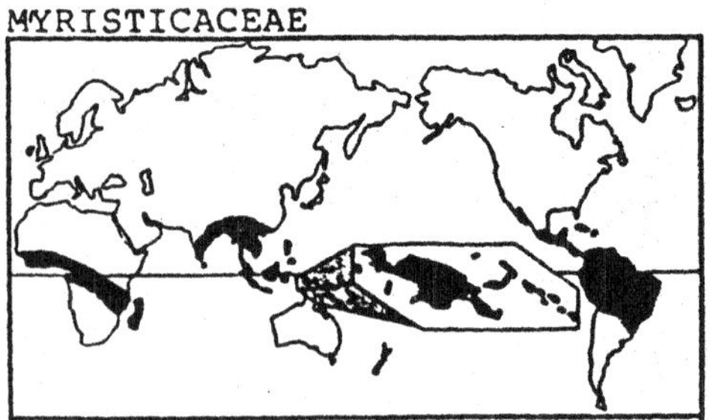

	WORLD	PAPUASIA
Number of genera :	17	5
Number of species :	350	70

Distribution : Tropical, mainly in lowland tropical rainforest.

Economic uses : Source of nutmeg and mace. Timber.

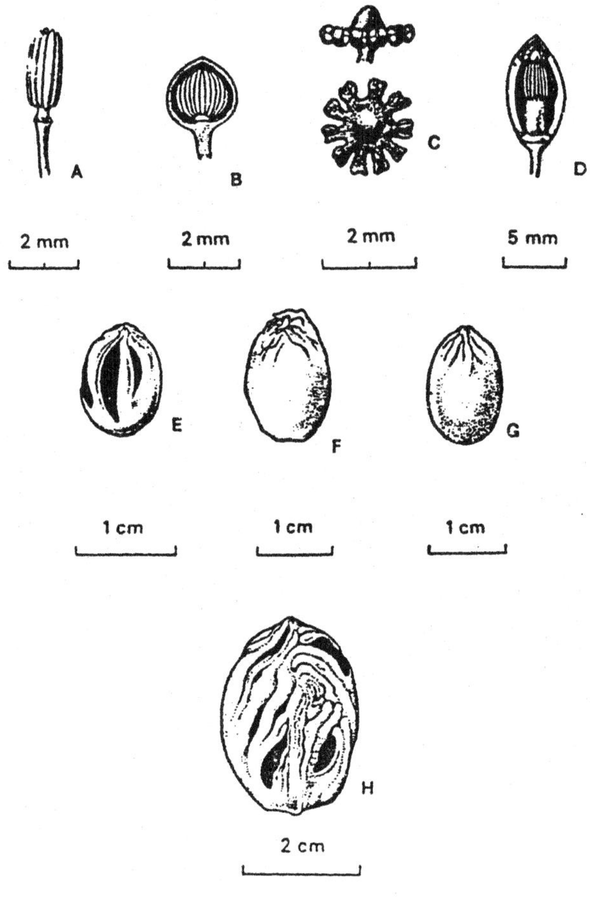

Gymnacranthera: (A) united filaments; (E) seed with aril divided almost to the base; Horsfieldia: (B) stamens free in the flower; (F) aril entire, completely covering the seed; Knema: (C) peltate disc below which the stamenss are attached; (G) aril almost completely covering the seed, lobed at apex; Myristica: (D) flower with column to which the stamens are dorsally attached. (H) seed with laciniate aril. (Source: Foreman (1978).

Sinclair, J. (1974). Horsfieldia. Gardens Bull. Sing. 27: 133 - 141.

Sinclair, J. (1975). Horsfieldia.. Gardens Bull. Sing. 28: 1 - 181.

de Wilde, W.J.J.O. (1979). New account of the genus Knema. Blumea 25: 321 - 478.

de Wilde, W.J.J.O. (1981). Supplementary data on Malesian Knema including three new taxa. Blumea 27: 223 - 234.

de Wilde, W.J.J.O. (1984). Endocomia a new genus in the Myristicaceae. Blumea 30: 173 - 196.

de Wilde, W.J.J.O. (1984). A new account of the Genus Horsfieldia. Pt. 1. Gard. Bull. Sing. 37: 115 - 179.

de Wilde, W.J.J.O. (1985). A new account of the Genus Horsfieldia. Pt. 2. Gard. Bull. Sing. 38: 55 - 144.

de Wilde, W.J.J.O. (1987). Further supplementary data on Malesian Knema. Blumea 32: 115 -141.

KEY TO THE GENERA IN PAPUASIA

(Based on Male Flowering Specimens)

(Source: de Wilde (1984)

1. Inflorescences short woody tubercle -like protuberances with the scars of previous pedicels and bracts. Bracteoles present............. 2

1. Inflorescence a branched panicle, the upper parts occasionally woody with a few scars. Bracteoles present or absent................. 3

2. Androecium a stalked disc with the anthers usually largely free, stellately attached by their bases. Bracteoles at the base of the flower or median on the pedicel. .... 4. Knema

2. Androecium a stalked column with the anthers completely fused to it by their backs, the apex of the column often sterile. Bracteoles embracing the base of the perianth on one side. .... 5. Myristica

3. Bracteoles present, at the base of the perianth. .... 5. Myristica

3. Bracteoles absent.............. 4

4. Androecium various, with the central column solid or excavated, usually considerably broader than the basal part or androphore. Perianth inside glabrous, the lobes or tepals in anthesis more or less erect, not reflexed. .... 3. Horsfieldia

4. Androecium elongated to depressed globose, the central column at apex not excavated, narrow, about as wide as or narrower than the androphore. Perianth inside (papillary) hairy, the lobes or tepals erect or reflexed in anthesis........... 5

5. Anthers elongate, mutually free in the apical portions. Perianth lobes suberect.. 2. Gymnacranthera

5. Anthers short, completely sessile, without apically free portions. Perianth lobes spreading or reflexed........ 1. Endocomia

KEY TO THE GENERA IN PAPUASIA

(Based on Female Flowering and Fruiting Specimens)

(Source: de Wilde (1984)

1. Aril completely covering the seed, laciniate to or nearly to the base, into many usually narrow segments. Hairs of leaf bud when dry flat and with a glassy appearance, each cell with 2 arms................... 2

1. Aril completely or but partly covering the seed, entire or at apex convoluted or laciniate up to about half-way. Hairs not particularly flat, not glassy, usually stellate, or dendroid, or filiform, each cell with 1 arm.................... 3

2. Twigs in older parts striate. Leaves above often with venation rather laxly reticulate. Bracteole present. Arms of hair cells of equal lengths... 5. Myristica

2. Twigs not striate. Reticulation of leaves indistinct above. Bracteole absent. Arms of hair cells of different lengths. .... 2. Gymnacranthera

3. Aril entire or shallowly laciniate or convoluted at apex. Seed not variegated, not pointed at one end. 4

3. Aril usually coarsely laciniate in the upper 1/3. Seed often bluntly pointed at one end. Bracteole absent. Stigma narrowly 2 - few-lobed. Leaves not whitish beneath (without alveolar tissue); reticulations lax. .... 1. Endocomia

4. Bracteole absent. Stigma usually minutely 2-lobed or 2-lipped. Leaves usually not whitish beneath; reticulations usually lax above. .... 3. Horsfieldia

4. Bracteole present. Stigma few- to many-lobed. Leaves usually whitish beneath; reticulations usually forming a dense network above. .... 4. Knema

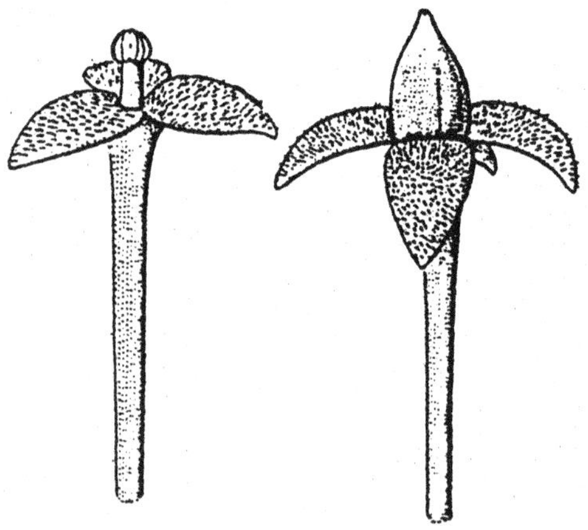

Endocomia macrocoma subsp. longipes (A) male flowers; (B) female flowers both at anthesis. (Source: de Wilde (1984)

1. ENDOCOMIA

• • • • • •

This genus was recently published by de Wilde (1984) to include 4 species previously described under the 'aggregate' species Horsfieldia macrocoma. The genus Endocomia can be distinguished from Horsfieldia (in brackets) by the following characters. Monoecious (dioecious); twigs without lenticels (usually with distinct lenticels); colour of perianth inside greenish-creamy to bright red and papillary-hairy to various degrees (perianth inside yellow and glabrous). de Wilde (1984: 174-175) lists several other character states to distinguish between the two genera.

A genus of 4 species, one with 3 subspecies, distributed from S China to New Guinea. In Papuasia a single species E. macrocoma (Miq.) de Wilde with 1 subspecies: subsp. prainii (King) de Wilde. This is fully described in de Wilde (1984: 187-190). The New Guinea collections are from the Vogelkop, Jayapura, Sepik, S Highlands and N Province. The subspecies occurs throughout the range of the genus.

Monoecious trees to 5-50 m tall. Leaf bud and twigs at apex rusty pubescent; hairs 0.2-0.5 mm. Inflorescence much branched, condensed, to 6-15 cm long. Flowers densely rusty pubescent. Pedicels 3-4 mm long. Male flowers: perianth 3-valved, cleft to 4/5, inside yellowish green; anthers 4, androphore 0.4-0.6 mm long. Female flowers: ovary ovoid; stigma broadly 2-lipped, finely lobulate. Fruits variable in shape; 1.7 -3.5 by 1.2-1.9 cm. Aril at apex laciniate from 1/3 to 1/2-way. Seed ± ellipsoid.

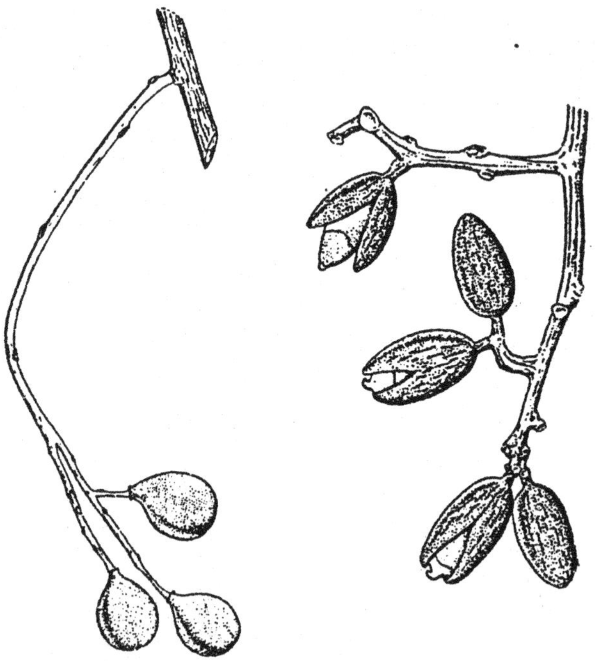

Endocomia macrocoma subsp. prainii (A) infrutescence of plant from Irian Jaya; Endocomia macrocoma subsp. macrocoma (B) infrutescence. (Source: de Wilde (1984)

Key to Species in Papuasia

1 widespread species:

Endocomia macrocoma subsp. prainii

Notes:

1. In Papuasia the species occurs in lowland and hill forest, alluvial and swamp forest. In the Sepik area the trees are used for house-building.

2. The subspecies is quite variable throughout its range. The nature of the variation is discussed by de Wilde (1984: 189 - 190).

2. GYMNACRANTHERA

A genus of 7 species (Schouten 1986) distributed from S India to Papuasia. In Papuasia a single species of tree: from 10-25m tall. The species occurs throughout the mainland, Papuan Islands and the Bismarck Arch. It has not been collected in the Solomon Islands. Foreman (1978) followed Sinclair (1958) in that he treats the New Guinea taxon as a variety of Gymnacranthera paniculata. As with other genera it has characteristic red sap when cut. The aril is segmented almost to the base.

Small to medium sized trees, 10-25 m tall. Twigs smooth and glabrous, not striate. Leaves glabrous, whitish or glaucous beneath, few almost invisible reticulations, nerves arching near margins. Inflorescence paniculate, axillary or in axils of fallen leaves; bracts falling at early stage; no bracteoles. Perianth 3-toothed, teeth acute. Male flowers with filaments fused into a column, anthers elongate and covering column. Female flowers: stigma minutely bilobed, sessile. Fruit small. Aril segmented almost to base.

Key to Species in Papuasia

1 widespread species;

Gymnacranthera farquhariana
var. zippeliana

Notes:

1. This is the most widespread taxon in the genus being distributed from Malaya to Papuasia.

2. It is a common understorey tree in secondary and primary forest, growing along streams and also on ridges at low altitudes (sea level to c. 1000m).

3. In New Guinea the bark is used with lime to prepare the skins of birds.

3. HORSFIELDIA

The genus Horsfieldia extends from Sri Lanka to the Solomon Islands. Included in the most recent revision by de Wilde (1984) are 100 species placed into three sections. In Papuasia all species belong to section Irya. Almost all the 40 species in this section have a 2-valved perianth: but some are aberrant and have a 3-4 or 2-4-valved perianth. Most species are local or restricted in distribution; the major exception is H. irya which covers almost the whole range of the species. The major centers of species diversity are New Guinea and Borneo.

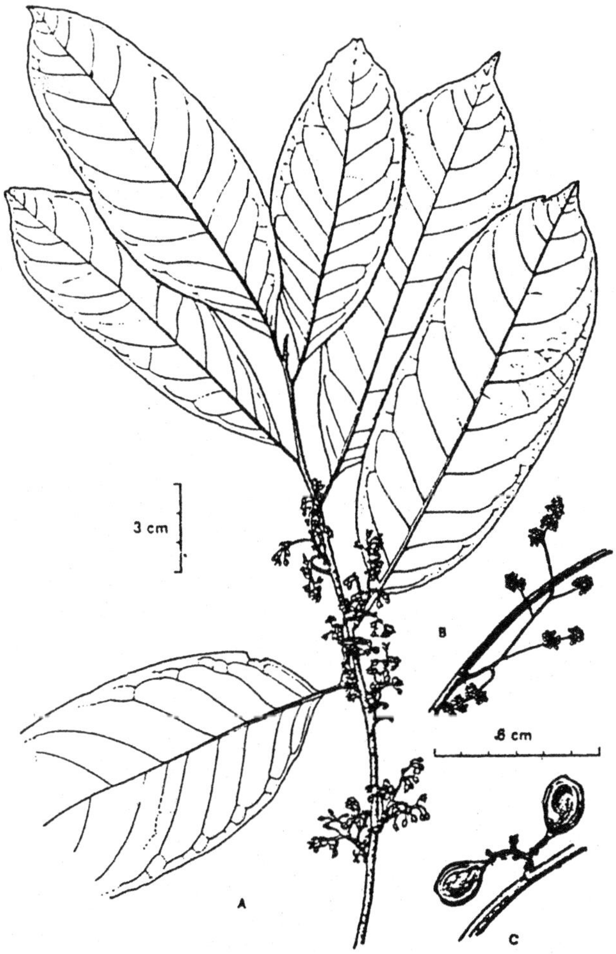

Gymnacranthera paniculata var. zippeliana (A) flowering twig, female flowers; (B) male flowers; (C) fruits. (Source: Foreman (1978).

de Wilde (1984) segregates some (4) species previously placed in the H. macrocoma complex into a new genus Endocomia.

Trees from 40-60m tall, dioecious. twigs early glabrescent, terete or angular or with two raised lines from petiole to petiole, bark striate. Leaves distichous, petioled, to 45 cm long, reticulations never forming a dense, close network (as in Knema), lower surface not glaucous. Inflorescences axillary, rarely on old wood, paniculate usually several times branched; female inflorescences usually smaller than male. Perianth thin-leathery to succulent; either 2- or 3- or 4-lobed, inside glabrous, greensih to yellowish, never red. Male flowers: perianth usually small, anthers 2-c.25. Female flowers: perianth usually larger than male; style absent; stigma 2-lobbed or 2-lipped, usually small. The fruits ellipsoid or globose, perianth sometimes persistent, characterised by the aril completely covering the seed, entire or shallowly lobed at apex.

Key to Species in Papuasia

(Based on male flowering specimens)

(Source: de Wilde (1984)

1. Phyllotaxis of plagiotropic (fertile) twigs distichous. Leaves membranous, usually with largish whitish marks of irregular shape and size. Male perianth 2-valved, globose, c. 1-1.5 (-2.0) mm diam. Androecium not or hardly laterally compressed; androphore distinct, tapering, broad above and hollowed to a saucer-shape, shallowly to deeply; anthers 6-10, their apices free to only c. 0.3 mm and ± incurved. (Sri Lanka to Solomon Is., generally not too far from the coast)..... H. irya

1. Phyllotaxis of plagiotropic twigs either distichous (alternate), or dispersed (spirally), or mixed in the same specimen. Leaf consistency various, usually, without whitish marks. Male perianth 2 - 4-valved, various in shape and size. Androecium various, but if hollow, deeply cup- or saucer-shaped then the anther-apices either mutually free for at least 1/2-way, or the anthers strongly inflexed into the cup (anthers connate or mutually free, number of anthers few to many)........................ 2

2. Perianth predominantly 2-valved, or sometimes a few flowers in an inflorescence 3- or 4-valved... 3

2. Perianth predominantly 3-valved, or predominantly 4-valved, or both 3 and 4 valves present; sometimes the odd flower in an inflorescence has a 2-valved perianth........ 28

3. Androecium ± cup-, or bowl-, or saucer-shaped, moderately laterally compressed; the anthers at one or both sides of the androecium distally distinctly incurved or inflexed into the cup-shaped central cavity. (N W New Guinea)....... 4

3. Androecium laterally flattened or not, the central column either (1) solid, or at apex, (2) broadly but shallowly hollowed only up to c. 1/3, or (3) narrowly channelled like a slit to various depths; the anthers straight or only slightly incurved, never inflexed into the cup-shaped central cavity. Moluccas to Solomon Is.................. 5

4. Leaves below with regularly spaced reddish-brown to brown-black dots presumed to be non-traumatic cork warts originating from hair-scars (use lens), not to be confused with a much finer punctation if also present. Twigs angular or ridges especially in the upper portion. .... H. inflexa

4. Leaves without dots. Twigs either terete, or angular, or winged. Male perianth together with the pedicel ± pear-shaped, the pedicel tapering. Petiole relatively long, c. 1.0-2.6 cm, leaf blade (6-)8 - 22(-25) cm. Twigs at apex generally terete, not angular. .... H. moluccana

5. Male perianth angular before anthesis, arranged into dense semi-globose clusters. Androecium much longer than wide, not or but slightly laterally compressed. Leaves lanceolate to lanceolate-linear, usually ± parallel-sided. .... H. sylvestris

5. Flowers not densely clustered, not angular in mature buds.......... 6

6. Twigs sharply angular, or markedly ridged, or winged from petiole to petiole at the apex of the twig as well as lower down in between the older leaves and in older wood; i.e., twigs not merely more or less distinctly lined.............. 7

6. Twigs not winged, i.e. terete or only somewhat angular at apex, lower down terete or merely provided with lines from petiole to petiole (the bark striate or not). 9

7. Male perianth 2- 3- or 4-valved, subspherical, hardly or not laterally compressed, drying ± glossy, not collapsing on drying. (New Guinea: Vogelkop Penin.) .... H. angularis

7. Male perianth predominantly 2-valved, little or much laterally compressed, drying matt, slightly or strongly collapsing on drying... 8

8. Leaves thinly coriaceous. Pedicels of male flowers about as long as or longer than the perianth. Perianth at anthesis cleft almost to the base. Hairs of inflorescence and pedicels c. 0.2 - 0.3 mm. Anthers 10-14; the central column of androecium at apex hollow for c. 1/4. .... H. iriana

8. Leaves membranous. Pedicel shorter than the perianth. Perianth at anthesis cleft to c. 2/3 - 3/4. Inflorescences and pedicels almost glabrous, with hairs of c.0.1 mm or less. Anthers (12-)14 - 18; the central column of androecium solid or almost so. .... H. aruana

9. Male inflorescence large, 25 - 35 cm long. Male perianth ± pear-shaped. Androecium longer than broad; androphore c. 0.5 mm long or more, about half as long as the anthers or longer.............. 10

9. Male inflorescences c.20cm long or less. Male perianth of various shapes. Androecium longer or shorter than broad; androphore short or long........................ 11

10. Male perianth glabrous (?), c. 4 x 2 mm. Anthers 10, androphore nearly as long as the anthers. Inflorescence glabrescent, to 25cm long. .... H. ampla

10. Male perianth pubescent, c. 3 by 3 mm. Anthers 7, androphore about half as long as the anthers. Inflorescence pubescent, c. 25 - 35cm long....... H. ampliformis

11. Male inflorescence delicate, 2 - 5 (-8)cm long 1 or 2 (or 3) times ramified. (New Guinea, including Aru Is.)..................... 12

11. Male inflorescences generally stouter, c. 5 - 20cm long, not or 1 - 4 times ramified (inflorescences of H. sinclairii and H. basifissa sometimes small). Celebes to Pacific.......................... 16

12. Male perianth pubescent or late glabrescent, distinctly or only somewhat longer than broad.. Anther-bearing part of the androecium much shorter than the elongate club-shaped androphore......... 13

12. Male perianth glabrous or early glabrescent, about as long as or shorter than broad. Androphore much shorter than the anthers.. 15

13. Leaves broadly obovate or elliptic to oblong, c. 12-17 x 5-11cm. Male perianth gradually passing into the thick and tapered pedicel, both together c. 10-12mm long. Androphore glabrous. .... H. crux-melitensis

13. Leaves elliptic to lanceolate, 4.5-20 x 0.7-6cm. Male perianth with pedicel 4-6(-9)mm 14

14. Leaves drying olivaceous. Male perianth with pedicel together club-shaped, c.5-5.5 x 2-2.2mm. Androphore glabrous. .... H. clavata

14. Leaves drying dark brown. Male perianth broadly ellipsoid, c.2.0-2.4 x 1.8-2.2(-2.4)mm, at base narrowed into a slender pedicel 2-3.5(-6)mm. The surface of the androphore in the upper part wrinkled-bullate, towards the base either densely, minutely scaly-hairy, or striate but glabrous. .... H. squamulosa

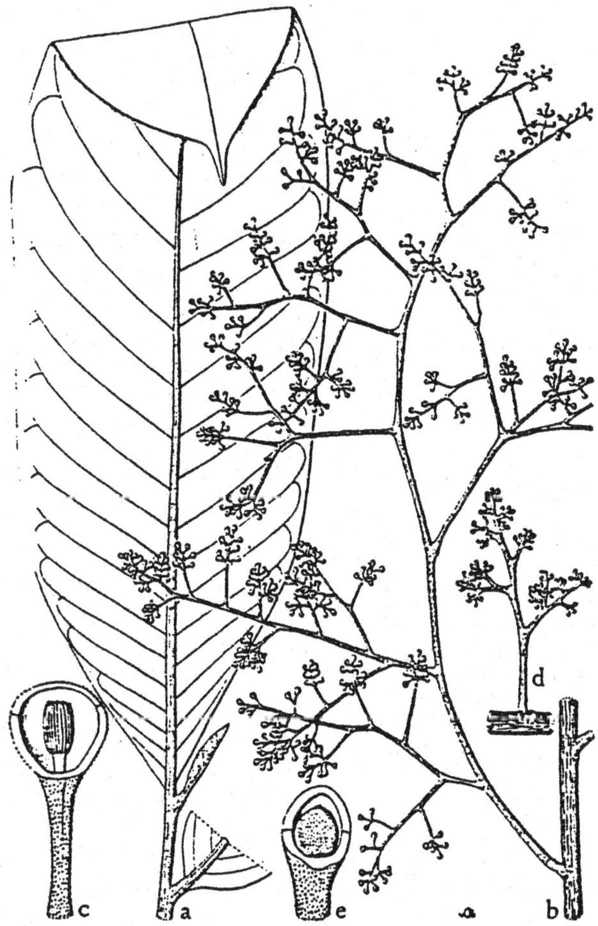

Horsfieldia ampliformis (A) twig with leaves; (B) male inflorescence; (C) male flower; (D) female inflorescence; (E) female flower.
(Source: de Wilde (1984).

15. Male perianth ± laterally compressed, generally ± obtriangular in lateral view, 1.8 - 3mm wide, usually ± collapsing on drying. Androphore slender, c.0.2.-0.5mm long, usually much shorter than the anthers... H. subtilis

15. Male perianth generally subglobose, 1-2mm wide, not or but slightly compressed, the surface wrinkled on drying but not collapsing. Androphore 0.4.-0.5mm long, about half the length of anthers. .... H. schlechteri

16. Male perianths together with the pedicels generally ± pear-shaped; the upper part of the perianth broadly rounded in lateral view, the lower (1/4-) 1/3-1/2 tapered and gradually passing into the ± tapered pedicel (these characters not always clear in certain specimens of H. tuberculata). Perianth glabrous, pubescent or glabrescent 17

16. Male perianth in lateral view of various shapes, either circular, or ovate, or obovate, or elliptic, or transversely elliptic, or reniform; at base either short-attenuate, or rounded, or truncate; not tapered, the pedicels generally not thickened towards the perianth. Perianth either hairy, at least at the base, or in H. psilantha, H. basifissa, H. sinclairii glabrous, subglabrous or glabrescent..... 19

17. Perianth glabrous. A variable species...... H. tuberculata

17. Perianth minutely pubescent, or early glabrescent in H. corrugata 18

18. Male pedicel stoutish, 2-4mm long. Male perianth ± membranous, glabrescent, with or without a few scattered blackish warts. Anthers c.8-12........ H. corrugata

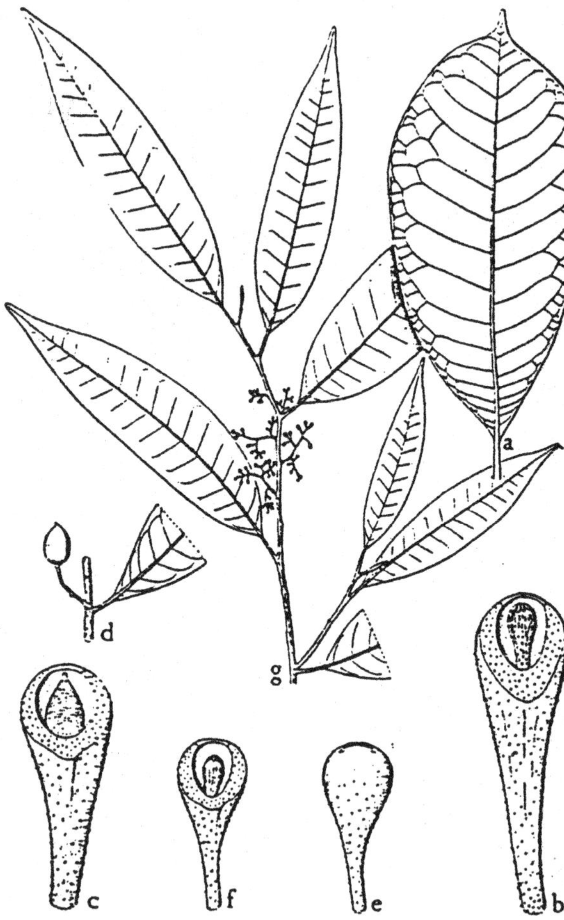

Horsfieldia crux-melitensis (A) leaf; (B), (E) & (F) male flower; (C) female flowers; (D) mature fruit.
H. squamulosa (G) twig with male inflorescences.
(Source: de Wilde (1984).

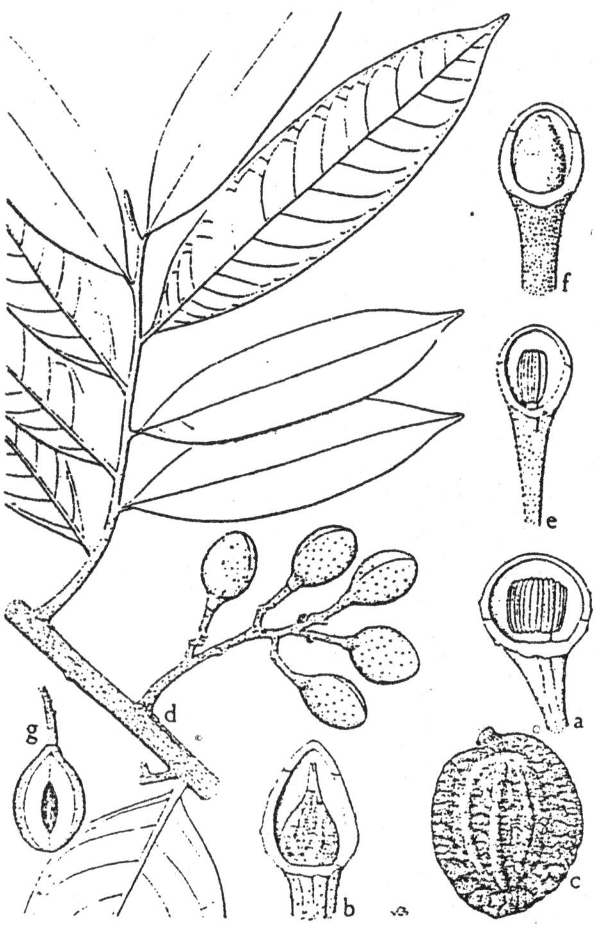

Horsfieldia corrugata (A) male flower; (B) female flower, pubescent ovary and narrow 2-lobed style; (C) fruit.
H. pachycarpa (D) leafy twig with infructescence; (E) male flower; (F) female flower; (G) almost mature fruit.
(Source: de Wilde (1984).

18. Male pedicel stout or rather slender, 2-5mm long. Male perianth ± fleshy, pubescent, without blackish warts. Anthers 5-10.
.... H. pachycarpa

19. Inflorescence moderately pubescent to nearly glabrous. Tomentum on leaf buds and apical portions of twigs short, rusty or greyish, the hairs 0.1-0.4(-0.5)mm long. Leaf largely or entirely (early) glabrescent, or with scattered stellate hairs beneath especially the younger ones 20

19. Inflorescence generally thick-woolly tomentose. Tomentum on leaf bud and apical portion of twig usually rust-coloured, conspicuous, its hairs coarse and long, (0.3-)0.5-1.5mm. Leaves with (sub) persistent tomentum, at least on and near the midrib beneath............ 25

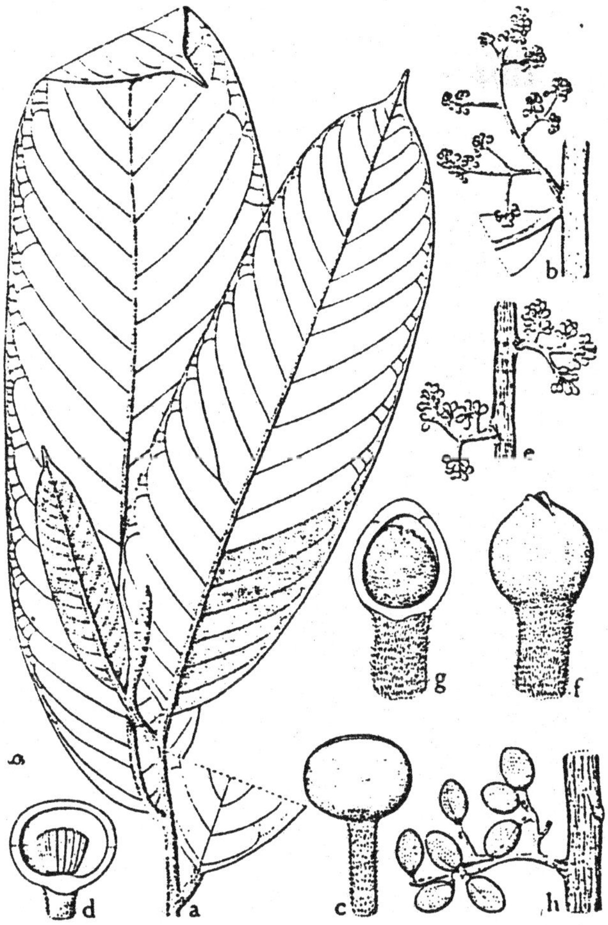

Horsfieldia hellwigii var. hellwigii (A) leafy twig; (B) twig with male inflorescence; (C) & (D) male flower; (E) twig with female inflorescences; (F) & (G) female flower; (H) mature fruits. (Source: de Wilde (1984).

20. Male perianth largely glabrous, subglobose, 2-3mm diam., at anthesis cleft to the base, not collapsing on drying.
.... H. basifissa

20. Male perianth glabrous or hairy, of various sizes and shapes, at anthesis cleft to c.1/2-way to near the base; perianth collapsing on drying or not........................ 21

21. Male perianth wholly or almost wholly clothed with ± persistent tomentum, the hairs may be scattered and very minute. Leaves drying olivaceous to brown, without a reddish tinge; marginal nerve not conspicuously regularly looped. Hairs of leaf bud c. 0.1-0.2mm long, usually greyish. Moluccas, New Guinea, New Britain; not in the Solomon Is.................... 22

21. Male perianth either glabrous, or early or late glabrescent, at least the upper 4/5. E. New Guinea to Solomon Is.................... 23

22. Mature male perianth small, c. 1.2-1.9mm diam. A variable species.
.... H. pilifera

22. Mature male perianth 2.0 - 3.3mm diam. A variable species.
.... H. laevigata

23. Perianths / pedicels glabrous; male perinth c. (2.0-)2.5 - 3.5(-4.0)mm diam. Leaves rather large, 20-40cm long, drying olivaceous to brown, the marginal nerve rather indistinct and not very regularly looped. Bagabag Is., Long Is., New Britain, New Ireland.
.... H. psilantha

23. Perianths and pedicels glabrous or pubescent; male perianth (1.0-) 1.5 - 2.0(2.5 mm diam. Leaves drying generally with a reddish tinge, especially the midrib and nerves........................ 24

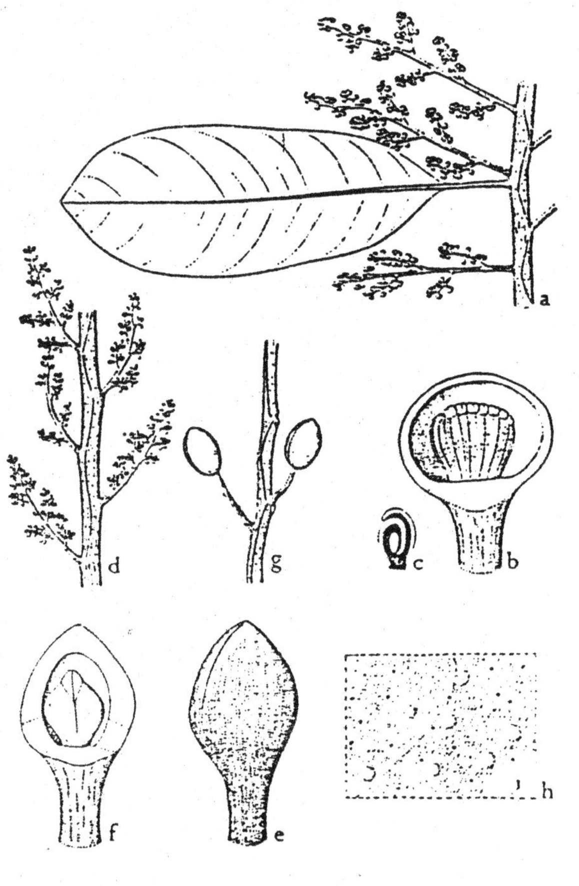

Horsfieldia inflexa (A) lined twig with male inflorescences; (B) male flower; (C) androecium; (D) twig with female inflorescences (E) & (F) female flower; (G) twig with infructescences; (H) lower leaf with cork warts. (Source: de Wilde (1984).

24. Leaves usually large, 9 - 30(-40)cm long, the marginal nerve very distinct and conspicuously regularly looped. Hairs on leaf bud 0.1-0.4mm, average c. 0.3mm long, usually reddish brown, sometimes greyish. Pedicel and base of perianth minutely pubescent. Male perianth c. 1.5 - 2.0mm diam., cleft at anthesis to c. 9/10. Solomon Is. .... H. whitmorei

27. Male perianth obovoid or ellipsoid. Anthers c.10(-12). Leaves generally oblong-lanceolate, at apex caudate....... H. ralunensis

28 Twigs distinctly ridged or nearly winged in between the petiole insertions, also in the older wood. 29

28. Twigs not ridged; sometimes twigs faintly ridged, or line or angular in the apical portion only. .... H. sepikensis

29. Male perianth slightly broader than long, short-pubescent in the lower half. Androecium slightly broader than long; anthers erect, not incurved.... H. angularis

29. Male perianth subglobose to broadly ellipsoid, glabrous. Androecium longer than broad, ± obovoid, the anthers with apex free and incurved, those of one side of the androecium clasping the others: .... H. olens

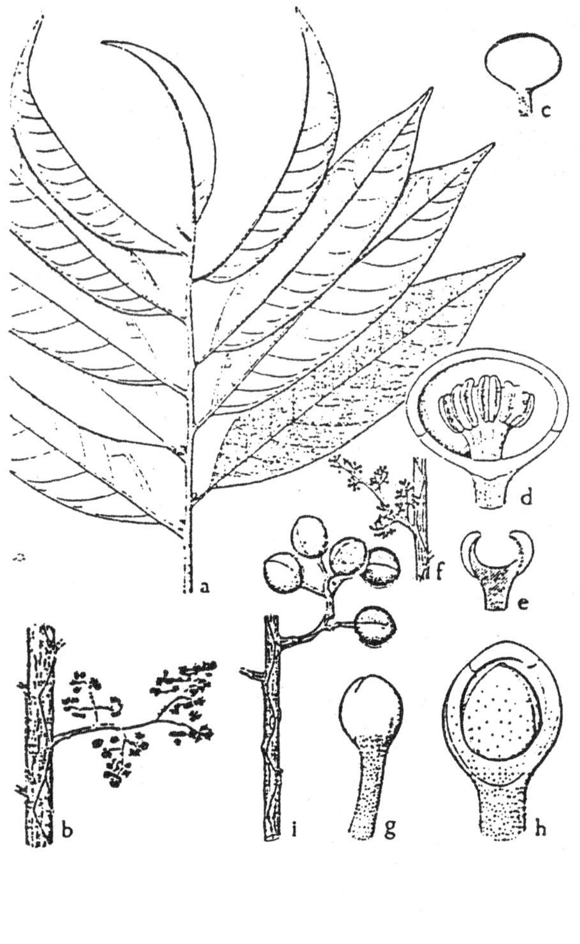

24. Leaves smaller, 6 - 20cm long, the marginal nerve indistinct and not very regularly looped. Hairs of leaf bud to c.0.1mm long, greyish brown. Pedicel and perianth glabrous. Male perianth c. 1.1 - 2.0mm diam., at anthesis cleft to c.1/2-way; perianth larger and pedicel pubescent in deviating specimens. E New Guinea. .... H. sinclairii

25. Perianth in the male largely pubescent, towards the base thick-walled and coriaceous, the remainder collapsing on drying; both male and female opening at apex by small pore-like slit less than 1mm long. Androecium subellipsoid, mainly consisting of the column with 2 minute anthers situated at the top, just below the pore. Leaves usually coriaceous, ± bullate; clothed with harsh hairs which when shed leave rough thickened bases. .... H. pulverulenta

25. Perianth glabrous or pubescent, membranous or chartaceous.

Slit to at least c.1/3. Androecium mainly consisting of 10-16 sessile anthers. Hairs on lower leaf surface not harsh, when shed not leaving rough thickened bases...... 26

26. Perianth pubescent, in male at anthesis cleft to c.3/4-5/6. Anthers 10-14. Leaves chartaceous or membranous. H. leptantha

26. Perianth glabrous, except at the very base; at anthesis cleft to c.1/2-way or less. Leaves generally membranous................. 27

27. Male perianth subglobose. Anthers 12-16. Leaves generally oblong to oblong-lanceolate, at apex not-caudate (always?). .... H. hellwigii

Horsfieldia irya (A) leafy twig; (B) ridged twig with male inflorescence; (C) male flower bud; (D) male flower; (E) androecium; (F) twig with female inflorescence; (G) & (H) female flower. (I) spherical fruits. (Source: de Wilde (1984).

Key to the Species in Papuasia

(Based on Female Flowering and Fruiting Specimens).
(Source: de Wilde (1984)

1. Leaves membranous, often with irregularly shaped whitish blotches here and there. Female perianth 2-valved; ovary glabrous. Fruit perfectly globose, c. 1.5-2.0cm diam., glabrous; pericarp c. 1-2mm thick; seed globose. Plant usually growing not too far from the coast. .... H. irya

1. Leaves of various consistency, generally without whitish blotches. Fruit either ellipsoid or globose; if globose either only c. 1cm diam. (H. subtilis), or the pericarp more than 2mm thick, at least at one side; seed mostly ellipsoid. Fruit glabrous or pubescent. Plant coastal or not.................. 2

2. Twigs towards apex distinctly angled or ridged from petiole to petiole. Aru Is. New Guinea...... 3

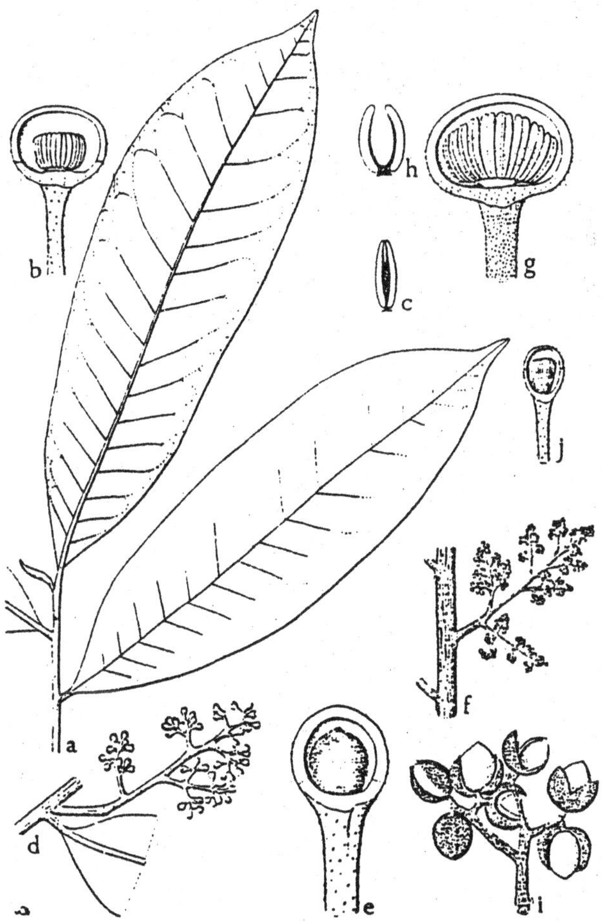

Horsfieldia laevigata var. laevigata (A) twig with leaves; (B) mature male flower; (C) androecium, longtitudinal section; (D) female inflorescence; (E) female flower, finely pubescent ovary, minute lobed stigma.
var. novobritannica (F) twig with male inflorescence; (G) male flower; (H) androecium, longitudinal section; (I) infructescence with mature fruits.
H. pilifera (J) male flower.
(Source: de Wilde (1984).

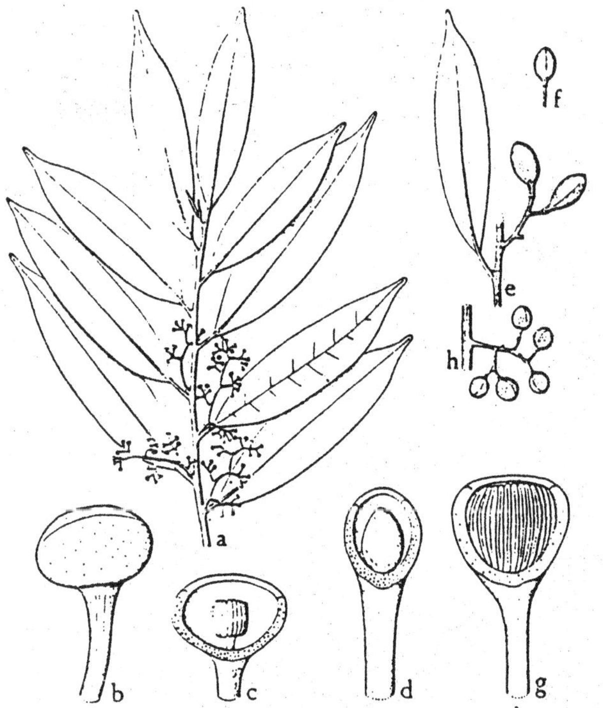

Horsfieldia (A) male inflorescence; (B) male flower; (C) stiped androecium; (D) female flower, glabrous ovary; (E) & (F) mature fruits.
H. subtilis var. subtilis (G) male flower; (H) twig with mature infructescence.
(Source: de Wilde (1984).

2. Twigs terete, sometimes lined in between the bases of petioles but neither angled nor ridged. New Guinea to Solomon Is............ 8

3. Leaves on lower surface with regularly scattered blackish-brown non-traumatic dots or cork warts originating from hair bases (lens!). Perianth 2-valved. Vogelkop to W. Sepik P. .... H. inflexa

3. Leaves without blackish-brown dots. 4

4. Perianth 3-(or 4-) valved 5

4. Perianth 2-valved.............. 6

5. Ovary glabrous (?). Fruit 10-16mm long, glabrous. Leaves chartaceous, 7-14cm long; petiole relatively long and slender, 11-20mm long. SW and S New Guinea (Digul W.P.)... .... H. olens

5. Ovary pubescent. Fruit 17 - 20mm long, pubescent at base. Leaves membranous to thinly chartaceous, 10-27cm long; petioles 7-15mm. Vogelkop penins. .... H. angularis

6. Female perianth depressed-globose, valves nearly 1mm thick; ovary pubescent. Fruit 17-20mm long, pubescent. Leaves membranous to thinly chartaceous. Vogelkop Pen. .... H. angularis

6. Female flower and fruit not known. 7

7. Leaves membranous. S W New Guinea; .... H. aruana

7. Leaves thinly coriaceous. S W New Guinea.... H. iriana

8. Perianth 3- (or 4-) valved; at anthesis cleft almost to the base. East Sepik P. .... H. sepikensis

8. Perianth 2-valved 9

9. Ovary and fruit glabrous 10

9. Ovary and fruit pubescent; hairs on fruit either distinct or small and inconspicuous and only to be seen remaining at the base of the fruit near the insertion of the stalk (lens!). Fruit ellipsoid, sometimes in H. sinclairii globose. 16

10. Tomentum of leaf bud, apex of twig, and inflorescence, with hairs 0.3-1.5mm long. Leaves large, 17-45cm long often ± parallel-sided, nerves 30-40 pairs. Fruit ellipsoid, 3.4-5.5cm long, glabrous. W and C New Guinea.... H. sylvestris

10. Tomentum with hairs c.0.2mm long or less; hairs in H. moluccana and H. tuberculata c. 0.1 - 0.3mm long. Leaves generally smaller, nerves fewer......................... 11

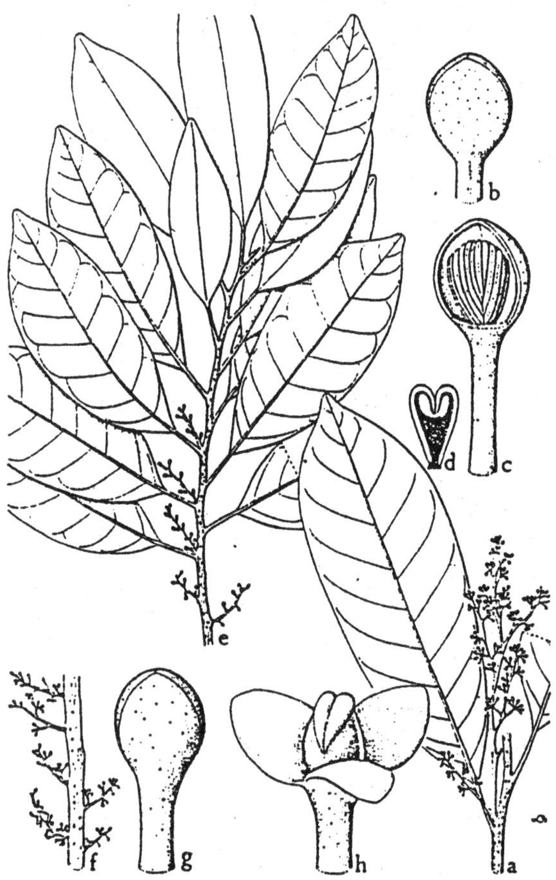

Horsfieldia sepikensis (A) twig portion with male inflorescence; (B) mature male flower, lateral view; (C) ditto, opened, showing androecium; (D) androecium, longitudinal section. (E) & (F) portions of female flowering twig, inflorescences axillary to leaves and fallen leaves; (G) mature female flower bud; (H) ditto, at full anthesis, showing glabrous ovary and large broadly 2-lipped stigmas.
(Source: de Wilde (1984).

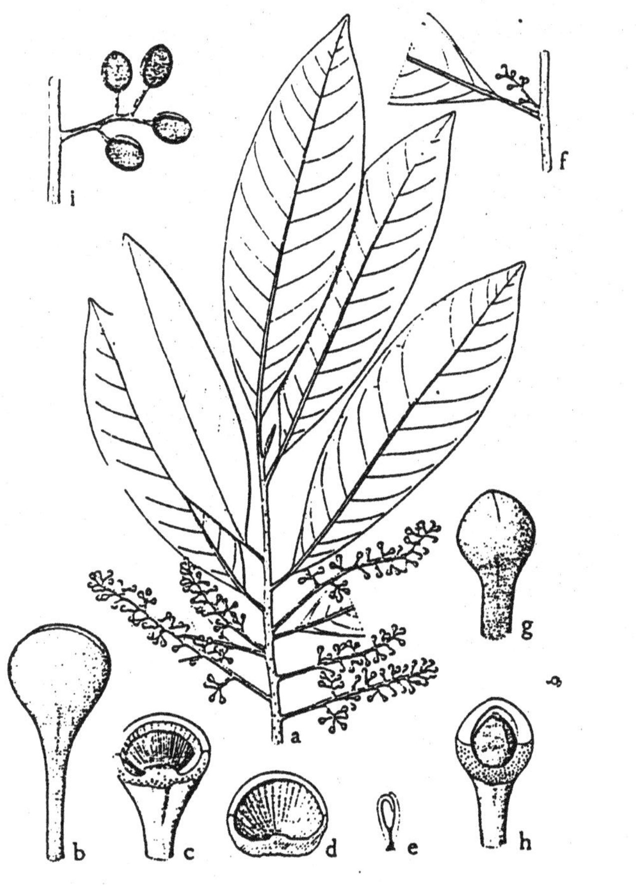

Horsfieldia spicata (A) leafy twig with male inflorescences; (B) mature male flower, lateral view; (C) male flower, opened, showing androecium; (D) male perianth, inner side, showing impression of androecium; (E) androecium, longitudinal section, schematic; (F) twig portion with female inflorescence; (G) mature female flower, lateral view; (H) ditto, opened, showing glabrous ovary and 2-lipped stigma; (I) twig portion with infructescence.
(Source: de Wilde (1984).

11. Fruit globose or subglobose, not beaked and without pseudostalk, c. 1.4cm diam. or less............ 12

11. Fruit ellipsoid, c. 1.3cm long or more.......................... 13

12. Fruit drying brown; dry pericarp c. 1.5 - 3mm thick. Female perianth at anthesis cleft nearly to the base. N Irian Jaya and N Papua New Guinea.... H. basifissa

12. Fruit drying blackish; dry pericarp c. 1mm thick. Female perianth at anthesis cleft to c. 1/3. Aru Is. New Guinea. H. subtillis var. subtilis

13. Fruit up to c. 2.0cm long, drying blackish; top pointed/beaked or not, base without or with long or short pseudostalk.............. 14

13. Fruit 1.3-3.7cm long, drying brown or dark brown; top rounded, base without pseudostalk............ 15

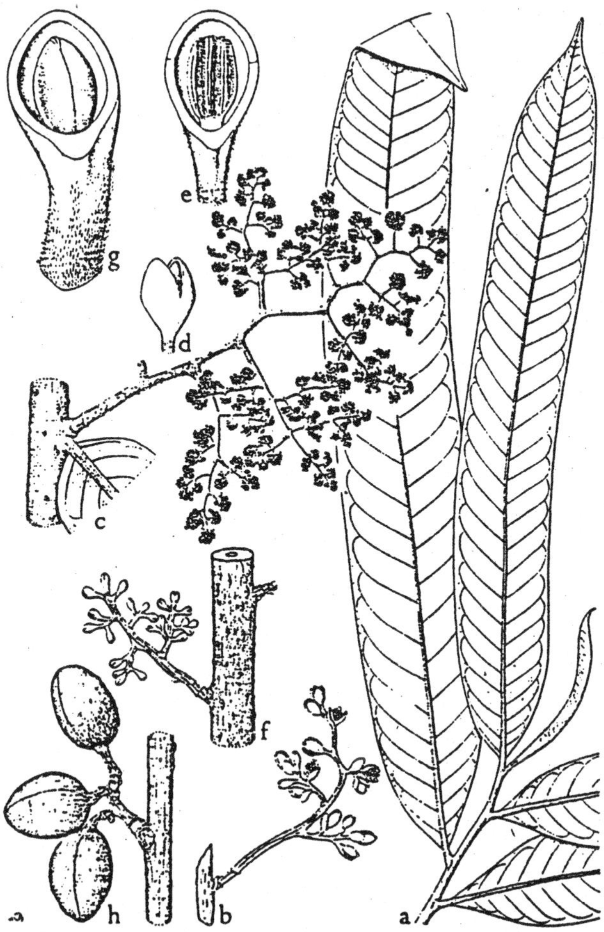

Horsfieldia sylvestris (A) leafy twig apex; (B) & (C) immature and full-grown male inflorescences, (D) & (E) male flower; (F) & (G) female inflorescence; (H) mature fruits.
(Source: de Wilde (1984).

14. Pseudostalk of fruit (1.5-)2 - 6mm long. Jayapura Dist. (Irian Jaya), W. Sepik P. H. schlechteri

14. Pseudostalk absent or up to 3mm long. Throughout New Guinea. .... H. subtilis

15. Female perianth c. 2mm long, cleft at anthesis to c. 1/2-4/5. Fruit 1.3 - 2.8cm long, dry pericarp 1-2mm thick. Irian Jaya. .... H. moluccana

15. Female perianth 2 - 3mm long, cleft at anthesis to c. 1.2 - 2/3. Fruits 1.5 - 3.7cm long, pericarp 1 - 8mm thick. Bismarck Arch., Papuan Is. and Milne Bay P. .... H. tuberculata

16. Arches of the submarginal nerve on the lower leaf surface distinct and very regularly looping; nerves often reddish brown. Leaves often ± parallel-sided. Tomentum of leaf bud composed of hairs c. 0.1 - 0.3 (-0.4)mm long. Female inflorescence and infructescence up to c. 7cm long. Fruit 1.7 - 2.5(-3.4)cm long, drying orange to brown; pericarp 1-2mm thick. Solomon Is. .... H. whitmorei

16. Marginal nerve not very regularly looping. Not from Solomon Is.. 17

17. Tomentum of leaf bud, apex of twigs and inflorescence short, composed of hairs c. 0.2mm long or less; in H. psilantha hairs c. 0.1 - 0.3mm 18

17. Tomentum long, with hairs c. 0.5 - 1.0mm long; hairs 0.2 - 0.5mm long in H. ampliformis; tomentum not known in H. ampla 28

18. Tomentum of leaf bud composed of hairs c. 0.1 - 0.3mm long. Infructescence (and inflorescence) large, much branched, 10-16cm long. Fruit 1.7-2.2cm long, pericarp thin, 1-2mm thick, brown. Bismarck Arch., Bagabag Is., Long Isl. .... H. psilantha

18. Tomentum of hairs 0.1 - 0.2mm long, or less. Female inflorescence and infructescence c. 10cm long or less. Fruits various.......... 19

19. Female perianth incl. pedicel long-obconical. Fruit stalk tapering, thickened towards the fruit. Mature fruit not known. E New Guinea: Morobe P.. H. crux-melitensis

19. Female perianth ± ellipsoid, well marked off from the slender pedicel. Fruit stalk not tapering. 20

20. Fruit 1.6 cm long or less; pericarp 1 - 3mm thick. Perianth pubescent. 21

20. Fruit 1.5cm long or more; pericarp 2mm thick or more; if fruit c. 1.5 cm long then almost globose and perianth glabrous............. 23

21. Fruit without pseudostalk, top rounded N New Guinea.
.... H. pilifera

21. Fruit with pseudostalk c. 1.5 - 5mm long, top rounded or usually acute, sometimes rostrate............ 22

22. Leaves drying olivaceous; midrib distinctly raised. Northern P.
.... H. clavata

22. Leaves drying dark brown; midrib slightly raised. Throughout Papua New Guinea (excl. Bismark Arch.).
.... H. squamulosa

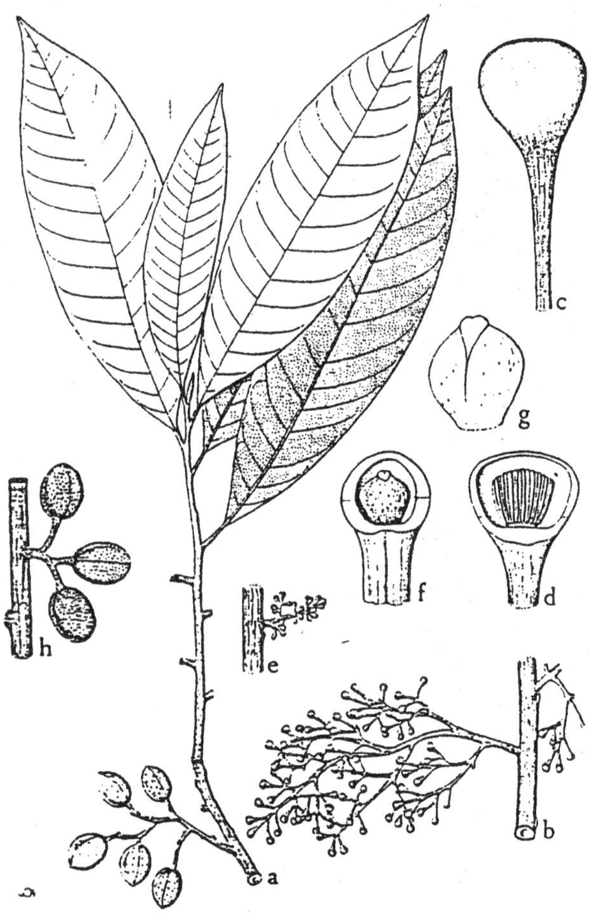

Horsfieldia tuberculata (A) leafy twig with infructescence; (B) twig with male inflorescences; (C) & (D) male flower; (E) twig with female inflorescence; (F) & (G) female flower; (H) mature fruits. (Source: de Wilde (1984).

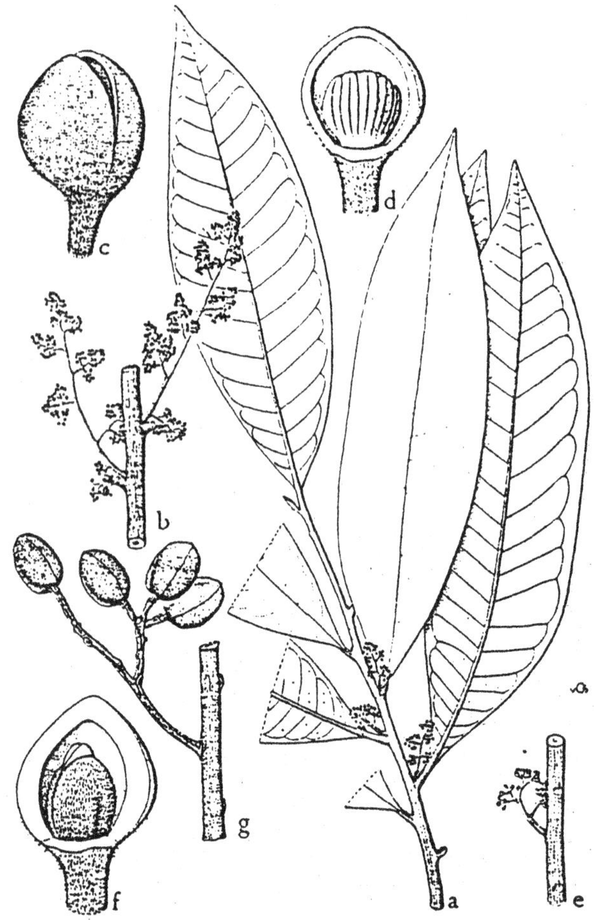

Horsfieldia whitmorei (A) twig portion with male inflorescences; (B) male inflorescences, lower down on the same twig; (C) mature male flower; (D) ditto, opened, showing androecium; (E) portion of twig with female inflorescence; (F) female flower longitudinally opened, showing pubescent ovary; (G) portion of twig with infructescence with mature fruits.
(Source: de Wilde (1984).

23. Female perianth c. 2 - 2.4mm long, glabrous. Fruit (sub)globose, or short-ellipsoid, or obovoid, c. 1.5-2.5 x 1.5-2.0cm; pericarp 4-6mm thick. Leaves 6-14cm long. Throughout Papua New Guinea.
.... H. sinclairii

23. Female perianth c. 2.5 mm long or more, pubescent or glabrescent. Leaves 10cm long or more....... 24

24. Fruit 1.6 - 3.0 cm long, usually with coarse pale-coloured wart-like lenticels; dry pericarp 2 - 6mm thick. Moluccas, New Guinea and Bismarck Arch.; 0-1000m alt. (sometimes much resembling small-fruited H. pachycarpa).
.... H. laevigata

24. Fruit (3.0-)3.5 - 7.5cm long; pericarp (4-)5mm thick or more. New Guinea; (450-)1000-2000m alt... 25

25. Perianth pubescent. Fruit 3.0 - 4.5cm long; pericarp 4-10mm thick. New Guinea Highlands.
.... H. pachycarpa

25. Perianth glabrescent. Fruit 6 - 7.5cm long; pericarp 10 - 20mm thick. Papua New Guinea Highlands. .... H. corrugata

26. Tomentum of leaf bud and inflorescence composed of hairs c. 0.2 - 0.5mm long. Female perianth c. 3mm long, at anthesis cleft to c. 2/3. Fruit not known. (Tomentum of leaf bud, female flower and fruit also not known in H. ampla)......... 27

26. Tomentum composed of hairs c. 0.5 - 1.0(-1.5)mm long. Fruit usually conspicuously pubescent........ 28

27. Inflorescence glabrescent. Sepik P. .... H. ampla

27. Inflorescence pubescent. Sepik P. Morobe P.. H. ampliformis

28. Leaves coriaceous, beneath with harsh hairs, when hairs shed leaving thickened scars. Female perianth c. 4mm long, at anthesis opening with a narrow pore-like slit at apex. Fruit 3.0 - 5.0cm long, pericarp 4 - 7mm thick. New Guinea.... H. pulverulenta

28. Leaves membranous or chartaceous. Perianth at anthesis cleft to c. 1/4 - 1/2..................... 29

29. Flowers (only the male known) entirely pubescent. Fruit 2.0 - 2.4 cm long, pericarp 4-7mm thick. New Guinea.... H. leptantha

29. Flowers largely glabrescent.... 30

30. Fruit 2.5 - 3.0cm long. Female perianth c. 4mm long. Leaves generally oblong-lanceolate, at apex caudate. New Britain. H. ralunensis

30. Fruit 1.2 - 2.8cm long. Female perianth c. 3mm long. Leaves oblong to oblong-lanceolate, at apex not caudate (always?). New Guinea and New Britain. .... H. hellwigii

4. KNEMA

A genus of 90 species (de Wilde 1987) distributed from India and Burma to Irian Jaya where it is represented by a single species, once collected from the Vogelkop at low altitudes (Foreman 1978). The species Knema tomentella also occurs in the Philippines, Celebes and the Moluccas.

A small to medium sized tree from 5-12m tall. When cut the trunk exudes copious red sap. Twigs striate at least in older parts. Leaves glaucous, nerves arching towards the margin, 7-25 by 2-7cm, lanceolate. Inflorescence axillary or in axils of fallen leaves, axis forms short, woody tubercles to 1.7cm long; bracts falling at early stage. Flowers in clusters or tubercles. Male flowers: globose; filaments joined, forming a peltate or triangular disc, short stalked or nearly sessile; anthers free attached around the disc by their bases. Female flowers: larger than male; stigma disc shaped, lobed. Fruit sessile to stalked, covered with rust-coloured tomentum, 2-2.5 x 1.8 cm. Aril segmented at apex only.

Key to Species in Papuasia

1 species in Vogelkop:

Knema tomentella

Note:

1. The single collection (Kostermans 4745) from the Vogelkop Penin. (Ransiki) Irian Jaya is from along a stream at 10m altitude.

5. MYRISTICA

A genus of circa. 70-125 species occurring from India and Indo-China to the Pacific. In Papuasia some 37 species. Several widely dispersed but with 3, 14 and 2 respectively endemic in Irian Jaya, Papua New Guinea and the Solomon Is. Most species occur in the lowland rainforest but a few are found at altitudes above 2000m. The seeds are commonly eaten by birds. The genus has the characteristic red exudate of all members of this family. Myristica fragrans is important as the source of mace (aril) and nutmeg (seed). Larger trees are used as sawlogs.

Trees 5 - 45 m tall. Twigs striate, sometimes inhabited by ants, then swollen. Stilt roots sometimes present. Leaves glabrous or with an easily removed indumentum of hairs and/or scales. Inflorescence axillary either a panicle or at the end of branches or a short woody knob, simple or branched. Bracts present, small and falling early. Bracteoles persistent and covering base of flower. Androecium forming column with 8 - 30 anthers joined by their backs to column. Ovary globose, glabrous or hairy; style absent; stigma minutely bilobed. Fruit various, glabrous or hairy, pericarp thick and fleshy. Aril divided to base of seed, red, yellow or orange. Seed brown to black.

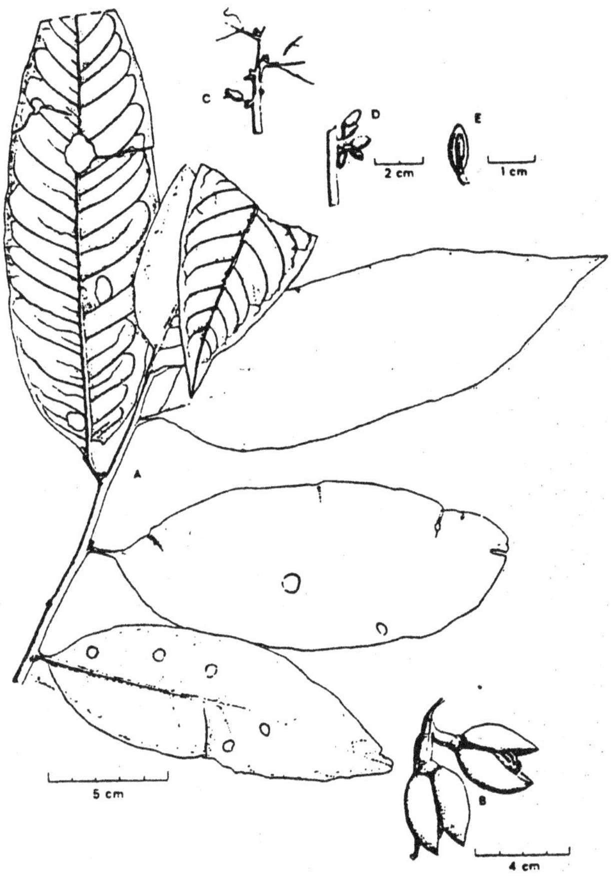

Myristica buchneriana (A) leafy twig; (B) fruit; (C) female flower; (D) male flower; (E) male flower with part of perianth removed to show staminal column. (Source: Foreman (1978).

Key to Species in Papuasia
(Source: Foreman 1978)

1. Leaves with easily removed indumentum of yellowish, cinnamon-coloured, silvery cf whitish-yellow scales on undersurface.......... 2

1. Leaves without easily removed indumentum of yellowish, cinnamon-coloured, silvery or whitish-yellow scales on undersurface.......... 21

2. Hairs as well as scales present on undersurface of leaves.......... 3

2. Scales only present on undersurface of leaves...................... 9

3. Inflorescence a panicle with several branches; scales on undersurface sometimes silvery as well as yellowish; hairs brown, persisting along midrib; male flowers + subglobose, large, to 7-10 x 5-7mm. Fruit oblong-ovoid, covered with short dark brown hairs. .... M. markgraviana

3. Inflorescence axis a woody, scar-covered knob, simple or bifurcate, scales usually a single colour, flowers mostly ellipsoid, not subglobose........................ 4

4. Reticulations on leaves absent or barely visible; nerves straight or curving slightly; secondary nerves absent, inconspicuous if present; tomentum wherever it occurs short, 1mm or less long................ 5

4. Reticulations on leaves very distinct, deeply impressed above or at least clearly defined on both surfaces; nerves curving widely, always prominent; secondary nerves present, numerous and conspicous; tomentum wherever it occurs dense, hairs 1.3mm long................ 6

5. Hairs on undersurface of leaves not abundant and often absent; scales sparse and soon disappearing, nerves on undersurface often not raised, usually reddish against paler background. .... M. buchneriana

5. Hairs on undersurface usually abundant but quite short, 0.5-1mm long; scales abundant, dark yellowish, usually persisting, nerves distinct and usually raised beneath. .... M. fatua and vars

6. Hairs wherever they occur dark brown or rust-coloured; flowers large, to 10-17 by 3-7mm, stalked; hairs on fruit usually short, 0.5-2mm long........................ 7

6. Hairs wherever they occur yellowish or golden brown, darkening with age, hairs persistent or not; male and female flowers smaller than above, ± sessile; hairs on fruit 3-5mm long.
.... M. chrysophylla and vars

7. Leaves ± subcordate or rounded at base, densely tomentose on undersurface; twigs hairy for considerable distance down twigs, to 20cm; male flowers 15-17 by 5mm, female flowers 10 x 7mm, densely hairy, ± inflated; fruit 7 by 3.3 - 4.4cm, densely hairy.
.... M. fusca

7. Leaves rounded at base but not subcordate, tomentose on undersurface, becoming ± glabrous; twigs becoming glabrous or hairy for short distance down twigs only; male flowers 10 x 3-4mm, female flowers 10 x 6mm, not inflated, not so densely hairy (those of M. womersleyi not yet seen); fruit 6-9cm in diameter, tomentum darker than above and less dense........... 8

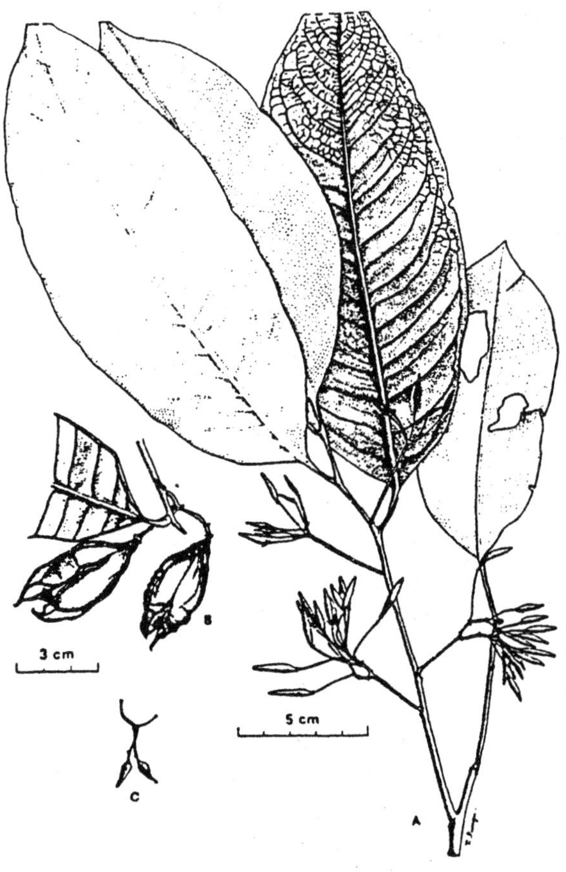

8. Hairs on undersurface of leaf < 1mm long; confined to nerves and midrib, disappearing with age, scales yellowish or sometimes silvery, tending to disappear with age; petiole 1.7 - 2.5cm long; seed not aromatic... M. sphaerosperma

8. Hairs on undersurface of leaf 1-2mm long, more dense and uniformly spread over entire undersurface of leaf, persistent; scales dark brown, becoming greyish with age; petioles 1 - 1.3cm; seed strongly aromatic... M. womersleyi

9. Reticulations numerous, usually on both surfaces of leaves, impressed above; scales yellowish, sparse, often absent altogether, nerves curving widely................. 10

9. Reticulations less numerous, often absent especially above; scales yellowish, usually present; nerves ± straight................... 11

10. Fruiting axis often forked; reticulations fine, visible on both surfaces but not raised beneath, secondary nerves numerous; twigs blackish; male flowers tubular.
.... M. cucullata

10. Fruiting axis not forked; reticulations distinct, often raised beneath, nerves raised beneath, secondary nerves may or may not be present; twigs usually not blackish; male flowers ± ellipsoid.
.... M. sphaerosperma

11. Leaves usually small, 8-15 x 3-7cm.
.... 12

11. Leaves usually larger than above, 11-32 x 4-15cm................ 15

12. Fruit 2.5-3 x 1-1.5cm, male flowers small, 3-5 x 2-2.5mm.
.... M. lepidota

12. Fruit and male flowers larger than above......................... 13

13. Fruit thick-walled; leaves ± elliptic......................... 14

13. Fruits thin-walled; leaves narrowly elliptic or lanceolate.
.... M. fatua var. papuana

Myristica cornutiflora (A) twig with male flowers; (B) fruit; (C) female flowers. (Source: Foreman (1978).

14. Leaves 4-15 by 3-5cm; petiole 10-15 by 1-2mm; fruit ellipsoid, 5.5-7 by 3.3-4cm, apex oblique. .... M. crassipes

14. Leaves 13-15 by 6.5-7cm; petiole 25-40 by 2mm; fruit obovoid, 7 by 3.3-4cm, apex not oblique. .... M. archboldiana

15. Petiole 6-15mm long........... 16

15. Petiole 2-6cm long............. 18

16. Venation prominent and well defined on undersurface of leaf; petiole 1-1.5cm long.................... 17

16. Venation not prominent or well defined on undersurface of leaf; petiole 6-15mm long. .... M. buchneriana

17. Fruit ovoid, 3 by 2.5cm, base narrowed into short pseudostalk. .... M. flosculosa

17. Fruit ± ellipsoid, 3.3 by 6x3-5cm, base rounded or slightly narrowed, pseudostalk absent. .... M. fatua var. papuana

Myristica fatua var. papuana: (A) leafy twig; (B) fruit; (C) fruit with part of pericarp removed to show seed surrounded by aril; (D) male flowers. (Source: Foreman (1978).

18. Fruit small, ± subglobose, 2 by 1.5 cm, stalk very slender, 15 by 2mm. .... M. pedicellata

18. Fruit larger than above, fruiting stalk stouter................. 19

19. Nerves fine, only slightly raised beneath; indumentum usually present on young leaves, older leaves becoming ± glabrous; blade 10-20 by 4-7 cm; petiole usually 2 - 2.5cm long....... M. tenuivenia

19. Nerves usually prominent and raised beneath; indumentum covering entire surface and persistent or may be present on young leaves only; petioles to 6cm long; leaves usually larger than above............... 20

20. Fruits covered with short, dark brown hairs; nerves fine but quite prominent; petioles to 6cm long; indumentum usually on young leaves only or persistent along midrib. .... M. petiolata

20. Fruit covered with yellowish or medium brown indumentum; nerves raised beneath and prominent; petiole usually < 4cm long; indumentum usually present and covering entire undersurface. .... M. fatua and vars

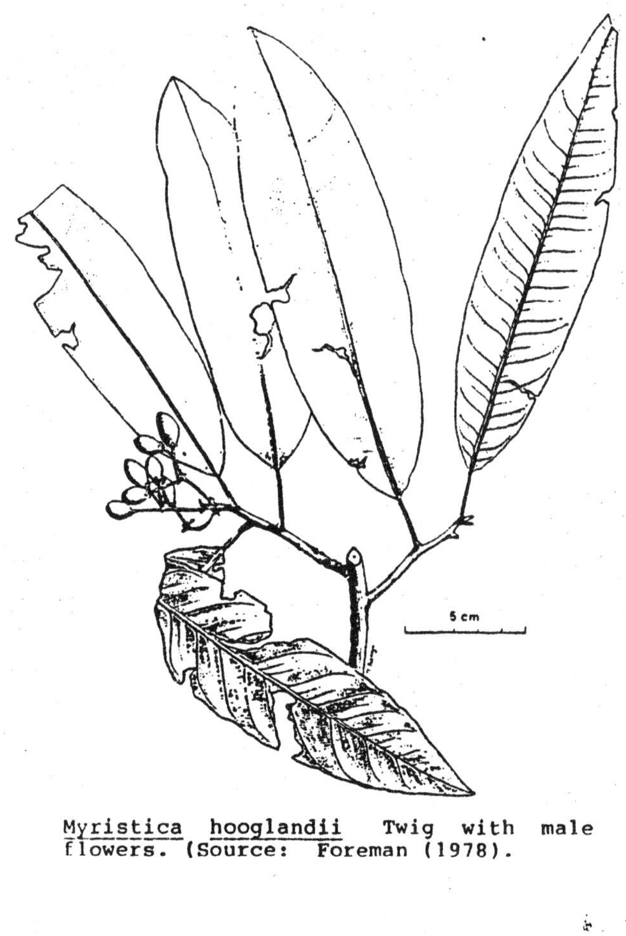

Myristica hooglandii Twig with male flowers. (Source: Foreman (1978).

1. Hairs sometimes present on young leaves.......................... 22

1. Hairs absent.................... 23

2. Fruit subglobose, 4 - 5 cm in diameter, covered with dark brown hairs; leaves whitish beneath, hairs confined to midrib.
.... M. uncinata

2. Fruit ovoid, flattened at base, 3x2.5cm, covered with medium brown hairs; leaves not whitish beneath, hairs covering entire undersurface at first, later ± glabrous.
.... M. inopinata

3. Inflorescence axis not woody, 1-8cm long, sometimes flattened, elongated, producing flowers only once, female inflorescence less branched and usually shorter than male one.
.... 24

3. Inflorescence axis rarely > 3cm long, main axis a short, woody, scar-covered knob, producing flowers more than once, sometimes basal portion smooth and upper parts scar-covered, female inflorescence similar to male one........... 34

4. Male inflorescence axis not branched.......................... 25

4. Male inflorescence axis branched.
.... 26

5. Leaves 6-13cm long, not silvery beneath, aromatic in all parts, twigs ± smooth, not native, always cultivated. M. fragrans

5. Leaves usually larger than above, silvery beneath, seed only slightly aromatic, twigs rough with pustular lenticels, native of New Guinea but also planted.
.... M. argentea

6. Leaves 6-12cm long, fruit 2-3.5cm long, pseudostalk usually present.
.... 27

6. Leaves and fruit larger than above, pseudostalk present or not..... 28

7. Leaves oblong to ovate base rounded to subcordate or cordate, 8-32 x 5-12cm; nerves ± distinct; fruit oblong, to 4 together, 3.5 x 1.5cm, pseudostalk 3 - 4mm long.
.... M. schleinitzii

7. Leaves lanceolate or narrowly elliptic, base acute or slightly rounded, 6-16 x 1.8-4cm; nerves faint on both surfaces; fruit not known.
.... M. rosselensis

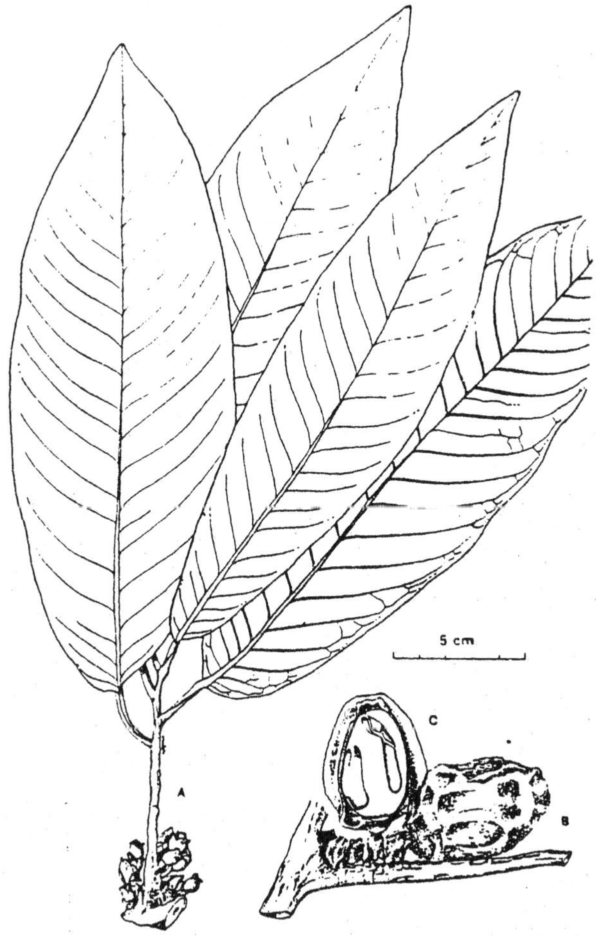

Myristica kajewskii (A) twig with male flowers; (B) fruit; (C) fruit with part of the percarp removed to show aril surrounding the seed.
(Source: Foreman (1978).

28. Leaves mostly 16-30cm long..... 29

28. Leaves mostly larger than above. 31

29. Leaves 18-30 by 5.5-10.5 cm; often pale grey beneath, fruit with distinctive uncinate tip, 4-6cm long, covered with dark brown hairs.
.... M. uncinata

29. Leaves mostly 16-30 by 5-11 cm; not pale grey beneath; fruit not as above.......................... 30

30. Leaves mostly panduriform; primary and secondary nerves faint; fruit oblong, becoming glabrous, base narrowed into short pseudostalk, with small cup-like receptacle at top of pedicel where it joins fruit...... M. garciniifolia

30. Leaves oblong-lanceolate or ovate-lanceolate; primary nerves more distinct, secondary nerves few; fruit ovoid, 3x2.5cm, densely hairy, flattened at base, no pseudostalk or cup-like receptacle present........ M. inopinata

31. Primary nerves prominent, especially beneath, secondary nerves mostly absent........................ 32

31. Primary nerves more slender than above, secondary nerves usually present....................... 33

32. Leaves 23-54x8-16cm, leaves drying reddish-brown above, whitish beneath, fruit 6.5-9x4.5-6cm, covered with fine short uniform medium brown hairs.
.... M. umbrosa

32. Leaves 21-45 x 5-17cm, drying olivaceous above, light brown beneath, nerves distinctive reddish-brown, fruit 5-5.5 x 2.3-2.5cm, with dense reddish-brown tomentum with rather shaggy appearance.
.... M. sepicana

33. Leaf base rounded or subcordate, blade 32-56 by 11-6.5cm, nerves raised upper surface, slightly raised beneath, mostly faint, midrib broad and well defined, raised on both surfaces.. M. neglecta

33. Leaf base mostly acute or bluntly acute, blade 22-42 x 6.5-11cm, nerves not raised above, slightly raised beneath, usually drying with glossy waxy appearance above, midrib sunken above, raised below.
.... M. hooglandii

34. Fruit with dense tomentum; leaves 12-24 x 3.5-10 cm; nerves 12 - 18 pairs......................... 35

34. Not as above.................. 36

35. Leaves glaucous beneath; nerves drying reddish-brown often faint beneath, impressed above; fruit 4 x 1.8-2 cm, covered with short medium to dark brown hairs.
.... M. buchneriana

35. Leaves drying ± same colour above and below; nerves well defined and raised beneath; fruit to 2.5-3.3 x 2 cm, covered with golden brown hairs 3 - 5mm long.
.... M. chrysophylla var. entrecasteauxensis

36. Leaves with cordate, subcordate or emarginate bases (especially oldest and largest leaves)........... 37

36. Leaves not cordate, subcordate or emarginate at base............ 39

37. Twigs with 2 lines running from petiole base to petiole base expanded into raised wings or not, twigs sometimes swollen and inhabited by ants; fruits to 3.5 x 2.8cm.... 38

37. Twigs without 2 lines running from petiole base to petiole base, not inhabited by ants; fruit 7-8.5 by 5.5-7.5cm.. M. kajewskii

38. Leaf-blade 20-35 by 5-13cm; 2 lines on twigs not expanded into wings, twigs not swollen and inhabited by ants; fruit minutely hairy at first, later ± glabrous, 3-6.5 by 2-4.5cm.... M. hollrungii

38. Leaf-blade 20-40 by 7-20cm; 2 lines on twigs raised into wings, twigs swollen and inhabited by ants; fruits minutely hairy, 1.5 by 1.3-2.5cm...... M. subalulata

39. Fruit globose or ellipsoid, tip ± rounded or pointed, base rounded, pseudostalk absent; inflorescence mostly simple, less often 1-3 short branches; bracteole or its scar at junction of pedicel and perianth; male flowers not usually tubular.
.... 40

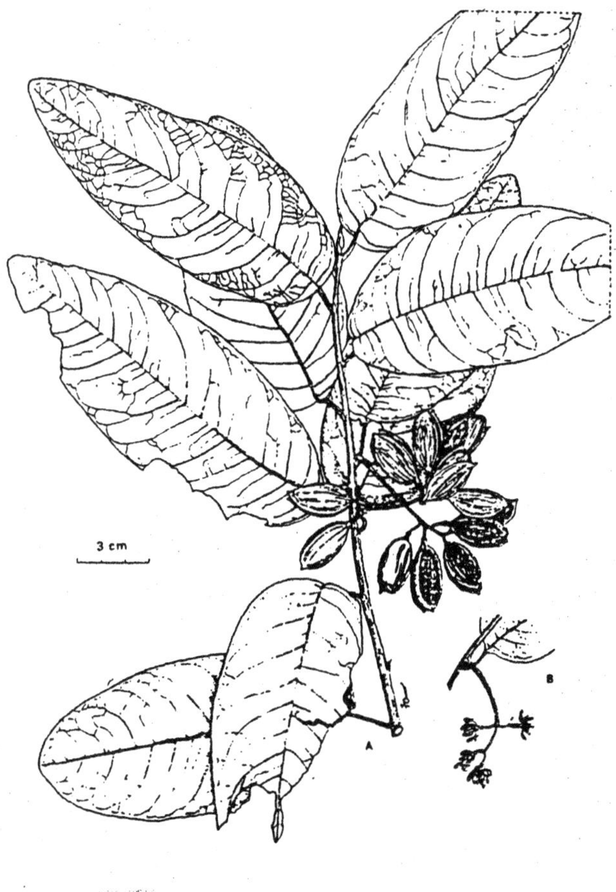

Myristica schleinitzii (A) twig with mature fruit; (B) male flowers. (Source: Foreman (1978).

39. Fruit ovoid, narrowly ellipsoid, ellipsoid or obovoid-ellipsoid, often ± pointed at both ends, pseudostalk usually present; inflorescence simple or split into 2 equal branches; bracteole or its scar 1-3mm down pedicels; male flowers tubular................ 48

40. Leaves mostly 5-15(-20) by 1-7.5cm; veins usually not prominent; 2 raised lines not present on apical portion of twigs; male flowers small, 2-5 by 2-3mm............ 41

40. Leaves large, to 15-40 by 3.5-15cm, veins prominent; apical portion of twigs with or without 2 raised lines; male flowers large, 5-15 by 3-5mm......................... 44

41. Primary nerves slender, fine, usually slightly raised beneath, most visible beneath, noticeably curved. 42

41 Primary nerves very fine, not raised and often invisible beneath, usually straight.
.... M. lancifolia and vars.

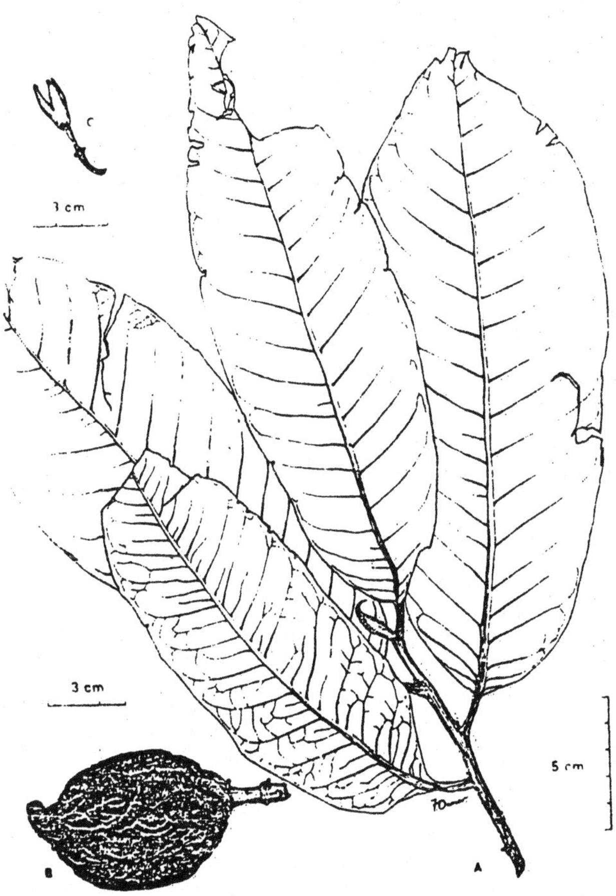

Myristica uncinata (A) leafy twig, note well-defined apical bud; (B) fruit; (C) female flower.
(Source: Foreman (1978).

42. Leaves small and narrow, 5-11 x 1-3cm; fruit ellipsoid, 1.8-2 x 1-1.2 cm......... M. concinna

42. Leaves larger and broader than above, 8-20 by 2.5-7.5 cm; fruit subglobose, globose or ellipsoid, 1.5-5 by 1.5-2.5cm............. 43

43. Fruit globose or subglobose, 1.5-2.5cm diameter; leaves 8-17 by 3-5.5cm, usually drying darker above than beneath.
.... M. globosa

43. Fruit globose when young, becoming ellipsoid, 5 by 2.5cm; leaves 10-20 by 2.5-7.5 cm, usually drying pale above and beneath.
.... M. insipida

44. Twigs with 2 raised lines running from petiole base to petiole base; fruits 1.5-6.5 by 1.2-4.5cm.... 45

44. Twigs without 2 raised lines running from petiole base to petiole base; fruit large, 6-8 by 5.5-7.5 cm............................ 47

45 Twigs swollen and inhabited by ants 2 lines raised into wings; fruit 1.5-5 by 1.2-2.5 cm ± sessile or with stalk to 15 mm long.
.... M. subalulata

45. Twigs not swollen nor inhabited by ants, 2 lines clearly defined but not raised into wings; fruit usually larger than above, 3-6.5 by 2-4.5 cm always stalked.......... 46

46. Leaf-blade 20-35 by 5-13cm, base rounded or sometimes acute (especially young leaves), usually drying a light colour above and beneath, not glossy; fruit ± ellipsoid, apex often pointed; male flowers ± glabrous.. M. hollrungii

46. Leaf-blade 11-26(-32) by 3.5-9cm, base rounded or often acute, usually drying darker above than beneath, often glossy; fruit subglobose, 3.8-6 by 3-5.5cm, apex not pointed; male flowers densely hairy.
.... M. sulcata

47. Fruit spherical, 6cm or more in diameter, covered with short rust-coloured hairs, pericarp smooth; male flowers ± ellipsoid, pedicels 8-10 by 1mm.
.... M. sphaerosperma

47. Fruit ± subglobose, 7-8.5 x 5.5-7.5cm, covered with short pale brown hairs, becoming glabrous, pericarp often warted; male flowers ± subglobose, pedicel 3 by 1-1.5mm, flattened.. M. kajewskii

48. Leaves very narrowly lanceolate, 17 -22cm by 2-3mm, nerves distinctly looped near margin; fruit fusiform, 4.5 x c.1.3cm pointed at both ends, base narrowed into pseudostalk. .... M. ensifolia

48. Leaves mostly elliptic or elliptic-lanceolate or oblong, interarching at margins distinct or not; fruit not as above.................. 49

49. Fruiting stalk slender (5-)10-35 x 0.5-2mm.................. 50

49. Fruiting stalk stout, 5-17 x 4-7mm. 53

50. Fruit cylindrical, 3 x 1cm; leaves covered beneath with closely appressed whitish scales; pseudostalk absent..... M. cylindrocarpa

50. Fruit not cylindrical; scales absent; pseudostalk present or not. 51

51. Twigs slender, 1mm thick at top; leaves ± elliptic, often with distinct drawn-out tip; petiole slender, 1.5-2mm thick; fruit pendulous, mostly single or in pairs, 4-7 x 1.3-2.5cm, pericarp thin, stalk 5 - 25 mm long. .... M. tubiflora

51. Twigs stouter, 2mm thick at top; leaves variable, apex bluntly acute or shortly acuminate; petiole stouter, 2 - 3mm thick; fruit not pendulous, mostly in pairs on forked peduncle, 3-7 by 1.5-3cm, pericarp thick, stalk 10-23 x 2mm.. 52

52. Male flowers to 25 x 3 mm; leaves 8 -23 by 6-10cm, nerves usually quite prominent; fruit ± subglobose or broadly ellipsoid, to 7 x 3cm, not noticeably drawn out at ends, short pseudostalk may be present. .... M. cornutiflora

52. Male flowers to 10 x 3-4mm; leaves 6-15 by 3-6cm, nerves faint; fruit narrowly ellipsoid, 3-5 x 1.5-2.3 cm, drawn out at both ends, pseudostalk usually present; fruit may be hairy when young. .... M. longipes

53. Fruit ellipsoid, thick-walled, with oblique apex; leaves elliptic, 4-15 by 1.5 - 5 cm. .... M. crassipes

53. Fruit not as above; leaves broadly elliptic to oblong, 11-23 by 4-9 cm........................... 54

54. Leaves broadly elliptic, less often elliptic-lanceolate, drying reddish-brown, secondary nerves neither numerous nor prominent; fruit ± ovoid, 3 by 2-2.5cm, with short pseudostalk. .... M. flosculosa

54. Leaves oblong or oblong-elliptic, drying pale yellowish or blackish-brown, secondary nerves numerous and usually well defined but not raised; fruit broadly to narrowly ellipsoid 3-6 by 2-3.5 cm, rounded at both ends. .... M. cucullata

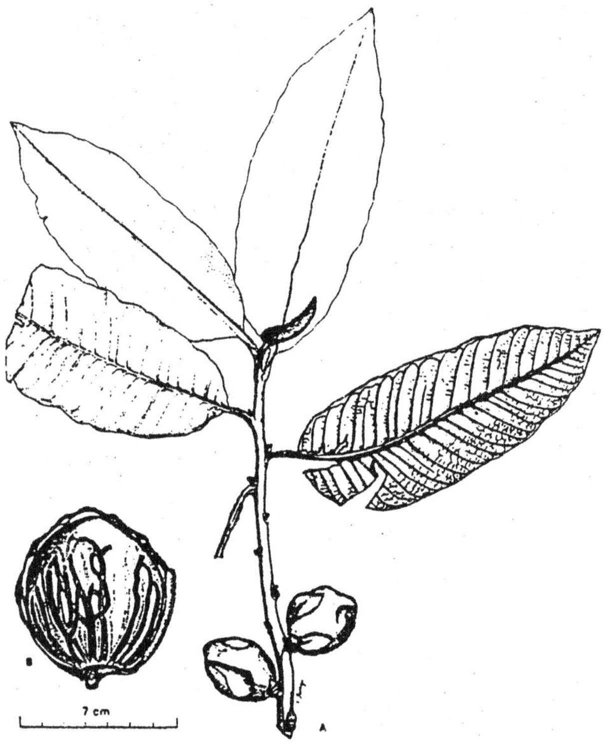

Myristica womersleyi (A) twig with young fruit; (B) fruit with part of pericarp removed to show seed and highly divided aril. (Source: Foreman (1978).

LAURALES

The order Laurales includes several small families of limited distribution. These small 'primitive' families are located around the southern margins of the Pacific Ocean. They include the Austrobaileyaceae (1 genus 2 species: Qld.); the Lactoridaceae (monotypic: Juan Fernandez Is.); Gormortegaceae (monotypic: Chile) and the Eupomatiaceae (1 genus 2 species: P.N.G. and E. Australia). In addition the order includes two large tropical and subtropical families: Monimiaceae and Lauraceae, and the Chloranthaceae which is tropical in distribution but does not occur in Africa.

EUPOMATIACEAE

	WORLD	PAPUASIA
Number of genera:	1	1
Number of species:	2	1

Distribution: Rainforest in E Australia and New Guinea.

Economic uses: Limited use for carving.

EUPOMATIACEAE

• • • • • •

Includes a single genus and two species of shrubs or small trees. The genus Eupomatia occurs in rainforest with E. laurina occuring throughout the range of the family. E. bennettii is restricted to Queensland. The decorative wood of E. laurina is used locally for carving. Both species have limited uses as ornamental plants.

Diagnositic features:

Leaves: alternate, simple, entire, glabrous, exstipulate.

Flowers: showy, solitary, bisexual, actinomorphic, perigynous without any clear distinction between petals and sepals. The perianth is fused into one mass, the upper part forming a conical lid; the lower filled with the receptacle sur ounded by stamens. The inner (?) stamens are sterile and petaloid. Receptacle top-shaped.

Fruit: flat-topped berry. Seeds with copious endosperm and very small embryo.

References

Croft, J.R. (1978). Eupomatiaceae. Handbooks of the Flora of Papua New Guinea. 1: 123 - 126.

Hotchkiss, A.T. (1955). Geographical distribution of the Eupomatiaceae. Journ. Arn. Arb. 36: 385 - 396.

KEY TO GENERA IN PAPUASIA

1 genus 1. Eupomatia

Eupomatia laurina: (A) twig with flowers and flower buds; (B) fruit.
(Source: National Herbarium, Lae).

1. EUPOMATIA

A family of 1 genus and 2 species. Shrubs or small trees to 10m tall, usually under 5m. Twigs striate. Leaves entire, 7-25 by 2.5-6 cm, base attenuate, apex acuminate, 7-17 lateral nerves anastomosing at margin. Flowers pale green, scented, 10-15mm diameter; pseudo-perianth whitish or pinkish; stamens 6-9mm long, anthers linear, filaments dilated at base, 2-3mm long; staminodes 3-7mm long, broadly attached to receptacle. Fruit pale green, succulent, 15-20mm diameter.

Key to Species in Papuasia

1 species:

Eupomatia laurina R. Br.

Notes:

1. The southern specimens of E. laurina, which include the type locality, differ from the New Guinea and N Queensland specimens. This requires further study.

2. This species has not been collected in Irian Jaya, the New Guinea Is. or the Solomon Is.

3. Locally common in understorey of lowland tropical rainforest as a subcanopy shrub. Coastal in distribution.

TRIMENIACEAE

A small family of rainforest trees, shrubs and woody climbers. Distributed through tropical and subtropical S W Pacific. Genera in this family are sometimes included in the family Monimiaceae. Rodenburg (1971) places both Trimenia and Piptocalyx in the Trimeniaceae. Philipson (1986) notes that while the habit of the two genera differs, the floral characters are similar. Philipson (1986) places species previously described in Piptocalyx into the genus Trimenis.

Diagnostic features

Leaves: exstipulate, opposite, subopposite or apparently whorled. Entire or glandular

TRIMENIACEAE

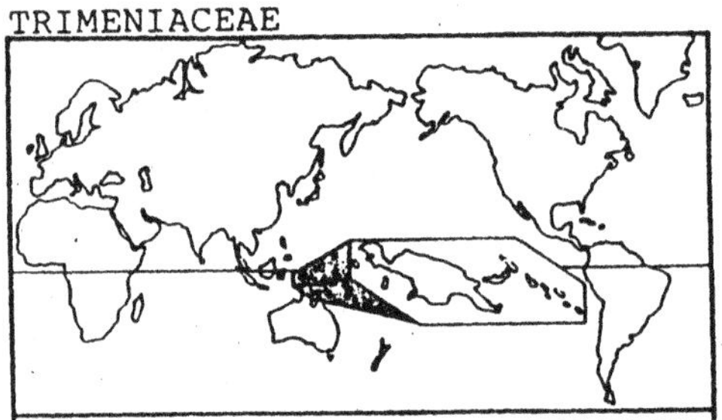

	WORLD	PAPUASIA
Number of genera:	3	1
Number of species:	12	3

Distribution: Tropics and subtropics of S W Pacific.

Economic uses: None known.

Flowers: unisexual or bisexual, receptacle continuous with pedicel; tepals caducous before or at anthesis, spirally arranged, imbricate, 10-38. Stamens 7-25, spirally arranged. Carpel solitary, rudimentary or absent in male flowers, superior; stigma tufted-papillose, 1-celled.

Fruit: small succulent berry. Seeds hard, smooth or ridged; embryo small with copious endosperm.

References

Bass, P. (1975). Vegetative anatomy and the affinities of Aquifoliaceae, Sphenostemon, Phelline and Oncothea. Blumea. 22: 311 - 407.

Hutchinson, J. (1964). The Genera of Flowering Plants. 1: 123 - 125.

Philipson, W.R. (1986). Trimeniaceae. Flora Malesiana Ser.1. 10: 327 - 333.

Rodenberg, W.F. (1971). Revision of Trimenia. Blumea 19: 3 - 15.

Royen, P. van (1982). The Alpine Flora of New Guinea. 3: 1251 -1254.

KEY TO GENERA IN PAPUASIA

1 genus 1. Trimenia

1. TRIMENIA

A genus of the Eastern Pacific Basin from Celebes and Moluccas, throughout Papuasia to Micronesia, New Caledonia and E Australia. Small trees or climbers in forests on ridges or reasonably exposed places. The fruit is probably bird dispersed (dark red to purple). Trimenia was placed in the Monimiaceae by Bentham and Hooker but since the 1950's has been placed in its own family.

Shrub or tree up to 20m tall; young branches reddish-brown villose, with age glabrous. Leaves 2-9 by 0.7-3.5cm, elliptic to lanceolate, base cuneate, apex acuminate, margin serrate or entire. Inflorescence axillary or terminal. Flowers unisexual or bisexual; receptacle continuous with the pedicel, glabrous,; tepals falling before or at anthesis, spirally arranged, imbricate, 10-38, to 3-5mm long. Stamens 7 - 25, spirally arranged, connective at apex. Carpel solitary (rudimentary or absent in male flowers), superior; stigma sessile, tufted-papillose. Fruit berry, small, spherical, succulent.

Key to Species in Papuasia

(Source: Philipson (1986)

1. Woody climber. Tepals 10 or fewer. .... T. macrura

1. Trees or shrubs. Tepals 13 or more. 2

2. Filaments relatively long and thin; mature anthers once or twice as long as the filaments. Seeds smooth... T. papuana

2. Filaments relatively short and broad; mature anthers at least 4 times as long as filaments. Seeds with longitudinal or reticulate ridges... T. weinmanniifolia

Notes

1. Philipson (1986) follows Rodendurg (1971) in treating T. papuana as a diverse species including both T. arfakensis and T. myricoides. A. C. Smith (1978) is critical of this approach and has a narrower circumscription of the species.

2. T. papuana is common on ridges from 1000 - 2700 m altitude. Its wood is used for fence posts and buildings and the leaves provide a treatment against dysentry. No uses are recorded for the other species.

3. More material is required from New Ireland as the identification of the species there is doubtful.

4. T. bougainvilleensis (Rodenburg) A. C. Smith (Allertonia 1 (1978) 154) is the only subspecies of T. weinmanniifolia Seeman found in Papuasia (Philipson, 1986).

Trimenia papuana (A) habit; (B) bud; (C), (D) & (E) outer, middle and inner tepals; (F) flower, after removal of tepals; (G) stamen; (H), (J) & (K) ovary; (L) fruit.

(Source: Perkins (1919).

SPHENOSTEMONACEAE

A small family of 8 species of shrubs or small trees, distributed from New Caledonia and Queensland, through New Guinea to the Moluccas. In montane rain-forest. The taxonomic position of the family has been disputed. It has been placed in the Aquifoliaceae, Icacinaceae, the Guttiferae and in the Trimeniaceae. Bass (1975) concluded that it was closely related to the Aquifoliaceae and Icacinaceae. The genus Sphenostemon includes both Nouhuysia and Idenburgia.

Diagnostic features

Leaves: simple, exstipulate, coriaceous from almost entire to dentate, subopposite to almost whorled, articulate at base.

Flowers: Racemes terminal or axillary. Bracts at base, early caducous. Flowers bisexual, pedicel articulate at base. Sepals 4, free, decussate, mostly hooded. Petals absent. Stamens 6 or 9 - 13, sessile. Ovary superior.

Fruit: Drupe, ellipsoid to globular. Stigma persistent. Pericarp fleshy, black.

SPHENOSTEMONACEAE

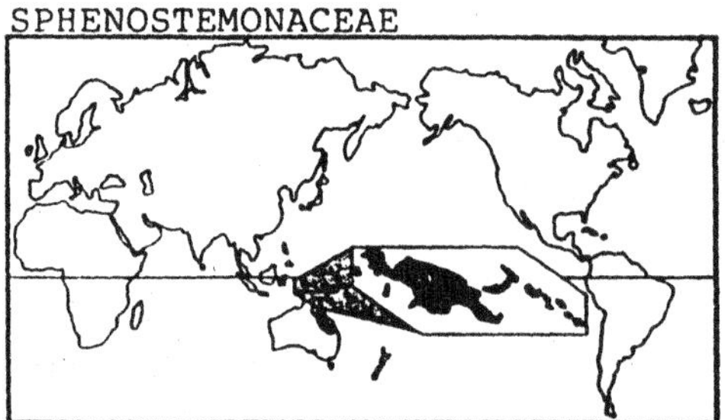

WORLD PAPUASIA

Number of genera : 1
Number of species : 3
Distribution : W. Pacific: Moluccas to New Caledonia.

Economic uses : None known.

References

Airy Shaw, H.K. (1972). Validation of the family name Sphenostemonaceae. Kew Bull. 27:

Bass, P. (1975). Vegetative anatomy and the affinities of the Aquifoliaceae, Sphenostemon, Phelline and Oncothea. Blumea 22: 311 - 407.

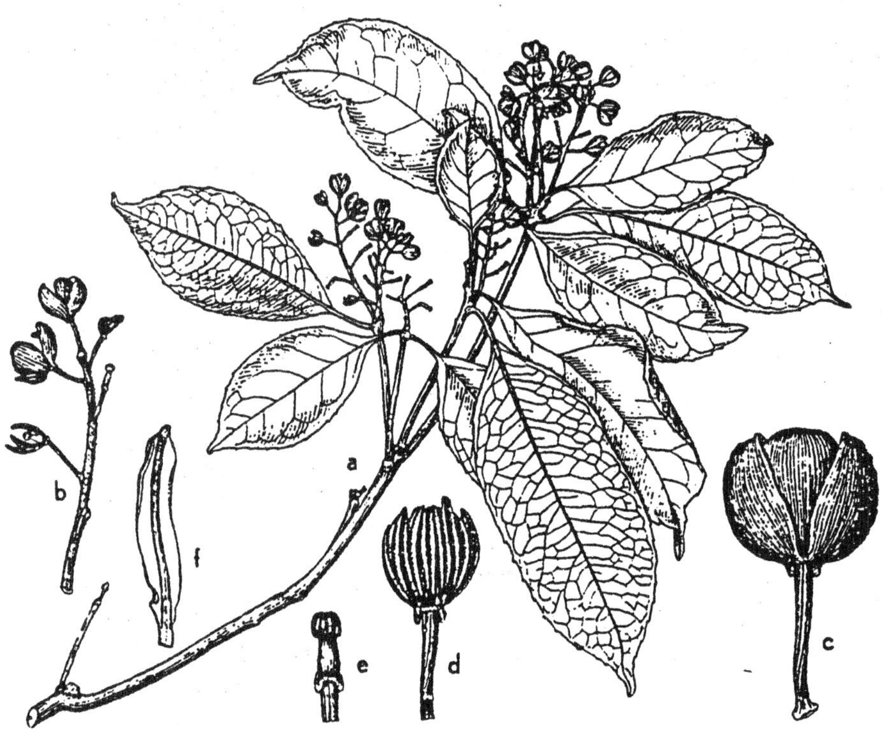

Sphenostemon papuanus (A) habit; (B) inflorescence; (C) ripe bud; (D) flower; (E) pistillum; (F) anther,

van Steenis, C.G.G.J. (1955). Some notes on the flora of New Caledonia and reduction of Nouhuysia to Spenostemon. Swensk. Bot. Tidskr. 49: 19 - 23.

van Steenis C.G.G.J. (1986). Sphenostemonaceae. Flora Malesiana Series 1. 10: 145 - 149.

van Royen, P. (1982). Alpine Flora of New Guinea. pp. 1250 - 1254.

KEY TO GENERA IN PAPUASIA

1 genus 1. Sphenostemon

1. SPHENOSTEMON

● ● ● ● ● ●

Distribution and description as in family discussion.

Key to Species in Papuasia
(Source: van Steenis (1986)

1. Stamens (5-)6. Inflorescences short to 1-2 cm. Pedicels even in fruit only c. 1/2cm. Leaves usually not conspicuously toothed to almost entire........................ 2

1. Stamens 8-13. Inflorescences long, to 11 cm in fruit. Pedicels to 2cm, in fruit up to 2cm. Stigma sessile, cap-shaped, appressed to fruit apex, 2-3mm diameter. Leaves almost always conspicuously toothed.. .... S. papuanus

2. Racemes puberulous, few-flowered. Stigma grooved, cupular, appressed to fruit apex, 2-3 mm diameter. Habit rather delicate, twigs thin. Leaves ovate to obovate-lanceolate, acute-acuminate to caudate, 5-11 by 1.75-3.25 cm; petiole 4-7 mm. .... S. arfakensis

2. Racemes mostly glabrous. Stigma punctate in flower, flat and ± elevated in fruit, hardly 3/4 to 1mm wide, not cap-shaped and grooved. Habit not delicate, twigs rather sturdy. Leaves mostly oblanceolate, rather short- and mostly blunt-acuminate, 5-13 by 2.5-4cm; petiole 10-20 mm. .... S. lobosporus

Notes:

1. Small trees in montane forest from 800 to 3000 m altitude.

2. S. papuanus is widespread: from Celebes to New Ireland. S. arfakensis is known from only 2 collections in the Arfak Mts of Irian Jaya. S. lobosporus occurs in S E New Guinea, the Jimi Valley and N Queensland.

Sphenostemon papuanum (A) flowering branch. (Source: van Royen (1982).

MONIMIACEAE

• • • • • •

A family of tropical to subtropical distribution in the southern hemisphere. Includes some 30 genera, several of which are monotypic including Lauterbachia (P.N.G.). Most are trees or shrubs and many produce aromatic oils, a few species are vines. The family is Austral in distribution.

Diagnostic features

Leaves: opposite, exstipulate, evergreen, entire or serrate, leathery and often containing aromatic oils.

Flowers: perigynous, symmetrical, solitary or in racemes or cymes; small, often unisexual or bisexual. Perianth of 4 to many (sometimes absent), segments in 2+ overlapping series; sometimes outer sepal-like and inner petaloid, and borne on rim of inflorescence. Stamens 1 to many: in 1 or two series or in male flowers scattered over inner receptacle surface. Rarely absent. Carpels separate, ovaries superior.

Fruits: an achene or drupe. Sometimes enclosed by recep~tacle.

References:

Hutchinson, J. (1964). The genera of Flowering Plants. 1: 110 - 123.

Perkins, J. (1925). Gattungen der Monimiaceae.

Philipson, W.R. (1980). The genus Wilkiea in Papua New Guinea. Blumea 26: 365 - 366.

Philipson, W.R. (1980). Kairoa a new genus from Papua New Guinea. Blumea 26: 367 - 372.

Philipson, W.R. (1980). A revision of Levieria. Blumea 26: 373 - 385.

Philipson, W.R. (1982). A revision of the Malesian species of Palmeria. Blumea 28: 85 - 101.

Philipson, W.R. (1984). A synopsis of the Malesian species of Steganthera. Blumea 29: 481 - 497.

Philipson, W.R. (1985). A synopsis of the Malesian species of Kibara. Blumea 30: 389 - 415.

Philipson, W.R. (1985). Faika and Parakibara: two new genera of Monimiaceae from Malesia. Blumea 30: 417 - 423.

Philipson, W.R. (1985). Monimiaceae. Flora Malesiana. Series I. 10: 255 - 326.

van Royen, P. (1982). Alpine Flora of Papua New Guinea. 3: 1259 - 1265.

MONIMIACEAE

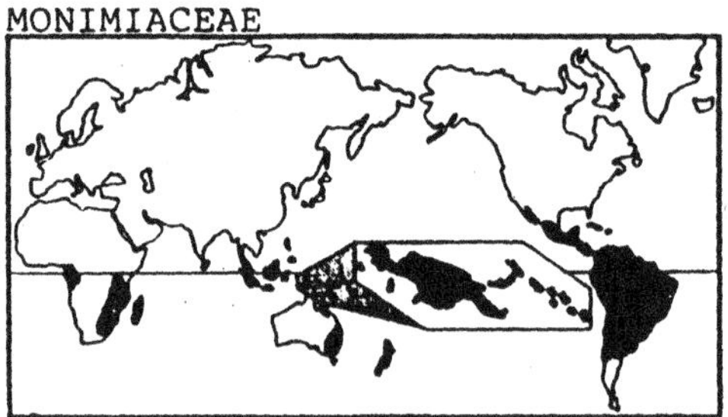

	WORLD	PAPUASIA
Number of genera:	25	9
Number of species:	102	82

Distribution: Tropical: Polynesia, Australasia, to America.

Economic uses: Timbers, fruits perfumes, medicines and dyes.

KEY TO GENERA IN PAPUASIA

1. Flowers hermaphrodite; anthers valvate; filaments of segments with a pair of glands; fruits with persistent elongated plumose style. .... 1. Dryadodaphne

1. Flowers unisexual; anthers dehiscing by slits; filaments without glands; style not persistent in fruit........................... 2

2. Lianes: receptacle splitting irregularly when ripe to reveal drupes dispersed over its inner surface. .... 6. Palmeria

2. Trees or shrubs, or if scandent (rarely) then fruit not as above. 3

3. Male flowers with a small, flat receptacle bearing relatively large rotund tepals; after anthesis female receptacle splitting irregularly from the ostiole; fruit a head of sessile drupes with the irregular rim of the receptacle reflexed below them. .... 5. Levieria

3. Male flower with a globose or flask-shaped receptacle, either with a small ostiole or (Kairoa) splitting open at anthesis to form a cup with sharply lobed margin; after anthesis upper half of female receptacle abscissing as a calyptra. Fruit: a head of sessile or stipitate drupes with the annular scar of the calyptra below them.. 4

4. Female receptacle with thickened glands inside the ostiole....... 5

4. Female receptacle without glands inside the ostiole.............. 7

5. Style-stigma subulate; anthers dispersed irregularly within the receptacle..................... 6

5. Stigma sessile, obtuse; anthers disposed regularly in 2 whorls of up to 4 stamens.
.... 4. Kibara

6. Anthers dehiscing by a horseshoe-shaped slit.
.... 9. Wilkiea

6. Anthers dehiscing by two vertical slits..... 2. Faika

7. Male receptacle at anthesis a fleshy cup with rim split into acute lobes; stamens numerous (over 100); stigma short, sessile; stems and branches with prominent irregular longitudinal ridges of cork...... 3. Kairoa

7. Male receptacle at anthesis with an ostiole surrounded by minute tepal; stamens few; branches ± smooth.... 8. Steganthera

Notes:

1. Barker (mans.) distinguished the genus Lauterbachia from other trees and shrubs (at 2) by the presence, in that genus, of a rim of tissue (velum) between the ostiole of the female receptacle and the upper most tepals. It does not occur in the other genera.

1. DRYADODAPHNE

A genus of 3 species in Queensland (1) and Papuasia (2). A common tree in montane rainforest from (500-)600-2800m altitude. Tall trees. D. crassa grows in Nothofagus forest on well drained sites usually on slopes or ridges. D. novoguineensis is represented by 2 subspecies. It occurs throughout the mountains of Papua New Guinea from the Hunstein and Torricelli Mts to Owen Stanley Range.

In 1969 Schodde 1969 considered Dryadodaphne as a member of the family Atherospermataceae (Philipson, 1986). This is not followed here.

Tall trees, young branches usually glabrous. Leaves dentate or almost entire. Inflorescence axillary; bracteoles enclosing the flowerbud, early caducous. Flowers bisexual, perianth and androecium on the rim of the hypanthium; tepals 8, in 2 whorls. Stamens in several whorls. Carpels free, styles terminal. Fruit dehiscing into 4 equal valves. Nutlets silky on one side, terminal styles.

Dryadodaphne novoguineensis (A) stem with flower; (B) flower; (C) fruits; (D) achene. (Source: Johns (1983).

Key to Species in Papuasia
(Source: Philipson (1986)

1. Stamens 8 (rarely 4); outer stamens broader than long with widely divergent ± acute glands. Leaves obovate (rarely elliptic) with rounded, obtuse, frequently retuse apices. Tepals in ± equal whorls, usually pale cream-green.
.... D. crassa

1. Stamens 4; including appendages longer than broad, with slightly divergent ± obtuse glands. Leaves lanceolate to elliptic with attenuate to acuminate (rarely rounded) apices. Tepals in unequal whorls, usually pale maroon-brown, pinkish or purplish.
.... D. novoguineensis

Notes:

1. Two subspecies *D. novoguineensis* occur in Papua New Guinea: ssp. *novoguineensis* and ssp. *occidentalis*. Within subspecies *novoguineensis* 2 varieties are recognised: var. *novoguineensis* and var. *macra*.

2. Subspecies *occidentalis* is known only from collections in Irian Jaya, some 200-250 miles to the west of ssp. *novoguineensis*.

2. FAIKA

• • • • • •

A monotypic endemic rainforest genus found only in Irian Jaya from the Vogelkop Peninsula to the Cyclops Mts. A species of shrub or small tree to 3m tall. Conspicuous villose stems and leaves. Flowers solitary in leaf axils; (?) dioecious. Male flowers: yellowish, receptacle ovoid, tepals in 3 decussate pairs. Stamens c. 24; filament short. Female flowers: orange, receptacle turbinate, with c. 5 decussate pairs of tepals and with 2-3 pairs of bracts at base. Carpels numerous, sessile; stigma long, subulate. Drupes ovoid, black when ripe.

Key to Species in Papuasia

1 endemic species:

Faika villosa

Notes:

1. It occurs in the undergrowth of rainforest between 100 and 1250m altitude.

Faika villosa (A) & (B) female flower and LS; (C) & (C) male flower and LS. (Source: Philipson (1986).

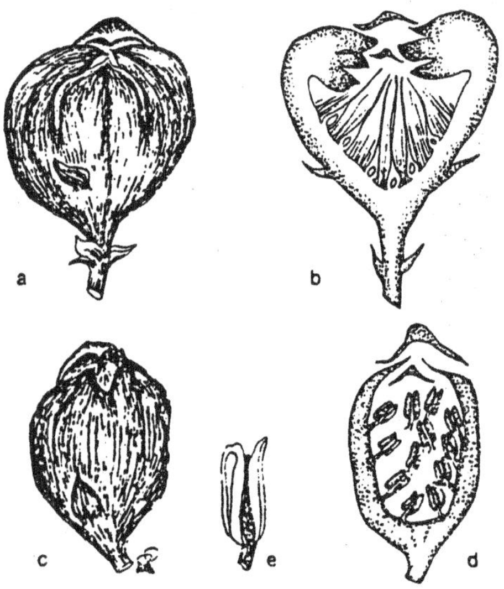

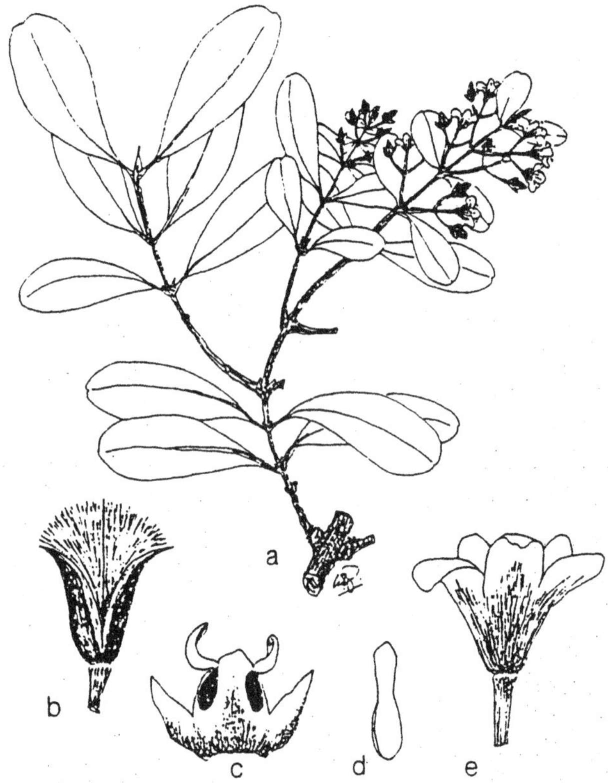

Dryadodaphne crassa (A) habit; (B) fruiting hypanthium; (C) anther of outer whorl; (D) embryo; (E) open flower. (Source: Philipson (1986).

Faika villosa (A) young infructescence; (B) (C) LS seed. (Source: Philipson (1986).

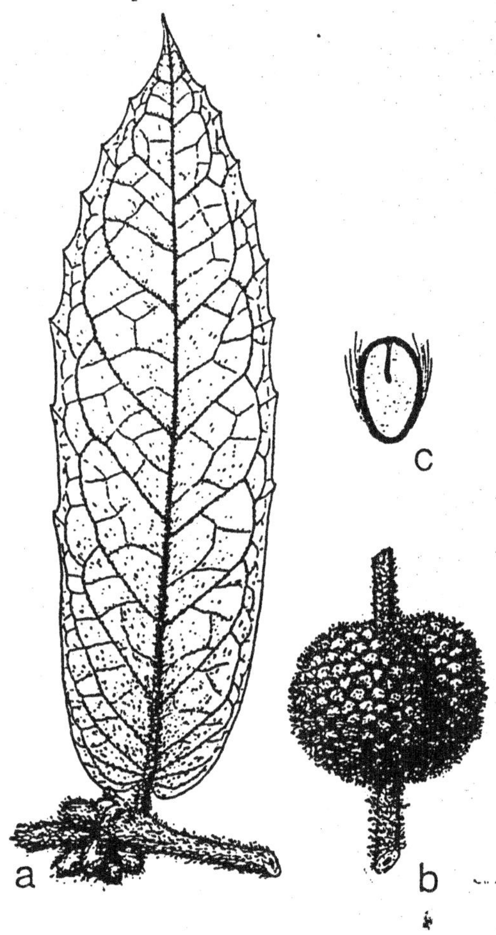

3. KAIROA

A monotypic understorey tree occuring in the Morobe and Northern Provinces of Papua New Guinea. Shrub or small tree with stout, terete branches. Leaves coriaceous, toothed, apex acute, secondary and tertiary veins form a distinct reticulation; petiole stout, 5-15mm long. Flowers axillary, monoecious. Male flowers: receptacle globose with 8 tepals; at anthesis splitting to form bowl-shaped flower with 4-6 radiating lobes, fleshy. Stamens numerous (+100), over surface of receptacle; filaments very short. Female flowers: receptacle oblate, 4 tepals, splitting into broad segments. Carpels numerous, sessile, style very short. Fruits numerous, ovoid, sessile on enlarged fleshy receptacle.

Key to Species in Papuasia

1 local, endemic species:

Kairoa suberosa

Notes:

1. The bark on the older stems is deeply fissured, a distinctive field character.

2. Locally frequent in lowland rain forest, occasionally with *Castanopsis* and *Lithocarpus* up to 1200 m altitude.

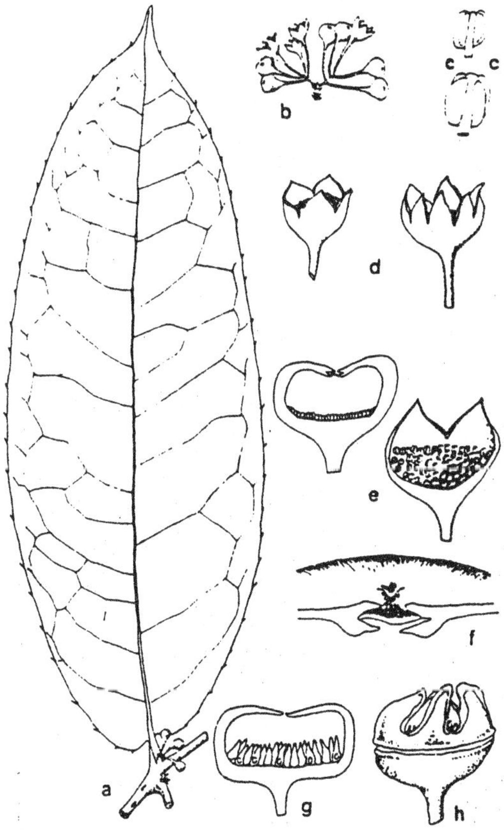

Kairoa suberosa (A) habit; (B) male inflorescence; (C) stamens; (D) & (E) male flowers and LS; (F) ostiole of male flower; (G) female flower. (Source: Philipson (1986)).

4. KIBARA

A genus of 43 species of which 37 occur in Papuasia. Most are understorey small trees or shrubs from 0 to 2800m altitude. Some occur on sandy or coral beaches. As noted by Philipson (1986) most species are regional and this should be used when making identifications. In consequence the distribution for species is given in the key.
Trees or shrubs. Leaves usually pubescent, becoming glabrous, entire or dentate, main secondary veins arched and meeting with margin. Monoecious. Inflorescence lateral or terminal, cymose (racemose in *K. streimannii*), pedicels usually thickened distally into receptacle. Male flowers usually smaller than female; ostiole minute surrounded by 2-4 decussate pairs of tepals; 4 larger outer stamens and to 4 smaller inner stamens or staminodes, may be rudimentary and ± connate. Female flowers: ostiole surrounded by 5 decussate pairs of tepals, inner pairs thickened and glandular. carpels numerous. Drupes sessile or stipitate.

Key to Species in Papuasia
(Source: Philipson 1986)

1. Leaves linear (leaf index 6). (S W Irian Jaya, Misool?). .... *K. roemeri*

1. Leaves broader (leaf index less than 4)........................ 2

2. Leaves shorter than 5 cm........ 3

2. Leaves longer than 5cm.......... 4

3. Leaves coriaceous, apex obtuse or retuse (Vogelkop). .... *K. oligocarpella*

3. Leaves membranaceous, tapering to a long apiculum (Sepik Region). .... *K. myrtoides*

4. Inflorescence shorter than, or little exceeding the petiole (c.2cm or less)......................... 5

4. Inflorescence considerably longer than the petiole (3cm or usually much longer).................. 16

5. Stems greatly expanded at the nodes (usually inhabited by ants)..... 6

5. Stems not markedly expanded at nodes.......................... 7

6. Leaves sessile, amplexicaul, margin strongly and sharply dentate (E. PNG).
.... K. ferox

6. Leaves petiolate, base cuneate or truncate, margin entire (S. PNG).
.... K. archboldiana

7. Inflorescence pubescent (young shoots also pubescent and hairs +- persisting on mature foliage)... 8

7. Inflorescence and mature foliage glabrous (buds and young foliage sometimes pubescent)........... 11

8. Leaves stiffly chartaceous, teeth sharply spinulose, often +- bullate (W.H.P.).
.... K. karengana

8. Leaves softy chartaceous, dentations not spinulose............ 9

9. Leaves bullate (M.P.).
.... K. bullata

9. Leaves flat................... 10

10. Pubescence of fine, pale, appressed hairs (SE PNG).
.... K. leachii

10. Pubescence of loose, curled, brownish hairs (Vogelkop).
.... K. versteeghii

11. Leaf blade larger than 10cm..... 12

11. Leaf blade shorter than 8cm..... 15

12. Leaf margins entire (Normanby Is.)
.... K. oblongata

12. Leaf margins dentate............ 14

13. Leaf margins spinulose-dentate (Vogelkop).
.... K. warenensis

13. Leaf margin serrate........... 14

14. Leaf base tapering to the petiole, blade narrowly obovate (Rossel I.).
.... K. rosselensis

14. Leaf base broadly cuneate, blade oblong-elliptic (Vogelkop).
.... K. royenii

15. Leaf apex with a slender apiculum (E.H.P.).
.... K. hartleyi

15. Leaf apex acute or obtuse, without an apiculum (Mt Shungol. M.P.).
.... K. shungolensis

16. Leaf blade small, membranaceous (usually under 10x3cm). Inflorescence branches very slender (Torricelli Mts.).
.... K. microphylla

16. Leaf blade larger (usually over 10cm long, if not, then coriaceous) 17

17. Male flowers racemosely arranged along the inflorescence branches.
.... K. streinannii

17. Male flowers cymosely arranged or solitary......................... 18

18. Flowers 6mm long at anthesis or longer, densely pubescent (E.H.P. and M.P.).
.... K. macrantha

18. Flowers less than 5mm long (or if 6mm, then glabrous)............ 19

19. Leaf margin dentate............ 20

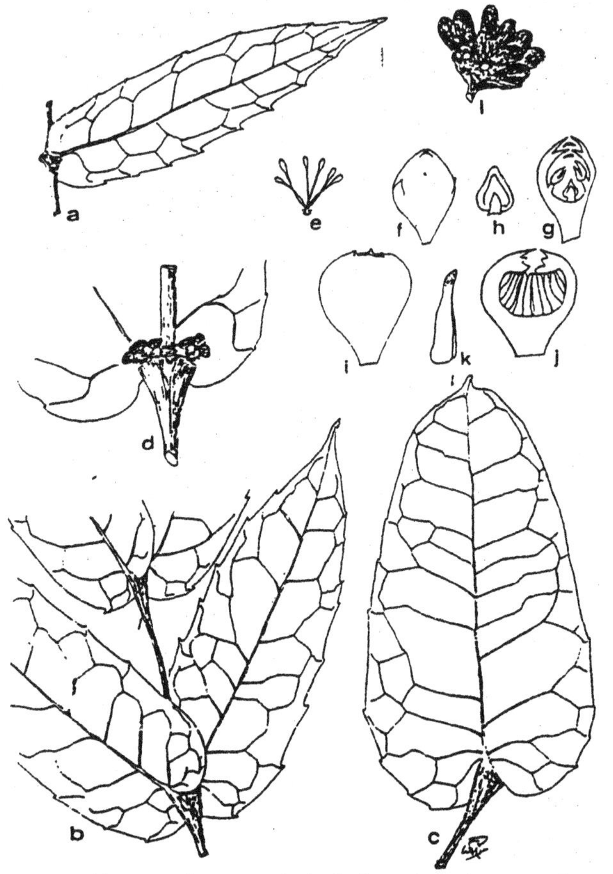

Kibara ferox (A) (B) & (C) leaves from upper, middle and lower branches; (D) female flowers; (E) male inflorescence; (F) & (GS) male flower and LS; (H) stamen; (I) & (J) female flower and LS; (K) carpel; (L) fruit.
(Source: Philipson (1986).

19. Leaf margin entire (occasionally some leaves with a few teeth near the apex)...................... 26

20. Stems greatly expanded at the nodes (usually inhabited by ants).
.... K. carrii

20. Stems not markedly expanded at nodes.......................... 21

21. Dentations of leaf margins sharply spinulose.
.... K. elongata

21. Dentations of leaf margin not spinulose....................... 22

22. Pedicels strong................ 23

22. Pedicels delicate.............. 25

23. Underside of mature leaves pubescent. Inflorescence a few-flowered cyme (S.H.P. and M.B.P.).
.... K. katikii

23. Underside of mature leaves glabrous or, if pubescent, inflorescence +-paniculate...................... 24

24. Leaf blade usually broadly elliptic -ovate, more than 10cm long, rounded to broadly cuneate base. Pedicels rigid (throughout region).
.... K. coriacea

24. Leaf blade usually elliptic, usually less than 10cm long, base cuneate. Pedicels less rigid (SE P.N.G.).... K. papuana

25. Inflorescence of paniculate cymes. Stamens inserted near base of receptacle (NW Irian Jaya).
.... K. fragrans

25. Inflorescence of simple pleiochasia. Outer stamens inserted at widest part of the receptacle. (Star Mts. Telefomin area).
.... K. nitens

26. Undersurface of mature leaf pubescent........................... 27

26. Undersurface of mature leaf glabrous........................... 28

27. Pubescence on inflorescence soft, not closely appressed (S.H.P. M.B.P.).... K. katikii

27. Pubescence on inflorescence stiffly appressed (throughout Malesia).
.... K. coriacea

28. Leaves stiffly coriaceous, broadly elliptic to subrotund (NG Highlands)..... K. laurifolia

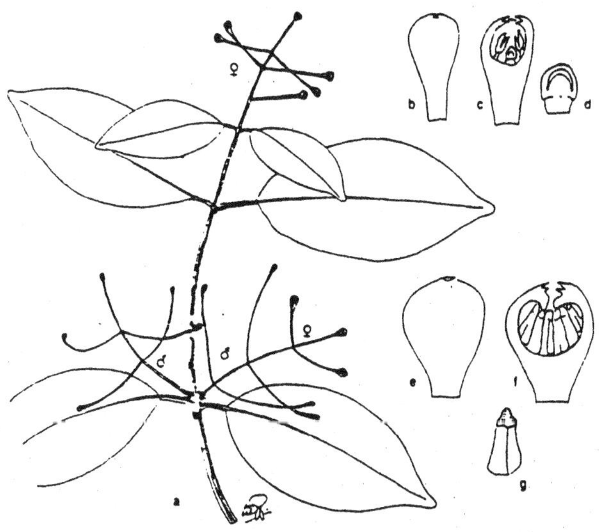

Kibara fugax (A) leafy shoot bearing male, female and mixed inflorescences; (B) & (C) male flower and LS; (D) stamens; (E) & (F) female flower; (G) carpel. (Source: Philipson (1986).

28. Leaves not as above............. 29

29. Inflorescence pubescent......... 30

29. Inflorescence glabrous (when young $\pm$ setulose in sleumeri).... 31

30. Leaves acuminate, usually broadly elliptic (throughout Malesia).
.... K. coriacea

30. Leaves obtuse, usually narrowly elliptic (Biak I.).
.... K. obtusa

31. Pedicels of male flowers elongated (20-45mm)...................... 32

31. Pedicels of male flowers shorter than 20mm...................... 34

32. Male flowers in separate inflorescences (Irian Jaya).
.... K. flagelliformis

32. Male flowers on lower branches of inflorescences................. 33

33. Leaf narrowly elliptic (S.P.).
.... K. chimbuensis

33. Leaf broadly elliptic or elliptic. (M.P. and C.P.).
.... K. fugax

34. Pedicels of female flowers noticeably thickened for c.10mm below the receptacle (N.B. and N.I.P.).
.... K. novobritanica

34. Pedicels of female flowers not strikingly thickened.......... 35

35. Male receptacle 4-6mm long; anthers kidney shaped. Leaf rigidly coriaceous........................... 36

35. Male receptacle 1.5-2.5mm long; anthers triangular. Leaf not rigidly coriaceous.................. 37

36. Leaf oblong, apex rounded or retuse with a minute mucro, often 3-whorled (W.P. and Queensland).
.... K. rigidifolia

36. Leaf narrowly elliptic, apiculate, opposite (Sudest Is.).
.... K. sudestensis

37. Leaves oblong, apex and base rounded (Vogelkop).
.... K. sleumeri

37. Leaves elliptic................. 38

38. Leaves membranaceous (Sepik).
.... K. monticola

38. Leaves firmly herbaceous or chartaceous (SE P.N.G.).
.... K. papuana

5. LAUTERBACHIA

A monotypic genus known only from the type collection from Ramu in the Madang Province. The single collection was destroyed in World War II. Philipson (1986) considers the interpretation by Perkins doubtful because of her description of a velum in the Mollinedieae. Until further collections of this plant are made it is not possible to accurately interpret this genus.
Trees or shrubs with entire leaves, 13-21 by 5-8cm, base cuneate or rounded, apex acute or acuminate, glabrous except for sparse hairs on the nerves, below clothed with long greyish hairs; petiole 10 - 12 mm tomentose. Inflorescence axillary or terminal. Male flowers: not known. Female flowers: turbinate, 8-9mm diameter, tepals 4, minute; carpels c. 38, narrowed above into a long subulate style.

Key to Species in Papuasia

1 endemic species:

Lauterbachia novoguineensis

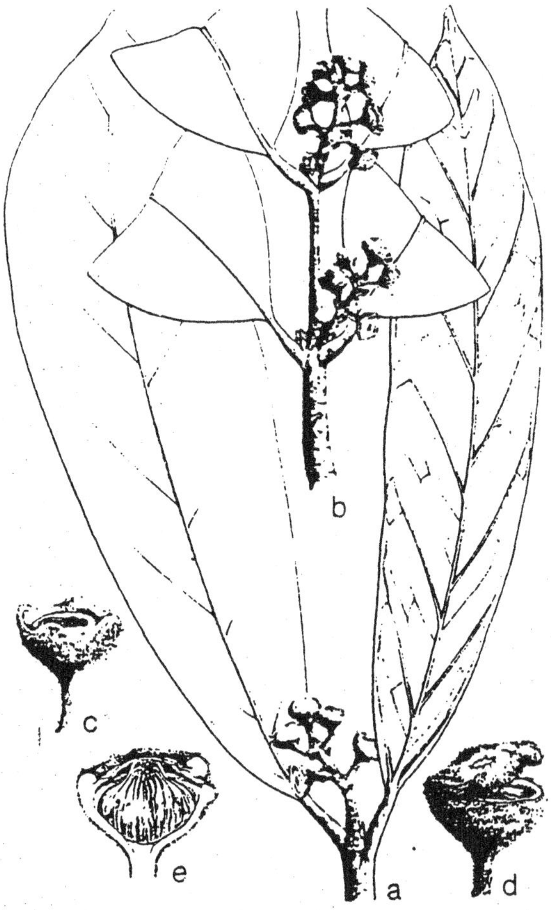

Lauterbachia novoguineensis (A) & (B) habit, with female flowers; (C) (D) & (E) open flowers and LS.
(Source: Philipson (1986).

6. LEVIERIA

A genus of some 7 species from the Celebes and Moluccas to Queensland. Common in the montane zone from 1200 to 3000m but descending to sea-level.
Trees or shrubs, rarely climbing. Leaves exstipulate, glabrous, entire or dentate. Dioecious with terminal or lateral cymose inflorescences. Male flowers: receptacle small, tepals 8, rounded enclosing numerous almost sessile stamens. Female flowers, ± globose with small ostiole bounded by 4 irregular tepals. Margin of receptacle reflexed to expose the ovoid sessile drupes.

Key to Species in Papuasia
(Source: Philipson 1986)

1. Undersurface of mature leaves glabrous.......................... 2

1. Undersurface of mature leaf with some indumentum (at least a few hairs at the base of the midrib) 4

2. Leaves obovate or narrowly elliptic .... L. squarrosa

2. Leaves elliptic to broadly elliptic 3

3. Leaf about 5cm long, or shorter. .... L. orientalis

3. Leaf about 7cm long, or longer. .... L nitens

4. Indumentum on lower leaf surface confined to the midrib.......... 5

4. Indumentum on lower leaf surface extending to lateral veins (and sometimes to the whole lamina)... 7

5. Liane with small leaves (3cm long, or less)... L. scandens

5. Trees or shrubs, leaves longer than 3cm............................... 6

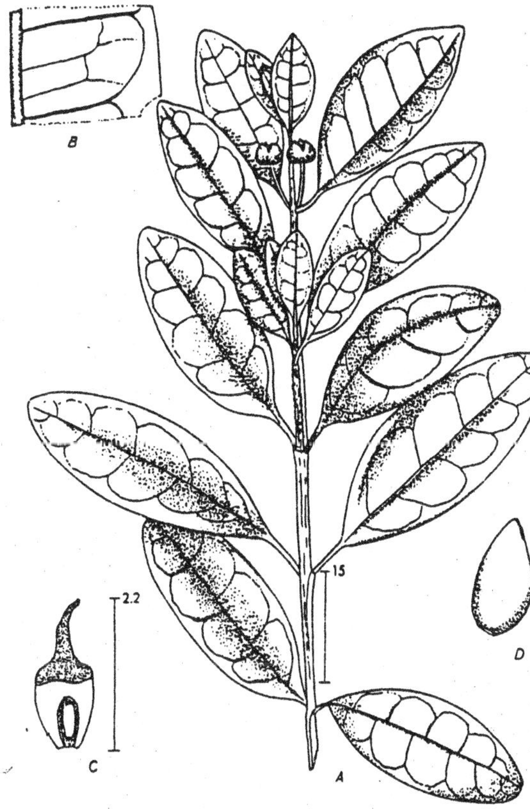

Levieria squarrosa (A) female branch; (B) limb of leaf; (C) female flower; (D) seed. (Source: van Royen (1982).

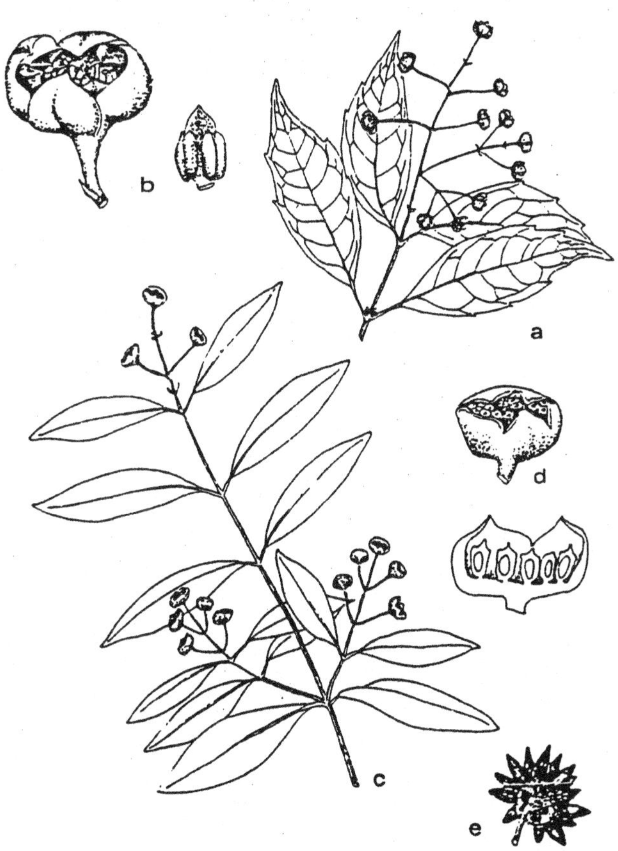

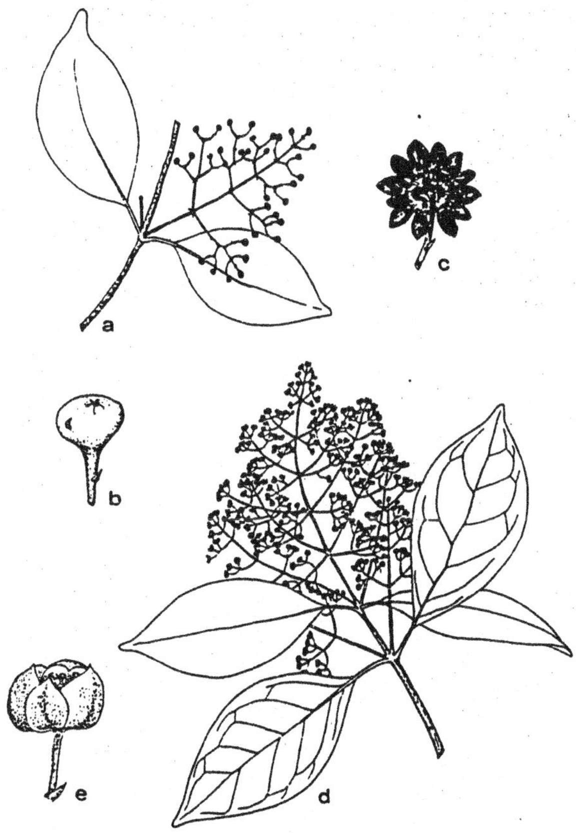

Levieria nitens (A) twig of female plant; (B) female flower; (C) fruit; (D) twig of male plant; (E) male flower. (Source: Philipson (1980).

6. Lateral veins clearly defined, blade usually rhombic or obovate. .... L. squarrosa

6. Lateral veins indistinct, blade narrowly elliptic. .... L. acuminata

7. Leaf with a long, narrow, apiculum; blade small (usually 7 x 2.5cm). .... L. acuminata

7. Apiculum, if present, short; blade larger (usually over 8cm long)... 8

8. Leaf oblong, hairs mostly straight and often sparse. .... L. montana

8. Leaf obovate, hairs crisped and usually copious. .... L. beccariana

Notes:

1. A frequent component of the sub-canopy of rainforest, often in regrowth areas. Most species are montane. All seven species occur in Papuasia.

Levieria beccariana (A) habit of male plant; (B) male flower; (C) twig of female plant; (D) female flowers; (E) fruits. (Source: Philipson (1986).

7. PALMERIA

● ● ● ● ● ●

A genus of 14 species of which 11 occur in New Guinea: 3 of these extend to Australia and one to the Celebes. In addition 3 species occur only in Australia. All are woody lianes with terminal panicles. No collections are recorded from the Solomon Islands.
Woody lianes with opposite, entire leaves, usually with stellate hairs. Dioecious, cymose panicles, inflorescences lateral or terminal. Male flowers: bowl- or saucer-shaped, tepals 4-7, incurved or spreading at anthesis to expose numerous stamens; stamens ± sessile on surface of receptacle. Female flowers: globose or flask shaped, tepals 5, small, surrounding a minute ostiole; carpels sessile, interspersed with numerous bristles. Receptacle enlarging to become a ± globose fruit which splits irregularly to exposed the drupes.

Key to Species in Papuasia (Source: Philipson 1986)

1. Outer surface of receptacle glabrous....... P. hooglandii

1. Outer surface of receptacle bearing hairs............................ 2

2. Undersurface of leaf +- glabrous. .... P arfakiana

2. Undersurface of leaf with indumentum............................. 3

3. Undersurface of leaf bearing numerous simple hairs (and usually also some stellate ones). .... P. gracilis

3. Undersurface of leaf bearing stellate hairs (occasionally also with a few simple hairs)......... 4

4. Mature fruits c. 7-8mm diam (before dehiscing).. Achenes about 5 mm long when dry (Irian Jaya: Angi Lake). .... P. angica

4. Mature fruits c. 15 - 20 mm diam (before dehiscing). Achenes about 7 mm long or longer when dry....... 5

5. Stellate hairs of undersurface of leaf laxly or densely distributed but not matted to form a close felt............................. 6

5. Stellate hairs densely distributed over the undersurface of the leaf and matted together to form a close felt............................ 7

6. Stellate hairs of the underside of the leaf large (0.6-0.9mm), generally distributed over the leaf surface, usually rather dense but the leaf surface clearly visible through them.
.... P. brassii

6. Stellate hairs of the underside of the leaf small (0.2-0.25mm), occasionally generally distributed and dense, but often confined to leaf base and petiole.
.... P. arfakiana

7. Felt on underside of leaf with a smooth even surface.............. 8

7. Felt on underside of the leaf with some larger hairs protruding above the general surface............. 10

8. Leaves large (most over 12cm long). from M.P. and C.P.
.... P. incana

8. Leaves smaller.................. 9

9. Leaves small, less than 5cm long. (Murray Pass).
.... P. montana

9. Leaves larger. (Tari and West).
.... P. hypargyrea

10. Veins on undersurface of the leaf fringed with long hairs.
.... P. clemensae

10. Veins on the underside of the leaf bearing tufted stellate hairs above the felt....................... 11

11. Leaves large, over 9cm long.
.... P. womersleyi

11. Leaves smaller, less than 6cm.
.... P. schoddei

Notes:

1. Most are montane species, growing from 1200 to 3200m altitude.

2. Several have traditional uses: employed as lashing materials and for wrapping tobacco. Several species are used as a source of salt.

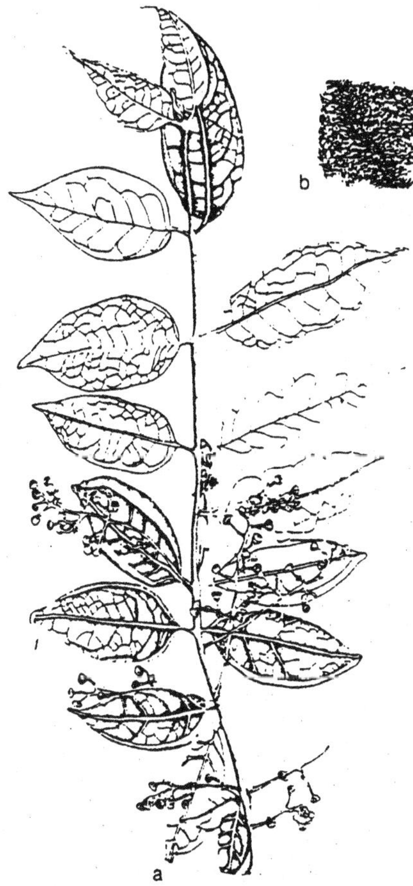

Palmeria clemensae (A) habit of male twig; (B) portion of lower surface of leaf. (Source: Philipson (1986).

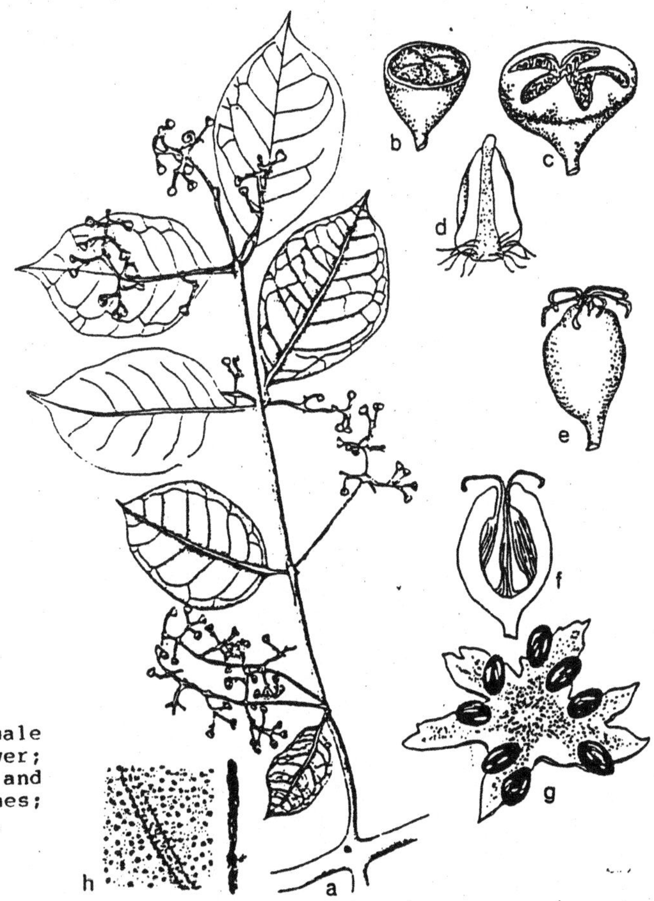

Palmeria womersleyi (A) habit of male twig; (B) & (C) bud of male flower; (D) anther; (E) & (F) female flower and LS; (G) receptacle, bearing achenes; (H) portion of lower surface of leaf. (Source: Philipson (1986).

8. STEGANTHERA

A genus of trees and shrubs extending from the Celebes and Moluccas, throughout Papuasia to Queensland. Several species are associated with ants. As noted by Philipson (1986) these species have enlarged nodes with distinct pores leading into them. This association is most common in S. hospitans but also occurs in S. moszkowskii, S. ledermannii and S. royenii. There are several species which are so poorly known that Philipson did not include them in his key.

Trees or shrubs. Leaves pubescent when young, becoming glabrous, entire or dentate, main secondary veins arched and joining margin. Monoecious or dioecious. Inflorescences lateral, cymose or fasciculate. Male flowers: turbinate; 4 small tepals; stamens (3-)4(-5), rarely with a second inner whorl of 1-4 stamens. Female flowers: similar to male but larger; ostiole surrounded by 4 tepals; carpels numerous; ovary pubescent; style glabrous, awl-shaped. Drupes sessile or long-stipitate.

Key to Species in Papuasia
(Source: Philipson 1986)

1. Female inflorescence subsessile. .... S. insculpta

1. Female inflorescence pedunculate. 2

2. Male inflorescence umbellate..... 3

2. Male inflorescence of dichasia, pleiochasia or panicles.......... 6

3. Leaves elliptic to broadly elliptic (mostly under 10cm long: W.S.P.). .... S. hentyi

3. Leaves oblanceolate to narrowly elliptic (mostly longer than 10cm). 4

4. Male receptacle globose (P.N.G. east of 146.30). .... S. fasciculata

4. Male receptacle discoid.......... 5

5. Leaves narrower than 4.5cm (Irian Jaya)...... S. cyclopensis

5. Leaves wider than 7cm (Irian Jaya). .... S. dentata

6. Pedicels of male flowers mostly in irregular clusters.............. 7

6. Pedicels of male flowers solitary.. 10

7. Nodes dilated, with a pore inhabited by ants (Irian Jaya).......... 8

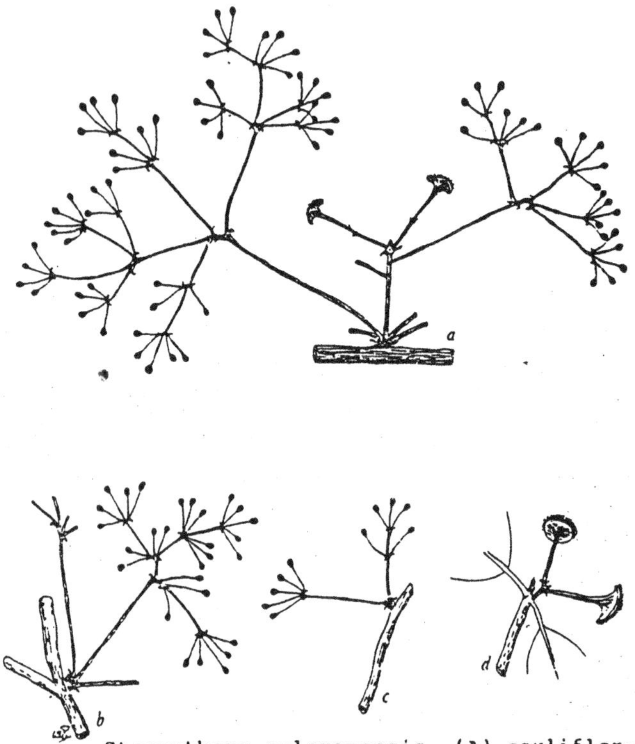

Steganthera salomonensis (A) cauliflorous panicle, with two female flowers; (B) (C) male inflorescences; (D) female inflorescence.
(Source: Philipson (1986).

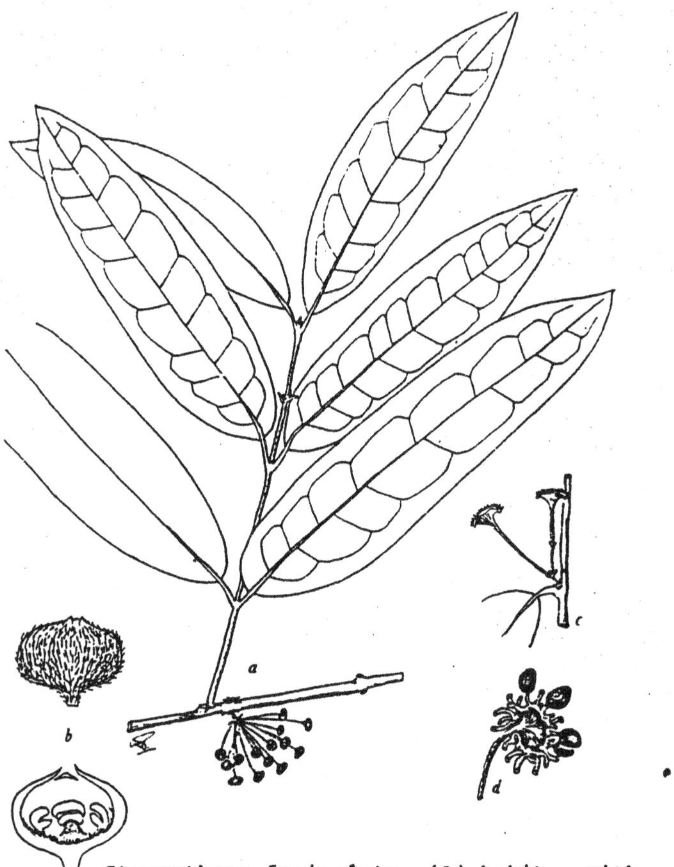

Steganthera faciculata (A) habit, with male flowers; (B) male flowers; (C) female inflorescence; (D) infructescence with 4 stipitate achenes.
(Source: Philipson (1986).

7. Nodes not harbouring ants........ 9

8. Male receptacle globose.
.... S. royenii

8. Male receptacle turbinate witha 4-lobed apical disc.
.... S. moszkowskii

9. Leaves large (rarely in Solomon Is leaves small on stunted plants), Infloresecence large with stout pedicels (Solomon Is. and P.N.G. east of 48.30).
.... S. salomonensis

9. Leaves smaller (less than 10cm long). Inflorescence small, with slender pedicels (M.P. and C.P.).
.... S. australiana

10. Receptacles discoid or turbinate...
.... 11

10. Receptacles globose (or only slightly depressed................. 14

11. Leaves 20cm, or more, long (widely distributed).
.... S. hospitans

11. Leaves considerably shorter..... 12

12. Male inflorescence 1-2 flowered, Leaves 6cm long, or shorter. (Irian Jaya, Idenburg R.).
.... S. myrtifloia

12. Male inflorescence with more flowers. Leaf longer than 6cm... 13

13. Male receptacle turbinate (Irian Jaya, W.S.P.).
.... S. myrtifolia

13. Male receptacle patelliform (C.P.).
.... S. oligantha

14. Leaves usually moderate to large (10-32cm long), and entire (widely distributed at low altitudes, but up to 2450m).
.... S. hirsuta

14. Leaves variable in size (usually less than 15cm long, but up to 21 cm), usually coriaceous, often dentate, but when narrow also entire (widely distributed above 1200m).
.... S. ilicifolia

14. Leaves 5-10cm long, coriaceous, +-rugose, entire (S.P. E.H.P. above 2400m).
.... S. chimbuensis

14. Leaves 5-10cm long, chartaceous, usually narrowly elliptic, acuminate, entire or dentate (M.P. C.P.).
.... S. australiana

Notes:

1. They occur from low altitudes to as high as 3,250m, in both primary and secondary forest, most species being subcanopy shrubs or small trees to 20m tall.

2. Steganthera is often confused with Kibara but they can be distinguished because Kibara has more tepals and a greater number of stamens (excepting S. salomonensis). The pedicel is also useful: in Steganthera the pedicel is slender and clearly disjunct from the receptacle; in Kibara the pedicel is thickened distally and has not a clearly defined junction with the receptacle. Also useful is the colour on drying: Steganthera drying brown and Kibara greenish or yellowish green (Philipson (1986).

3. The association with ants is striking. Some species have enlarged nodes with well-defined pores leading into the hollow stem. The cavities are lined with scale insects. This is most common in the abundant species S. hospitans but also in moszkowskii, ledermannii and royenii.

Steganthera ilicifolia (A) habit; (B) male flowers; (C) anther. (Source: Philipson (1986).

9. <u>WILKIEA</u>

• • • • • •

A genus confined to Papuasia (C.P. in P.N.G.) and to Queensland and New South Wales. Philipson (1986: 288) includes a discussion on the separation of Wilkiea and Kibara. In Papuasia this genus is represented by a small regrowth tree which occurs in mid-montane forest. It is characterised by light green, very hairy leaves (Philipson 1980).
Shrubs or trees with opposite, entire or serrate leaves, elliptic, base and apex rounded, petiole to 7-10mm. Densely hairy in all parts. Flowers axillary, singly or in few-flowered cymes; pedicels stout, to 4cm long, often with scars of 2 bracts below the flower. Male flowers: globose, ostiole surrounded by 6 tepals; stamens c. 18-30. Female flowers: globose, leathery, hairy within, tepals 4, carpels numerous (c. 75-100), sessile on lower half of receptacle; style slender. Ripe fruit unknown.

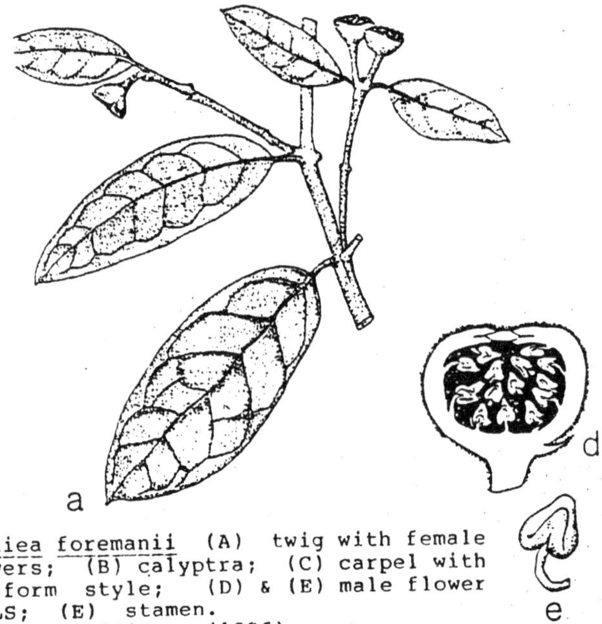

Wilkiea foremanii (A) twig with female flowers; (B) calyptra; (C) carpel with filiform style; (D) & (E) male flower in LS; (E) stamen.
(Source: Philipson (1986).

Key to Species in Papuasia

1 endemic species:

Wilkiea foremanii Philipson

Notes:

1. Known from Goilala region of the Central Province of Papua.

2. A small tree in mid montane forest and regrowth, 2400 - 2600 m altitude.

CHLORANTHACEAE

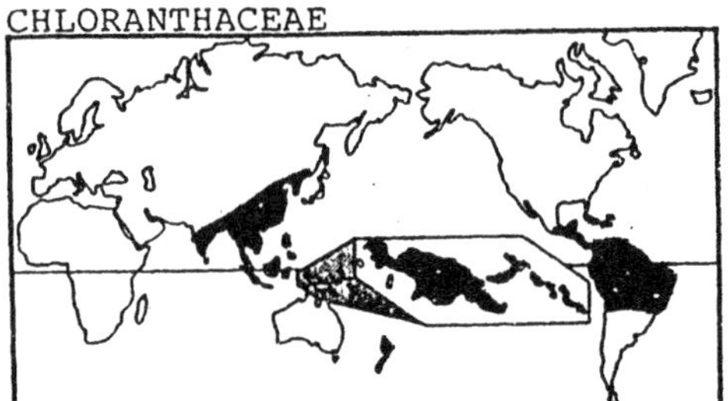

	WORLD	PAPUASIA
Number of genera:	4	3
Number of species:	80	6

Distribution: Tropics: America, E. Asia and Pacific.

Economic uses: Ornamentals, medicines and beverages.

CHLORANTHACEAE

A small family of herbs, shrubs and trees native to the tropics and southern and northern temperate regions. It includes 4 genera and some 80 species Limited economic use as ornamentals and locally used in Malesia to make beverages. One species of Hedyosmum is used in South America to treat stomach complaints. All moist evergreen forest species from sealevel to 3300m in Papuasia.

Diagnostic features

Leaves: opposite, simple, toothed, with small stipules.

Flowers: bisexual or unisexual, minute, devoid of petals. Male flowers in large numbers on a racemose inflorescence: no sepals and either 1 stamen or three fused stamens. Female and bisexual flowers in small numbers on a short raceme or spike: three-toothed calyx, fused to unilocular, inferior ovary.

Fruit: small ovoid or globose drupe, outer wall more or less succulent. Single seeded with minute embryo and copious endosperm.

References

Keng, H. (1969). Orders and Families of Malayan Seed Plants.
Smith, A.C. (1976). Ascarina in the South Pacific. J. Arn. Arb. 57: 405-424.
Verdcourt, B. (1985). Kew Bull. 40: 213.
Verdcourt, B. (1986). Chloranthaceae. Flora Malesiana Series 1. 10: 123 - 144.

KEY TO GENERA IN PAPUASIA
(Source: Verdcourt (1986)

1. Flowers truly unisexual, or apparently so.. 1. Ascarina

1. Flowers essentially bisexual, the male parts adnate to the ovary. 2

2. Anther-loculi 4 on a 3-lobed organ interpreted as 3 fused stamens with central 2-locular anther and a 1-locular anther on each side. Leaves of widespread species usually finely serrate. Fruit white, greenish or yellowish, rarely tinged violet or pinkish, the tip sometimes dark violet-purple.
.... 2. Chloranthus

2. Anther-loculi two on a narrowly clavate organ, i.e. apparently 1 stamen with a 2-locular anther. Leaves usually very coarsely serrate. Fruit red, rarely black.
.... 3. Sarcandra

1. ASCARINA

● ● ● ● ● ●

A genus of 12 species: from Madagascar through Malesia to the Pacific Islands and New Zealand. In Papuasia mostly in montane forests between 450 and 3300 m altitude. In Papuasia 4 species.

Shrubs or trees, apparently dioecious or monoecious. Leaves decussate, serrate, serrations often gland tipped; petiole sheath short; stipules very small. Flowers in condensed cymes borne in axils of bract with abaxial male flower and 1-2 adaxial female flowers (variable) but with most inflorescences reduced to a single male flower or 1-2 female flowers and appearing monoecious or dioecious. Male flowers bracteate, reduced to 1-2 subsessile anthers. Female flowers of a naked sessile ovary with outer bracts or 2 caducous inner bracts or if flowers paired also bracteoles; stigma sessile, truncate ot 2-lipped. Fruits purplish-grey turning black, obovoid; exocarp thin, succulent, endocarp stony. Seeds ovoid with smooth testa.

Key to Species in Papuasia
(Source: Verdcourt (1986)

1. Leaves rounded or only very shortly acuminate at apex. Flowers with a single subtending bract; male flowers with 2 stamens. Endocarp smooth........................ 2

Ascarina philippinensis (A) male flowers; (B) stipules; (C) male inflorescence; (D) base of anthers showing bract; (E) apiculate apex of anther; (F) TS anther; (G) female flowering branchlet; (H) female inflorescence; (I) portion of older female inflorescence; (J) fruiting branchlet; (K) LS ovary; (L) fruits; (M) seed; (N) leaf. (Source: Verdcourt (1986).

1. Leaves distinctly long-acuminate at apex. Flowers with 3 subtending bracts; male fowers with 1 stamen. Endocarp warty-papillate.
.... A. diffusa

2. Leaves subsessile, the petioles short and thick, about 2mm long; blades narrowed to rounded or subcordate at base.
.... A. subsessilis

2. Leaves distinctly petiolate, the petioles slender, 0.4 - 1.5 cm long; the blades cuneate at base.
.... 3

3. Fruits 1.2-2 mm long, with thinner more finely wrinkled exocarp (dry material), in more lax terminal and axillary inflorescences, the distances between the fruits greater and inflorescence branches more slender... A. maheshawrii

3. Fruits 2-3 mm long, with thicker more coarsely wrinkled exocarp, in lax to usually very densely congested terminal inflorescences, the distances between the fruits shorter and inflorescence - branches usually much stouter.
.... A. philippinensis

Notes:

1. A. philippinensis is a widely distributed high altitude species, in New guinea usually in Nothofagus forest or upper montane forest from 450 to 3,000 m altitude.

2. A. maheshwarii is a low altitude species endemic throughout Papuasia. In primary rainforest and open forest from 450 - 750 m altitude.

3. A. subsessilis is a local endemic in the Highlands and Morobe P of Papua New Guinea. In montane rainfoest and secondary forest on limestone from 1830 - 3200m altitude.

5. A. diffusa is a widespread Pacific species found in the Solomon Islands, Bougainville and New Britain.

2. CHLORANTHUS

• • • • • •

A genus of 20 species distributed from Japan and China to Papuasia. In New Guinea a single species.

Shrub 0.3 - 3m tall, rarely epiphytic, leaves ± when crushed, nodes distinctly swollen, sometimes purplish. Leaves oblong to elliptic, 8-29 by 3-13 cm, usually acuminate at apex, cuneate at base, margin glandular-serrate, sometimes purple beneath. Inflorescence scented, with 5-13 spikes each to 2.5-5cm long; bracts sheathing. Anther bearing organ yellow, 3-lobed, to 1.6mm long. Ovary enclosed by male part. Fruit white, cream or tinged violet or pinkish, succulent t0 5-7mm diameter. Seeds yellowish-white.

Key to Species in Papuasia

1 widespread species:

Chloranthus erectus

Notes:

1. A species distributed from Nepal to New Britain and New Ireland. Grows in primary and secondary forest in lowlands but up to 2535m altitude.

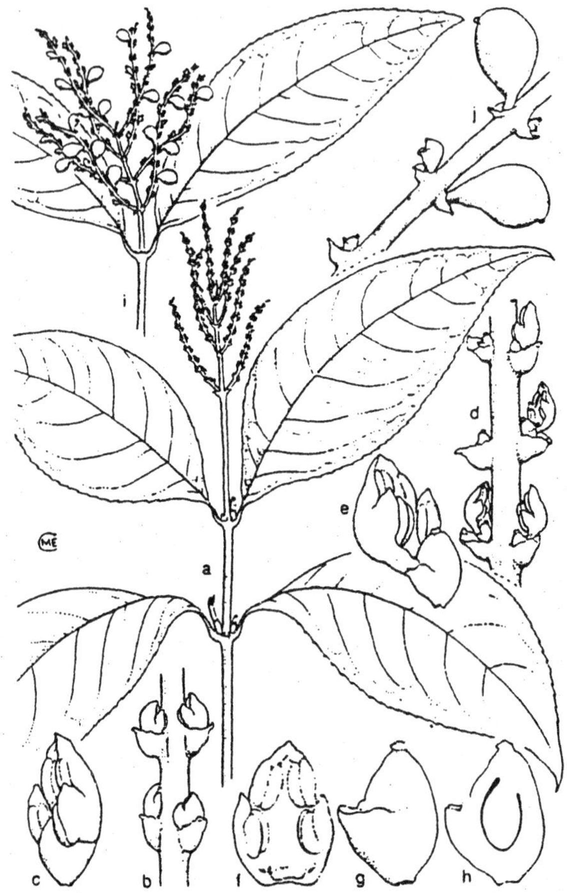

Chloranthus erectus (A) flowering branchlet; (B) young inflorescence; (C) young flower; (D) older inflorescence; (E) mature flower; (F) anthers; (G) and (H) ovary and LS; (I) fruiting branchlet; (J) fruits. (Source: Verdcourt (1986).

3. SARCANDRA

● ● ● ● ● ●

A genus of 2 species distributed from Sri Lanka and China to New Guinea. In Papuasia a single widespread species with a single variety in Papuasia.

Shrub or half-woody herb. Xylem lacks vessels. Stems with swollen nodes. Leaves 2-20 by 1-8 cm, acuminate at apex , cuneate at base, very sharply serrate or dentate-serrate; stipules small. Inflorescence greenish or white, the spike dense to 3.5(-5) cm long; paired bracts to 3 mm long; bracteoles oblong. Male parts to 2 mm long, anthers extending from half to whole length. Female flask shaped to 1.5 mm long, stigma to 1mm wide. Fruit yellow to red or bright orange when ripe, rarely black; to 7mm diameter. Seed pale, yellowish or cream.

Key to Species in Papuasia

1 widespread species:

Chloranthus glabra
ssp. brachystachys
var. brachystachys

Notes:

1. In lowland and lower montane forest and also stream banks from 130 - 2550 m altitude.

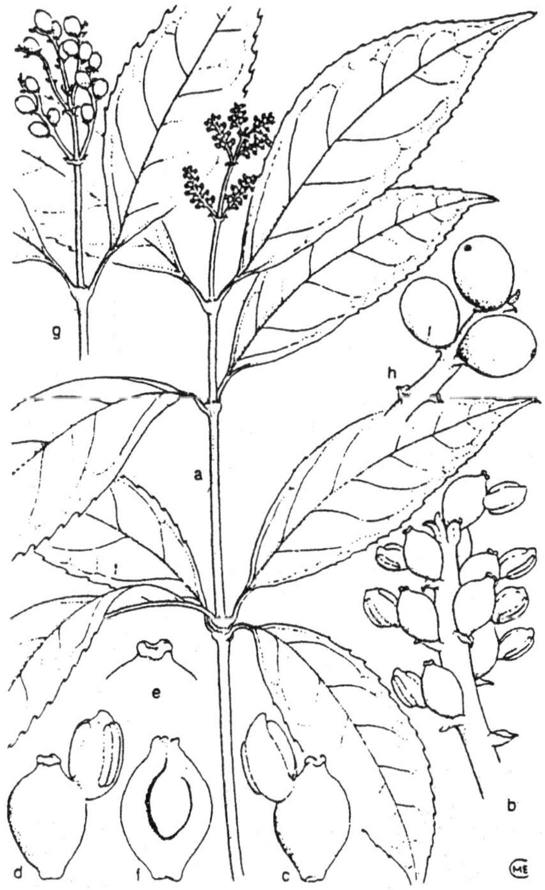

Sarcandra glabra (A) flowering branchlet; (B) inflorescence; (C) and (D) flower, back view, front view; (E) stigma; (F) LS ovary; (G) fruiting branchlet; (H) fruits.
(Source: Verdcourt (1986).

HERNANDIACEAE

● ● ● ● ● ●

A pantropical family of 4 genera and 76 species. Most are coastal. The timber of Hernandia is used locally.

Diagnostic features

Habit: trees, shrubs or lianes.

Leaves: large, alternate, exstipulate, simple or 3-5 palmately compound.

Flowers: regular, small, unisexual with flowers on same plant (in PNG), cymose inflorescences. Perianth of 3-10 sepal-like segments, 3-5 stamens in a single whorl, and (minute or lacking) 1 to 2 outer whorls of staminodes. Ovary inferior, uniloculate.

Fruit: dry, indehiscent achene, 2-4 winged, or enclosed by fleshy cup. Seed solitary, without endosperm.

HERNANDIACEAE

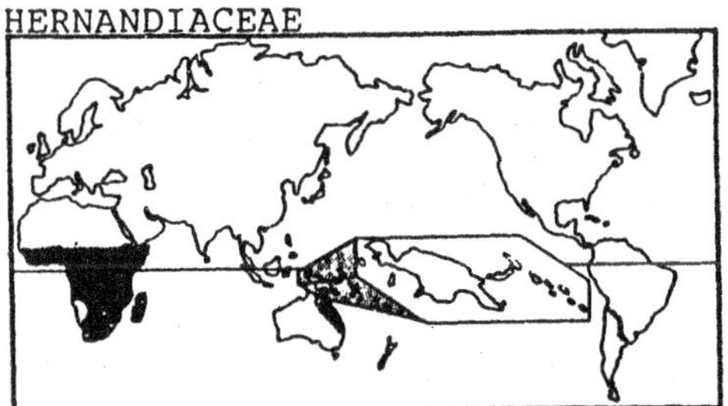

	WORLD	PAPUASIA
Number of genera:	4	3
Number of species:	76	7

Distribution: Pantropical mainly coastal.

Economic uses: Timber.

References

Croft, J.R. (1981). Hernandiaceae. Handbooks of the Flora of Papua New Guinea. 2: 190 - 201.

Kubitzki, K. (1969). Monographie der Hernandiaceen. Bot. Jahrb. Vol. 89: 78 - 209.

White, C.T. (1929). J. Arn. Arb. 10: 197 - 274.

KEY TO GENERA IN PAPUASIA

1. Deciduous trees: flowers minute, buds less than 1.5mm diameter; inflorescence an ebracteate thyrse; fruits with 2 apical wings; leaves simple..... 1. Gyrocarpus

1. Evergreen trees or lianes; flowers conspicuous, buds more than 2mm in diameter; inflorescence a bracteate thyrse; fruits with 2-4 broad lateral wings or surrounded by an expanded fleshy cupule: leaves simple or digitately compound....... 2

2. Trees; leaves simple; fruits surrounded by an inflated receptacle; flowers unisexual, rarely bisexual. .... 2. Hernandia

2. Lianes; leaves 3 - 5 digitately compound; fruit with 2 - 4 lateral wings; flowers bisexual. .... 3. Illigera

1. GYROCARPUS

A genus of 3 species with 11 subspecies found in lowland monsoon forests of both hemispheres. In Papuasia a single species in lowland areas to 400m altitude. Tree to 30m tall with a cream exudate. The timber is of no commercial value.

Deciduous trees. Leaves simple, entire, ovate to lanceolate, venation pedate. Inflorescence corymbose thyrse, with leaves towards branchlet apices. Flowers minute, numerous, polygamous, mostly male; stamens 4-5; staminodes clavate, alternating with stamens or partly connate. Male flowers with 4 - 8 -lobed perianth. Bisexual flowers with minute (6-)7(-8) -lobed perianth. Fruit a samara; nut ovoid, with 2 apical wings.

Key to Species in Papuasia

1 species:

Gyrocarpus americanus Jacq.

ssp. americanus

Notes:

1. Common in savannah areas around Port Moresby and also reported from Ramu Valley.

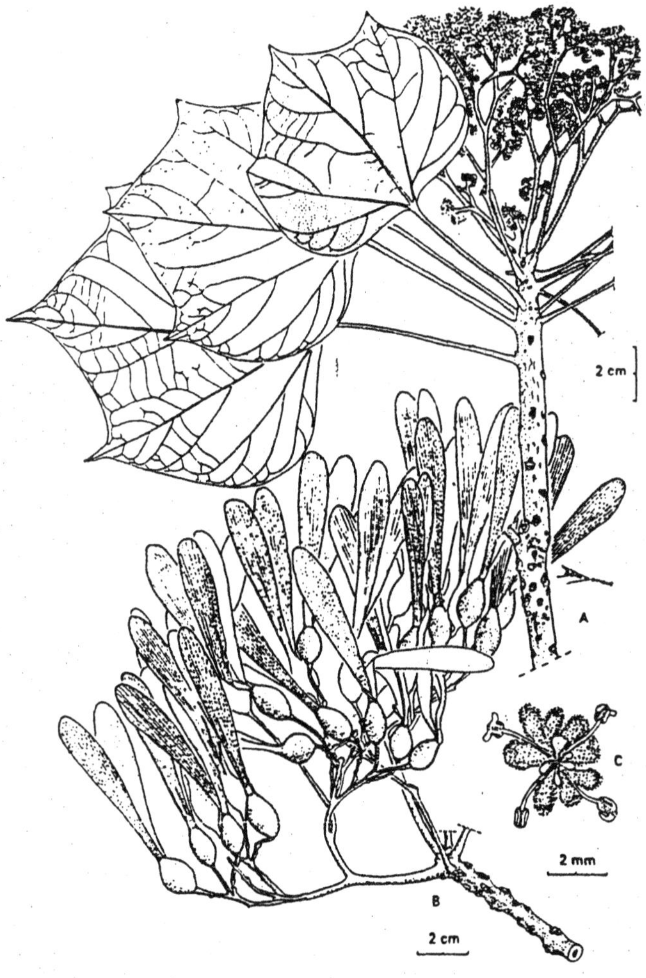

Gyrocarpus americanus ssp. americanus (A) twig with inflorescence; (B) fruit; (C) flower. (Source: Croft (1981).

2. HERNANDIA

A genus of 24 species mainly in coastal areas. Pantropical. In the region there are 4 species. A key based on vegetative material is included. Croft (1981) includes a key based on flowering material.

Evergreen trees. Leaves simple, peltate or not, venation pedate with 3-7 pairs of lateral veins. Inflorescence a thyrse in upper leaf axils, with foliaceous bracts subtending each branch, with single upper female flower and 2 lower male flowers; densely, finely, tomentose. Male flowers 3-5(-6)-merous; pistillode usually absent; stamens 3-5(-6), filaments free to partly connate, paired glands at base. female flowers in small cupule, 4-6-merous, staminodes lacking; glands 4-5(-12) free or connate around the style. Nut ovoid, enclosed by expanded fleshy cupule. Seed with hard testa.

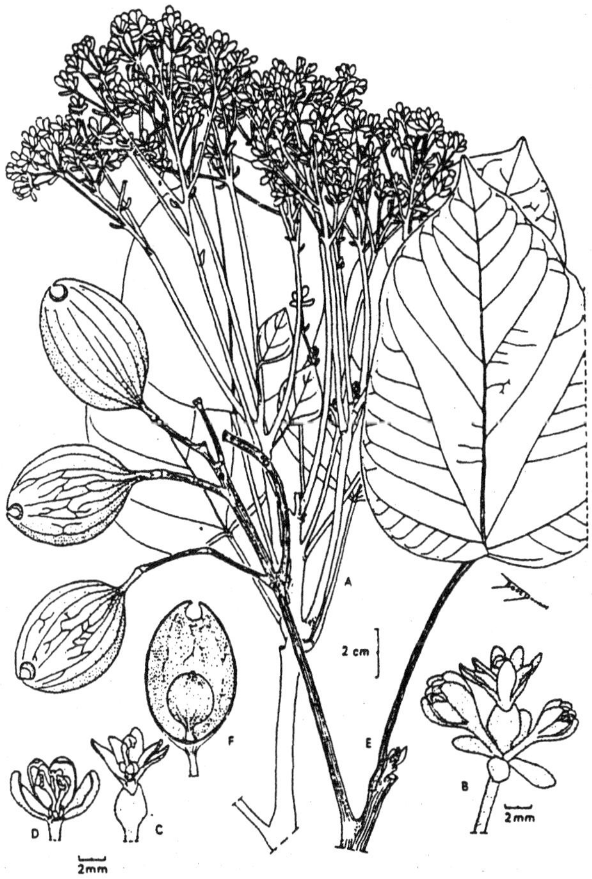

Hernandia ovigera (A) twig with inflorescence; (B) dichasium of flowers; (C) and (D) flowers; (E) fruit; (F) cut -away view of fruit within calyx. (Source: Croft (1981).

Key to Species in Papuasia
(Source: Croft (1981)

1. Leaves peltate, strand species. .... H. nymphaeifolia

1. Leaves not peltate, lowland rain-forest species.................... 2

2. Leaves ± cuneate-obovate, apex broadly rounded. .... H. rostrata

2. Leaves broadly ovate to cordate, apex ± acute or acuminate....... 3

3. Hairy domatia present in axils of lower veins; leaves ± coriaceous. .... H. moerenhoutiana ssp. samoensis

3. Hairy domatia absent; leaves ± chartaceous. .... H. ovigera

Notes:

1. The timber of H. moerenhoutiana is widely used for canoe building and for mouldings and interior finish. Trees of this species grow up to 900m altitude on ridges or well drained slopes.

2. H. ovigera has a light timber suitable only for light construction. H. nymphaeifolia is a tree of poor form traditionally used in ENB for making drums; the bark was also used with pig food for fattening. H. rostrata is only known from Santa Isabel in the Solomon Islands.

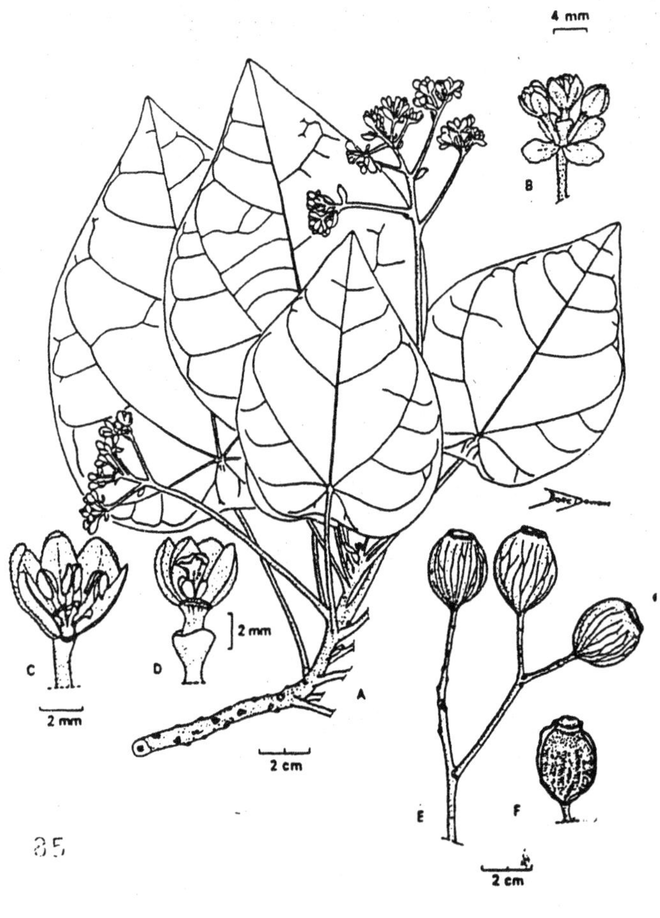

Hernandia nymphaeifolia (A) twig with inflorescences; (B) dichasium of central flower and 2 lateral flowers; (C) and (D) flowers; (E) fruit; (F) cut -away view of fruit within calyx. (Source: Croft (1981).

3. ILLIGERA

A genus of 18 species distributed from W Africa to Papuasia. Very rare lianes in Papua New Guinea.

Lianes or scandent shrubs. Leaves 3-foliolate (rarely simple, 3-lobed or 5-foliolate). Inflorescence terminal and axillary, bracteate thyrses; 1-3 bracteoles present at base of flowers. Flowers bisexual, 5-merous; tepals in 2 whorls of 5, deciduous, valvate in bud; stamens 5, opposite external tepals; ovary ovoid, 1-locular; ovule solitary, pendulous. fruit samara with 2-4 longitudinal membranous wings.

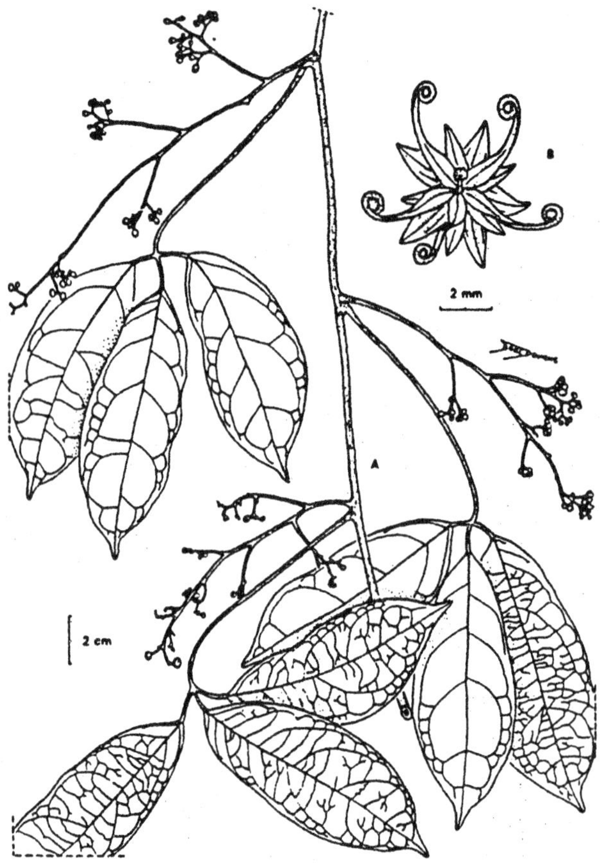

Illigera celebica (A) inflorescences on stem; (B) flower.
(Source: Croft (1981).

Key to Species in Papuasia
(Source: Croft (1981)

1. Filaments very flattened and circinate, staminodes clavate, 0.6mm long, interstaminal glands absent; leaves glabrous with 5-7 pairs of lateral veins, domatia absent; samara 4-winged.
.... *I. celebica*

1. Filaments only slightly flattened, straight to curved, staminodes urceolate, 2mm long, interstaminal glands present; leaves pilose on midrib above and on lower surface of blade, 4(-5) pairs of lateral veins, domatia present; samara 3(-4) winged.
.... *I. novoguineensis*

Notes:

1. *I. celebica* is locally common in the Vogelkop up to an altitude of 100m. *I. novoguineensis* Kubitzki is endemic to a restricted area of the lower Ramu in the Madang Province. It is apparently very rare as no collections have been made of this species since c. 1900.

LAURACEAE

A large tropical and subtropical family of mostly trees and shrubs with 32 genera and some 2500 species. Most authorities also include Cassytha which is a group of parasitic climbers without true leaves. The group includes several important commercial species: Cinnamomum the source of cinnamon and camphor; Persea americana the avocado pear; Litsea a source of many local medicines and many valuable timber trees. Several genera in Papua New Guinea are very diverse with many species.

LAURACEAE

	WORLD	PAPUASIA
Number of genera:	32	12
Number of species:	2500	274
Distribution:	Tropics and subtropics.	S.E. Asia and Brazil

Economic uses: Important timber trees, avocado and spices. Some ornamental species.

Diagnostic features

Leaves: alternate or opposite (radially arranged in Actinodaphne), exstipulate, leathery, containing numerous oil cavities.

Flowers: in racemes or cymes, regular, bisexual or unisexual on the same plant. Flower parts in multiples of three. The petals and sepals form an inconspicuous perianth, the sepals and petals being poorly distinguished from each other. Ovary superior.

Fruit: a berry or drupe-like, often enclosed by the perigynous parts of the flower which can become more or less fleshy. Seeeds lack endosperm.

References

Allen, C.K. (1942). Studies in Lauraceae IV. Preliminary study of the Papuasian species collected by Archbold Expeditions. Journ. Arn. Arb. 23: 112 - 155.

Kostermans, A.J.G.H. (1950). Misc. Notes on New Guinea Plants. I. Misc. Notes on the Lauraceae. Bull. Jard. Bot. Buitenz. Ser.III. 18: 435 - 443.

Kostermans, A.J.G.H. (1952). The genera of the Lauraceae.

Kostermans, A.J.G.H. (1956). New and Critical Malesian Plants. IV. Djalan Perniagaan 44: 1 - 40.

Kostermans, A.J.G.H. (1957). Lauraceae. Reinwardtia 4: 193 - 256.

Kostermans, A.J.G.H. (1960). New and Critical Malesian Plants. VI. Reinwardtia 5: 341 - 369.

Kostermans, A.J.G.H. (1962). Misc. Bot. Notes. III. Reinwardtia 6: 155 - 187.

Kostermans, A.J.G.H. (1962). Misc. Bot. Notes. IV. Reinwardtia 6: 281 - 325.

Kostermans, A.J.G.H. (1965). New and Critical Malesian Plants. VII. Reinwardtia 7: 19 - 46.

Kostermans, A.J.G.H. (1965). Misc. Bot. Notes. V. Reinwardtia 7: 141 - 146.

Kostermans, A.J.G.H. (1968). Materials for a revision of the Lauraceae. I. Reinwardtia 7: 291 - 356.

Kostermans, A.J.G.H. (1969). Materials for a revision of the Lauraceae. II. Reinwardtia 7: 451 - 536.

Kostermans, A.J.G.H. (1969). A New Guinea cinnamon used as a contraceptive. Reinwardtia 7: 539 - 541.

Kostermans, A.J.G.H. (1970). Materials for a revision of the Lauraceae. III. Reinwardtia 8: 21 - 196.

Kostermans, A.J.G.H. (1974). Materials for a revision of the Lauraceae. IV. Reinwardtia 9: 97 - 115.

Weber, J.Z. (1981). A taxonomic revision of Cassytha in Australia. J. Adelaide Bot. Gard. 3: 187 - 262.

White, C.T. (1950). Ligneous Plants from the Solomon Islands (and New Guinea). Journ. Arn. Arb. 31: 81 - 116.

Notes:

1. Some authorities exclude the parasitic climbers in the genus Cassytha from the Lauraceae. They place them in their own family: the Cassythaceae. This is not followed here. Kostermans (1952) notes that despite their aberrant habit and ecology, the floral characters are not different from other Lauraceae. They are closest to Cryptocarya.

KEY TO GENERA IN PAPUASIA

(Source: Kostermans (1957)

1. Twining parasites with reduced leaves.... 3. Cassytha

1. Trees with normal leaves....... 2

2. Flower umbels surrounded by an involucre of large, decussate, persistent bracts. Fruits ± embedded in a cup....................... 3

2. Flowers not surrounded by a persistent involucre............... 5

3. Flowers dimerous. .... 10. Neolitsea

3. Flowers trimerous. 4

4. Anthers 2-celled. .... 8. Lindera

4. Anthers 4-celled. .... 9. Litsea

5. Fruit completely enclosed by adnate enlarged flower tubes with usually only a small hole at the apex. Anthers 2-celled. .... 5. Cryptocarya

5. Fruit on a naked pedicel or only partly covered by a shallow or deep cup developed from the flower tube. Anthers 2- or 4-celled......... 5

6. Fruit seated on and partly covered by a shallow or deep cup developed from the flower tube. Anthers 4-celled........................ 6

6. Fruit on a naked pedicel or perianth persistent, but no cup formed. Anthers 2- or 4-celled......... 7

7. Inflorescence covered by long-persistent, large, non-decussate bracts. Leaves markedly 3-veined, opposite. Bark with cinnamon smell. .... 4. Cinnamomum

7. Bracts of inflorescence small, soon deciduous. Leaves penniveined, in whorls.... 1. Actinodaphne

8. Perianth in the fruit hardened, clasping the base of the fruit. Anthers 4-celled. .... 12. Phoebe

8. Perianth deciduous in fruit, or if persistent, not hardened and not clasping fruit. Anthers 2- or 4-celled........................ 8

9. Fruit pedicel strongly thickened, fleshy, often highly coloured. Anthers 2-celled. Only in Irian Jaya...... 6. Dehaasia

9. Fruit pedicel not or hardly thickened: if thickened anthers 4-celled. Anthers 2- or 4-celled. 9

10. Anthers 4-celled. (Avocado). .... 11. Persea

10. Anthers 2-celled............... 10

11. Fertile stamens 3. Leaves aerolate. .... 7. Endiandra

11. Fertile stamens 6 or 9. Leaves reticulate. Outer 2 staminal whorls facing outwards. .... 2. Beilschmiedia

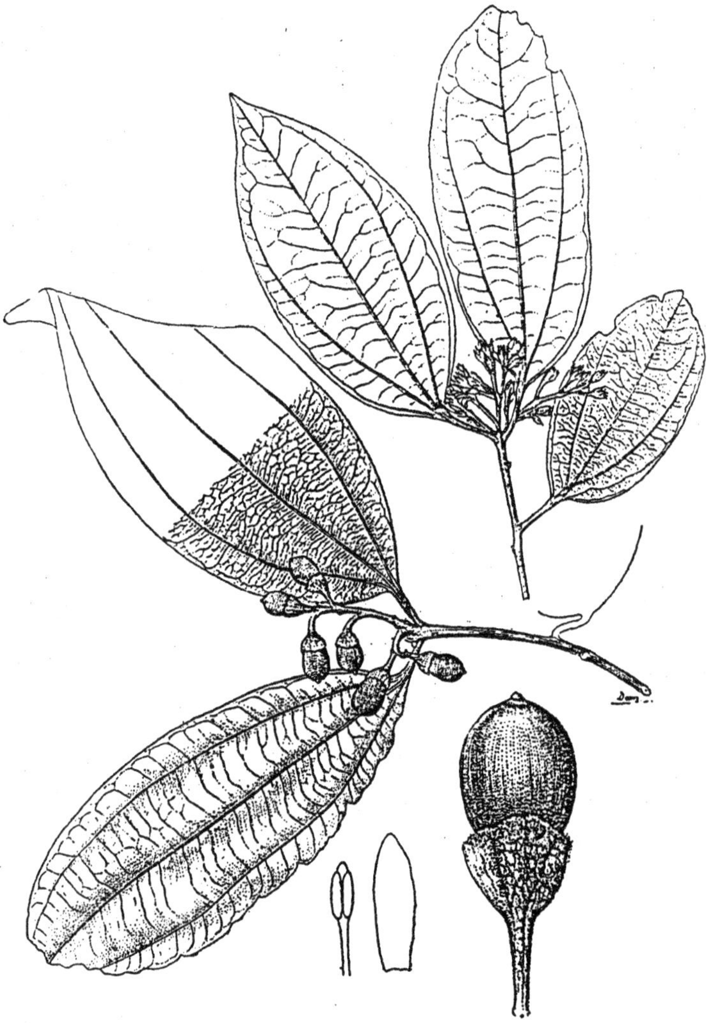

Cinnamomum eugeniliferum
(Source: Kostermans (1970)

1. ACTINODAPHNE

● ● ● ● ● ●

A genus of 60 - 70 species restricted to the region from the Asiatic mainland to Papuasia. It is closely related to Sassafras, a genus of 2 species with which it might be combined (Kostermans 1952, 1957). Allen (1942) described 3 species of Actinodaphne from Papuasia: A. brassii, described from Santa Isabel, is a synonym of A. multiflora (Kostermans 1969: 453). Allen also described A. archboldii from Lake Daviumbu in the Western Province and A. solomonensis from Bougainville.

Leaves usually verticillate. Flowers dioecious; tepals 6, subequal. Fertile stamens in male flowers usually 9; inner whorl biglandular. Anthers 4-celled. Staminodes in female flowers usually 9; inner whorl biglandular. Stigma peltate. Fruit on a shallow or deep cup.

List of Species in Papuasia

Actinodaphne archboldii
Actinodaphne macgregorii
Actinodaphne multiflora
A. brassii
Actinodaphne solomonensis

Notes:

1. A multiflora is common in lowland tropical rainforest. the Type collection was made in the Japen Is. group in Irian Jaya.

2. Teschner (1923) described several new species of Actinodaphne. These were described as having a 6-lobed perianth and consequently most belong in Neolitsea rather than in Actinodaphne. The species described by Teschner were: caesia; densiflora, engleriana, latifolia, ledermannii, nitida, novoguineensis, obtusa, pulchra and tomentosa.

2. BEILSCHMIEDIA

● ● ● ● ● (●

A pantropical genus of over 200 species: well represented in Africa. In Papuasia some 15 species. Collections of several undescribed species are in the National Herbarium at Lae.

Trees. Leaves often subopposite; reticulation lax, usually conspicuous. Flowers bisexual. Tepals equal or subequal. Fertile stamens 9 or 6, 2-celled; two outer whorls introrse, glandless; third whorl extrorse, flanked by glands; fourth whorl consisting of conspicuous, usually stipitate staminodes. Stigma inconspicuous. Fruit on a not swollen (or hardly swollen) bare pedicel.

List of Species in Papuasia

Beilschmiedia acutifolia
B. archboldiana
B. tessendorffiana
Beilschmiedia dielsiana
Beilschmiedia dilmyana
Beilschmiedia ledermannii
Beilschmiedia longifolia
Beilschmiedia morobensis
Beilschmiedia myrmecophila
Beilschmiedia novae-britanniae
Beilschmiedia podagrica
Beilschmiedia pubescens
Beilschmiedia pullenii
Beilschmiedia pustulata
Beilschmiedia rosseliana
Beilschmiedia sericea
Beilschmiedia triplinervis

Notes:

1. One of the few examples of myrmecophilly in the Lauraceae occurs in the species Beilschmiedia myrmecophila, collected by L.J.Brass (13693) along the Idenburg River.

2. Trees of Beilschmiedia are locally common on areas dominated by ultrabasic soils and also in the Lower Montane zone.

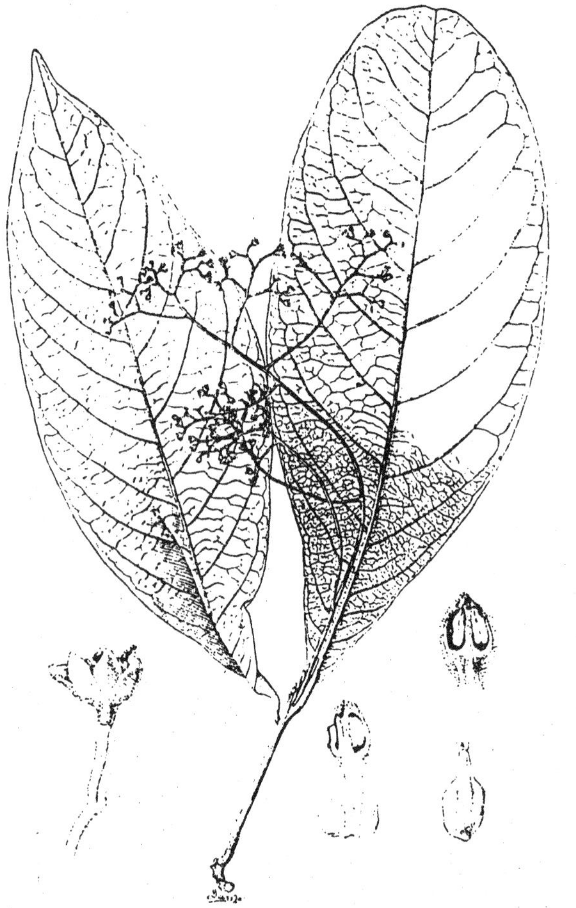

3. CASSYTHA

A genus of 15 - 20 species mainly Australian in distribution but one species, C. filiformis is pan tropical. Although aberrant in their habit and ecology, being parasitic twiners with small haustoria, the floral characters hardly differ from the Lauraceae. Consequently they are not separated into a the distinct family: Cassythaceae.

Two species were described from Papuasia by Allen (1942), but both of these are synonymous with earlier described species.

Key to Species in Papuasia

(Source: Weber (1981)

1. Fruit glabrous, globular, greenish (drying black). Peduncle +0.5 mm thick..... C. filiformis (syn. C. archboldiana)

1. Fruit strigose, ovoid, reddish (never drying black). Peduncles 0.3 - 0.4 mm thick. .... C. capppillaris (syn. C. tenuis)

Notes:

1. C. filiformis is distributed from New Guinea to San Cristobal in the Solomon Islands. C. archboldiana described from the Balim Valley in Irian Jaya in a synonym. C. tenuis described from the Western Province, is a synonym of C. capillaris.

2. C. filiformis var. pubescens is not accepted as , with age, the species becomes pubescent.

3. C. novoguineensis was described by Kanehira and Hatusima (Bot. Mag. Tokyo 57 (1943) 190). No material has been seen of this species.

4. CINNAMOMUM

A genus of 391 described species but will reduce considerably in numbers when revised. Extends from Asia and Formosa to Australia and the Pacific. Throughout Papuasia where it is represented by some 21 species. The main leaf characters used in identification are the length of the basal or subbasal, ascendent, lateral nerves and the indumentum. Fertile material is usually required for accurate identification.

Trees. Leaves usually opposite and triplinerved. Flowers bisexual, rarely polygamous; fertile stamens 9 (or 6) in 3 whorls; outer 2 whorls introrse, glandless; inner whorl extrorse, flanked by glands. Anthers 4-celled, very rarely 2-celled. Fourth whorl consisting of conspicuous, stipitate, glandless staminodes. Stigma discoid or peltate.

List of Species in Papuasia

Cinnamomum archboldianum
Cinnamomum clemensii
Cinnamomum culilaban
Cinnamomum englerianum
Cinnamomum eugenoliferum
Cinnamomum frodinii
Cinnamomum gigaphyllum
Cinnamomum gracillimum
Cinnamomum hentyi
Cinnamomum kami
Cinnamomum ledermannii
Cinnamomum massoia
Cinnamomum piniodorum
C. gilgianum
Cinnamomum podagricum
Cinnamomum rosselianum
Cinnamomum singularis
Cinnamomum sleumeri
Cinnamomum salomonense
Cinnamomum splendens
Cinnamomum verum
Cinnamomum xanthoneurum

Cinnamomum hentyi
(Source: Kostermans (1970)

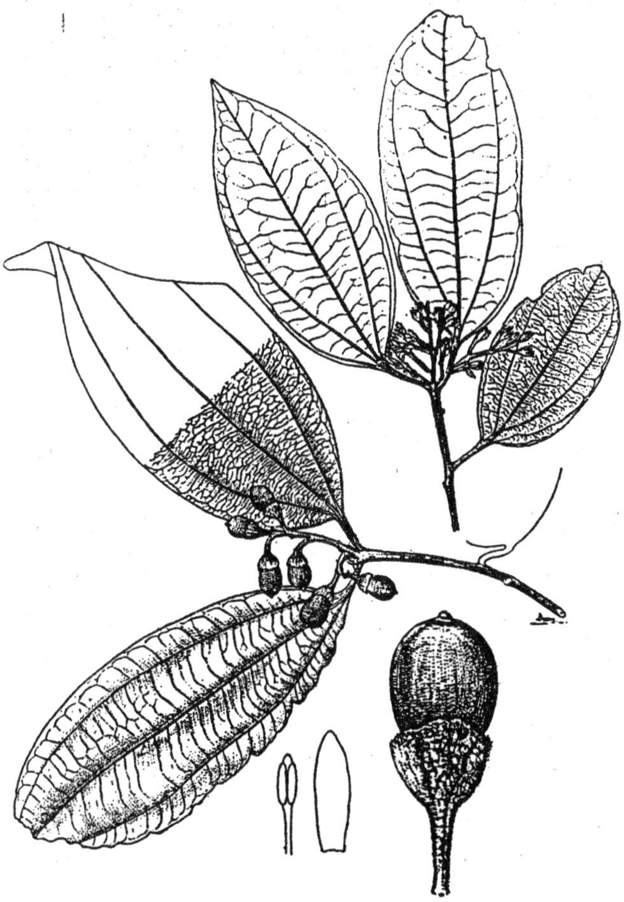

Cinnamomum eugenoliferum
(Source: Kostermans (1970)

Notes:

1. C. zeylanicum Bl. has commonly been considered the proper name for the common cinnamon tree. The oldest valid name is C. verum J. S. Presl published in 1825 (Kostermans 1965: 141).

5. CRYPTOCARYA

● ● ● ● ● ●

A pantropical genus of 200 - 250 species centered in Malaysia, only lacking from C Africa. In Papuasia some 96 species mainly growing in the lowland and lower montane zones. This genus requires detailed revision. Several species are logged in commercial operations.

Leaves alternate or opposite. Flowers bisexual; tube slender, conspicuous; tepals 6, equal. Fetile stamens 9, 6 or 3, anthers 2-celled; anthers of two outer whorls introrse, of third whorl extrorse and filaments flanked by glands. Forth whorl consisting of conspicuous stipitate staminodes. Stigma small or inconspicuous, rarely peltate. Fruit entirely included in the enlarged flower tube leaving only a minute orifice at apex. Endocarp and exocarp often bony and ribbed (Kostermans 1957).

List of Species in Papuasia

Cryptocarya ainikini
Cryptocarya albida
Cryptocarya alleniana
Cryptocarya alticola
Cryptocarya apamaefolia
Cryptocarya arfakensis
Cryptocarya argyrophylla
Cryptocarya aromatica
Cryptocarya atra
Cryptocarya aurea
Cryptocarya aureosericea
Cryptocarya boumiensis
Cryptocarya brassii
Cryptocarya brevipes
Cryptocarya bullata
Cryptocarya burckiana
Cryptocarya cagayanensis
Cryptocarya caloneuria
Cryptocarya camptodroma
Cryptocarya caudata
    C. bernhardiensis
Cryptocarya cinnamomifolia
    C. tetragona
Cryptocarya cordifolia
Cryptocarya depressa
Cryptocarya endiandraefolia
Cryptocarya engleriana
Cryptocarya exfoliata
Cryptocarya fagifolia
Cryptocarya filicifolia
Cryptocarya flumiensis
Cryptocarya forbesii
    C. myrcioides
Cryptocarya fusco-pilosa
    C. impressivena
Cryptocarya gigantocarpa
Cryptocarya globosa
Cryptocarya graebneriana
Cryptocarya hartleyi
Cryptocarya idenburgensis
    C. gonioclada
Cryptocarya ilocana
Cryptocarya invasiorum
Cryptocarya iridescens
Cryptocarya vridescens
Cryptocarya kajewskii
Cryptocarya kamahar
    C. aureobrunnea
Cryptocarya laevigata
Cryptocarya lanuginosa
Cryptocarya ledermannii
    C. angica
Cryptocarya longepetiolata
Cryptocarya macrophylla
Cryptocarya magnifolia
Cryptocarya massoy
Cryptocarya medicinalis
Cryptocarya minutifolia
Cryptocarya multinervis
    C. sulcata
Cryptocarya multipaniculata
    C. cordata
Cryptocarya nothofagetorum
Cryptocarya novoguineensis
    C. archboldiana
Cryptocarya ovata
Cryptocarya ovato-caudata
Cryptocarya pachyphylla
Cryptocarya palmerensis
Cryptocarya pauciflora
Cryptocarya parallelinervia
Cryptocarya patentinervis
    var. glabra
Cryptocarya percrassa
Cryptocarya pergamentacea
Cryptocarya pergracilis
Cryptocarya perlucida
Cryptocarya pulchella
Cryptocarya pullenii
Cryptocarya pusilla
Cryptocarya rarinervia
Cryptocarya renicarpa
Cryptocarya resinosa
Cryptocarya rhizophoretum
Cryptocarya roemeri
Cryptocarya ruruvaiensis
Cryptocarya scalariformis
Cryptocarya schlechteri
Cryptocarya schoddei
Cryptocarya sericeo-triplinervia
Cryptocarya sleumeri
Cryptocarya splendens
Cryptocarya subbullata
Cryptocarya subfalcata
Cryptocarya subtunervis
Cryptocarya sulcata
Cryptocarya tannaensis
Cryptocarya tebaensis
Cryptocarya tesselata
Cryptocarya unilonata
Cryptocarya verrucosa
Cryptocarya viridiflora
Cryptocarya weinlandii
Cryptocarya whiteana
Cryptocarya wilsonii
Cryptocarya womersleyi
Cryptocarya xylophylla

6. DEHAASIA

A genus of some 20 species closely related to Persea subgenus Nothaphoebe. The genus is restricted to Malaysia with 1 species in Papuasia. Airy-Shaw (1940) proposed the fusion of Alseodaphne with Dehaasia. In contrast Kostermans (1957) maintained the genus Dehaasia.

The genus is characterised by sessile, thick, 2-celled anthers, minute staminodes and a ± permanently semi-closed perianth with smaller outer tepals. The fruit-pedicel is strongly enlarged, fleshy, and often highly coloured.

Key to Species in Papuasia

1 endemic species:

Dehaasia novoguineensis

Notes:

1. A tree to 8m tall with alternate leaves. Growing in secondary forest at 1200m altitude (Kostermans 1956).

2. Known only from the type collection (Carr 15627) from Isuarava.

7. ENDIANDRA

A genus of some 95 species distributed from Malaysia to Australia and the Pacific. Closely related to Beilschmiedia, but Endiandra has a third whorl of 3, often sessile stamens (one species has 6 stamens). In Papuasia some 35 species.

Trees. Leaves alternate or opposite with areolate reticulation. Floral characters as in Beilschmiedia, but in the third whorl only 3 sessile stamens. Staminodes are minute or absent.

List of Species in Papuasia

Endiandra acuminata
E. brassii
E. clemensii
Endiandra aggregata
Endiandra albiramea
Endiandra aneityensis ?
Endiandra archboldiana
Endiandra arfakensis
Endiandra asymmetrica
Endiandra beccariana
Endiandra bullata
Endiandra carrii
Endiandra eyphellophora
Endiandra clememsiae
Endiandra dielsiana
E. glandulosa
Endiandra djamuensis
E. rubescens non Bl.
Endiandra engleriana
Endiandra euadenia

Endiandra macrostemon
(Source: Kostermans (1970)

Endiandra faceta
Endiandra ferruginea
Endiandra flavinervis
Endiandra forbesii
Endiandra formicaria
E. brassii
Endiandra fragrans
Endiandra fulva
Endiandra gem
Endiandra gemopsis
Endiandra glandulosa
Endiandra glauca
Endiandra grandifolia
Endiandra havelii
Endiandra impressicosta
Endiandra inaequitepala
Endiandra invasiorum
Endiandra latifolia
Endiandra ledermannii
Endiandra macrostemon
Endiandra magnilimba
Endiandra micrantha
Endiandra microphylla
E. merrilliana
Endiandra minutiflora
Endiandra montana
Endiandra multiflora
Endiandra novoguineensis
Endiandra oblonga
Endiandra ochracea
Endiandra papuana
Endiandra parviflora
Endiandra pilosa
Endiandra recurva
Endiandra schlechteri
Endiandra sericea
Endiandra sleumeri
Endiandra solomonensis
Endiandra spathulata
Endiandra sphaerica
Endiandra spathulata
Endiandra teschneriana
Endiandra versteeghii
Endiandra whitmorei
Endiandra xylophylla

Endiandra minutiflora
(Source: Kostermans (1970)

8. LINDERA

A widely distributed genus of some 100 species, not represented in Africa. It is represented by a single species in Irian Jaya.

Slender tree to 8m tall. Leaves alternate, thin, papery. Petiole slender, densely tomentellous. Flowers and fruits as in Litsea, but anthers 2-celled (very rarely part 4-celled; fruit cup usually shallow, surrounded by remnants of perianth.

Key to Species in Papuasia

1 species:

Lindera novoguineensis

Notes:

1. Growing in mid-montane forest. In Irian Jaya associated with conifers and Nothofagus between altitudes of 1200 and 1900 m. In Moluccas at 1000 m altitude.

9. LITSEA

A genus with some 474 described species, including several which are synonyms. The number will be reduced to circa. 400 when critically revised. Tropical and subtropcal in distribution. Around 65 species in Papuasia, but material of many undescribed species appears to exist in the National Herbarium at Lae. a widely dispersed genus only missing in Africa and Europe. Most species however tropical.

A genus of woody trees. Flowers dioecious, surroinded by an involucre of usually persistent, large, decussate bracts. Tepals 6 or 0. Male flowers with 9 or 12 fertile stamens. Outer 2 whorls usually lacking glands, the third and inner whorl usually flanked by glands. Filaments slender; anthers 4-celled, introrse. Oval stipitiform or absent. Female flowers, staminodes equal in number to stamens in male flower; stigma peltate, conspicuous. Fruit seated in dveloped cup or disc; perianth cauducous.

List of the Species in Papuasia

Litsea acuminata
Litsea alba
Litsea alveolata
Litsea anamo
Litsea amicorum
Litsea aneityensis
Litsea aureo-sericea
Litsea bainingensis
Litsea brassii
Litsea breviumbellata
Litsea buinensis
Litsea calophyllantha
Litseaacaroli Litsea caroli
Litsea carrii
Litsea chinensis
Litsea chrysoneura
Litsea clarissae
Litsea collina
Litsea complanata
Litsea crenata
Litsea densiflora
Litsea dielsiana
Litsea domarensis
Litsea elliptica
Litsea engleriana
L. bernhardensis
Litsea exsudens
Litsea filipedunculata
Litsea fo
Litsea fulvo-sericea
Litsea gigaphylla
Litsea globosa
Litsea grandifolia
Litsea habbemensis
Litsea henricksonii
Litsea kauloensis
Litsea latifolia
Litsea ledermannii
Litsea longipedicellata

Litsea longipedicellata
(Source: Kostermans (1970)

Litsea psilophylla
(Source: Kostermans (1970)

Litsea maluensis
L. gilgiana
Litsea maluensis
Litsea melchioriana
Litsea minor
Litsea morobensis
Litsea novoguineensis
Litsea papuana
Litsea perlucida
Litsea psilophylla
Litsea pumila
Litsea racemosa
Litsea rigidifrons
Litsea schlechteri
Litsea sebifera
Litsea sepikensis
Litsea solomonensis
Litsea spathulata
Litsea subauriculata
Litsea thulatasp
Litsea subauriculata
Litsea tannaensis
Litsea teshneri
Litsea timonioides
Litsea trichophylla
Litsea versteeghii
Litsea zeylanica

10. NEOLITSEA

A genus of some 80 species from Asia to Malaysia, 6 species in Papuasia. Several undescribed species occur in Papuasia, collections of these are in the herbarium at Lae.

Flowers dioecious, with large, decussate, persistent, involucral bracts. Flowers dimerous; flowers dimerous; fertile stamens in male flowers and staminodes in female flowers 6 in 3 whorls; the inner whorl with glands. Anthers 4-celled, introrse. Stigma in female flowers conspicuous, peltate. fruit on disc-like cup; pedicel only slightly thickened.

List of Species in Papuasia

Neolitsea archboldiana
Neolitsea arfakensis
Neolitsea brassii
Neolitsea novoguineensis
Neolitsea pubescens
Neolitsea teschneriana

Notes:

1. The type of Neolitsea brassii was collected at Wassi Kussa in the Western Province along grassland margins. the same species is also known from the Morobe Province as a small tree in grasslands at 1350m, forest margins and as a tree on a cliff overhanging L Wanum.

2. Neolitsea pubescens has been collected from high altitudes (2250-2550m) in the Bismark Ranges and the Star Mountains. Two collections are known from Missima Island at 700m in the understorey of tree fern glades.

Litsea carrii
(Source: Kostermans (1970)

11. PERSEA

A genus with some 239 described species, pantropic but mainly in S America. P. americana is introduced to Papuasia and widely cultivated. The inclusion of Nothaphoebe and Alseodaphne must await satisfactory revisions of these genera to conclude whether both should be incorporated in Persea. In this treatment the specimens are listed in the genus Nothaphoebe, with the exception of the common avocado.

Trees or shrubs. Leaves alternate, chartaceous to rigidly coriaceous. Panicles axillary or sub-terminal. flowers bisexual; tepals 6, the outer whorl usually smaller than the inner one, deciduous or persistent; tube very shallow. Stamens 9 or 6 (2 or 3 whorls) with long slender filaments in Persea s.s. or stamens sessile in Nothaphoebe; Staminodes stipitate. Anthers 4-celled (rarely 2-celled in third whorl). Fruits on naked pedicel, or perianth persistent but not enlarged. Pedicel cylindrical or enlarged and fleshy.

Key to Species in Papuasia

1 introduced species:

Persea americana

ALSEODAPHNE and NOTHAPHOEBE

Both generic names have been applied to the material from Papuasia. It is characterised by a swollen and enlrged fruit pedicel and distinct leaf reiculation. Kostermans (1960) noted that they 'may be upheld as a distinct species'. In 1950, Kostermans treated N. archboldiana as a variety of A. umbelliflora; in 1950 he recognised it as a distinct species. He also described several other species from Irian Jaya.

List of Species in Papuasia

Nothophoebe archboldiana
Alseodaphne elata
Nothophoebe novae-brittanniae

Notes:

1. N. archboldiana is widespread in lowland tropical rainforest (Gogol, Sorong) and also in savannah regions (Oriomo).

2. N. novae-brittanniae has been collected at low altitudes in New Britain in regrowth and advanced secondary forest.

3. A distinct, unnamed species, from Irian Jaya is represented by BW 12839.

CARYODAPHNOPSIS

Airy Shaw (1940) includes the record of C. tonkinensis from Irian Jaya. Two collections are recorded bb. 32872 from Mimika, Aria; and Kanehira and Hatusima 12850 from Nabire.

List of Species from Papuasia

1 species:

Caryodaphnopsis tonkinensis

Persea americana (A) habit; (B) (C) flowers.
(Source: Purseglove (1977)

12. PHOEBE

A genus with some 174 described species (several synonyms), and some could be placed into Persea and perhaps Cinnamomum, depending on the delimitation of the genus. As constituted by Kostermans (1957) the genus is restricted to Asia. In Papuasia probably three species.

Genus with floral characters as in Persea. Perianth persistent, indurate, enlarged, clasping the basal part of the fruit.

Key to Species in Papuasia

Phoebe clemensiae
Phoebe forbesii
P. novo-guineensis

ARISTOLOCHIALES

A small order which includes two families: the Aristolochiaceae and the Nepenthaceae. By their 3-merous flowers, anther type and secreting cells related to the Magnoliaceae. In some publications authors consider the order contains a single family, the Aristolochiaceae.

ARISTOLOCHIACEAE

• • • • • •

This family includes 7 genera of tropical and temperate distribution. Many members of the family are twining lianas. A large number of species are cultivated because of their curious flower shapes and colourful variegated leaves. In Papua New Guinea several species act as host plants to species of Birdwing Butterflies. Farming of the butterflies depends on the cultivation of the host plants.

Diagnostic features

Leaves: alternate, palmately veined, entire or lobed, exstipulate.

Flowers: axillary, solitary or in racemes or cymes, bisexual, zygomorphic. Calyx regular 3-4 lobed or irregular and 1-3 lobed. Corolla usually absent, rarely 3 petals present. Stamens 6 (in New Guinea)-40 in number, free, conate or joined to a stylar column (gynostemium). Ovary inferior, 3-6 locules and 3-6 stigmas.

Fruit: a many seeded, 6-valved, capsule, seeds small with copious endosperm.

References

Backer, C.A. and Bakhuizen van den Brink, R.C. (1963). Flora of Java 1: 61

Ding Hou (1983). New Species of Malesian Aristolochiaceae. Blumea 28: 343 - 352.

Ding Hou (1986). Aristolochiaceae. Flora Malesiana. Series 1. 10: 53 - 108.

KEY TO GENERA IN PAPUASIA

1 genus: 1. Aristolochia

ARISTOLOCHIACEAE

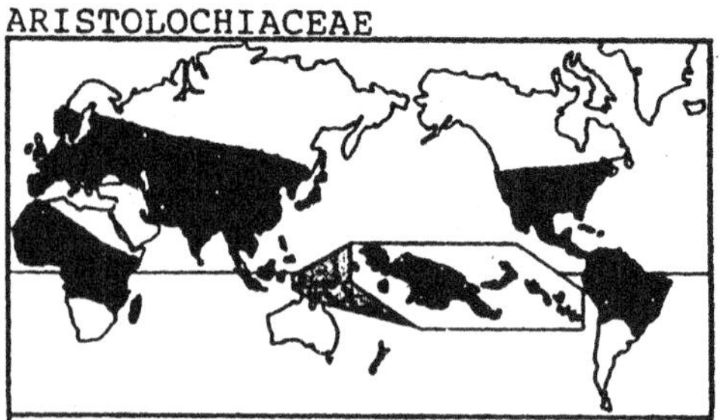

	WORLD	PAPUASIA
Number of genera:	7	1
Number of species:	630	9

Distribution: Tropical and temperate regions. Woody lianes.

Economic uses: Common garden plants. Host birdwing butterflies.

1. ARISTOLOCHIA

• • • • • •

Aristolochia is a genus of about 400 species distributed in tropical and subtropica areas but with some species in temperate regions such as Australia. It is represented by 9 species in Papuasia. It grows throughout the lowlands and montane areas of Papuasia up to an altitude of c. 2250m. It is rarely in primary forest, most species are in secondary regrowth. Usually flowering material is required for correct identification.

Herbaceous perennials, climbing, sometimes high lianes; often with tuberous rhizome or rootstock. Leaves entire, sometimes 3-lobed. petiole grooved above. Flowers usually zygomorphic, solitary or in inflorescences. Perianth inflated at basal part, then contracted into tube. Stamens 6, adnate to style column. Ovary oblong; style column 6-lobed. Fruits capsular. Seeds often winged, flat.

Key to Species in Papuasia
(Source: Ding Hou (1984)

1. Leaves deeply 3-lobed, in shape often similar to a 'W'. .... A. jackii

1. Leaves entire.................. 2

2. Leaf veins closely reticulate, prominent beneath. Fruits cylindric or oblong, 2.5-4.5 by 2-3cm, minutley granular. Leaves cordate, auricles usually much overlapping; apex acuminate.
.... A. crassinervia

2. Leaf veins loosely reticulate and some crossbar-like, often slightly elevated beneath................ 3

3. Perianth limb 6-lobed. (Leaf base cordate)... A. schlechteri

3. Perianth limb 3-lobed or 1-lipped.
.... 4

4. Perianth without a stipe; limb 3-lobed. Seeds not winged......... 5

4. Perianth with a distinct stipe; limb 1-lipped. Seeds distinctly winged......................... 7

5. Flower buds or perianth lobes with the apical 20-30mm strongly contracted or narrowed and often tail-like when dry (easily broken).... 6

5. Flower buds or perianth lobes without a tail-like apical part. Perianth lobes triangular, 5-6 by 8-10mm, apex obtuse or retuse.
.... A. engleriana

6. Perianth lobes suborbicular, c. 10 mm diameter, with an apical tail-like apex c. 20mm long.
.... A. momandul

6. Perianth lobes triangular to narrow -triangular, 35-50 by 12-14mm at base, apical 25-30mm often tail-like when dry.
.... A. dielsiana

7. Leaves glabrous. Perianth limb obovate or oblanceolate, 17-20 by 8-12mm. Seed (incl. wing) transverse-oblong, 6-11.5 by 12-16mm, smooth on both surfaces.
.... A. gaudichaudii

7. Leaves minutely hairy or sparsely short-hairy beneath. Seeds (incl. wing) triangular, 6-10 by 5-10mm, usually verrucose on both surfaces.
.... 8

8. Leaves traingular or deltoid in outline, minutely hairy beneath. Inflorescences 2-3(-15)mm long; internodes condensed, obscure or invisible; bracts minute, c. 1mm long. Perianth lobes oblong, c.20 mm long.... A. linnemannii

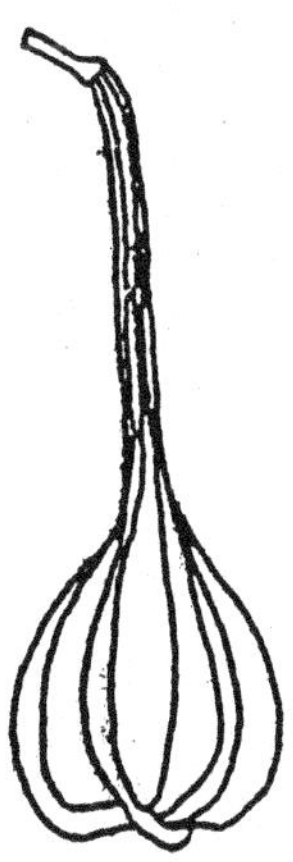

8. Leaves variable in shape and size, ovate, ovate-oblong, rarely suborbicular, sparsely shortly hairy beneath. Inflorescence 20-60mm long; internodes spacious, distinct: bracts up to 10mm long. Perianth lobes lanceolate, 20-30(-40)mm long.
.... *A. tagala*

Notes:

1. Most species are scattered, rarely abundant, in secondary or primary forest. Mostly at low or medium altitudes but occasionally up to 2250m in altitude.

2. Most species have very characteristic capsular fruits. They are usually 4-6-loculed and dehiscent. Seeds are usually numerous.

3. For proper identification it is essential to collect flowers. Note that flowers often vary in colour with age.

4. *A. schlechteri* is unique in Papuasia in that it has a 6-lobed perianth.

5. Ding Hou (1984: 108) includes notes on 2 species described from the Sepik. The type material of both species was collected at Lordberg. The types and only known collections were destroyed in Berlin during World War II. Ding Hou includes brief descriptions of these species in his revision in Flora Malesiana.

NEPENTHACEAE

● ● ● ● ● ●

This family of carnivorous herbs and shrubs occurs throughout S. E. Asia to New Guinea and northern Australia. It also occurs on Madagascar. The areas with most species are Borneo and in New Guinea there are 5 species. In Papuasia the species occur from sea-level to 2000m, usually in very moist conditions. One species is however common in lowland grassland situations. In Sri Lanka and Malaysia the stems of some species are used for basket making and making cord.

The common name for these plants is the Pitcher Plants. Many taxonomists place the Nepenthaceae in the same family as the other insectiverous plants the Sarraceniaceae and the Droseraceae. However they are usually placed in the Aristolochiales, Sarraceniales and the Rosales respectively.

Scandent herbs or shrubs, climbing by twirling leaf-tendrils. Leaves alternate. mature leaves consist of the lamina, tendril and the pitcher with an expanded lid. Flowers borne in racemes, unisexual and dioecious. Perianth: 2 whorls each two-merous. Stamens 4 to many; filamenst connate into a column. Fruit: capsule opening by 4-valves. Seeds minute, winged.

Diagnostic features

Stem: many species are epiphytic and have climbing stems to 3cm diameter.

Leaves: alternate, exstipulate, petiole indistinct winged. Midrib forms tendrils from apex of leaf. Ends of tendrils develop into pitchers, with a lip projecting over mouth.

NEPENTHACEAE

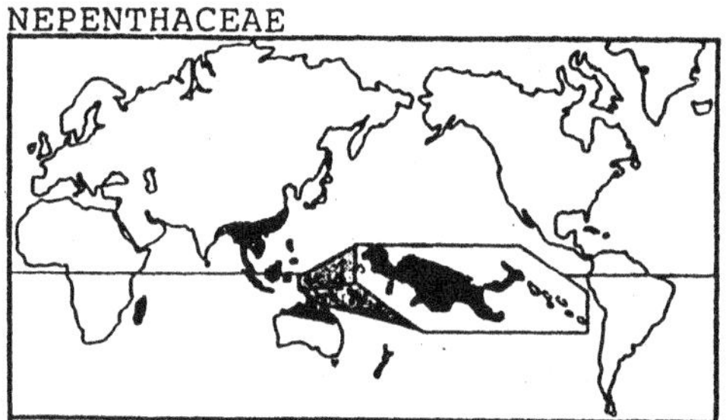

	WORLD	PAPUASIA
Number of genera:	1	1
Number of species:	70	10
Distribution: Old World tropics.		
Economic uses: Horticulture.		

Pitchers: edges curve inwards, spotted with honey glands to attract insects. Below the region of glands the surface is slippery so that when an insect falls into the pitcher it cannot climb out. The plant absorbs the products of decay.

Flowers: small, red, yellow or green, regular, unisexual, with male and female flowers on separate plants. Borne in spike-like inflorescence. Perianth of 3-4 sepals. Male flowers 4-24 Stamens, filaments united into column and anthers crowded into mass. Female flowers, discoid stigma, style short or absent, ovary superior, of four locules with numerous ovules.

Fruit: leathery capsule, seeds with hair-like projections on end. Embryo minute with fleshy endosperm.

References

Danser, B. (1928). The Nepenthaceae of the Netherlands Indies. Bull. Jard. Bot. Buitenz. Series III. 9: 249 - 438.

KEY TO GENERA IN PAPUASIA

1 genus: 1. Nepenthes

1. NEPENTHES

A genus of some 60 palaeotropical pitcher plants. Most are herbs in boggy places, climbing with the aid of tendrils developed from the midrib of the leaves. The end of the tendril usually develops into a pitcher. The top section of the pitcher has secreting glands which, together with the colour, attract insects. The insects drown in the pitcher and the plant absorbs the products of their decay. In Papuasia there are 10 species.

Key to Species of Nepenthes in Papuasia

(Source: Jebb (unpub.)

1. Lower pitchers urceolate. Lid narrow, oblong-linear in shape, upper pitchers rudimentary. .... N. ampullaria

1. Lower pitches tubular to ovoid. Lid rounded to ovate, upper pitchers always present.................. 2

2. Lid with 2 appendages on lower surface............................ 3

2. Lid with no appendages on lower surface......................... 4

3. Mouth of pitchers hooded, pointing forwards, the apex being almost horizontal. .... N. klossii

3. Mouth oblique, the apex vertical. .... N. maxima

4. Leaf base winged and decurrent to half way down the internodes..... 5

4. Leaf distnctly petiolate......... 6

5. Peristome over 1cm in width. .... N. insignis

5. Peristome less than 0.5cm in width. .... N. vieillardii

6. Upper pitchers infundibulate, widest at mouth.................... 7

6. Upper pitchers cylindrical, narrowed towards the mouth............ 8

7. Pitchers small, less than 10cm long, leaves with 2 to 3 longitudinal veins in outer third of blade. Inflorescence a panicle. .... N. paniculata

7. Pitchers large, to 25cm long, leaves with 2 to 3 longitudinal veins in outer two thirds of the lamina, some arising from the midrib. Inflorescence a raceme. .... N. treubiana

8. Margin of lower leaves fimbriate, upper pitchers not winged. .... N. mirabilis

8. Margin of lower leaves entire, upper pitchers commonly winged...... 9

9. 2 to 3 longitudinal veins in outer third of lamina only, numerous pinnate nerves. .... N. neoguineensis

9. 5 to 7 longitudinal veins throughout lamina, pinnate nerves obscure. .... N. papuana

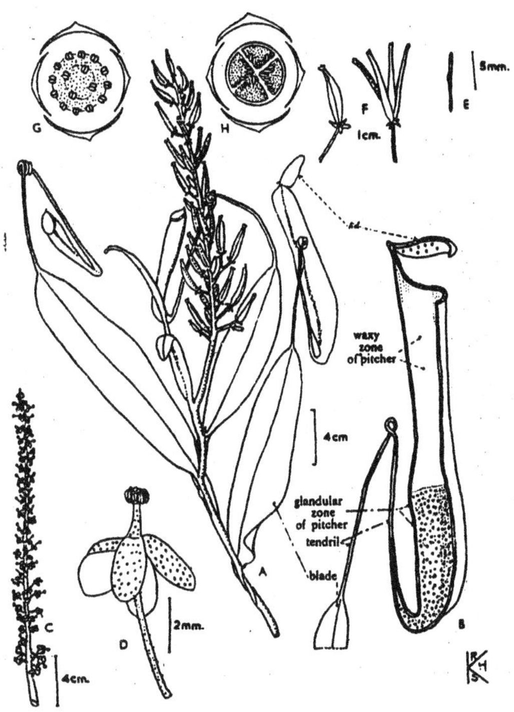

Nepenthes gracilis (A) branch with pitcher and fruiting inflorescence; (B) pitcher; (C) (D) male: inflorescence, flower; (E) (F) seed, fruit; (G) (H) male and female flowers.
(Source: Keng (1969)

PIPERALES

The order Piperales includes three families, two of which occur in Papua New Guinea. The families Piperaceae and Peperomiaceae are pantropical in distribution, extending also into subtropical and temperate regions, particularly in the southern hemisphere. A third family, the Saruraceae, is restricted in its distribution to eastern Asia and North America.

In the following key the genera are included for both the families Piperaceae and Peperomiaceae. This approach has been adopted because of the very close relationship between the two families.

KEY TO GENERA IN PAPUASIA

1. Flowers bisexual, in erect dense spikes. Spikes several to numerous, +- umbellate on stout axillary peduncles. Peduncles 1-3 together. Leaves heart-shaped or kidney-shaped, palmately nerved: base of petiole broad, sheathing, leaving an annular scar on stem. Erect, robust shrub.
   .... Pothomorphe

1. Flowers bisexual or unisexual, in solitary spikes or racemes....... 2

2. Epiphytic or terrestrial herbs, not climbing. Leaves exstipulate, base of petioles not winged. Flowers bisexual, very small. Stamens 2, filaments very short. Stigma 1.
   .... Peperomia

2. Climbing or erect shrubs or trees, rarely herbs. Leaves stipulate or with winged base to petiole. Flowers bisexual or unisexual. Stamens 2 or more, Stigmas 2-5.
   .... Piper

PIPERACEAE

● ● ● ● ● ●

This family of small trees, shrubs and woody climbers is tropical in distribution. Many authorities consider it as comprised of a single genus Piper with some 2,000 species. Some workers split off some 70-80 Central and South American species into closely related genera. The Peperomiaceae, a closely related family, is sometimes treated as a part of this family. Peperomia differs from Piper in being herbaceous, with flowers reduced to a single carpel and two anthers subtended by one bract.

Economic Uses

Piper nigrum is the source of pepper. The other species in common use in Papua New Guinea is Piper betle which is eaten with betel nut. The roots of Piper methysticum are used for making kava.

Diagnostic Features

Stem: conspicuously jointed. Anatomy is unusual in that some vascular bundles are scattered in stem like in monocotyledons.

Leaves: alternate, simple, entire, dotted with glands which contain aromatic oils, with winged petioles which sheath the stem.

Flowers: in a raceme or spike, borne opposite the leaves; flowers very small, bisexual or unisexual, no perianth, stamens 2 to several, ovary superior with 2-4 fused carpels.

Fruit: fleshy single-seeded drupe.

References

Backer, C.A. and Bakhuizen van den Brink, R.C. (1963). Piperaceae. Flora of Java 1: 167 - 174.

Chew, W.L. (1972). The genus Piper in New Guinea, Solomon Islands and Australia. J. Arn. Arb. 53: 1 - 25.

van Royen, P. (1982). Piperaceae. The Alpine Flora of New Guinea. 3: 1267 - 1288.

PIPERACEAE

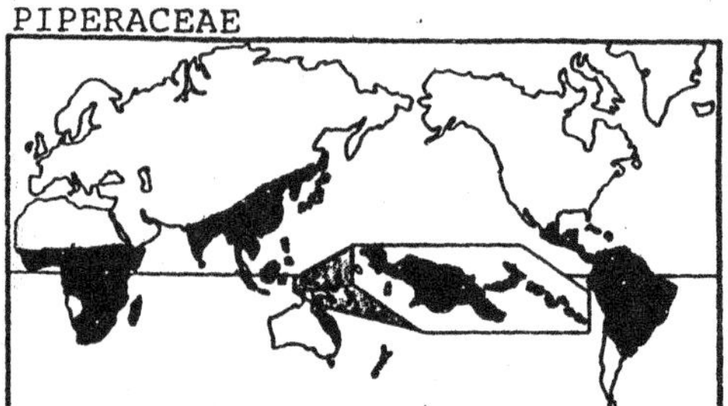

	WORLD	PAPUASIA
Number of genera:	5	2
Number of species:	2100	100

Distribution: Pantropical, rainforest species.

Economic uses: Pepper and kava.

KEY TO GENERA IN PAPUASIA

1. Inflorescence leaf-opposed, solitary spike.
.... 1. Piper

1. Inflorescence axillary, spikes in an umbellate cluster.
.... 2. Pothomorphe

1. PIPER

• • • • • •

A tropical genus of some 700 species, mostly in C and S America. Climbing shrubs. In New Guinea many species, some 8 above 3000m altitude. The genus is much in need of revision. Some 100 species are described from Papuasia.
Small creeping herbs to shrubs or trees, sometimes epiphytic. Monoecious or dioecious. Stipulate, stipules small or distinct. Leaves distichous or spirally arranged, simple, usually entire. Inflorescence usually terminal, sometimes axillary, leaf-opposed, solitary, stalked or sessile. Flowers unisexual or bisexual, perianth lacking. Stamens 1 - 10. Ovary sessile, stigmas 2 - 5, style very short or absent. Berry sometimes immersed into bract or fleshy rachis.

Key to Species in Papuasia
(Source: Jebb, M. (unpub.)

1. Trees (10 species).............. 2

1. Climbers or scandent shrubs (26 species)...................... 11

2. Inflorescences paired at node, and branched, with 4-5 inflorescences.
.... P. umbellatum

2. Inflorescences solitary at node, unbranched...................... 3

3. Leaf markedly triangular, the leaf margins straight, the base +- cordate........ P. triangulare

3. Leaf otherwise.................. 4

4. Plant glabrous in all parts.. 5

4. Plant pubescent or hispid........ 6

5. Lateral veins all arising from the base of the leaf.
.... P. gibbilimbum

5. One pair of veins arising from the midrib..... P. plagiophyllum

6. Leaves roughly hispid above and below...... P. aduncum

6. Leaves usually glabrous above, pubescent below, especially on the mid rib and veins.................. 7

7. Lateral veins in 2-4 pairs, all arising from the base of the leaf.
.... 8

7. Lateral veins in 4 - 7 pairs, the most distal pair arising from the midrib, away from the base...... 10

8. Leaves cordate, asymetric....... 9

8. Leaves rounded, ± symetric.
.... P. gibbilimbum

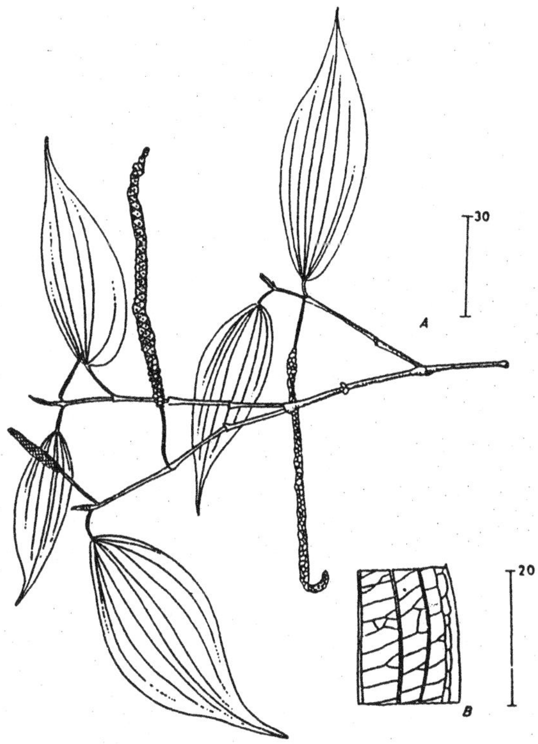

Piper breviantherum (A) flowering branch; (B) part of leaf enlarged. Source: van Royen (1982).

9. Veins hispid - pubescent below, the hairs <1mm long.
.... P. bolanicum

9. Veins densely pubescent below, the hairs >2mm long.
.... P. wabagense

10. Inflorescence <5cm long.
.... P. methysticum

10. Inflorescence >7cm long.
.... P. wichmanii
or .... P. grandispicum

11. Plant pubescent, at least in part. 12

11. Plant glabrous in all parts, leaves sometimes hispid............... 23

12. Veins all arising from the base of the leaf........................ 13

12. Some veins arising from the midrib, away from the base of the leaf.. 15

13. Base of leaf flattened to cordate. .... P. arfakianum

13. Base of leaf rounded to cuneate. 14

14. Inflorescence to 13-20cm, peduncle to 3cm..... P. macropiper

14. Inflorescence to 11cm, peduncle to 1cm........ P. breviantherum

15. Lateral veins in 4-6 pairs...... 16

15. Lateral veins in 2-4 pairs...... 18

16. Base cordate.................... 17

16. Base rounded to cuneate. .... P. celtidiforme

17. Lateral veins arising from midrib up to half way along the leaf. .... P. majusculum

17. Lateral veins arising from midrib, but always within 3cm of the base. .... P. pseudoamboinense

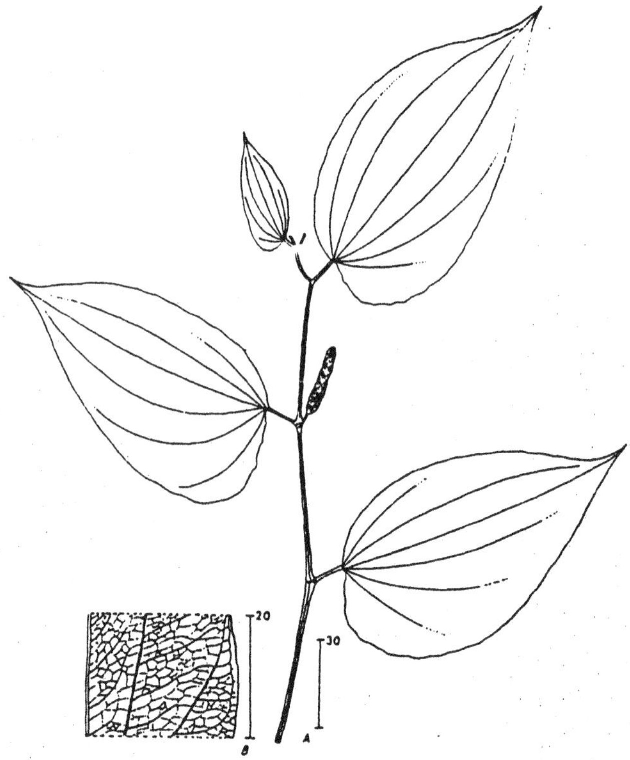

Piper gibbilimbum (A) branch with flowers; (B) part of leaf enlarged. (Source: van Royen (1982).

18. Base rounded to cuneate......... 19

18. Base flattened to cordate....... 20

19. Inflorescence shorter than the leaf .... P. caninum

19. Inflorescence longer than the leaf. .... P. novoguineense

20. Ratio of leaf length : breadth <= 3:2............................ 21

20. Ratio of leaf length : breadth >= 2:1............................ 22

21. Veins densely and shaggily pubescent below, the petiole <1.5 cm. .... P. wilhelmense

21. Veins + pubescent below, petiole >= 2 cm........ P. sclerophloeum

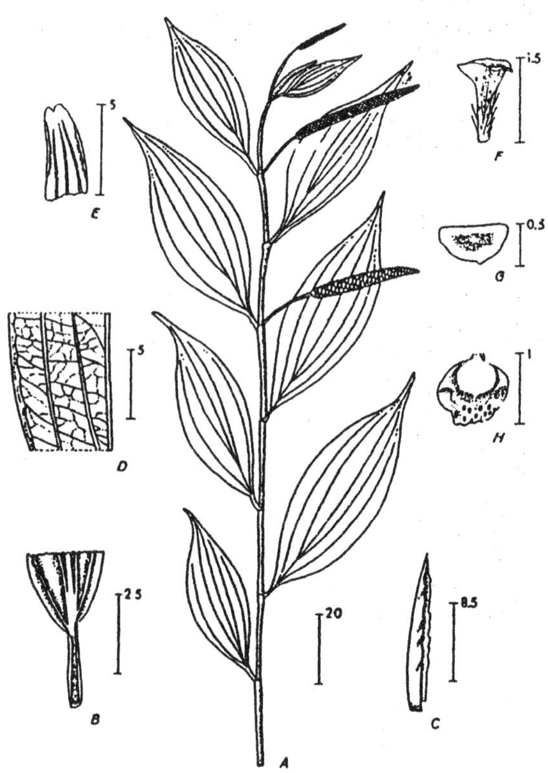

Piper caninum (A) female flowering branchlet; (B) base of leaf; (C) stipule; (D) venation; (E) apex of leaf; (F), (G) and (H) female flowers. (Source: van Royen (1982).

22. Base of leaf markedly assymetric and auriculate, leaves 3:1.
.... P. lessertianum

22. Base of leaf assymetric, the base of the leaf uneven in its insertion to the petiole, veins pubescent below...... P. pullibaccum

22. Base of leaf ± assymetric, cuneate to cordate.
.... P. caninum

23. Inflorescences paired at each node, with large, acute basal bracts >1 cm long, leaves ± cordate.
.... P. vertseegii

23. Inflorescence solitary at each node, lacking bracts; base of leaves cordate to cuneate............. 24

24. Lateral veins all arising from near the base of the leaf............ 25

24. At least some veins arising from the midrib away from the base of the leaf....................... 28

25. Leaf base cordate............... 26

25. Leaf base rounded to cuneate.... 27

26. Petiole to 1 cm long.
.... P. bosniacum

26. Petiole 2-6 cm long.
.... P. truncatibaccum

27. Leaf base cuneate, lamina with 2 prominent pairs of lateral veins.
.... P. fragile

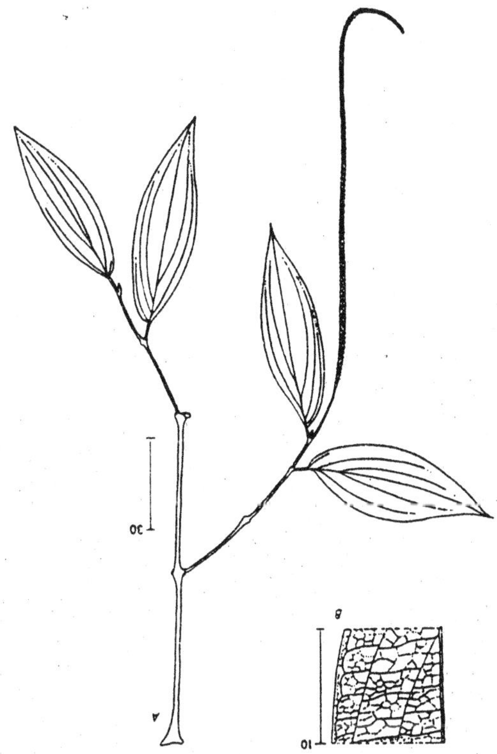

Piper mestonii (A) flowering branch; (B) part of leaf enlarged.
(Source: van Royen (1982).

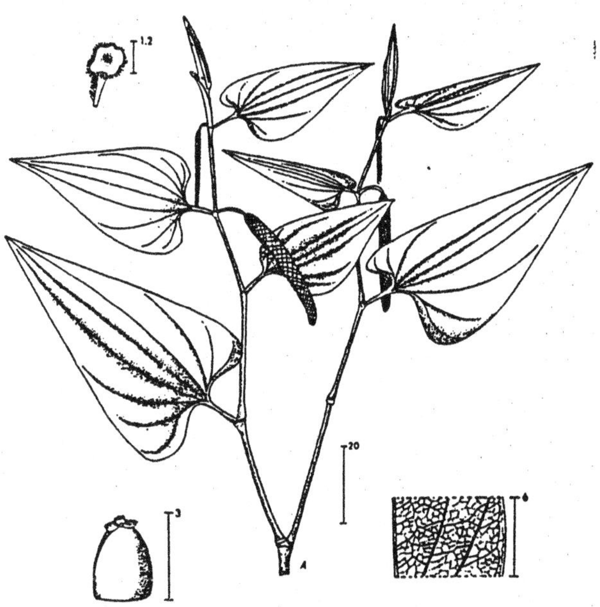

Piper triangulare (A) flowering branch, (B) part of leaf enlarged; (C) bract of female flower; (D) female flower.
(Source: van Royen (1982).

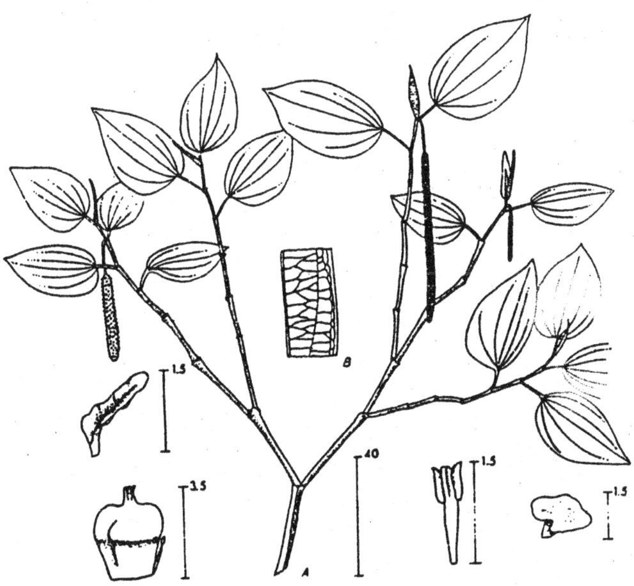

Piper interrupta (A) flowering branch; (B) part of leaf enlarged; (C) bract; (D) stamen; (E) young fruit with staminode; (F) stigma.
(Source: van Royen (1982).

27. Leaf base rounded, lamina with 1 prominent pair of lateral veins. .... *P. macropiper*

28. Leaf base rounded to cuneate, never cordate........................ 29

28. Leaf base cordate 36

29. Petiole 2-3 cm long............. 30

29. Petiole <2 cm long.............. 31

30. Leaf apex acute. .... *P. dumilimbum*

30. Leaf apex acuminate. .... *P. stenocarpum*

31. Inflorescence and peduncle >15 cm long 32

31. Inflorescence and peduncle <15 cm long 33

32. Leaves silvery-grey glaucous below. .... *P. interruptum*

32. Leaves green below.
.... *P. stenocarpum*
or .... *P. mestonii*

33. Distal pair of veins opposite. .... *P. subcanirameum*

33. Distal pair of veins alternate.. 34

34. Lateral veins all prominent to tip. .... *P. radatzii*

34. Lateral veins not all prominent to tip............................ 35

35. Leaves more slender than 2:1. .... *P. abbreviatum*

35. Leaves less slender than 2:1.
.... *P. austro-caledonicum*
or ...... *P. novoguineense*

36. Base of leaf auriculate, lamina rugose, swelling upwards between veins, the secondary veins prominent, and giving a net-like reticulation, leaf very large. .... *P. decumanum*

36. Base of leaf cordate, lamina flat, not as large................... 37

37. Leaf regularly cordate, Length: breadth ratio <1.8 : 1. .... *P. trombek*

37. Leaf lanceolate, ovate or irregularly cordate.
.... *P. mestonii*
or .... *P. betle*
or .... *P. amboinense*
or .... *P. singkojang*

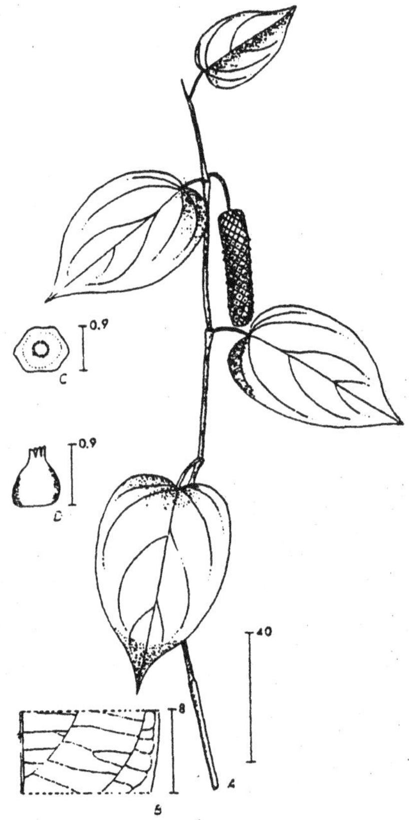

Piper trambek (A) flowering branch; (B) part of leaf enlarged; (C) bract of female flower; (D) female flower. (Source: van Royen (1982).

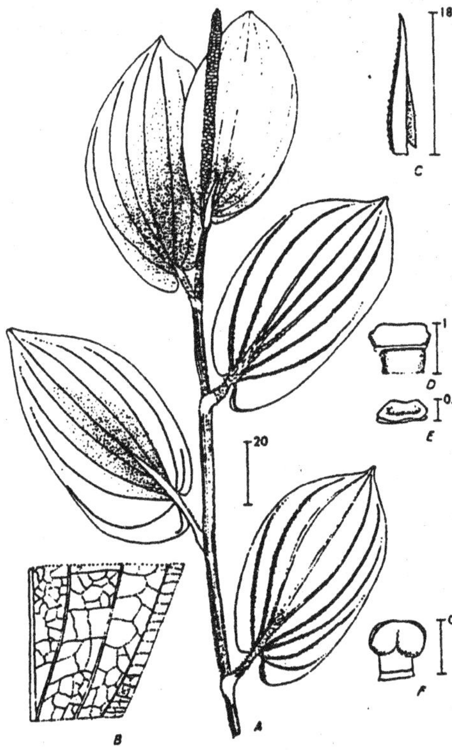

Piper wilhelmense (A) part of flowering branch; (B) part of leaf enlarged; (C) stipule; (D) and (E) bract of female flower; (F) stamen. (Source: van Royen (1982).

2. POTHOMORPHE

A pantropical genus represented in Papuasia by 2 species. The genus can be distinguished from *Piper* because it has axillary inflorescences and spikes in an umbellate cluster. *Piper* always has a solitary spike.

A pantropical genus of terrestrial, woody plants, with large, alternate leaves. The flower spike is an umbellate cluster.

Shrubs, usually glabrous, except for leaves. Leaves ovate, often unequally cordate, to 20 cm long: several nerves radiating from base. Spikes crowded on axillary branchlets, pendulous, white, to 7 cm long, rather thick. Flowers bisexual. Bracts usually peltate. Stamens 2 - 3. Ovary free with 3 stigmas. Fruit minute.

Notes:

1. No published records of identified species have been seen for Papuasia.

2. Only two collections of *Pothomorphe* are in the National herbarium at Lae. Schodde and Craven 677 was collected along a forest margin at an altitude of 75 m in the Gulf Province. Kairo, NGF 24395, was a small tree collected at an altitude of 600 m in grassland. The collections appear to represent two distinct species.

3. Ridley (1924) in *Flora of Malay Peninsula* places the single Malayan species of *Pothomorphe* in *Piper* Section *Heckaria* which he characterised by: spikes several, umbelled flowers , bisexual, erect shrub; leaves very large. fruit very small.

PEPEROMIACEAE

A small family of succulent herbs (rarely sub-shrubs), most epiphytic or living in rocky habitats. Many cultivated due to their attractive foliage. Some 1000 species in 4 genera, very common in tropical America and the West Indies. In Papua New Guinea the family is represented only by *Peperomia*.

Some species have economic potential as ornamental plants. In America the leaves and stems of *P. vividispica* are eaten.

Many botanists still include the 4 genera of the Peperomiaceae in the Piperaceae, but the lack of stipules, 2 stamens per flower, unique pollen grains as well as anatomical differences are sufficient to warrant a separate family.

PEPEROMIACEAE

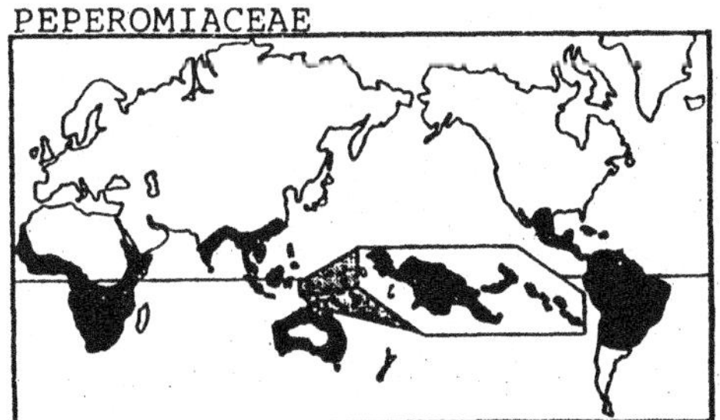

	WORLD	PAPUASIA
Number of genera:	4	1
Number of species:	1000	23
Distribution: Pantropical, common in disturbed rainforest.		
Economic uses: Garden ornamentals.		

Diagnostic features

Stem: in *Peperomia* the outer vascular bundles of the stem are scattered in the manner characteristic of many monocotyledons.

Leaves: without stipules, large, alternate (rarely opposite or whorled).

Flowers: in dense spikes, bisexual, very small. Subtended by fleshy or succulent bracts. The flowers have no sepals or petals and only two anthers which are fused.

Fruit: a berry with a succulent, thin or dry pericarp. It contains a small seed which consists of a minute embryo surrounded by mealy endosperm.

References

see under Piperaceae.

KEY TO GENERA IN PAPUASIA

1 genus. 1. Peperomia

1. PEPEROMIA

• • • • • •

A tropical and subtropical genus of some 500 species mainly in America. Many are epiphytes with creeping stems, adventitous roots and fleshy leaves.

Succulent small herbs, often epiphtic or on rocks. Leaves alternate, opposite or whorled, exstipulate. Flowers axillary or terminal, minute, perfect, in slender bracted spikes. Bracts minute, peltate. Perianth lacking; stamens 2, subsessile, anther-cells confluent. Ovary sessile, 1-celled; stigma penicillate; ovule solitary, erect. Fruit minute, indehiscent.

List of Species in Papuasia

Peperomia adscendens
Peperomia baueriana
Peperomia angustilimba
Peperomia argyroneura
Peperomia bamleri
Peperomia bismarckiana
Peperomia bryophila
Peperomia lasiorhachis
Peperomia lauterbachii
Peperomia ledermannii
Peperomia linearifolia
Peperomia lonchophylla
Peperomia microstachya
Peperomia muscicola
Peperomia obliqua
Peperomia peekelii
Peperomia pellucida
var. minor
Peperomia purpurea
Peperomia reineckei
Peperomia rubrivaculata
Peperomia schlechteri
Peperomia teniopila
Peperomia udisilvestris
Peperomia versteegii

Notes:

1. Peperomia pellucida is a common tropical weed, naturalised in gardens and on waste ground. It is a native of S America.

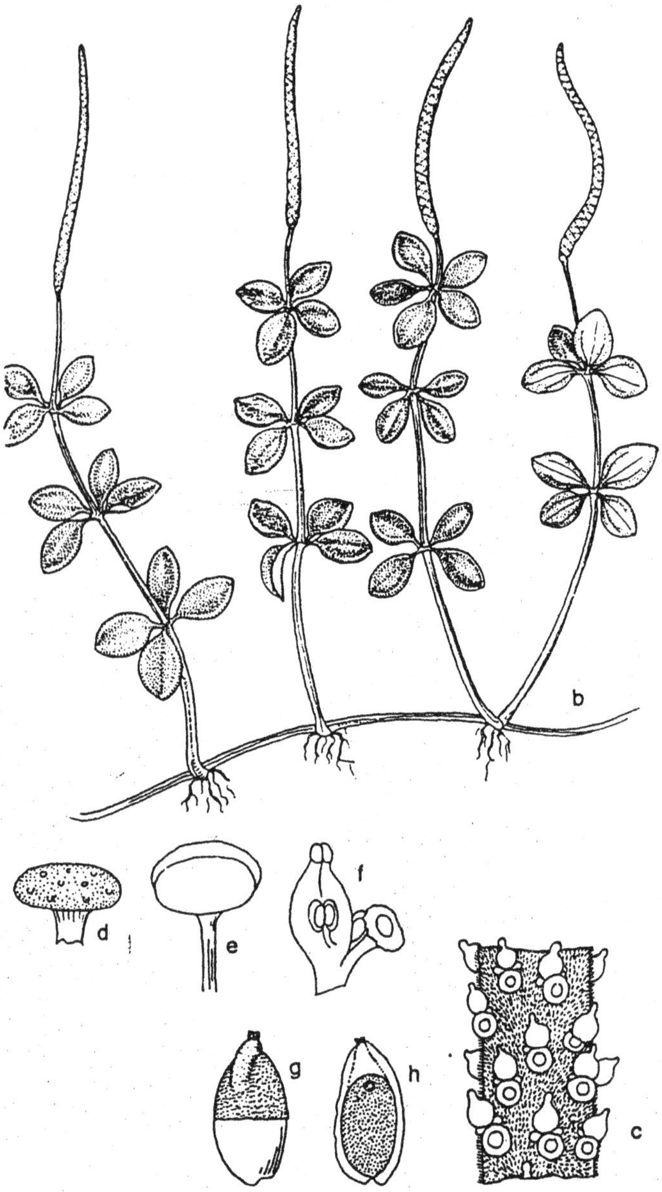

Peperomia tetraphylla (A) part of flowering branch; (B) flowering plant; (C) flowering inflorescence; (D) bract; (E) stamen; (F) flower; (G) fruit showing pseudocupule; (H) LS fruit. (Source: Morley and Toelken (1983).

NYMPHAEALES

This order consists of two families and some 65 species of aquatic plants. The water lilies (Nymphaeaceae) are usually considered allies of the Ranunculaceae, but some authors have placed the family in the monocotyledons with which it shares many characteristics. Members of this family are widely cultivated. Ceratophyllum is often grown in fish tanks. The genus Nelumbo is often treated as a distinct family, the Nelumbonaceae.

CERATOPHYLLACEAE

● ● ● ● ● ●

A cosmopolitan family (world wide in distribution) of aquatic herbs, lacking roots. A single genus, Ceratophyllum with several species. It can provide valuable cover to fish fry, but also provides protection for malaria-carrying mosquito larvae. It can reproduce rapidly by vegetative propagation and the floating masses can close waterways.

Ceratophyllum is very variable and is a difficult genus in which to recognise the distinct aquatic species.

CERATOPHYLLACEAE

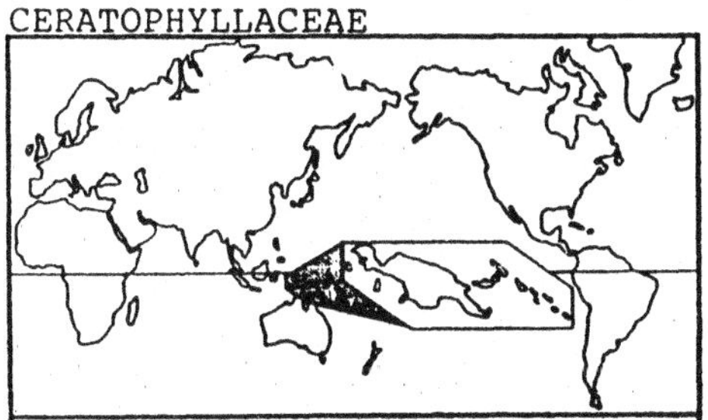

	WORLD	PAPUASIA
Number of genera:	1	1
Number of species:	2	2

Distribution: Worldwide in fresh water.

Economic uses: Protection of fish fry.

Diagnostic features

Leaves: lack stomata and a cuticle, whorled with 3-10 per node. Each leaf is rigid, often brittle and 1-4 times forked, ultimate leaf-segments bear two rows of minute teath and are tipped by two bristles.

Flowers: unisexual, solitary in axils of one leaf in whorl. Male and female flowers usually on alternate nodes. Perianth of 8-12 linear, bract-like segments, united at base. Stamens numerous. Ovary superior, solitary, comprising one carpel with a single pendulous ovule.

Fruit: one seeded nut, tipped by persistent spine like style and often additional spines at base or margin of the seed.

References

Leach, G.J. and Osborne, P.L. (1985): Freshwater plants of Papua New Guinea. 97 - 102.

van Steenis, C.G.G.J. (1949). Ceratophyllaceae. Flora Malesiana. Series 1. 4: 41 - 42.

KEY TO GENERA IN PAPUASIA

1 genus: 1. Ceratophyllum

1. CERATOPHYLLUM

● ● ● ● ● ●

A cosmopolitan genus, most diverse in temperate regions. 10 species worldwide but only 2 recognised in Papuasia. Usually growing in clear streams, ponds and water-courses. Occasionally from muddy and estuarine conditions.

Submerged, rootless, monoecious, perennial herbs. Stems either suspended in water or anchored by slender rhizoidal branches; branches irregular, 1-several at node. Leaves cauline, either simple or dichotomously divided, opposite or whorled; segments with 2, parallel, marginal rows or persistent thorn-like denticles. Male and female inflorescences greatly reduced, solitary, extra-axillary and alternating with the leaves, 1-several at a node. Male flowers: actinomorphic, naked; stamens 3-numerous, arranged spirally. Female flowers actinomorphic; naked, single carpel, tapers to persistent spine-like style, stigma at base of stylar groove; ovary superior. Fruit an achene.

Key to Species in Papuasia
(Source: Leach and Osborne (1986)

1. Fruits with a long (2-7mm) apical stylar spine and two basal spines (1-8mm);, fruit margins wingless or winged; leaves usually no greater than 2nd order forked (occasional individuals with 3rd-order forked leaves are found); teeth (denticles) along leaf-segment margins conspicuous and set in a broad base of tissue. .... C. demersum

1. Fruits with a short (1-3mm) apical stylar spine, basal spines absent; margin of fruit surrounded by a thin, membranous wing; leaves usually 3rd to 4th-order forked; teeth (denticles) along leaf-segment margins inconspicuous and lacking a broad base of tissue. .... C. submersum

Notes:

1. Ceratophyllum submersum is only known from low altitudes in Papuasia. C. demersum is known from sea level to 2500m altitude.

NYMPHAEACEAE
(Water Lilies)

A cosmopolitan family of fresh water plants with nine genera and over 90 species. Plants grow in a variety of freshwater habitats: ponds, streams and lakes. Many species are cultivated. The seeds and rhizomes of Nymphaea are sometimes eaten as are the roasted seeds of Victoria and Nelumbo. Arrowroot is refined from the seeds of Euryale.

Diagnostic features

Plants: aquatic, with large rhizomes and massive shoot apices.

Leaves: peltate or cordate, often floating (rarely submerged then finely dissected).

Flowers: solitary, 3-6 green or coloured sepals and 3 to many petals. Stamens numerous. Ovary superior or inferior, of 5-35 united or free ovules. Flowers pollinated by beetles (or bees).

Fruit: a spongy berry, dehiscing by swelling of mucilage within. Seeds usually with aril.

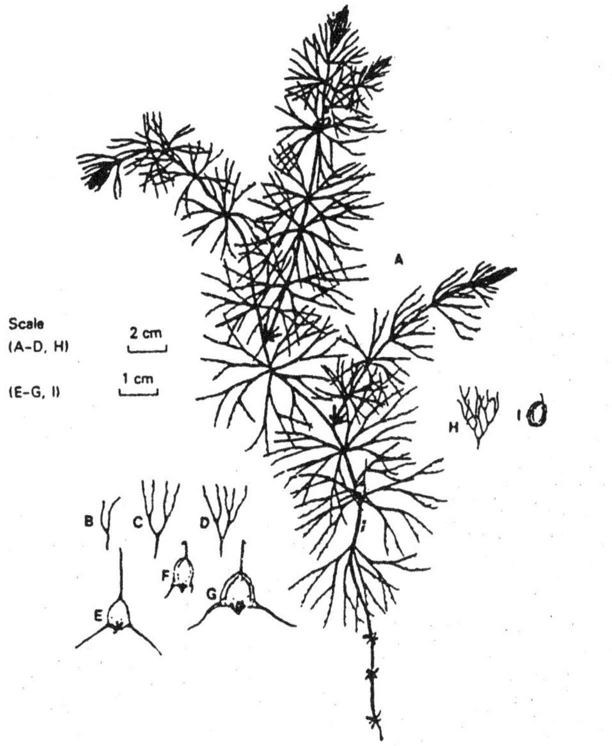

Ceratophyllum demersum (A) typical habit; (B), (C) and (D) 1st-, 2nd- and 3rd- forked leaves; (E) "typical" 3-spined fruit; (F) short-spined fruit; (G) alate fruit.
Ceratophyllum submersum (H) 4th-order forked leaf; (I) winged fruit.
(Source: Conn (1985).

NYMPHAEACEAE

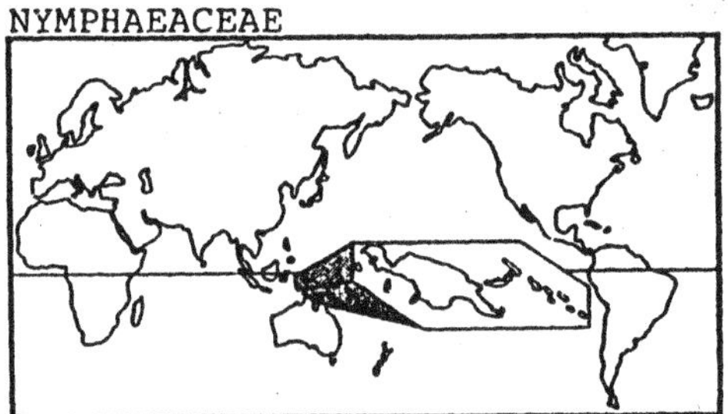

	WORLD	PAPUASIA
Number of genera :	9	3
Number of species :	90	8

Distribution : Consmopolitan in fresh water.

Economic uses : Ornamental aquatics: waterlilies, lotus. Some edible seeds and rhizomes.

References

Conn, B.J. (1985). Water Plants of Papua New Guinea. 191 - 201.

van Royen, P. (1962). Sertulum Papuanum 5: Nymphaceae. Nova Guinea Bot. N.S. 8: 103 - 126.

KEY TO GENERA IN PAPUASIA

(Source: Conn (1985)

1. Sepals and petals not distinctly different; carpels free, superior, all sunk into a common spongy receptacle; styles present; stigmas distinct; one ovule per carpel; fruit nut-like; lamina distinctly peltate, emergent; flowers and fruits emergent. .... 2. Nelumbo

1. Sepals and petals distinctly different; carpels connate, forming a many locular inferior or semi-inferior ovary; style absent; stigmas united to form a depressed stigmatic disc at the upper surface of the ovary, ovules numerous per carpel; fruit a spongy, mucilaginous berry; lamina peltate (but appearing to be cordate or not peltate), generally floating on the surface or submerged; flowers on or above the surface; fruits below the surface......................... 2

2. Sepals 5; petals connate, forming a corolla tube; mature lamina submerged. .... 1. Hydrostemma

2. Sepals 4; petals free, mature lamina usually floating on the water surface. .... 3. Nymphaea

1. HYDROSTEMMA

● ● ● ● ● ●

A genus of 3 species ranging from Asia to Papua New Guinea. Not recorded from the Solomon Islands. The single species in Papuasia is also recorded from Sumatra and Borneo. It is a perennial herb with a rhizome-like tuber.
Perennial herbs with tuber-like rhizomes. Lamina usually submerged, erect, simple, base cordate, margin entire. Flowers emerging, axillary, solitary, bisexual. Sepals 5, free unequal. Stamens adnate to corolla tube, incurved, numerous, outer series at apex staminodal, gradually changing into petals. Ovary superior; style absent; stigmas radiate with abaxial appendages which curve over the stigmas. Fruit many-seeded, surface usually with soft hairs.

Key to Species in Papuasia

1 widespread species:

Hydrostemma motleyi (Hk. f.) Mabb.

Notes:

1. Grows in shallow often muddy pools and in areas with sago swamps and small muddy or gravelly streams. Grows from sealevel to 1500m altitude.

2. In New Guinea occurs only in the West (W.H.P., W.S.P. and W.P.).

~~Barclaya~~ motleyi forma membranacea ~~and~~ Hydrostemma.
(Source: van Royen (1962).

2. NELUMBO

A genus of wide distribution with 2 species 1 of which occurs in Papuasia. The genus Nelumbo is often placed in its own family: the Nelumbonaceae. The fruits are used as a food for man and livestock. It grows in mud in sheltered waters of lakes at low altitudes. Only collected from the mainland of Papua New Guinea. The species in Papuasia is distributed from S. Russia through Asia and Malesia to Australia.
Robust perennial herbs with submerged creeping rhizomes; latex milky. Petioles terete; lamina above water surface. Flowers emergent, large, pink, fragrant. Tepals numerous, free. Stamens numerous, connective prolonged to form a large apical appendage. Carpels free, spirally arranged, sunken into a superior obconical, spongy receptacle; ovules one in each carpel. fruits hard, nut-like; one seed per fruit.

Key to Species in Papuasia

1 widespread species:

Nelumbo nucifera

Notes:

1. The fruits are used as food for both man and livestock.

Nelumbo nucifera in Lake Rombebai.
Nelumbo nucifera in Lake Wanum.
(Source: van Royen (1962).

3. NYMPHAEA

A genus of 50 species with 6 naturally occuring in Papuasia. The numerous cultivars are not included. They are perennial herbs with ascending or creeping rhizomes or have tubers with long runners. Two of the species are endemic.

Perennial herbs, rhizome ascending or creeping, or with tubers which form long runners. Petiole terete, distinctly thickened; lamina usually floating on or just below water surface, pelatate, margin entire or dentate, midrib and main veins arising from tubercle, veins anastomising near margin. Flowers usually floating on surface or emergent (to 10 cm), large, conspicuosuly coloured, fragrant. Sepals usually 4. Petals numerous, variable, free. Stamens numerous, adnate to or just below apical margin of ovary. Ovary superior; stigmas as many as locules, outer end usually raised and bilobed. Fruit ripening under water, retracted to muddy substrate, spongy, many-seeded, wall of ovary gradually dissolving. Seeds small.

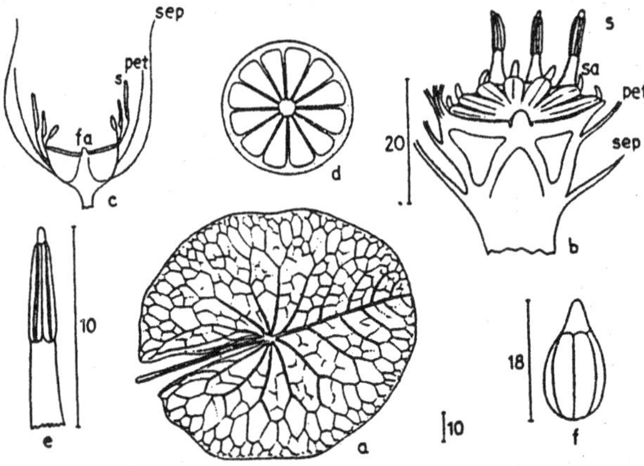

Nymphaea nouchali (A) underside of leaf; (B) LS androgynaecium: sep. sepals, pet, petals, s. stamens, sa. stigmatic appendages; (C) LS flower: sep., pet. and s. as before, fa. floral axis; (D) TS ovary; (E) stamen; (F) seed. (Source: van Royen (1962).

Key to Species in Papuasia

(Source: Conn (1985)

1. Lamina margin entire or with round or obtuse lobes (sometimes subpinnate in *N. nouchali*)............. 2

1. Lamina margin with acute teeth... 3

2. Stamens immediately above petals, filaments broadly lanceolate. .... *N. nouchali*

2. Stamens distinctly separate from petals, filaments filiform. .... *N. violacea*

3. Lower surface of lamina pubescent, sometimes hairs restricted to veins...... *N. pubescens*

3. Lower surface of lamina glabrous. 4

4. Anthers with an apical appendage 1-1.5mm long; sinus of lamina closed and margins overlapping. .... *N. gigantea*

4. Anthers with an apical appendage less than 0.5mm long or appendage absent; sinus of lamina open..... 5

5. Inner surface of sepals glabrous; number of petals 18-26; number of stamens 350-500; number of stigmas 13-15; flowers 7-9cm in diameter.; marginal teeth of lamina mostly 2-3mm long. .... *N. dictyophlebia*

5. Inner surface of sepals warty; number of petals 12-14; number of stamens 150-200; number of stigmas 6-10; flowers 4-7cm in diameter.; marginal teeth of lamina 1mm long. .... *N. macrosperma*

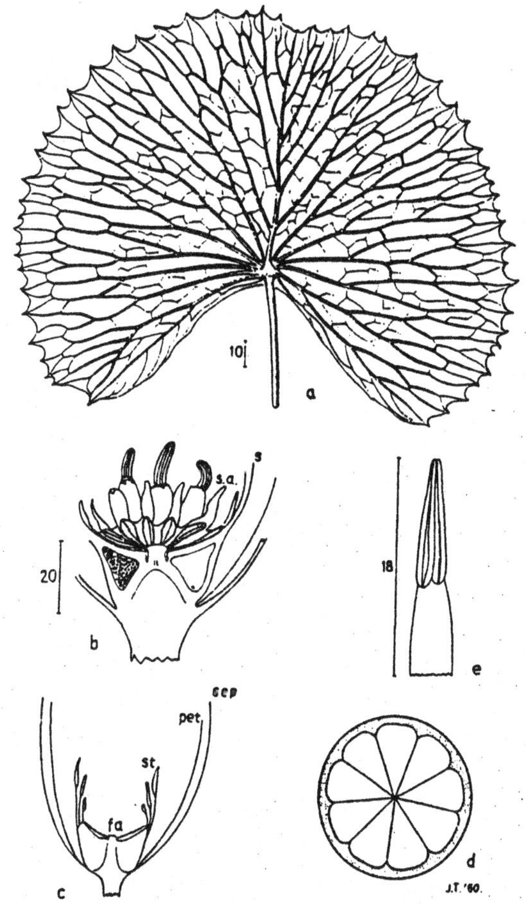

Nymphaea pubescens (A) underside of leaf; (B) LS ovary; (C) LS flower: sepals, pet. petals, s. stamens, sa. stigmatic appendages, fa floral axis; (D) TS ovary; (E) stamen.
(Source: van Royen (1962).

Notes:

1. Widely distributed throughout Papuasia but as with many plants poorly collected in the Solomon Islands.

2. Most species occur at low altitudes.

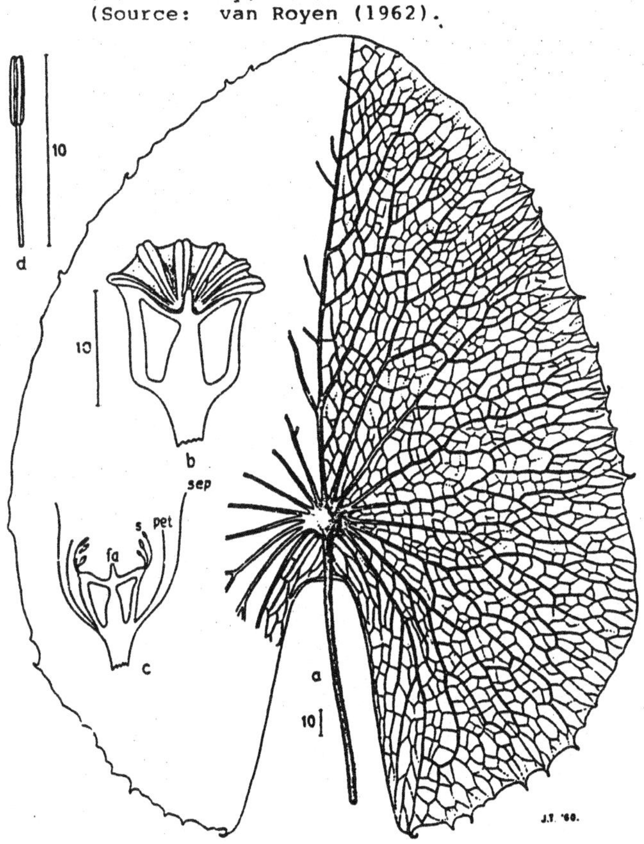

Nymphaea dictyophlebia (A) underside of leaf; (B) LS ovary; (C) LS flower: sep. sepals, pet. petals, s.l stamens, fa. floral axis; (D) stamen.
(Source: van Royen (1962).

RANUNCULALES

An order of 4 to 6 families and some 3,200 species. The largest families are the Ranunculaceae, the Berberidaceae and the Menispermaceae. Variable in habit from herbs to woody vines, shrubs and rarely small trees. Cronquist includes the Coriariaceae and the Sabiaceae in the order Ranunculales. This is not followed here.

RANUNCULACEAE

● ● ● ● ● ●

A family of world wide distribution with most species in temperate areas. It includes 50 genera and about 1,800 species. In Papua New Guinea there are two common genera at higher altitudes: species of small yellow flowered buttercups (Ranunculus) which grow in subalpine grassland and Thalitricum, a glabrous herb in swampy alpine areas. Several species of Clematis are common climbers in the montane zone: one species is common in the hoop and klinki plantations near Bulolo. The family includes many horticultural plants, particularly species and cultivars of Clematis, a woody climber. Most genera are herbaceous. Several genera are highly poisonous.

Diagnostic features

Leaves: alternate, opposite in Clematis. Exstipulate except in Ranunculus, palmately lobed or much divided, pinnately lobed in Clematis. In Clematis the petiole is sensitive to touch and supports the stem as a tendril.

Flowers: terminal, in racemes or cymes, rarely solitary. Petaloid with rarely a true calyx or corolla (except Ranunculus). Stamens numerous, spirally arranged. Carpels 1 to many, spirally arranged.

Fruits: an achene or many seeded follicle. Seeds with a small straight embryo and copious endosperm. Some seeds have styles which develop into long feathery structures adapted for wind dispersal.

RANUNCULACEAE

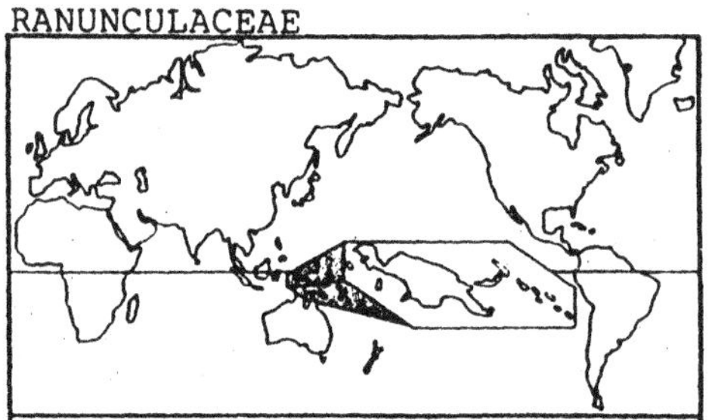

	WORLD	PAPUASIA
Number of genera:	50	3
Number of species:	1800	38

Distribution: Centred in N Hemisphere.

Economic uses: Ornamentals. Some narcotics.

Notes:

1. Members of the family are generally protandrous: stamens shedding their pollen before the ovaries mature: or protogynous: ovaries maturing before the stamens shed their pollen.

These processes favour cross pollination and outbreeding.

References

Diels, (1929). Bot. Jahrb. 62: 479 - .

Eichler, Hj. (1958). Revision der Ranunculaceen Malesiens. Bibliothec. Bot. 124:

Eichler, Hj. (1963). Flora of Java 1: 140 - 1

van Royen, P. (1982). The Alpine Flora of New Guinea. 3: 1289 - 1366.

van Steenis, C.G.G.J. (1934). Bull. Jard. Bot. Buitenz. 13: 238 - 240.

Merrill, E.D. and Perry, L.M. (1943). Journ. Arn. Arb. 24: 36 - 41.

Prantl, K. (1887). Beitrage zur Morphologie und systematik der Ranunculaceen. Bot. Jahrb. 9: 225 - 273.

Pulle, (1911). Nova Guinea 8: 631.

van Royen, P. (1982). The Alpine Flora of New Guinea. Ranunculaceae: 3: 1289 - 1366.

(1924). Trans. Roy. Soc. Edin. 29:

KEY TO GENERA IN PAPUASIA
(Source: Eichler (1958)

1. Leaves opposite. Woody climbers. .... 1. *Clematis*

1. Leaves alternate. Herbs not climbers, sometimes creeping......... 2

2. Leaves simple, enitire or deeply palmately lobed, or ternate. (Petals yellow). .... 2. *Ranunculus*

2. Leaves bi- or tri-ternate (at least lower ones) or repaetedly imparipinnate. Leaflets distinctly petiolate. Petals absent (sepals petaloid). .... 3. *Thali~~tricum~~ctrum*

1. CLEMATIS

A genus of over 200 species almost world wide in distribution. In Papuasia a genus of at least 7 species. The taxonomy is complicated as several species are monoecious: both male and female plants being described as distinct species.

Climbing or scrambling plants. Leaves opposite, mostly compound, with twining petioles and rachis. Inflorescence few to many flowered (rarely flowers solitary). Flowers actinomorphic, unisexual or bisexual. Sepals 4-6, subpetaloid, free or joined by interlocking hairs at base. Petals none. Stamens many (staminodes in female flowers). Carpels (3-)5, style persistent. Fruit a head of nutlets, each with an elongate, plumose style.

List of Species in Papuasia

- *Clematis archboldiana* Merrill & Perry *Journ. Arn. Arb.* 24 (1943) 35
- *Clematis clemensiae*
- *Clematis javana* D.C.
- *Clematis papuasica* Merrill & Perry *Journ. Arn. Arb.* 24 (1943) 36 Type: Brass 3168. Solomon Is.
- *Clematis phanerophlebia* Merrill & Perry *Journ. Arn. Arb.* 24 (1943) 35. Type: Brass 11235.
  - *C. perspicuinervia* Merrill & Perry *Journ. Arn. Arb.* 24 (1943) 155. Type: Clemens 4447.
  - *C. pickeringi* Gray var. *novoguineensis* Merrill & Perry *Journ. Arn. Arb.* 24 (1943) 154. Type: Clemens 4148.

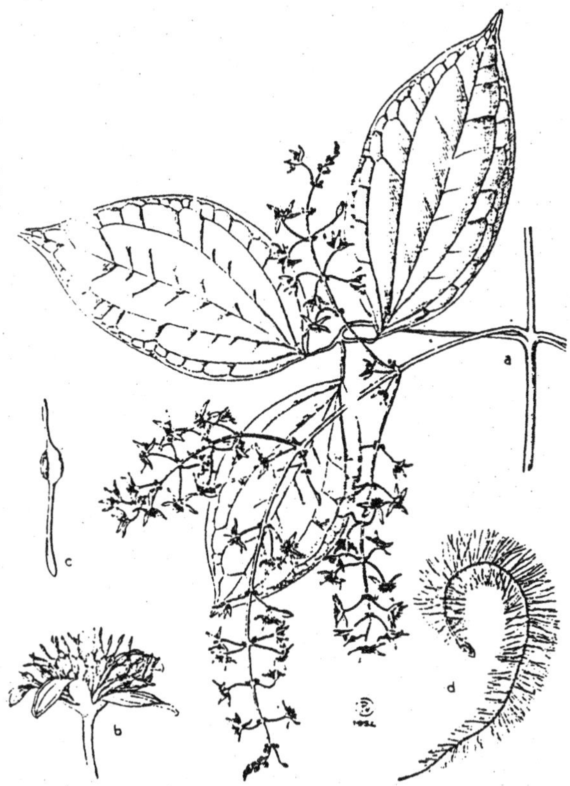

Clematis papuasica (A) flowering stem; (B) flower; (C) stamen; (D) seed. (Source: Eichler (1958)

Clematis pickeringii A. Gray
Clematis smilacifolia Wall.

Species listed as synonyms by Eichler are not included in this list.

2. RANUNCULUS

● ● ● ● ● ●

A large genus of between 300 and 800 species, mostly in the northern temperate regions, but also common at high altitudes in the tropics. Eichler (1958) revised the genus for Malesia but 'probably too many species were recognised' (van Royen, 1982). The key to species below is from van Royen (1982) and is based on a translation of the key proposed by Eichler. The group requires detailed studies to ellucidate intraspecific variation in Ranunculus and also a study of hybridisation is necessary as this appears common between Papuasian species. Thirty species are included in the following key.

Terrestrial or subaquatic small to large herbs. Stipules lacking, but leaf-sheaths may develop. Leaves basal, entire to much divided. Flowers terminal, soliatry or paniculate; bisexual. Sepals 3-5, free, deciduous. Petals (2-)5(-25), free, yellow or white, often dark brown or red on outside. Stamens many, free. Receptacle slightly raised. Style often curved. Nutlets laterally compressed.

Clematis archboldiana (A) fruiting stem; (B) dichasium; (C) seed.
(Source: Eichler (1958)

Key to Species in Papuasia
(Source: van Royen (1982)

Eichler (1958) stated that probably too many species are recognised. It appears that hybridisation occurs on a fairly large scale in New Guinea Ranunculus (van Royen 1982: 1298).

1. Leaves entire, coarsely dentate, or at the utmost lobed to the middle of the limb.................... 2

1. Leaves palmatilobed at least to the middle of the limb, or 3-partite, sometimes basal leaves entire and most apical leaves deeply 3-partite (see van Royen 1982 for notes under 3. R. capellaensis van Royen) 31

2. Leaves entire, rarely with 1 or 2 short teeth, one or either side. 3

2. Leaves either lobed or at least with one pair of coarse, apical and /or lateral, entire or lobed lobes 12

3. Petals narrowly linear-obovate, 3-nerved. Honeyglands more or less in the middle of petals. Sepals 1-nerved (always?). Flowers almost sessile in cushionforming rosettes of leaves.. R. bellus

3. Petals oblong, oblong-obovate to elliptic, 5- or more-nerved. Honeyglands inserted at base of petals. Sepals 3-nerved in basal part. Flowers distinctly stalked..... 4

4. Petioles above sheath much shorter than limb of leaves. .... R. angustipetalus

4. Petioles above sheath as long as or longer than limb.............. 5

5. Carpels 4 - 16. Style straight or almost so. Pedicels with thin, adpressed silky hairs.......... 6

5. Carpels 24 - 50. Style curved or or almost spirally curved (see van Royen 1982 for notes under 14. R. wahgiensis). Pedicels in upper part with scattered, relatively thick, adpressed hairs......... 7

6. Petioles densely hairy with long silky hairs. Receptacle densely hairy...... R. amerophylloides

6. Petioles densely hairy with long silky hairs. Receptacle sparsely bristly.... Ranunculus sp. nov. (Collecteed from Mt Scratchley by Gilianetti (s.n. K) see van Royen 1982: 1358 - 1359: see also couplet 35).

7. Limb of leaves gradually narrowed into petioles, oblong to narrowly elliptic...................... 8

7. Limb of leaves rounded or abruptly narrowed at base, narrowly ovate, elliptic-lanceolate, or oblong. 10

8. Limb of leaves glabrous above or with scattered (1.5 -)2 - 3 mm long hairs 9

8. Limb of leaves on either side more or less regularly covered with 1 - 1.5 mm long bristle which sit on small papillae. .... R. tridens

9. Limb of leaf ovate, ovate-elliptic, or elliptic, acute at tips, often weakly 3-nerved, usually glabrous above, sometimes with scattered long hairs. Petals 5 - 7, about 2 mm wide, Stamens 15. Sepals on outside densely hairy in median area....... R. wahgiensis

9. Limb of leaves oblong, obtuse at tips, 1-nerved, above with solitary hairs 1.5 - 3 mm long, parallel to margins, or rarely scattered over entire surface. Petals 5; 3 - 4 mm wide. Stamens 23 - 30. Sepals on outside only with hairs at tips. .... R. amerophyllus

10. Limb of leaves glabrous on either side, only along margins densely hairy, hairs short. Stamens 35 - 40. Sepals glabrous or subglabrous on outside. R. psilophyllus

10. Limb of leaves with scattered long hairs at least on upper side. Stamens 10 - 12. Sepals hairy on outside......................... 11

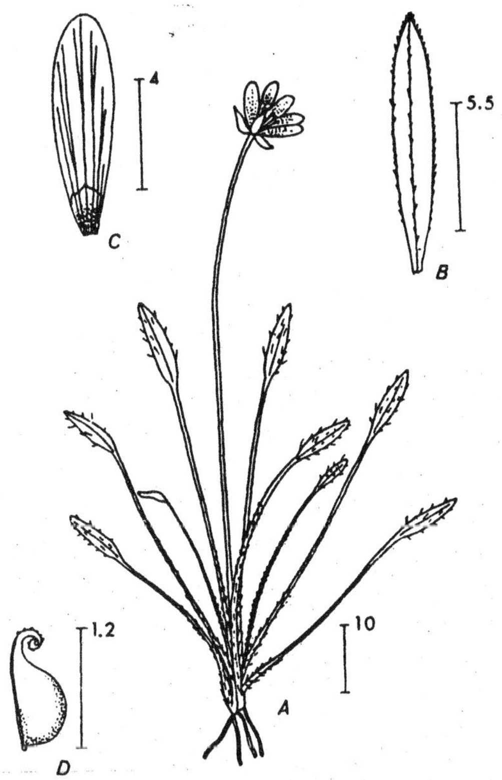

Ranunculus amerophyllus (A) habit; (B) sepals; (C) petals; (D) nutlet. (Source: van Royen (1982)

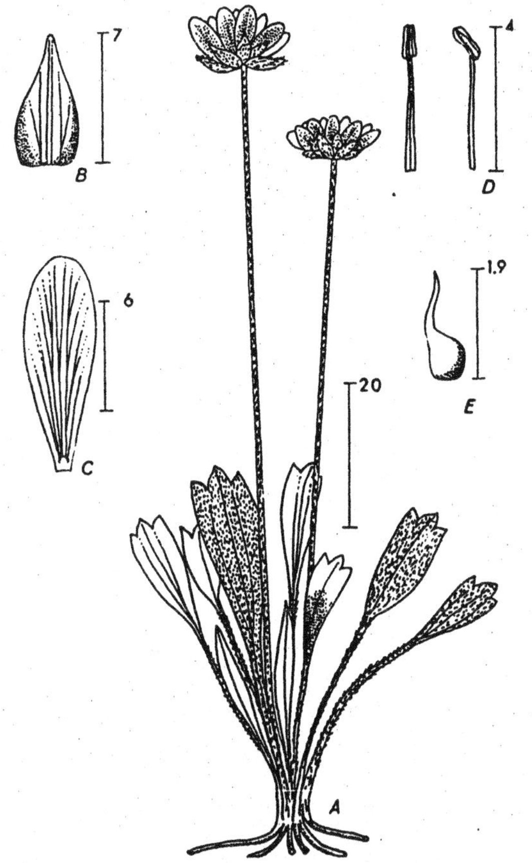

Ranunculus habbamensis (A) habit; (B) sepals; (C) petals; (D) stamens; (E) carpel. (Source: van Royen (1982)

11. Limb of leaves entirely glabrous below, narrowly ovate to oblong, 6 - 9 by 2.5 - 4 mm. Petioles 1 - 3 cm long Petals elliptic, 4 - 6.5 by 2 - 2.5 mm, distinctly clawed. .... R. minimifolius

11. Limb of leaves hairy on either side, elliptic or ovate, 11 - 15 by 7 - 9 mm. Petioles 5 - 8.5 cm long. Petals oblong-obovate or - ovate, 7.5 - 10 by 3 - 4 mm, gradually narrowed towards base. .... R. tridactylus

12. Limb of leaves distinctly longer than wide, coarsely dentate, teeth at tip only or along entire margin, median tooth 1/3 or less of length of limb...................... 13

12. Limb of leaves palmatilobed, median lobe at least 1/3 the length of limb, or coarsely dentate and then roundish or wider than long. 27

13. Limb of leaves more or less cuneately narrowed intopetiole, coarsely dentate only in apical 1/2 - 1/3. Pedicels distinctly longer than leaves (see notes in van Royen 1983 under R. consimilis where the pedicel is sometimes not or only slightly longer).............. 18

13. Limb of leaves more or less abrupt-ly narrowed into petiole, dentate in upper 1/3 or also in lower half. Pedicels at anthesis usually shorter or only slightly longer than leaves....................... 14

14. Limb of leaves on both sides densely to subdensely, sometimes sparsely hairy, on underside sometimes along nerves only............. 15

14. Limb of leaves on upper side only sparsely hairy, at tip with 3 coarse, forward directed teeth. Styles spirally recurved. .... R. tridactylus

15. Style hookshaped or almost straight, or if spirally curved limb of leaves only 3-(or 5-) dentate in upper half. Limb of leaves rarely entire or below or in the middle on both sides with another forward directed tooth, upper side densely hairy........................ 16

15. Style long and widely spirally re-recurved. Limb of leaves from base along both sides with 2 or 3 coarse teeth each, upper side densely furry hairy. .... R. peridutus

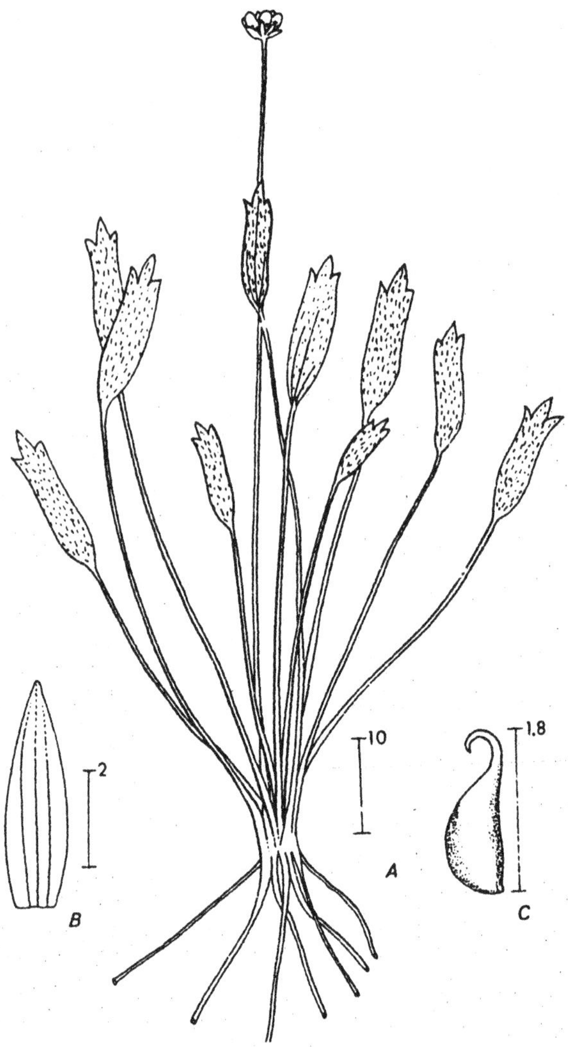

Ranunculus keysseri (A) habit; (B) sepals; (C) carpel. (Source: van Royen (1982)

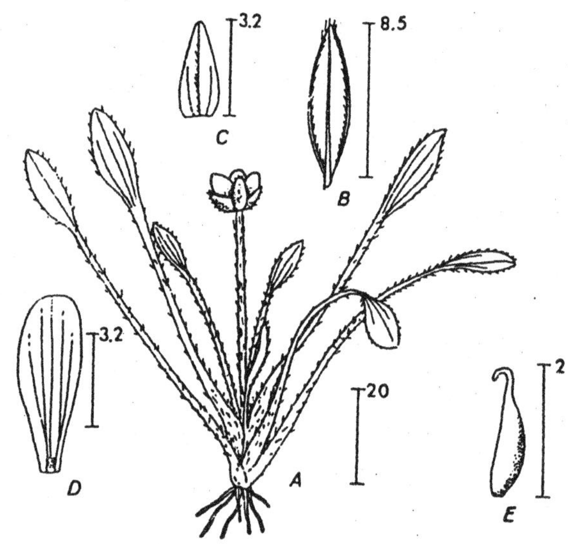

Ranunculus minimifolius (A) habit; (B) leaf; (C) sepals; (D) petal; (E) carpel. (Source: van Royen (1982)

16. Limb of leaves only with teeth in apical 1/2 - 1/3 17

16. Limb also with 2 teeth in basal part....... R. basilobatus

17. Stamens 8 - 12. Petals at base 5-nerved. Style of nutlets strongly spirally curved. .... R. ? celebicus

17. Stamens about 20. Petals at base 6 - 9-nerved. Styles of nutlets straight or almost so. .... R. saruwagedicus

18. Leaves (0.7-)1 - 10 cm long, herbaceous, thin. Petals narrowly elliptic-oblong or oblong-obovate, 4 - 11.5 by 1.5 - 2.2mm............ 19

18. Leaves (1 -)1.8 - 4(- 25) cm, fleshy or leathery. Petals obovate to oblong-obovate, rarely oblong, 6 - 14 by (2 -)2.5 - 6 mm.......... 21

19. Stamens 15 - 23. Plants 3 - 18 cm high.......................... 20

19. Stamens about 25. Plants up to 2cm high. Limb of leaves sparsely adpressed hairy on either side. Style spirally curved. .... R. gainanensis

20. Limb of leaves glabrous above or with a few hairs only, rarely finely adpressed villose, and on underside of nerves only. Styles almost straight or only slightly bent. .... R. wahgiensis

20. Limb of leaves sparsely antrorse strigose above, glabrous below. .... R. schoddei

21. Limb of leaves either 2- or 3-dentate at tips, with forward directed teeth, or 5 - 7-dentate in upper 1/2 - 1/3 22

21. Leaves in upper 1/3 coarsely 3-dentate, on either side with a sharply outward directed lateral tooth. .... R. tridens

22. Limb of leave oblong - lanceolate, acute, 3-dentate at tip, median tooth longer than lateral; limb 5 - 9(- 12)mm wide, sparsely adpressed hairy below. Sepals usually glabrous....... R. keysseri

22. Limb of leaves cuneate - spathulate to cuneate-obovate, usually obtuse, 2 - 5-dentate, if 3-lobed the median lobe not longer than lateral; limb (2-)12 - 20mm wide. Sepals on outside sparsely to densely hairy almost to the tip, or at tip only. 23

23. Leaves with an occasional hair only on upper side of limb, but denser so along the thickened margins and underside of midrib. .... R. bidens

23. Leaves at least on entire underside more or less densely or subdensely hairy.......................... 24

24. Sepals densely pilose on outside. Limb of leaves adpressed pilose above, glabrescent, on underside densely adpressed strigose. .... R. habbemensis

24. Sepals sparsely villose on outside mainly along median nerve or in a dorsal patch.................. 25

25. Limb of leaves sparsely adpressedly villose above, antrorse and usually with shorter hairs along margins, and on underside often mainly on the nerves antrorse adpressed villose.......................... 26

25. Limb of leaves densely antrorse, strigose mainly along margins and underside of nerves, less so on underside and sometimes even less so above... R. consimilis

26. Leaves 3-dentate with forward directed teeth. R. montiswilhelmii

26. Limb of leaves 5-dentate. .... R. perindutus

27. Lobes of limb of leaves or teeth entire........................ 28

27. Median lobe and often also the lateral lobes at tip 3-dentate, sometimes the younger leaves with entire median lobe............... 29

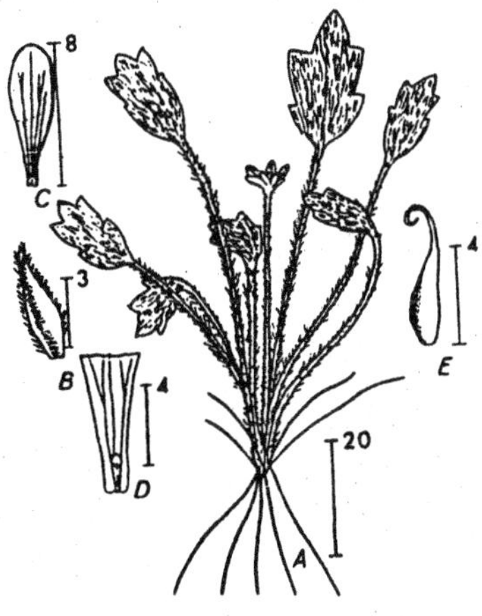

Ranunculus perindutus (A) habit; (B) sepals; (C) petal; (D) petal with honey gland; (E) carpel.
(Source: van Royen (1982)

28. Limb of leaves obovate to rarely broadly elliptic, only in apical half 3-lobed, but sometimes laterally on either side with a small tooth, rounded at base. .... R. psilophyllus

28. Limb of leaves more or less round to broadly elliptic, from the base upwards 5 - 7-lobed, subcordate at base....... R. pseudolowii

29. Limb of leaves 10 - 13(-25)mm long, shaggy hairy on either side, especially along margins. Petioles and pedicels shaggy hairy.......... 30

29. Limb of leaves over (12-) 15mm long upper side sparsely hairy or glabrous...... R. papuanus

30. Pedicels slender, patently hairy. Hairs light brown hirsute. Nutlets obliquely obovoid, about 2.5 by 1 mm, narrowed at base. Limb of leaves thin, subcoriaceous. .... R. coacervatus

30. Pedicels stout, adpressedly hairy. Hairs white. Nutlets subsemiorbicular, 1.2 - 1.5 by 0.7 - 1 mm, subrotundate at base to subtruncate. Limb of leaves thick, herbaceous. .... R. mentiens

31. Sepals 1-nerved. Petals 3-nerved. .... R. perpusillus

31. Sepals 3-nerved. Petals 3- or more -nerved....................... 32

32. Median lobe of limb of leaves narrowly lanceolate. .... R. capellaensis

32. Median lobe of limb of leaf broadly elliptic, 3-dentate at tip, sometimes in some leaves with a lanceolate median lobe, or leaves 3-foliolate or deeply 3-lobed.. 33

33. Hairs on upper side of limb of leaves 1.5 - 2 times longer than those on underside. Stamens 25-40. Sepals on outside subdensely long-hairy...... R. heterotrichus

33. Hairs on upper side of limb of leaves not or hardly longer than those on underside; sometimes glabrous or subglabrous on underside. Stamens 7 - 20. Sepals on outside only sparesely hairy........... 34

34 Longest petioles (4-) 8- 15cm long. Sepals broadly elliptic or ovate, 2.5 - 4 by 0.7 - 2 mm. .... R. uncostigma

34. Longest petioles 3cm long or shorter. Sepals ovate, 2.5-3.5 by about 1.5mm. 35

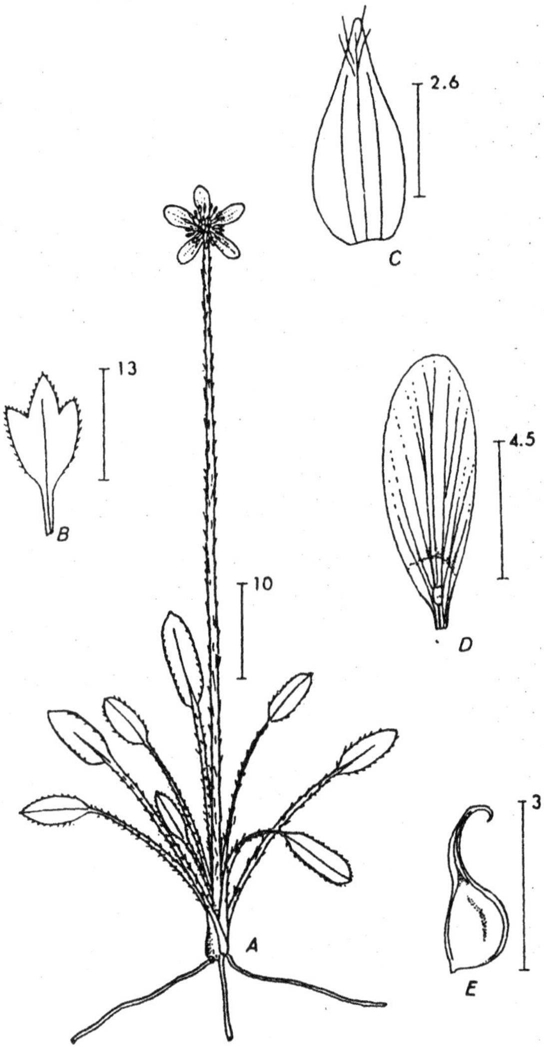

Ranunculus psilophyllus (A) habit; (B) leaf; (C) sepals; (D) petal; (E) nutlet. (Source: van Royen (1982)

35. Sepals sparsely hairy on outside, hairs about 1.5mm long. Petals 5-nerved. Hairs on petioles 1 - 1.2 mm long.... Ranunculus sp. nov. (Collected on Wharton Mts by Giulianetti & English (s.n. K) see van Royen (1982) 1308 - 1310; Eichler (1958) 101.

35. Sepals very sparsely hairy on outside, hairs 0.5 - 1mm long. Petals 7-9-nerved. Hairs on petioles about 0.5 mm long.
.... Ranunculus sp. nov. (see notes under couplet 6b which refers to the same species).

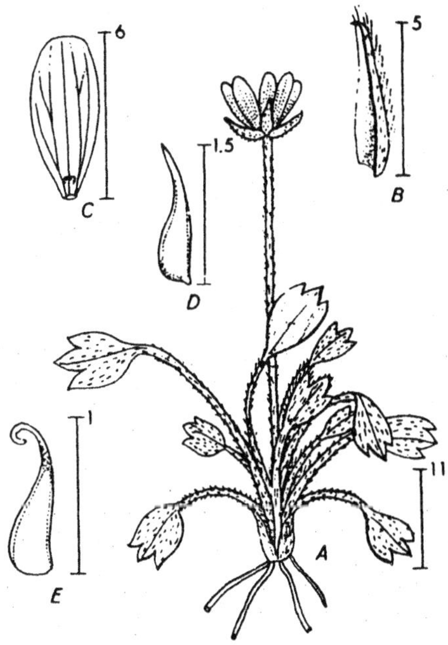

Ranunculus saruwagedicus (A) habit; (B) sepal; (C) petal; (D) carpel; (E) young nutlet.
(Source: van Royen (1982)

3. THALI~~TRICUM~~CTRUM

A family of some 250 species, most in the Northern Hemisphere. In Papuasia a single species (with 2 varieties) in high altitude regions. They are slender, glabrous herbs in swampy areas of alpine vegetation between 2950 and 3550 m altitude. Eichler (1963) described 2 varieties. The type was collected on Mt Jaya (Carstensz) by Boden Kloss and the species T. papuanum described by Ridley. Eichler described var. oranjense based on Versteegh 2474 collected on the Oranje Gebirge (see Steenis 1934) and a second, var. ceramense from Ceram. Both Papuasian varieties occur in Irian Jaya and extend only into the Star Mts. and Mt Capella.

Erect perennial herbs, with slender branched or unbranched stems. Leaves spirally arranged, repeatedly pinnate or ternate. Petiole sheath present. Inflorescence terminal or axillary, many-flowered. flowers small, usually bisexual. sepals 4, sometimes 3 or 5, often petaloid. Petals absent. Carpels many to few. Style spirally curved. Nutlets to 7mm long.

Key to Species in Papuasia

1 species:

Thali~~tricum~~ctrum papuanum

Notes:

1. Two varieties are recognised: var. papuanum and var. oranjense.

2. Distributed in Irian Jaya and the Star Mountains in subalpine shrubbery and alpine meadows. Altitudinal ranges 1200 - 3680 m and 2950 - 3550 m respectively.

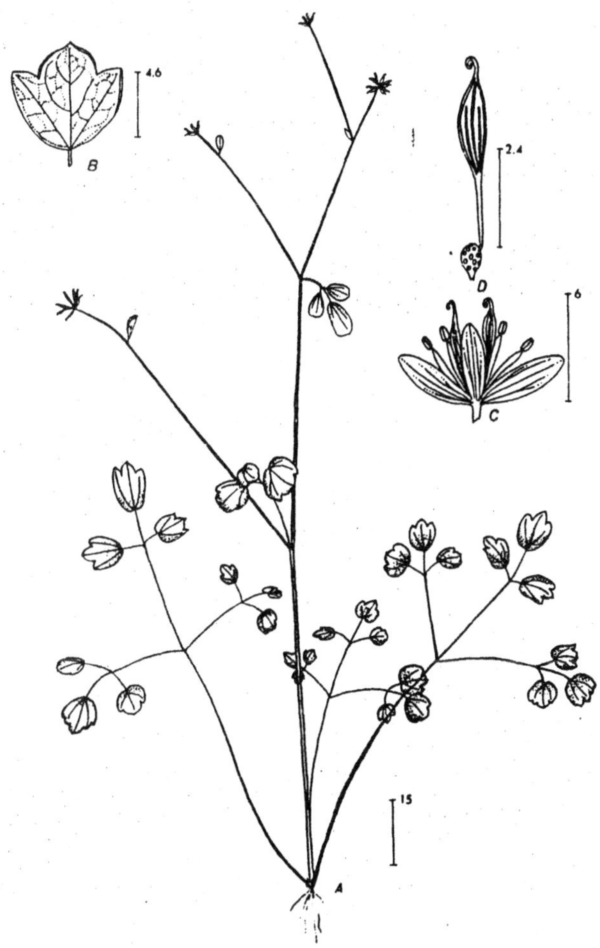

Thalictrum papuanum var. papuanum (A) habit; (B) leaflet; (C) flower; (D) nutlet. (Source: van Royen (1982)

MENISPERMACEAE

A family of mainly tropical lianes, with a few shrubs (rarely trees or herbs). It includes some 65 genera and about 350 species. Some species are subtropical or warm temperate. The family is a source of medicines and drugs. Best known is curare, an extract of Chondrodendron tomentosum used in surgical operations.

MENISPERMACEAE

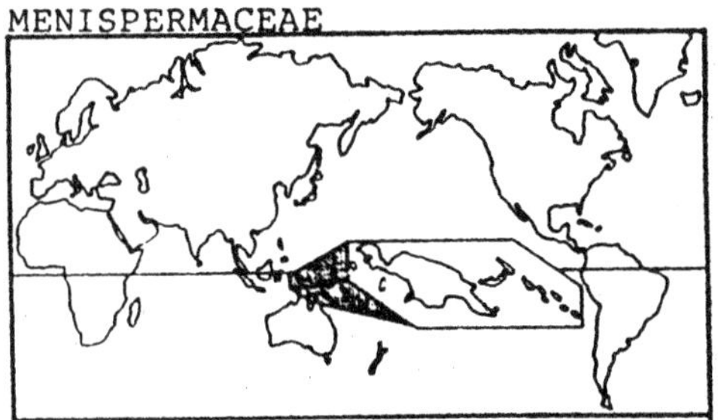

	WORLD	PAPUASIA
Number of genera:	65	19
Number of species:	250	33

Distribution: Tropical rainforest.

Economic uses: Medicines and drugs.

Diagnostic features

Leaves: exstipulate, alternate, sometimes peltate.

Flowers: very small, often greenish-white; male and female flowers on separate plants. Sepals two rows of three, petals in two or three rows. Stamens three to many, free, united or in bundles. Carpels one to 32, free. Stigma terminal.

Fruit: a drupe, usually curved, endocarp sculptured.

References

Diels, L. (1910). Menispermaceae. Das Flanzen. 46: 1 - 345.

Diels, L. (1915). Neue Menispermaceae von Papuasien. Bot. Jahrb. 52: 187 - 190.

Kundu, B.C. and Guha, S. (1976). New species of Stephania and Rhaptonema. Bot. Notiser. 129: 257 - 265.

Forman, L.L. (1956). The Menispermaceae of Malesia and adjacent regions I. Kew Bull. 11: 41 - 69.

Forman, L.L. (1957). The Menispermaceae of Malesia and adjacent regions II. Kew Bull. 12: 447 - 459.

Forman, L.L. (1961). The Menispermaceae of Malesia and adjacent regions IV. Cocculus. Kew Bull. 15: 479 - 487.

Forman, L.L. (1968). The Menispermaceae of Malesia and adjacent regions V. Kew Bull. 22: 349 - 374.

Forman, L.L. (1972). The Menispermaceae of Malesia and adjacent regions VI. Pycnarrhena, Macrococculus and Haematocarpus. Kew Bull. 26: 405 - 422.

Forman, L.L. (1972). The Menispermaceae of Malesia and adjacent regions VII. A revision of Legnephora. Kew Bull. 26: 275 - 280.

Forman, L.L. (1974). The Menispermaceae of Malesia and adjacent regions. Kew Bull. 29: 477 - 481.

Forman, L.L. (1975). The Menispermaceae of Malesia and adjacent regions VIII. The Tribe Triclisieae in Asia, the Pacific and Australia. Kew Bull. 30: 77 - 100.

Forman, L.L. (1978). The Menispermaceae of Malesia and adjacent regions IX A revision of the Tribe Coscineae. Kew Bull. 32: 323 - 338.

Forman, L.L. (1979). The Menispermaceae of Malesia and adjacent regions. X The identity of Batania. Kew Bull. 32: 323 - 338.

Forman, L.L. (1981). The Menispermaceae of Malesia and adjacent regions XI A revision of Tinospora in Asia to Australia and the Pacific. Kew Bull. 32: 323 - 338.

Forman, L.L. (1986). Menispermaceae. Flora Malesiana. Series 1. 10: 158 - 253.

KEY TO GENERA IN PAPUASIA
(Source: Forman (1986)

1. Seeds without endosperm 2

1. Seeds with endosperm 6

2. Condyle obsolete or septiform. TILIACOREAE.................... 3

2. Condyle hollow with a central aperture on both sides of endocarp. (LMENISPERMEAE). .... 12. Pachygone

3. Condyle obsolete. Seed straight, broadly ellipsoidal. Stamens often more than 6 and ± connate...... 4

3. Condyle an elongate septum around which the seed is sharply bent. Stamens 6, free. Inner sepals valvate. Leaves penninerved. .... 4. Carronia

4. Drupe with style-scar below the apparent apex. Sepals all free. 5

4. Drupe with style-scar near base. Inner sepals connate in a fleshy tubular calyx. Stamens c. 18 - 24, connate in a conical cluster. .... 1. <u>Albertisia</u>

5. Drupes not narrowed at base into a stipe, 3 - 6 on a ± glabose carpophore. Endocarp usually papyraceous to crustaceous and ± smooth, sometimes thick, bony and verrucose. Stamens 6-18, usually connate for most of length; connective not prolonged. 15. <u>Pycnarrhena</u>

5. Drupes narrowed at base into a stipe, 5 - 10 radiating from the margin of a disciform carpophore. Endocarp thick and bony, dorsally covered with a coarse open network of ridges, ventrally smooth. Stamens 9, 3 inner free surrounded by 6 slightly joined at base; connective prolonged into a tongue. .... 11. <u>Macrococculus</u>

6. Drupe with style-scar terminal. Embryo with foliaceous cotyledons. Petals 3 - 6. TINOSPOREAE (incl. Fibraureeae).................. 7

6. Drupe with style-scar near base or on one side.................... 10

7. Stamens free.................. 8

7. Stamens united in a peltate synandrium. Endocarp spiny, the condyle forming a ventral concavity or an enclosed inflated chamber. .... 13. <u>Parabaena</u>

8. Endocarp globose, covered with series of branched spines. Condyle clavate, 2-chambered, deeply intruding into the seed cavity. Stamens with flattened elliptic filaments broader than the anthers. .... 5. <u>Chlaenandra</u>

8. Endocarp not spiny. Stamens with ± terete filaments not broader than the anthers................... 9

9. Endocarp knobbly, papillose or smooth with a ventral aperture or elongate depression; condyle sometimes forming a ventral cavity. Endosperm usually ventrally ruminate. Plant without a laticiferous system. Leaves not striate above. .... 19. <u>Tinospora</u>

9. Endocarp rugose to rugulose, strongly compressed dorsiventrally without a ventral aperture or depression; condyle obsolete. Endosperm not ruminate. A laticiferous system present throughout the plant. Leaves finely and closely striate above. .... 18. <u>Tinomiscium</u>

10. Embryo with foliaceous, divaricate cotyledons. Seed broadly ellipsoidal or deeply cup-shaped. Petals 0. COSCINEAE..................... 11

10. Embryo elongate with narrow, semiterete or flattened cotyledons lying close together. Seed elongate, strongly curved, ± horseshoe-shaped. Petals present, at least in male flowers. Endosperm not ruminate. Condyle usually obovate in outline with the seed curved around its margin, sometimes hollow and 1-2 chambered. MENISPERMEAE (except Pachygone)............. 12

11. Condyle obsolete. Seed broadly ellipsoidal. Carpophore clavate. Anthers 9-12 in a sessile cluster. .... 3. <u>Arcangelisia</u>

11. Condyle deeply intruding into the seed-cavity. Seed deeply cup-shaped. Condyle deeply 2-lobed, hollow. Carpophore shortly branched below the drupes. Anthers 30-35 in a stalked cluster. Inflorescence in a panicle. .... 2. <u>Anamirta</u>

12. Sepals spirally arranged, imbricate, variable in number, 7 - 12. Stamens 9-40, free or connate. Carpels 2-3 (Mal. spp.). .... 8. <u>Hypserpa</u>

12. Sepals in whorls of 3 or in one united whorl................... 13

13. Carpels 3 - 6. Stamens free or only filaments connate 14

13. Carpel 1. Stamens completely connate in a peltate synandrium with the anthers horizontal around the margin........................ 18

14. Inner sepals valvate. Condyle hollow, 2-chambered. .... 10. <u>Limacia</u>

14. Inner sepals imbricate. Condyle ± lamelliform................... 15

15. Petals thick and fleshy, much larger than the sepals. Stamens with connate filaments.
.... 16. Sarcopetalum

15. Petals smaller than sepals or absent in female flowers. Stamens free. Petals entire, sometimes absent in female flowers....... 16

16. Condyle obovate to rotund in outline. Inflorescences axillary, cymes or thyrses............. 17

16. Condyle elongate, very narrow with the seed sharply bent around it. Inflorescences from old woody stems long racemes of cymes.
.... 7. Diploclisia

17. Petals absent in female flowers. Anthers dehiscing with transverse ± introrse slits, the cells separated abaxially by the thickened connective. Endocarp with an entire dorsal median wing or ridge and prominent lateral curved or cucullate crests bordering the condyle, the dorsal region smooth.
.... 9. Legnephora

17. Petals present in female flowers. Anthers dehiscing with longtitudinal latrorse slits, the cells parallel and close together. Endocarp without a dorsal median wing or ridge and lateral crests, dorsally covered with short pointed processes. 14. Pericampylus

18. Female inflorescences with accrescent bracts. Flowers 4 - merous. Petals connate.
.... 6. Cissampelos

18. Female inflorescences without accrescent bracts. Inflorescence composed of umbelliform cymes or disciform capitula. Condyle lamelliform. Flowers 3-4 merous. Petals free.. 17. Stephania

KEY TO GENERA IN PAPUASIA
(Based on Male Plants)

(Source: Forman (1986)

1. Upper surface of leaves very finely and closely striate. Inflorescences racemose on leafless stems, usually ferrugineous-tomentose.
.... 18. Tinomiscium

1. Upper surface of leaves not striate. Leaves not triplinerved.. 2

2. Inner 3 sepals connate in a fleshy tubular calyx, minute outer sepals free. Stamens c. 18-24, connate in a conical cluster.
.... 1. Albertisia

2. Sepals free or if connate then calyx of one whorl only........ 3

3. Petals thick and fleshy, much larger than sepals. Inflorescences pseudoracemose. Leaves peltate or subpeltate.
.... 16. Sarcopetalum

3. Petals smaller than sepals or absent......................... 4

4. Sepals 7-12, spirally arranged, not in distinct whorls. Stamens 9-40, free or connate. Inflorescences cymose or thyrsoid.
.... 8. Hypserpa

4. Sepals in one or more distinct whorls......................... 5

5. Connective prolonged into a flattened tongue. Stamens 9, the outer 6 joined at base, the inner 3 free. Inflorescences fasciculate on old stems..... 11. Macrococculus

5. Stamens without a prolonged connective.......................... 6

6. Stamens united or at least partly connate at the base, or only the inner 3 stamens connate........ 7

6. Stamens all free. Petals present. Stamens without a prominent collar below the anther. Wood not yellow.
.... 12

7. Stamens completely united in a peltate or clavate synandrium, the anthers in a horizontal ring.... 8

7. Stamens with the filaments ± connate or only the inner stamens connate. Inflorescence a panicle, cyme or fascile............... 10

8. Inflorescences composed of umbelliform cymes or disciform capitula. Leaves peltate.
.... 17. Stephania

8. Inflorescences cymose, thyrsoid or pseudoracemose. Leaves peltate or not.......................... 9

9. Inflorescence a peduncled, corymbose cyme. Sepals 4. Petals connate in a cup. Leaves peltate.
.... 6. Cissampelos

9. Inflorescences thyrsoid or pseudoracemose, or if cymose then sepals 6 free in 2 whorls and petals 6, free. Inflorescences thyrsoid or cymose. Leaves not peltate. .... 13. Parabaena

10. Lower surface of leaves with domatia in axils of main nerves. Leaves palmately nerved at base. Inflorescence a panicle. Anthers in a globose cluster........... 11

10. Leaves lacking domatia, penninerved or subpalmately nerved at base. Inflorescence a cyme or fascicle. Stamens 6-18 with the filaments ± joined.... 15. Pycnarrhena

11. Flowers pedicellate. Anthers 30 - 35 in a stalked cluster. .... 2. Anamirta

11. Flowers sessile or subsessile. Anthers 9-12 in a sessile cluster. .... 3. Arcangelisia

12. Inner 3 sepals valvate. Sepals tomentellous to densely pilose. Petiole not rugulose........... 13

12. Inner 3 sepals imbricate....... 14

13. Inflorescence a cyme or lax pseudopanicle... 10. Limacia

13. Inflorescence narrowly thyrsoid, the branches subspicate. .... 4. Carronia

14. Main basal nerves and their main outer lateral branches extending to the leaf-margin................ 15

15. Inflorescences on old leafless stems, long racemes of cymes. .... 7. Diploclisia

15. Inflorescences axillary........ 16

16. Anthers dehiscing with longitudinal latrorse slits, the cells parallel and close together. Leaves usually broadly and shallowly crenate. .... 14. Pericampylus

16. Anthers dehiscing with transverse ± introrse slits, the cells separated abaxially by the thickened connective. Leaves usually entire or subentire. 9. Legnephora

17. Inflorescence a large panicle up to c. 50 cm long with lateral branches up to 15 cm, borne on old leafless stems. Stamens with flattened elliptic filaments broader than the anthers... 5. Chlaenandra

17. Inflorescences much smaller and narrower, pseudoracemose, pseudospicate, narrowly thyrsoid or narrowly pseudopaniculate, axillary or on leafless stems. Stamens with filaments not broader than the anthers. Petals entire 18

18. Inflorescences yellowish tomentose to pubescent, always axillary and pseudoracemose. Leaf-base rounded to truncate or very obtuse (rarely subcordate). .... 12. Pachygone

18. Inflorescences glabrous, puberulous or hispidulous, axillary or on leafless stems, various in form. Leaf-base often cordate. .... 19. Tinospora

KEY TO GENERA IN PAPUASIA
(Based on Female Plants)

(Source: Forman (1986)

1. Upper surface of leaves very finely and closely striate. Inflorescences racemose on leafless stems, usually ferrugineous-tomentose. .... 18. Tinomiscium

1. Upper surface of leaves not striate, leaves not triplinerved.... 2

2. Inflorescences with accrescent rotund bracts becoming up to 1.5 cm long and persisting at fruiting stage, often obscuring the 5 mm long pubescent drupes. Leaves peltate, tomentose to puberulous. Slender climber. .... 6. Cissampelos

2. Inflorescences without accrescent bracts......................... 3

3. Inflorescences composed of umbelliform cymes or disciform capitula. Drupe with style-scar near base, 4-11 mm long. Endocarp compressed, ± obovate in outline with a sometimes perforate lamelliform condyle surrounded by a horseshoe-shaped seed cavity, dorsally ornamented with projections of various forms or transverse ridges. .... 17. Stephania

3. Inflorescences not of umbelliform cymes or disciform capitula.... 4

4. Seed and seed cavity straight or broad and concave or deeply cupular......................... 5

4. Seed and seed cavity elongate and strongly curved, semi-annular to horseshoe-shaped. Drupes not on a conspicuous carpophore. Carpels 1 - 6. Petioles not rugulose.... 12

5. Seed and seed cavity deeply cup-shaped, surrounding the thickly clavate, sometimes 2-lobed, deeply intrusive condyle. Drupes subglobose. Leaves palmately nerved at base...................... 6

5. Seeds and seed cavity either straight, ± ellipsoidal and endocarp with condyl absent, inconspicuous or forming a longitudinal groove, or seed dorsiventrally flattened and concave and endocarp with condyl forming a ventral cavity, concavity or inflated chamber..... 7

6. Lower surface of leaves with domatia in axils of main nerves. Drupes 9-11 mm... 2. <u>Anamirta</u>

6. Leaves without domatia. Drupes 40-45 mm , glabrous. Endocarp covered with branched spines. Leaves glabrous...... 5. <u>Chlaenandra</u>

7. Seed ± ellipsoidal filled by large embryo with thick cotyledons, endosperm absent. Leaves penninerved, sometimes with some subbasal nerves 8

7. Seed largely filled with endosperm which surrounds the very thin embryo. Leaves palmately nerved at base or with basal nerves diverging from midrib slightly above the base......................... 10

8. Drupes narrowed at base into a pronounced stipe, 5-10 radiating from margin of a disciform carpophore. Endocarp 3-6 cm long, bony, covered dorsally with a coarse open network of ridges, ventrally smooth. Inflorescences fasciculate on protuberances on old wood. Leaves densely reticulate.
.... 11. <u>Macrococculus</u>

8. Drupes with base not narrowed into a stipe, 3-6 on a knob-like carpophore. Endocarp either papyraceous to crustaceous and ± smooth to wrinkled or bony and verrucose. 9

9. Drupes subcompressed-ellipsoidal, tomentellous with style-scar near base, 2.5-6 cm long. Inner sepals connate into a fleshy tube.
.... 1. <u>Albertisia</u>

9. Drupes subglobose or broadly ellipsoidal, glabrous puberulous or tomentellous with style-scar below the apparent apex, 0.8 - 3cm long. Sepals all free.
.... 15. <u>Pycnarrhena</u>

10. Lower surface of leaves with hollow domatia in axils of main nerves, leaf papillose on upper surface around insertion of petiole. Drupes 2.2 - 5.5cm long with style-scar on one side, borne on a clavate carpophore up to 4 cm long. Endocarp covered with a fibrous layer, condyle inconspicuous or absent.... 3. <u>Arcangelisia</u>

10. Lower surface of leaves without domatia or only present in basal nerve-axils. Drupes with terminal style-scar. Basal nerves arising from the base of the leaf. Endocarp variously ornamented or smooth, often ± rotund in outline but sometimes subellipsoidal. Wood not yellow........................ 11

11. Endocarp spiny, the condyle forming a ventral concavity or an enclosed inflated chamber. Endosperm not ruminate.. 13. <u>Parabaena</u>

11. Endocarp rugose, papillose or smooth with the condyle forming either a ventral cavity with a central external aperture, or a ventral elongate depression. Endosperm usually ventrally ruminate.
.... 19. <u>Tinospora</u>

12. Leaves penninerved. Drupes and narrowly thyrsoid inflorescences tomentellous.
.... 4. <u>Carronia</u>

12. Leaves with base palmately 3 - 7 nerved. Drupes glabrous to pubescent......................... 13

13. Petals fleshy, larger than the sepals. Drupes semicircular in outline, (4-)6 mm long. Endocarp ornamented dorsally with pointed processes and/or transverse ridges, laterally concave. Seed and seed cavity semi-annular. Leaves subpeltate or peltate with petiole inserted up to 3 mm from basal margin, fine reticulation prominent especially on lower surface. Woody stems verruculose.
.... 16. <u>Sarcropetalum</u>

14. Inflorescences on old woody stems, consisting of very long racemes of cymes, up to c. 50 cm long. Drupes nearly obovate in outline, 15-30 mm long. Seed and seed cavity deeply horseshoe-shaped with the 2 arms ± parallel.. 7. Diploclisia

14. Inflorescences axillary. Drupes obovate to rotund in outline... 15

15. Endocarp with an entire dorsal median wing or ridge and laterally spreading curved or cucullate crests bordering the condyle, the dorsal region smooth. Petals absent..... 9. Legnephora

15. Endocarp without a dorsal median wing or prominent ridge and lateral crests, the dorsal region sometimes ornamented, e.g. verruculose, ridged or spinulate. Petals present. 16

16. Seed without endosperm, cotyledons large and thick, collateral. Endocarp rather smooth with a small central perforation on both lateral faces..... 12. Pachygone

16. Seed with endosperm surrounding the narrow embryo................. 17

17. Drupes 14 - 25 mm long. Dorsal region of endocarp smooth or only slightly rugose. Condyle hollow, 2-chambered. Inner sepals valvate. .... 10. Limacia

17. Drupes less than 12 mm long. Dorsal region of endocarp often ornamented. Condyle lamelliform or hollow. Inner sepals inbricate. 18

18. Leaves with main basal nerves and their main outer lateral branches leading direct to the margin which is usually broadly and shallowly crenate. Endocarp laterally concave, dorsally covered with short pointed processes, condyle lamelliform, imperforate. .... 14. Pericampylus

18. Leaves with main nerves breaking up well within the entire margin. Endocarp with thick, hollow condyle, often perforate. Condyle with 2 distinct lateral chambers each with a large lateral aperture. Endocarp not ventrally perforate. Carpels (2-)3 or 6. Sepals 6-12. Leaves never peltate. Drupes 6-12 mm long. Dorsal surface of endocarp rugose. Carpels (2-)3. Sepals spiral, 7-12. Petals entire.... 8. Hypserpa

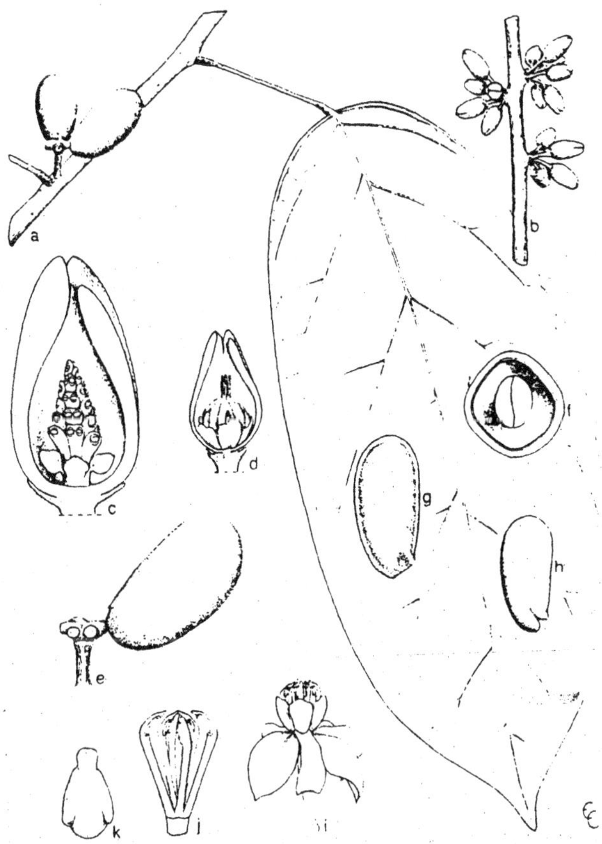

Albertisia papuana (A) - (C) male: habit, inflorescences, flower; (D) female flower; (E) (F) drupe; (G) endocarp; (H) seed. Macrococculus pomiferus (I) - (K) male: flower, stamens, anther. (Source: Foreman (1986)

1. ALBERTISIA

A small genus of 12 species from tropical and subtropical Africa to Indo-Malesia and Papuasia where it is represented by a single species.

Lianes, branches with prominent discoid petole-scars. Petioles conspicuously swollen at both ends. Male flowers: sepals (1-)2 outer whorls of 3 minute triangular sepals, inner 3 sepals connate, forming a fleshy corolla tube with minute apical opening; petals 3 or 6, minute fleshy; stamens 18 - 24, connate into a conical synandrium, anthers 2-celled. Female flowers: sepals and petals as in male; staminodes 6; carpels 4 - 6, style subulate. Drupes radiating from margins of carpophore terminating the peduncle. Seeds without endosperm; cotyledons very thick (Forman 1986).

Key to Species in Papuasia

1 widespread species:

Albertisia papuana

Notes:

1. This species is distributed from Thailand to Papuasia. Leaves and fruits are edible.

2. In primary rainforest at low altitudes.

2. ANAMIRTA

• • • • • •

A monotypic genus occuring from Sri Lanka and India to Papuasia. Large almost glabrous, lianes. Leaves palmately nerved at base, hairy domatia in axils of nerves. Inflorescences usually cauliflorous (male sometimes axillary), paniculate. Male flowers: 2 minute outer sepals, with 6 (3+3) larger inner sepals, concave, imbricate; petals absent; anthers c. 30-35 in globular cluster. Female flowers: sepals as in male fowers; petals absent; 6 minute staminodes; 3(-4) carpels arising laterally from a central conical gynophore; stigma recurved. Drupes: borne on short terminal divergent branches continuous with peduncle; style-scar lateral. Seeds with copious endosperm; cotyledons much larger than radicle.

Key to Species in Papuasia

1 widespread species:

Anamirta cocculus

Notes

1. Grows in both rainforest and monsoon forest. Also in coastal forest, savannah and riverine forest. Most common in savannah areas.

2. Fruits used as fish poison and also to kill lice. Fruits contain picrotoxin which has been used to treat schizophrenia.

3. ARCANGELISIA

• • • • • •

A genus of 2 species from China to Papuasia. One endemic species can only be recognised in fruiting condition.

Plants glabrous, apart from domatia on leaves. Stems exude yellow sap when cut, with prominent cup-like petiole scars. Petioles swollen at both ends, leaves ovate, apex acuminate; palmately 5-nerved at base. Inflorescence axillary or cauliflorous, paniculate, 10-15cm long. Male flowers sessile or subsessile, subtended by an ovate bracteole, which is c. 1mm long and thickened at base. Sepals 3-4, minute, less than 1 mm long; inner sepals 6 (3+3), elliptic to narrowly obovate, to 1.5 - 2.5 cm long; synandrium 0.5 - 1 mm long. Female flowers: sepals 6, narrowly oblong, apex becoming recurved, 2.5 - 4 mm long; staminodes scale-like, minute; carpels 3, to 1.5mm long; stigma broad. Infrutescences cauliflorous, branched to 7-30(-45) cm. Fruits on lateral branches, 1-3 borne together on a club-shaped, unbranched carpophore. Drupes yellow with a layer of radially arranged fibres.

Key to Species in Papuasia
(Source: Forman (1986)

1. Fruits transversely subovoid, 2.2 - 3 cm long by 2.5 - 3.3 cm broad; endocarp covered with a dense mat of radially arranged fibres.
.... A. flava

1. Fruits subglobose, 4.5 - 5.5 cm diameter; endocarp with interrupted layer of radially arranged fibres, which form a dense dorsal ridge and thin lateral transverse plates.
.... A. tympanopoda

Notes:

1. A. flava occurs at altitudes up to 1000 m. A tympanopoda occurs in secondary forest on clay soils to 350m altitude. Plants often overhanging water.

2. A. tympanopoda has two distinct forms of infrutescence which Forman (1986) suggests may relate to the position where they are borne: on older or younger stems.

4. CARRONIA

• • • • • •

A genus of 3 species: 2 in Australia and 1 in Papuasia. Lianes. Branchlets with prominent discoid petiole scars. Leaves elliptic. Inflorescence axillary, terminal or ramiflorous; flowers in congested clusters. Male flowers: sepals 9 - 30 in whorls of 3, inner ones larger, externally densely pilose, internally glabrous, inner whorl(s) glabrous; petals 6, minute; stamens 6, free, anthers introrse. Female flowers: perianth parts as in male; staminodes absent; carpels 6, densely pilose; styles erect or recurved, glabrous. Drupes narrowed at base into a stipe; seed horseshoe-shaped, endosperm absent, cotyledons thick.

Key to Species in Papuasia

1 low altitude species:

Carronia thyrsiflora

5. CHLAENANDRA

A monotypic endemic genus in New Guinea. Large, glabrous, woody climber. young stems prominent discoid petiole-scars; old stems with rough ridged bark. Petioles 5 - 9cm long, thickened at base and apex. Leaf ovate to 12-19 by 7 - 12 cm; palmately 3-5-nerved at base, finely reticulate. Male inflorescence: borne on old stems to 50cm long, with lateral branches to 15 cm; flowers: sepals, 2 series: outermost 6, minute less than 1mm long; inner sepals 6, 2-2.5mm long; petals 6, elliptic 1.3 mm long; stamens 6 to 2mm long, filaments broader than anthers. Female inflorescence unknown. Reddish, densely covered with branched spines. seed with copious uniform endosperm.

Key to Species in Papuasia

1 widespread species:

Chlaenandra ovata

Notes:

1. Up to 1200m altitude in lowland swamp forest and rainforest. Beccari stated that the seeds are avidly eaten by cassowaries.

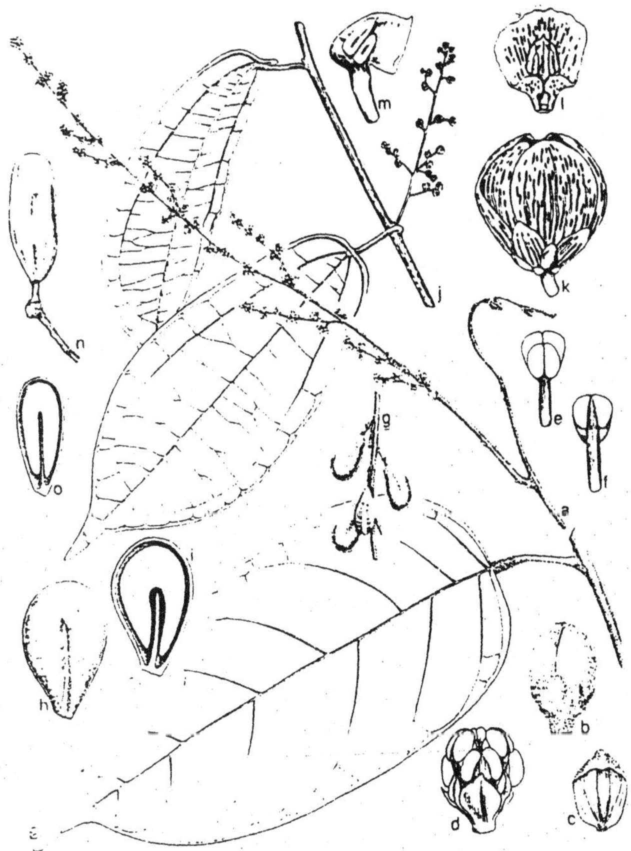

Carronia thyrsiflora (A) - (F) male: inflorescence, bud, inner sepal, flower, stamen; (G) drupes; (H) (I) endocarp. Haematocarpus subpeltatus (J) habit; (K) male bud; (L) (M) stamen; (N) drupe; (O) endocarp. (Source: Foreman (1986)

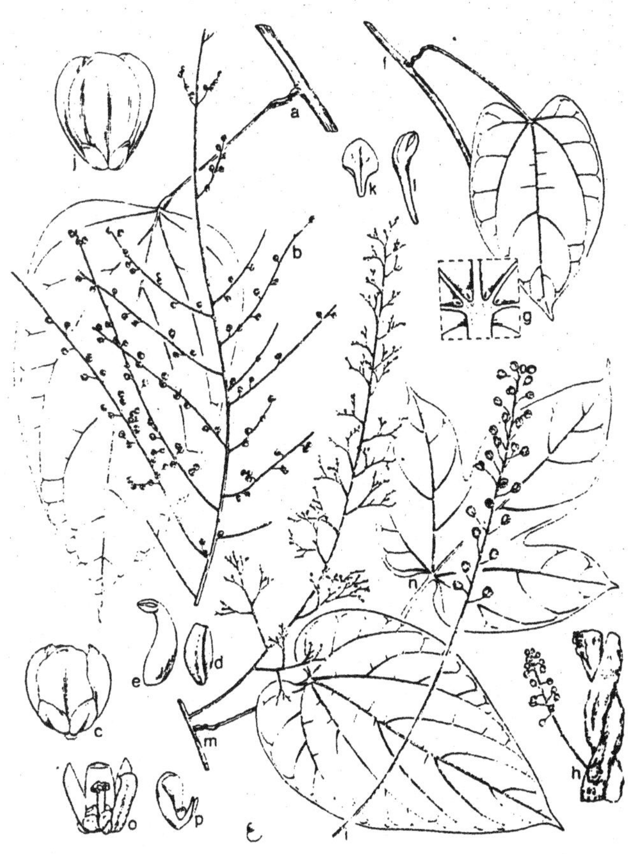

Chlaenandra ovata (A) leaf; (B) - (E) male: inflorescence, flower, petal, stamen. Tinospora glabra (F) leaf; (G) domatia; (H) - (L) male: inflorescence, flower, petal, stamen. Parabaena megalocarpa (M) male inflorescence; (N) leaf; (O) male flower; (P) petal. (Source: Foreman (1986)

6. CISSAMPELOS

A pantropical genus with 20 - 25 species. One widespread pantropical species in Papuasia where it is represented by one variety. Scandent shrub with woody old stems. Leafy stems slender, densely pubescent to glabrous. Petioles 2-9cm long, densely to sparsely pubescent, inserted 1-18mm from basal margin of lamina. Leaf variable, to 11 by 12 cm; palmately 5-7-nerved. Male flowers: green to yellowish, pedicels 1-2 mm; sepals 4, 1.2-1.5 mm long; corolla cup-shaped, to 0.5 mm long; synangium to 0.75 mm long. Female inflorescence to 18cm long, each fascicle in axil of bract to 1.5 cm long. Female flowers: sepals 1; petals 1; staminodes 0; carpel 1. Drupe orange or red, with 2 rows of 9 - 11 prominent ridges.

Key to Species in Papuasia

1 widespread species:

Cissampelos pareira
var. hirsuta

Notes:

1. Found in primary and secondary forest to 1300 m altitude. Often climbing over trees or river banks.

2. The brewed root is used as diuretic. Pounded leaves are used for snake bites and for scabies.

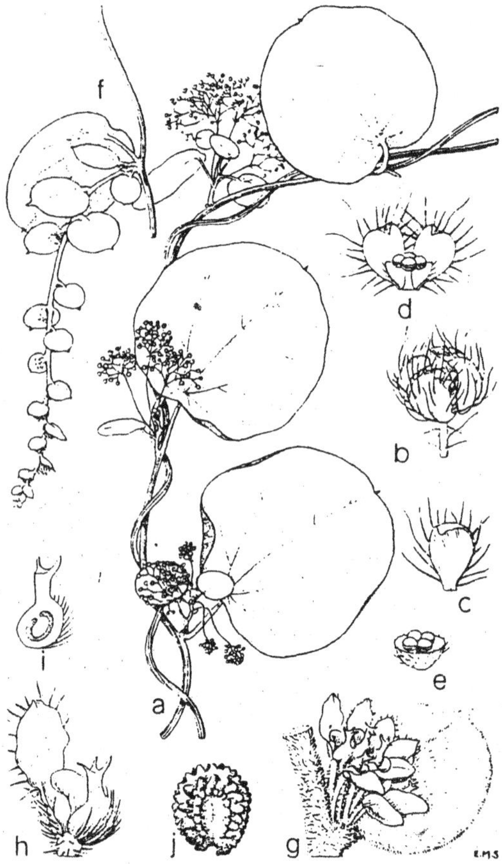

Cissampelos pareira var. hirsuta (A) - (C) male: habit, flower, sepal, petal; (F) - (J) female: habit, inflorescence, flower, carpel, endocarp.
(Source: Foreman (1986)

7. DIPLOCLISIA

A genus of 3 species from S E Asia to Irian Jaya. Glabrous, woody climbers to 30 m long. Stems to 5cm diameter. Leaves with petioles to 5 - 12 cm long; base rounded to cordate; peltate or non-peltate; to 6 - 11 by 6 - 11 cm. Inflorescences cauliflorous to 52 cm long and 7 cm wide. Male flowers pale yellow; pedicels to 2-4mm long; sepals 6, with dark brown reticulum; outer 3 elliptic, to 2.5mm long; inner 3 broadly elliptic, to 2.5mm long; petals 6, 1mm long; stamens 6, 2mm long. female flowers; sepals and petals as in male flowers, staminodes 6; carpels to 2mm long. drupe yellow to orange; endocarp elongate, to 20mm long, ornamented with transverse ridges.

Key to Species in Papuasia

1 widespread species:

Diploclisia glaucescens

Notes:

1. Growing in lowland forest up to 600m in altitude.

2. No fruiting material is known from Papuasia. This is needed to confirm the identification (Forman 1986: 230).

8. HYPSERPA

A genus of 6 species distributed from China and Sri Lanka to Australia and Polynesia; throughout Papuasia.

Scandent shrubs or woody climbers. Tip of stem often like a tendril. Leaf bases with 3 - 7 nerves. Inflorescences axillary or supra-axillary. Male flowers; sepals 7 - 12 spirally arranged; outer ones minute, bracteoliform; inner ones larger, imbricate; petals 5 - 9, fleshy; stamens 9 - 40, free or connate. Female flowers; sepals and petals similar to male;

staminodes 0 to several; carpels 2 - 3, stigma entire, reflexed. drupes curved, style scar near base, endocarp with 2 lateral cavities. Seed horseshoe-shaped, narrow, embedded in endosperm.

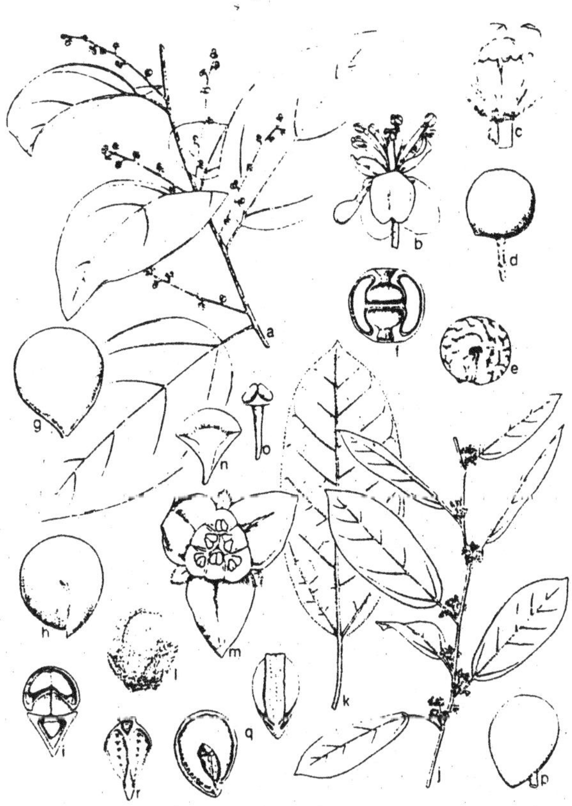

<u>Hypserpa</u> <u>nitida</u> (A) (B) male: habit, flower; (C) - (F) female: flower, drupe, endocarp. <u>Pachygone ovata</u> (G) drupe; (H) (I) endocarp. <u>Limacia scandens</u> (J) male: habit, leaf, bud, flower, petal, stamen; (P) drupe; (Q) (R) endocarp. (Source: Foreman (1986)

<u>Key to Species in Papuasia</u>
(Source: Forman (1986)

1. Inflorescence glabrous.
.... <u>H. laurina</u>

1. Inflorescence hairy............ 2
.... <u>H. polyandra</u>

2. Branches leaves and inflorescences glabrescent or minutely puberulous.
.... var. <u>polyandra</u>

2. Branches leaves and inflorescences tomentose. var. <u>tomentosa</u>

<u>Notes</u>:

1. <u>Hypserpa</u> <u>laurina</u> grows in Queensland and S W Papua in rainforest and <u>Imperata</u> grasslands.

2. H. <u>polyandra</u> grows in mangrove swamps and lowland mixed rainforest to 1200 m altitude. In Bougainville the bark is pounded and applied to the head to treat headaches. This species is widespread from W Sumatra to Australia, the Carolines and Vanuatu.

3. var. <u>tomentosa</u> (Forman 1968:360) grows near rivers and lakes in the Morobe Province.

9. LEGNEPHORA

• • • • • •

A genus of 5 species from the Sunda Is, Moluccas and Philippines, throughout Papuasia to E Australia.

Woody climbers. Leaves palmately 3-7-nerved at base, broadly ovate. Inflorescencs: pedunculate cymes. Male flowers sepals 6, outer 3 usually narrower than inner 3; petals 6 with sides folded inwards around the opposite stamens, glabrous; stamens 6, free. female flowers, sepals 6 - 9; petals 0; staminodes 6, claviform; carpels 3, stigmas recurved, flattened. Drupes curved with style scar near base; endocarp with entire dorsal wing and prominent lateral crests. Seed curved.

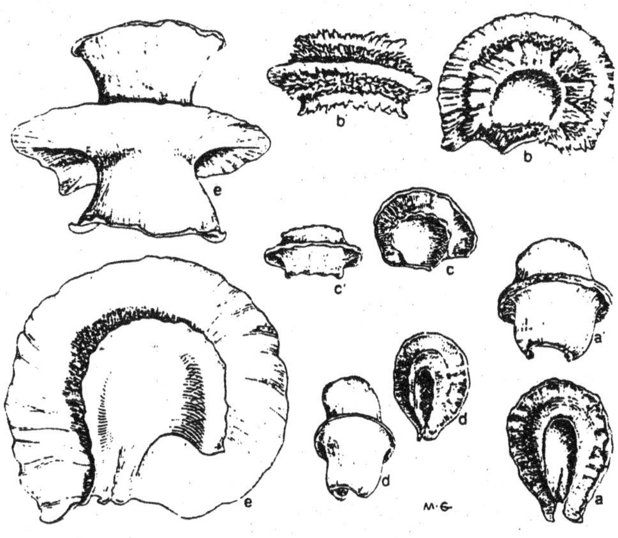

<u>Endocarps of Legnephora</u> (A) <u>L. acuta</u> (B) <u>L. moorei</u> (C) <u>L. phillipinensis</u> (D) <u>L. microcarpa</u> (E) <u>L. minutiflora</u> (Source: Foreman (1986)

<u>Key to Species in Papuasia</u>
(Source: Forman (1986)

1. Leaves long and finely acute at apex, entirely glabrous. Endocarp 8

- 9 mm long with a dorsal wing and lateral crests.
.... L. acuta

1. Leaves rounded or broadly pointed at apex, hairy, at least on petioles...................... 2

2. Endocarps up to 6 mm long.
.... L. microcarpa

2. Endocarps 10 - 18mm long.
.... L. minutiflora

Notes:

1. L. acuta is known only from W Highlands of Papua New Guinea. In primary or old secondary forest at 1750m.

2. L. microcarpa is known from 2 collections in Morobe Province between 800 and 1200 m altitude.

3. L. minutiflora occurs from Moluccas to New Ireland. A rainforest species growing from sealevel to 1800m altitude.

4. A single collection is known from Santa Cruz in the Solomon Islands. The species is uncertain as fruits are not known.

10. LIMACIA

● ● ● ● ● ●

A genus of 3 species from S E Asia to Celebes and Philippines. Reports from New Guinea are doubtful.

Woody climbers. Young growing tips sometimes like tendrils. Leaf base 3-5-nerved. Inflorescences axillary or supra-axillary. Male flowers: sepals 6 or 9 in whorls of 3, tomentose, inner whorl valvate and larger than outer whorls; petals 6, concave; stamens 6, free; rudimentary carpels 3 or absent. Female flowers: sepals and petals as in male flowers; staminodes 6 or absent; carpels 3, tomentose, style reflexed, stigma entire. Drupes with style scar near base. Sometimes placed in Hypserpa (see Forman 1986: 222).

11. MACROCOCCULUS

● ● ● ● ● ●

A monotypic genus, endemic in New Guinea, excluding Solomon Islands. Closely related to Pycnarrhena but retained as distinct by Forman (1986: 178).

Liane. Older branchlets producing conspicuous swollen cups around the base of petioles. Leaves: petioles swollen at base and apex; lamina not or slightly peltate, penninerved. Inflorescences cauliflorous. Male flowers: sepals 5 to 6 whorls of 3, imbricate; outer 3-4 whorls of sepals minute; inner 2-3 whorls much larger; petals 6, thick, fleshy; stamens 9, ring of 6 slightly joined at base surrounding 3 inner ones, filaments flattened. Female flowers unknown. Drupe red, globose, up to 10 by 8 cm, glabrous. Endocarp 3 - 6 cm diameter, with 3-5 longitudinal ridges connected by anastamising ridges forming a loose network.

Key to Species in Papuasia

1 endemic species:

Macrococculus pomiferus

Notes:

1. In forest from sea-level to 1000 m altitude. Beccari states the fruits are often eaten by cassowaries.

12. PACHYGONE

● ● ● ● ● ●

Genus of 11 species from China and S E Asia to Australia and Polynesia: throughout Papuasia. Probably too many species are described.

Woody climbers. Leaves ovate, base 3-5-nerved. Inflorescences axillary. Male flowers: sepals 6(-12), inner 3(-6) to 1.5 - 2.5mm long, outer 3, 1.5-2mm long, imbricate; petals 6, articulate towards base; stamens 6. Female flowers: sepals and petals as in male; staminodes 6; carpels 3, glabrous, style reflexed, stigma entire. Drupes on pedicels to 3 - 5mm long, glabrous.

Key to Species in Papuasia

1 widespread species:

Pachygone ovata

Notes:

1. Distributed from Sri Lanka and India to Papuasia. Along seashores and in lowland rainforest up to an altitude of 100m. Also in areas with seasonal climates.

2. Diels (1910) maintained 4 distinct

species in the above, but Pachygone is highly variable in leaf shape and leaf hairiness. Flowers and fruits are rather constant (Forman 1986: 218).

13. PARABAENA

A genus of 6 species from continental S E Asia to Papuasia where it is represented by a single species.

Slender woody climbers. Leaves swollen and geniculate at base, lamina cordiform to rotund, base cordate, apex acuminate, to 10-22 by 9-22 cm. Inflorescence axillary, to 18-35cm long. Male flowers: pedicels 1-2 mm; sepals white or yellow, 2.5mm, outer 3 elliptic, inner 3 broader and concave; petals to 1mm long, fleshy, synandrium to 1mm long. Female flowers: on pedicels to 3-4mm long, sepal and petals as in male; staminodes minute; carpels to 1mm long, stigma with 3-6 reflexed lobes. Drupes white (or red), radiating from carpophore; endocarp elliptic in outline, ventral surface with incurved spines.

Key to Species in Papuasia

1 Papuasian species:

Parabaena tuberculata

Notes:

1. Common in secondary rainforest (Pometia-Intsia), river banks and cane - breaks along floodplains; up to 1000m.

2. Forman (1984) provides a full discussion of this species. Beccari discussed anomalous bisexual flowers in material from Timor.

14. PERICAMPYLUS

A genus of 2-3 species in Asia and Malesia. One species in Papuasia, distributed through range of genus.

Woody climber. Older stems longitudinally ridged. Leaves: petioles to 3-7cm, traingular-ovate, margin crenate, 5-10cm long and broad, base palmately 5-nerved. Inflorescence 2-4cm long, yellowish tomentose. Male flowers: white or yellow with pedicels; sepals 9, hairy outside, , outer 3 minute, narrow, to 0.5mm long, middle 3 oblanceolate, inner 3 obovate, 1mm long; petals 6, glabrous, 0.5mm long; stamens 6, o.75mm long. Female flowers: sepals and petals as in male flowers; staminodes 6; carpels 0.5mm long; stigma recurved, bifid. Drupes purple to black, glabrous; endocarp round in outline, 5mm diameter, with 2 dorsal rows pointed projections.

Key to Species in Papuasia

1 widespread species:

Pericampylus glaucus

Notes:

1. In primary and secondary rainforest and thickets up to 1700m in altitude.

2. Stems are used for tying and basket making. Sap used as eye medicine.

3. Field notes taken for plants of this species from Kinabalu record a broad tuber to 30cm long.

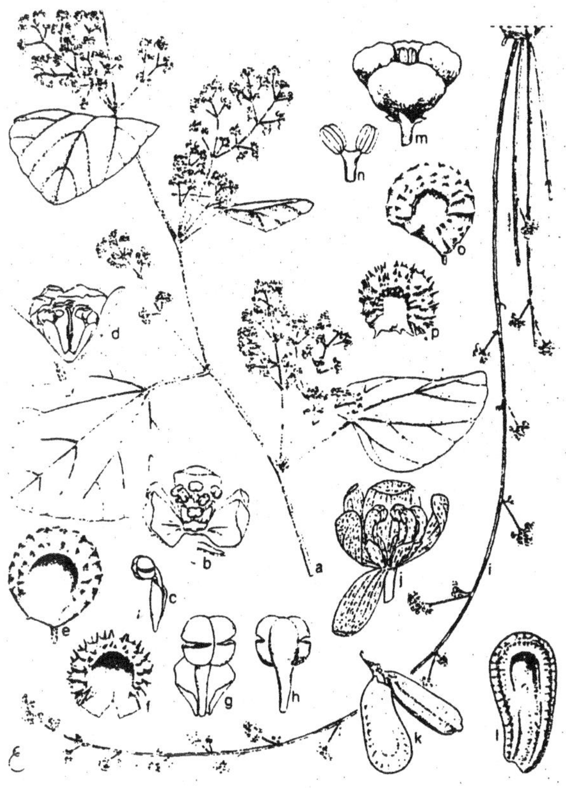

Pericampylus glaucus (A) - (C) male: habit, flower, stamen; (D) - (F) female: flower; (E) drupe; (F) endocarp. Legnephora minutiflora (G) (H) stamen; Diploclisia kunstleri (I) (J) male: inflorescence, flower; (K) drupe; (L) endocarp. Sarcopetalum harveyanum (M) (N) male: flower, stamens; (O) drupe; (P) endocarp. (Source: Foreman (1986)

15. PYCNARRHENA

A genus of 9 species distributed from S China to Queensland; 3 species occur in Papuasia. Forman (1986) notes that most collections are incomplete and in consequence the genus was difficult to revise.

Lianes or scandent shrubs. Leaves with petioles swollen at the base and usually also the apex, leaving a prominent cup-like scar on the stem. Lamina elliptic. Inflorescence axillary or ramiflorous, cymose, peduncles sometimes 1-flowered. Male flowers: sepals 6 - 15 in whorls of three, imbricate, the outer ones minute, inner ones larger (except sometimes in P. ozantha); petals 3 - 6 (or absent in P. ozantha, minute; stamens variable in number, 6 - 18, filaments usually clustered for most of their length. Female flowers: sepals and petals as in male; staminodes absent; carpels 3 - 6 stigmas recurved. Drupes subglobose; endocarp usually thin.

Key to Species in Papuasia

1. Clearly differentiated petals absent (occasionally 1). Carpels 4 - 6. Drupes 12 - 20 mm diameter, drying shrivelled, glabrous. endocarp thick, very hard, verucose...... P. ozantha

1. Petals present, clearly differentiated from the sepals. Carpels 3(-4). Endocarp crustaceous, smooth to wrinkled............ 2

2. Drupes 8 - 10 mm diameter. Infrutescences branched, slender, elongate up to 7 cm long. .... P. novoguineensis

2. Drupes 15 - 30 mm diameter, drying smooth when ripe. Infrutescences with rather thick peduncles, usually unbranched and fascicled when axillary, but branched when ramiflorous. .... P. tumefacta

Notes:

1. P. ozantha occurs throughout the Papuasian region in forest up to 1600 m altitude. Often on limestone but also on volcanic boulders.

2. P. novoguineensis occurs in Queensland and Papuasia. It is common in Irian Jaya but is rare in Papua New Guinea. At low altitudes (below 50 m). Male flowers are heavily scented.

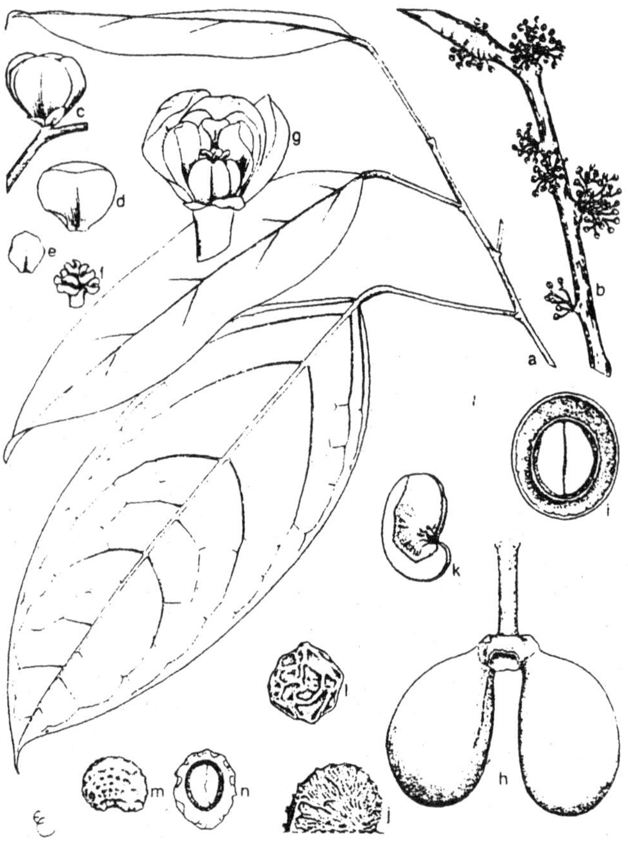

Pycnarrhena tumefacta (A) - (F) male: habit, inflorescences, flower, inner sepal, petal, stamens; (G) - (K) female: flower, infructescence, drupe, endocarp, seed. P. ozatha (L) (N) drupe; (M) endocarp. (Source: Foreman (1986)

3. P. tumefacta is a widespread species from Solomons to Java and Borneo. Usually below 400 m altitude in Papuasia. Includes several distinct forms which may be local species (Forman 1986). Flowers are very fragrant.

16. SARCOPETALUM

A monotypic genus in E Australia and S New Guinea.

Woody climbers. Leaves with petioles to 4 - 10 cm long, inserted up to 3 mm from basal margin of lamina, lamina ovate, base cordate, apex acuminate or obtuse, 9-15 by 7-12 cm, reticulations prominent, glabrous. Male inflorescence: 4cm long, puberulous, with minute, lanceolate bracteoles to 2mm long. Male flowers: pedicels 2-3mm long; sepals (2-) 3-5, minute, triangular, to 1mm long, margins dentate; petals 3-5, very fleshy to 2mm long, 1.5 mm thick; stamens to 1.5mm long with filaments connate into column, anthers 3-4 free, arising from top of column. Female flowers: sepals and petals as in male flowers; staminodes minute; carpels 3-6, stigma recurved, divided into 2 - 3 at apex. Drupe red, glabrous; endocarp 6x5 mm, dorsally covered scattered sharp points.

Key to species in Papuasia

1 species S New Guinea:

Sarcopetalum harveyanum

Notes:

1. In New Guinea known from L Daviumbu and Wassi Kussa both from seasonal rainforest areas at low altitudes.

2. Forman (1986:225) notes that the ornamentation of the endocarp in the Papuan specimens is slightly different from the Australian specimens. He considers it merely a regional variation.

17. STEPHANIA

• • • • • •

A genus of 35 species, mostly tropical, throughout Old World. Species have different life forms: *S. japonica* has annual or seasonal stems, *S. zippeliana* is perennial.

Slender climbers. Stems woody or herbaceous. Roots sometimes tuberous. Leaves ovate to suborbicular, palmately 8 - 13-nerved. Inflorecences axillary on old, leafless stems; all Papuasian species have inflorescences composed of umbelliform cymes, solitary or racemose. Male flowers: sepals free, imbricate, 6-8 in 2 ± whorls or symmetrical; petals free 3 or 4; stamens connate into peltate synandrium with 4-8 anther-cells. Female flowers: sepals 1-8; petals 2-4, both similar to male; carpel 1, style short or absent, stigma shortly lobed. Drupes ovoid with style-scar near base, glabrous; endocarp bony with 2 or 4 longitudinal rows of transverse ridges or processes. Seed horseshoe-shaped.

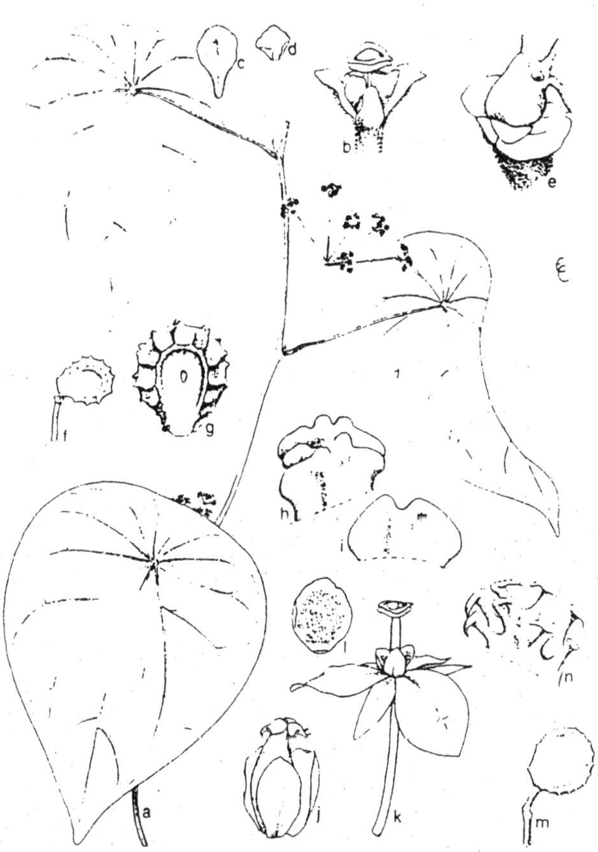

Stephania japonica var *timorensis* (A) habit; (B) - (D) male: flower, sepal, petal; (E) (F) female: flower, drupe, (G) - (I) endocarp.
Stephania corymbosa (K) (L) male: flower, petal; (M) drupe; (N) endocarp.
(Source: Forman (1986)

Key to Species in Papuasia
(Key to Male Plants)

(Source: Forman (1986)

1. Flowers sessile or subsessile in dense clusters. .... *S. japonica*

1. Flowers pedicellate........... 2

2. Inflorescences verruculose; peduncles of cymes arising retrorsely from main axis. Leaf margins with wide, shallow crenations: petiole inserted 2 - 3 mm from base. .... *S. grandiflora*

2. Inflorescences glabrous, papillose or puberulous; peduncles of cymes not retrorse along main axis. Leaf margins not widely crenate: petiole inserted more than 6 - 10 mm from base........................ 3

3. Leaves with very fine reticulation. .... *S. zippeliana*

3. Leaves with lax reticulation.. 4

4. Leaves triangular-orbicular to orbicular, apex rounded to very broadly obtuse. .... *S. salomonum*

4. Leaves triangular to ovate, apex acute or obtuse.............. 5

5. Umbelliform cymes 2 - 5 cm with primary rays less than 1 cm. Leaves ovate to ovate-triangular, apex obtuse.... S. moluccana

5. Umbelliform cymes 4 - 14 cm with primary rays more than 1 cm. Leaves triangular to narrowly triangular, apex very acute. .... S. montana

Key to Species in Papuasia
(Key to Female Plants)

(Source: Forman (1986)

1. Inflorescences verruculose. Leaf margins with wide, shallow crenations: petiole inserted 2 - 3 mm from base. S. grandiflora

1. Inflorescences glabrous, papillose or puberulous. Leaf margins not widely crenate: petiole inserted more than 6 - 10 mm from base.. 2

2. Drupes sessile or subsessile. Leaves with fine reticulations. .... S. japonica

2. Drupes pedicellate, apart from occasional sessile ones.

3. Leaves with very fine reticulation, visible on one or both surfaces. .... S. zippeliana

3. Leaves with lax reticulation, visible on both surfaces.

4. Leaves triangular-orbicular to orbicular, apex rounded to very broadly obtuse. .... S. salomonum

4. Leaves triangular to ovate, apex acute or obtuse.............. 5

5. Leaves ovate to ovate-triangular, apex obtuse. .... S. moluccana

5. Leaves triangular to narrowly triangular, apex very acute. .... S. montana

Notes:

1. S. japonica occurs from Japan and China to N and E Australia, Polynesia, and throughout Papuasia. Grows in secondary vegetation from sea-level to 2000 m. It is used medically for fever, diarrhoea, urinary diseases and stomachache. Crushed leaves in water are applied for breast infections.

Two varieties occur in Papuasia. They can be distinguished using the following key (Forman 1956: 49):

1. Leaves glabrous below. .... var. timorensis .

1. Leaves puberilous below. .... var. discolor

The regions around the insertion of the petiole can be variable. Both varieties are widespread from Asia to Australia and Papuasia.

2. S. grandiflora is endemic in the Highlands of Papua New Guinea. Occurs in regrowth areas between 2300 and 2700m altitude. The stem is used to make belts.

3. S. salomonum occurs in the Bismark Archipelago, Manus and the Solomon Is. including Bougainville. Grows in secondary regrowth up to 600m altitude.

4. S. moluccana is possibly represented from S E New Guinea by a fruiting collection, Jacobs 8682 (Forman 1986: 252).

5. S. zippeliana is a widespread Pacific species, occurring up to 1500m altitude in Papuasia.

6. Kunda and Guha (1976: 257) described S. neoguineensis based on a collection with a detached male inflorescence. Its status is doubtful (Forman 1986: 253)

18. TINOMISCIUM

A monotypic genus growing from Assam to Papuasia.

Large woody climber to 30m, exuding white latex when cut. Old stems coarsely striate, glabrous. Petioles to 6-21cm long, slender, sometimes swollen at base and apex. Lamina ovate, base truncate to cordate, apex acuminate, 11 - 25 by 4.5 - 20 cm, base 3 - 5-nerved with 2-3(-4) pairs of distal lateral nerves, nerves prominent on lower surface. Inflorescences arising from protuberences on old stems, racemose, 8-28 cm. Male flowers on pedicels 2.5-5cm long; sepals 3 outer triangular, 1-2mm long, 6 inner sepals white to yellow, elliptic, 4-5mm long; petals elliptic, 2.5-3.5 mm long, erect, contiguous at anthesis, concave, glabrous; stamens with or without apiculus, 2-2.5 mm. Female flowers: sepals and petals as in male flowers; staminodes acute, to 3 mm; carpels curved to 2mm, stigma multi-lobed. drupes green with white spots, latex white to yellow or orange, base narrowed into short stalk; endocarp compressed.

Key to Species in Papuasia

1 widespread species:

Tinomiscium petiolare

Notes:

1. Occurs in forests up to 1400m but mostly below 500 m altitude. The milky exudate is used against dental caries and to alleviate fever. Fruits are used as fish poison and also as a rat poison. Diels (1910) recognised 7 closely related species but these are considered by Forman (1986) as a single variable species.

Tinomiscium petiolare (A) (B) leaf; (C) - (F) male: inflorescences, flower, stamens; (G) - (L) female: flower, drupe, endocarps.
(Source: Foreman (1986)

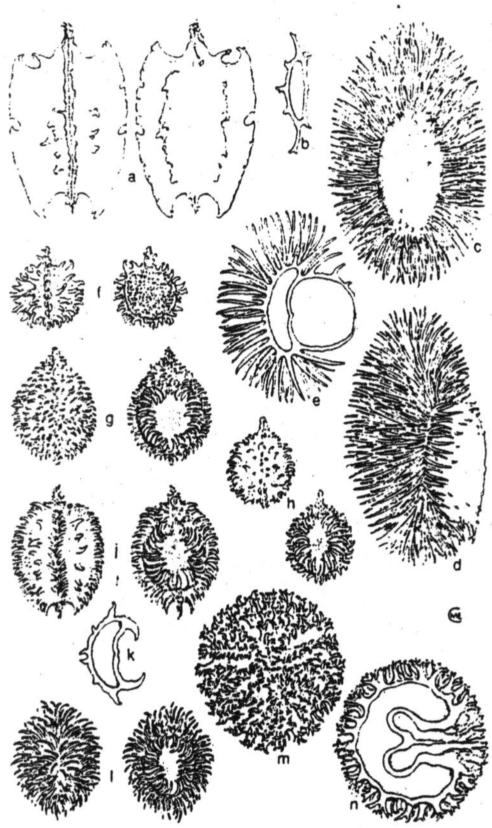

Endocarps of Tinosporeae (A) (B) *Aspidocarya uvifera*; (C) - (E) *Parabaena megalocarpa*; (F) *P. sagittata*; (G) *P. elmeri* (H) *P. denudata*; (J) (K) *P. tuberculata*; (M) (N) *Chlaenandra ovata*
(Source: Foreman (1986)

19. TINOSPORA

An Old World genus of some 32 species distributed from T Africa to Asia, Australia and the Pacific. In Papuasia 4 species.

Woody climbers. Petioles swollen at base, lamina cordate, 3-5 basal nerves and 1-3 pairs distal, lateral nerves, sometimes with domatia in basal, nerve-axils. inflorescences thyrsoid. Male flowers: sepals usually free, rarely joined at base, 6, outer 3 smaller, occasionally with 1 - 3 additional minute outer ones present; petals 6 (occasionally 3), lateral edges inrolled, fleshy; stamens 6, free. Female flowers: sepals and petals similar to male flowers; staminodes 6, subulate; carpels 3, stigma reflexed with short, pointed lobes. Drupes on arpophore, style scar terminal; endocarp bony; seed with endosperm, usually ruminate.

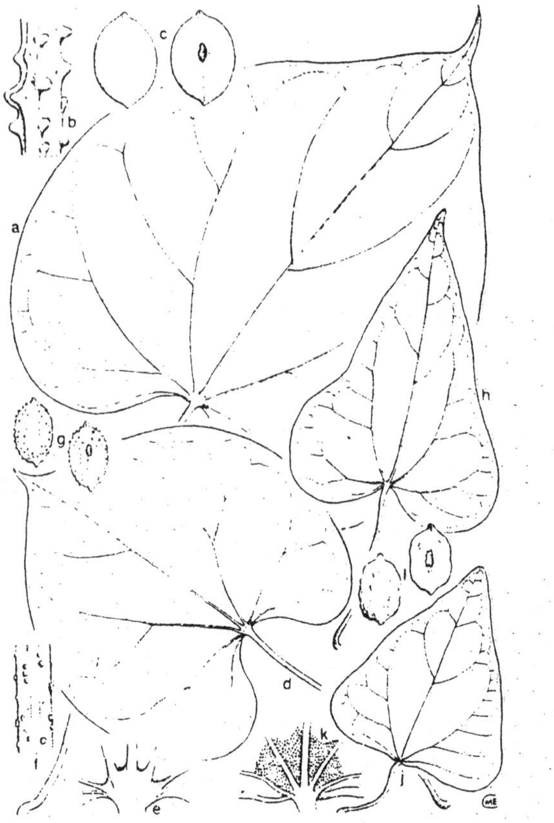

Tinospora crispa (A) - (C); T. baenzigeri (D) -(G); endocarp. T. smilacina (H) - (L). (Source: Foreman (1986)

Key to Species in Papuasia

(Source: Forman (1986: 190)

1. Leaves 18-28 by 11-23 cm, thinly coriaceous, lacking domatia or glandular patches in basal nerve-axils. Endocarp 1.7-2.4 cm long, surface very smooth, whitish. .... T. arfakiana

1. Leaves smaller and thinner, sometimes with domatia or glandular patches in basal nerve axils. Endocarps not as above.......... 2

2. Inflorescences, at least towards base, with lateral branches. .... T. dissitiflora

2. Inflorescences unbranched..... 3

3. Flower pedicels 8 - 12 mm. .... T. glabra

3. Flower pedicels 4 - 5 mm. .... T. subcordata

Notes:

1. Species occupy various forest types: T. dissitiflora grows in lowland swamp and rain forest below 300m; T. arfakiana in lower montane forest from 600 to 1500 m; T. glabra in littoral and mangrove vegetation but also in secondary growth and seasonal areas to 500 m altitude and T. subcordata in seasonally dry vegetation.

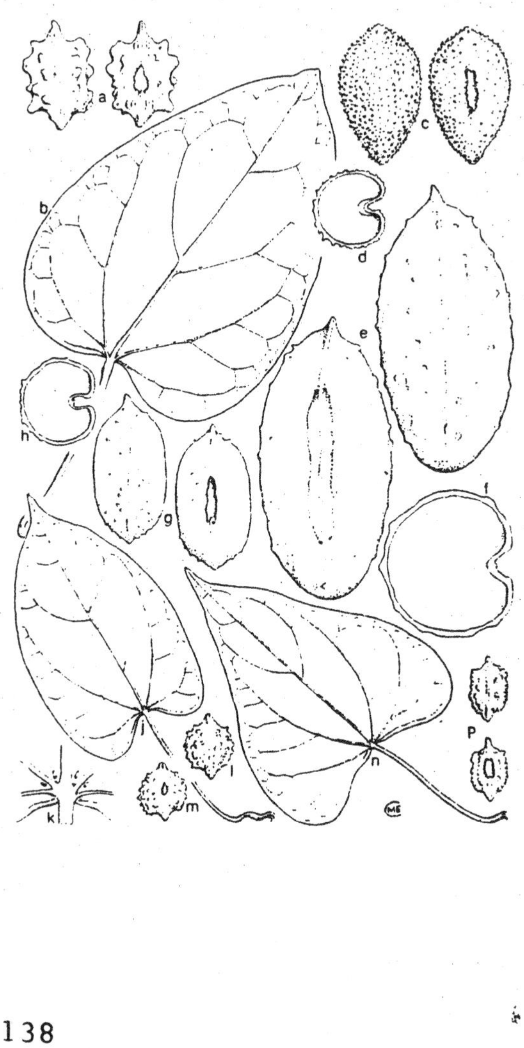

Tinospora dissitiflora (A) endocarp. T. tinosporoides (B) leaf; (C) - (F) endocarp. T. teijsmannii (G) (H) endocarp. T. glabra (J) leaf; (K) leaf base with domatia; (L) (M) endocarp. T. subcordata (N) leaf; (P) endocarp. (Source: Foreman (1986)

PAPAVERALES

PAPAVERACEAE

Distributed mainly in the north temperate zone. A family of herbaceous, latex producing, plants in 26 genera with some 250 species. The most important species is the opium poppy *Papaver somniferum*; many species are cultivated as ornamentals.

Most parts of the plant contain secretory canals which produce a yellow, milky or watery latex.

Diagnostic features

Leaves: alternate, exstipulate, entire but often lobed or deeply dissected.

Flowers: large, conspicuous, solitary or in a cymose or racemose inflorescence. Regular, bisexual, with two free sepals which fall off before the flower opens. Two whorls each of two free petals and several whorls of numerous stamens. Ovary superior with a single locule.

Fruit: a capsule, opening by valves or pores. Seeds with a small embryo and copious mealy or oily endosperm.

References

van Steenis, C.G.G.J. (1954) *Flora Malesiana* Series 1. 5: 114 - 117.

Several introduced species are cultivated.. None are native in Papuasia.

PAPAVERACEAE

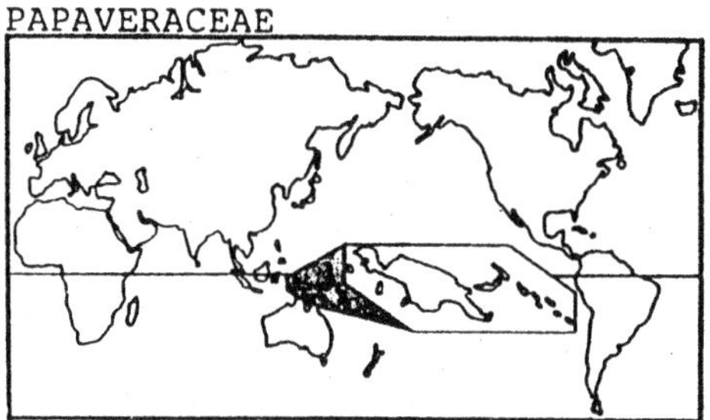

	WORLD	PAPUASIA
Number of genera:	26	(1)
Number of species:	250	(1)

Distribution: Mainly Northern Hemisphere.

Economic uses: Includes opium poppy. Many ornamentals.

INDEX TO FAMILIES AND GENERA

Actinodaphne	89
Albertisia	127
Alphonsea	20
Alseodaphne	97
Anamirta	128
Annona	21
Annonaceae	16
Arcangelisia	128
Aristolochia	98
Aristolochiaceae	98
Artabotrys	22
Ascarina	81
Beccariodendron	24
Beilschmiedia	89
Belliolum	14
Bubbia	14
Cananga	24
Carronia	128
Caryodaphnopsis	97
Cassytha	90
Ceratophyllum	109
Ceratophyllaceae	109
Chlaenandra	129
Chloranthus	82
Chloranthaceae	80
Cinnamomum	91
Cissampelos	130
Clematis	115
Cyathocalyx	25
Cyathostemma	26
Cryptocarya	92
Dasymaschalon	27
Dehaasia	93
Desmos	27
Diploclisia	130
Drimys	10
Dryadodaphne	69
Elmerrillia	8
Endiandra	93
Endocomia	44
Eupomatia	64
Eupomatiaceae	63
Faika	70
Fissistigma	27
Friesodielsia	28
Galbulimima	16
Goniothalamus	28
Gymnocranthera	45
Gyrocarpus	84
Hernandia	85
Hernandiaceae	83
Himantandraceae	15
Horsfieldia	45
Hydrostemma	111
Hypserpa	130
Illigera	86
Kairoa	71
Kibara	71
Knema	55
Lauraceae	87
Lauterbachia	74
Legnephora	131
Levieria	74
Limacia	132
Lindera	94
Litsea	95
Macrococculus	132
Magnolia	8
Magnoliaceae	7
Meigoyne	29
Melodorum	29
Menispermaceae	122
Michelia	9
Miliusia	30
Mitrella	30
Mitrephora	30
Monimiaceae	68
Myristica	56
Myristicaceae	42
Neolitsea	96

Nelumbo 112
Nepenthaceae 100
Nepenthes 101
Nothophoebe 97
Nymphaea 112

Nymphaeaceae 110
Oncodostigma 31
Oreomitra 32
Orophea 31, 32
Pachygone 132

Palmeria 76
Papaveraceae 139
Papualthia 33
Parabaena 133
Peperomia 108

Peperomiaceae 108
Pericampylus 133
Persea 97
Petalophus 34
Pheanthus 34

Phoebe 97
Piper 103
Piperaceae 102
Platymitra 35
Polyalthia 35

Polyaulax 36
Popowia 37
Pothomorphe 107
Pseuduvaria 31, 37
Pycnarrhena 134

Ranunculus 116
Ranunculaceae 114
Rauwenhoffia 39
Saccopetalum 30
Sarcandra 83

Sarcopetalum 134
Schefferomitra 39
Sphenostemon 67
Sphenostemonaceae 66
Steganthera 78

Stelechocarpus 40
Stephania 135
Thalitricum 121
Tinomiscium 137
Tinospora 138

Trimenia 65
Trimeniaceae 64
Uvaria 40
Wilkiea 80
Winteraceae 9

Xylopia 41
Zygogynum 14

INDEX TO FAMILIES

Acanthaceae VIII
Aizoaceae III
Alangiaceae VI
Alseuomiaceae IX
Amaranthaceae III
Anacardiaceae VII
Annonaceae **I**
Apocynaceae VIII
Aquifoliaceae VI
Araliaceae VII
Aristolochiaceae I
Asclepiadaceae VIII
Asteraceae
(see Compositae) IX
Avicenniaceae VIII
Balanophoraceae VI
Balsaminaceae VII
Basellaceae III
Batidaceae III
Begoniaceae IV
Bignoniaceae VIII
Bixaceae IV
Bombacaceae IV
Boraginaceae VIII
Burseraceae VIII
Byblidaceae VIII
Campanulaceae IX
Capparaceae IV
Caprifoliaceae IX
Cardiopteridaceae VI
Caricaceae IV
Caryophyllaceae III
Casuarinaceae II
Celastraceae VI
Ceratophyllaceae **I**
Chenopodiaceae III
Chloranthaceae I
Chrysobalanaceae V
Clethraceae IV
Cochlospermaceae IV
Combretaceae V
Compositae IX
Connaraceae VII
Convolvulaceae VIII
Coriariaceae VII
Cornaceae VI
Corynocarpaceae VI
Crassulaceae V
Cruciferae IV
Cucurbitaceae IV
Cunoniaceae V
Daphniphyllaceae VI
Datiscaceae IV
Dichapetalaceae VI
Dilleniaceae IV
Dipterocarpaceae IV
Droseraceae V
Ebenaceae IV
Elaeagnaceae VI
Elaeocarpaceae IV
Elatinaceae IV
Epacridaceae IV
Ericaceae IV
Erythroxylaceae VII
Euphorbiaceae VI
Eupomatiaceae I
Fabaceae
(see Leguminosae) V
Fagaceae II
Flacourtiaceae IV
Gentianaceae VIII
Geraniaceae VII
Gesneriaceae VIII
Gonostylaceae VI
Goodeniaceae IX
Guttiferae IV
Haloragaceae V
Hamamelidaceae II
Hernandiaceae I
Himantandraceae I
Hypericaceae IV
Icacinaceae VI
Juglandaceae VII
Labiatae
(see Lamiaceae) VIII
Lamiaceae VIII
Lauraceae I
Lecythidaceae IV
Leeaceae VI
Leguminosae V
Lentibulariaceae VIII
Linaceae VII
Lobeliaceae IX
Loganiaceae VIII
Lophopyxidaceae VI
Loranthaceae VI
Lythraceae V
Magnoliaceae I
Malphigiaceae VII
Malvaceae IV
Melastomataceae V
Meliaceae VII

Index to Families (contd.)

Menispermaceae	I
Monimiaceae	I
Moraceae	IV
Moringaceae	IV
Myoporaceae	VIII
Myricaceae	II
Myristicaceae	I
Myrtaceae	V
Nepenthaceae	I
Nyctaginaceae	III
Nymphaeaceae	I
Ochnaceae	IV
Olacaceae	VI
Oleaceae	VIII
Opiliaceae	VI
Onagraceae	V
Orobanchaceae	VIII
Oxalidaceae	VII
Pandaceae	VI
Papaveraceae	I
Passifloraceae	IV
Pentaphragmataceae	IX
Peperomiaceae	I
Phytolaccaceae	III
Piperaceae	I
Pitttosporaceae	V
Plantaginaceae	VIII
Plumbaginaceae	III
Podostemaceae	V
Polygalaceae	VII
Polygonaceae	III
Portulacaceae	III
Primulaceae	IV
Proteaceae	VI
Rafflesiaceae	VI
Ranunculaceae	I
Rhamnaceae	VI
Rhizophoraceae	V
Rosaceae	V
Rubiaceae	IX
Rutaceae	VII
Sabiaceae	VII
Santalaceae	VI
Sapindaceae	VII
Sapotaceae	IV
Saurauiaceae	IV
Saxifragaceae	V
Scropulariaceae	VIII
Simaroubaceae	VII
Solanaceae	VIII
Sonneratiaceae	V
Sphenostemonaceae	I
Stackhousiaceae	VI
Staphyleaceae?	VII
Sterculiaceae	IV
Stylidaceae	IX
Styracaceae	IV
Symplocaceae	IV
Theaceae	IV
Thymelaeaceae	V
Tiliaceae	IV
Trimeniaceae	I
Tropaeolaceae	V
Ulmaceae	IV
Umbelliferae	VII
Urticaceae	IV
Valerianaceae	IX
Verbenaceae	VIII
Violaceae	IV
Viscaceae	VI
Vitaceae	VI
Winteraceae	I
Zygophyllaceae	VII

THE FLOWERING PLANTS OF PAPUASIA

DICOTYLEDONS

PART 2
HAMAMELIDAE

R. J. Johns

FORESTRY DEPARTMENT

1988

P.N.G. UNIVERSITY OF TECHNOLOGY
LAE
PAPUA NEW GUINEA

Christensen Research Institute

Publication Number 1

REGIONS OF PAPUASIA

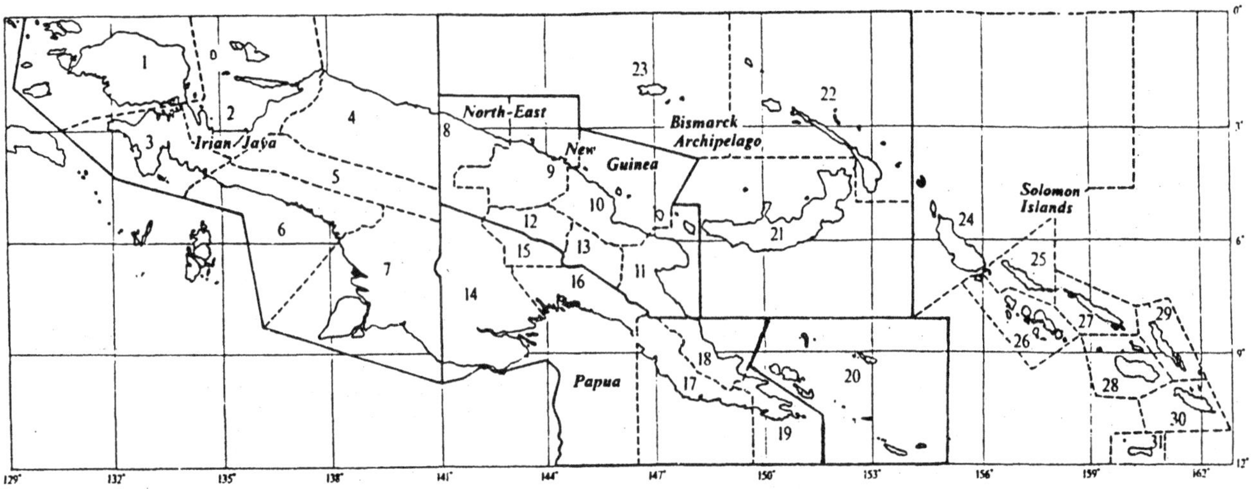

o o o o o o
I II III IV V VI

I. Irian Jaya

1. Vogelkop
2. Geelvink Bay
3. Fakfak
4. Jayapura
5. Snow Mountains
6. Mimika
7. Digul

II. North-east New Guinea

8. West Sepik
9. East Sepik
10. Madang
11. Morobe
12. Western Highlands
13. Eastern Highlands

III. Papua

14. Western
15. Southern Highlands
16. Gulf
17. Central
18. Northern
19. Milne Bay

IV. Papuan Islands

20. Papuan Islands

V. Bismarck Archipelago

21. New Britain
22. New Ireland
23. Manus

VI. Solomon Islands

24. Bougainville
25. Choiseul
26. New Georgia
27. Santa Isabel
28. Guadalcanal
29. Malaita
30. San Cristobal
31. Rennell Group

V

FLOWERING PLANTS OF PAPUASIA

PART TWO: THE HAMAMELIDAE

R. J. JOHNS

Johns, R. J. (Robert J.)

The flowering plants of Papuasia: Dicotyledons
9 volumes

1. Botany -- Papua New Guinea.
2. Botany -- Indonesia -- Irian Jaya.
3. Botany -- Solomon Islands.

I. Title. II. Papua New Guinea University of Technology.
Dept. of Forestry

583.0995
ISBN 9980-56-010-X

INTRODUCTION

The first volume of the Flowering Plants of Papuasia was published in 1987 and included the members of the subclass Magnoliidae. Unfortunately some of the families included were not in strict accordance with the system proposed by Cronquist in 1981 and some additional families were included. This will be corrected in a subsequent edition of Part 1.

As indicated in the introduction to Part 1 the authorities will be included for all species in this and subsequent volumes. In contrast to the first volume, dot distributions are given for all species. This should prove particularly useful for people working in the different regions of Papuasia but it should be noted that collecting is incomplete in many areas so additional records will come from any expedition.

ACKNOWLEDGEMENTS

I wish to express my sincere thanks to the Christensen Research Institute which has generously given a grant towards the production of the plates and the printing costs of the second part of the Flowering Plants of Papuasia. Mrs M. Popeu has typed the key to Ficus prepared by Mr E. E. Henty. Also included are translations by Dra C. M. H. Lapre of selected keys prepared to the species of the Urticaceae by Winkler (1922). These were included where not superseded by more recent revisions. I am most grateful for this assistance. I also wish to express my gratitude to Mr. T. Iwagu and Mrs. R. Bialostocki for their careful drawings of selected species to continue the objective of illustrating each genus included in these publications.

The Staff of the National Herbarium have been of assistance and I am grateful for access to their collections and Library. Considerable assistance has also been provided by the Staff of the Herbarium at the University of Papua New Guinea. I wish to express my gratitude to the staff of Herbarium Australiense in Canberra for access to the Card Index compiled by Dr. R. Hoogland. The Herbarium staff at Brisbane also generously provided a Check List of their collections from the Solomon Islands. These have prooved invaluable in compiling this volume.

This volume includes 10 families (including 1 introduced), 50 genera and around 490 species. Several large genera in the Urticaceae are in urgent need of revision. I would be grateful to receive any corrections, reprints of relevant articles, keys, or illusrations which could be included in revision editions of Parts 1 or 2, or in the preparation of subsequent parts.

R. J. Johns
Department of Forestry
P. N. G. University of Technology
P.O. Private Bag
LAE
PAPUA NEW GUINEA.

SUB-CLASS HAMAMELIDAE

The sub-class Hamamelidae consists of 11 orders, 24 families and some 3400 species. Most species belong to the Urticales (2/3) and a further quarter to the Fagales. The other orders are all small. The Hamamelidae has a long fossil history back into the Lower Cretaceous, more than 100 million years ago. Most members are woody or herbaceous dicotyledons, often with tannins. Most flowers are inconspicuous and small, with some exceptions in the Hamamelidaceae.

The families in Papuasia are:

		Genera	Species
	Casuarinaceae	3	14
	Cannabidaceae*	1	1
	Daphniphyllaceae	1	1
	Fagaceae	3	25
	Hamamelidaceae	1	1
	Juglandaceae	1	2
	Moraceae	14	188
	Myricaceae	1	1
	Ulmaceae	5	15
	Urticaceae	20	244
Totals	10	50	492

Cronquist (1981) includes the following illustration to show the putative evolutionary relationships amongst the orders of the Hamamelidae.

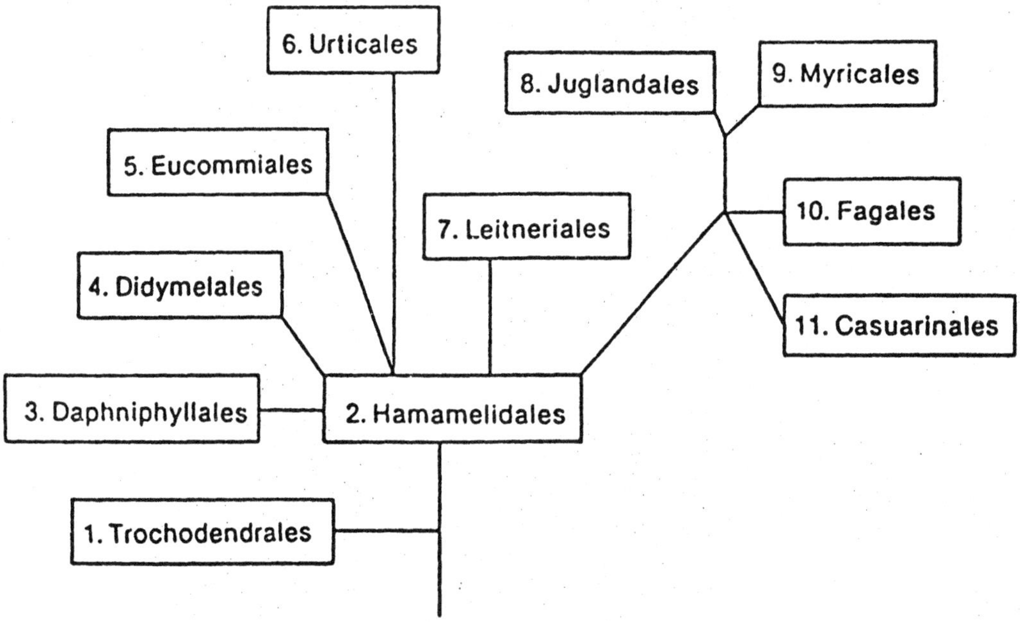

CLASSIFICATION OF THE HAMAMELIDAE

Orders and families underlined occur in Papuasia.

Trochodendrales
- Tetracentraceae
- Trochodendraceae

Hamamelidales
- Cercidiphyllaceae
- Eupteleaceae
- Platanaceae
- Hamamelidaceae
- Myrothamnaceae

Myricales
- Myricaceae

Didymelales
- Didymelaceae

Eucommiales
- Eucommiaceae

Urticales
- Barbeyaceae
- Urticaceae
- Cannabidaceae*
- Moraceae
- Cecropiaceae^
- Ulmaceae

Leitzneriales
- Leitzneriaceae

Juglandales
- Rhoipteleaceae
- Juglandaceae

Daphniphyllales
- Daphniphyllaceae

Fagales
- Dalanopaceae
- Fagaceae
- Betulaceae

Casuarinales
- Casuarinaceae

Alphabetical List of Families in Papuasia

Casuarinaceae
Cannabidaceae*
Daphniphyllaceae
Fagaceae
Hamamelidaceae
Juglandaceae
Moraceae
Myricaceae
Ulmaceae
Urticaceae
(including Cecropiaceae)

* Families Introduced into Papuasia.

^ The family Cecropiaceae, described by Berg in 1978 (Taxon 27: 39-44) included six genera. It includes Poikilospermum, here treated in the Urticaceae, which is the only genus occuring in Papuasia.

HAMAMELIDALES

HAMAMELIDACEAE

● ● ● ○ ● ○

This family has a very discontinuous distribution in both subtropical and temperate zones. The family includes 23 genera and some 100 species. A revision of the family in Malesia was published in 1951. A single genus Sycopsis, occurs in New Guinea.

The family includes many isolated, perhaps relict, genera. Liquidamber and Altingia are sometimes placed in a separate family, Altingiaceae. von Mueller (1877) recorded Liquidamber altingia Blume from Papua. No subsequent collections of this genus have been made in Papuasia, and the record by von Mueller is probably a misidentification or the locality was incorrectly recorded for Papua.

HAMAMELIDACEAE

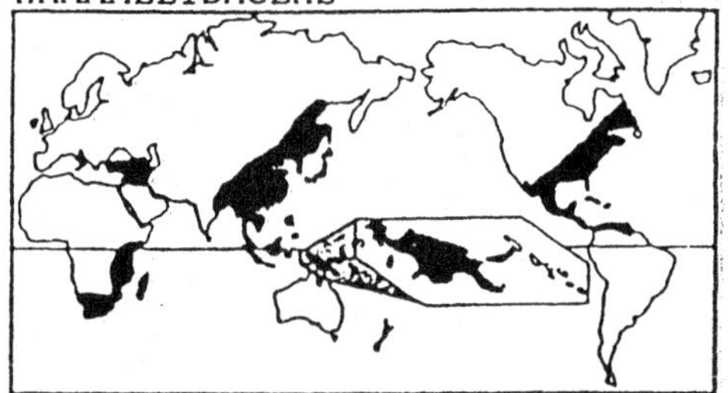

	WORLD	PAPUASIA
Number of genera:	23	1
Number of species:	100	1

Distribution: Discontinuous in tropics and subtropics.

Economic uses: Timber, gum, perfumes, medicines.

Diagnostic features

Leaves: simple, alternate, entire, stipules usually present.

Flowers: unisexual, calyx and petals lacking (in N.G.): male flowers: 4-10 free stamens, anthers 2-celled, connective often extended; female flowers: semi-inferior ovary, composed of 2 carpels each containing 2 locules; styles 2, free.

Fruit: 4 valves; 1 seed per locule.

References

Croft, J.R. (1981). Handbooks of the Flora of Papua New Guinea. 2: 186 - 189.

Endress, P. K. (1970). Die Infloreszenzen der apetalen Hamamelidaceen. Bot. Jahrb. 90: 1 - 54.

Hutchinson, J. (1967). The Genera of Flowering Plants 2: 93 - 103.

von Mueller, F. (1877). Descriptive Notes on Papuan Plants V: 99.

van Steenis, C.G.G.J. (1972). Addenda. Flora Malesiana Series I. 6: 952.

Vink, W. (1951). Hamamelidaceae. Flora Malesiana Series I. 5: 363 - 379.

Walker, E. H. (1944). A revision of Distylium and Sycopsis. J. Arn. Arb. 25: 319 - 341.

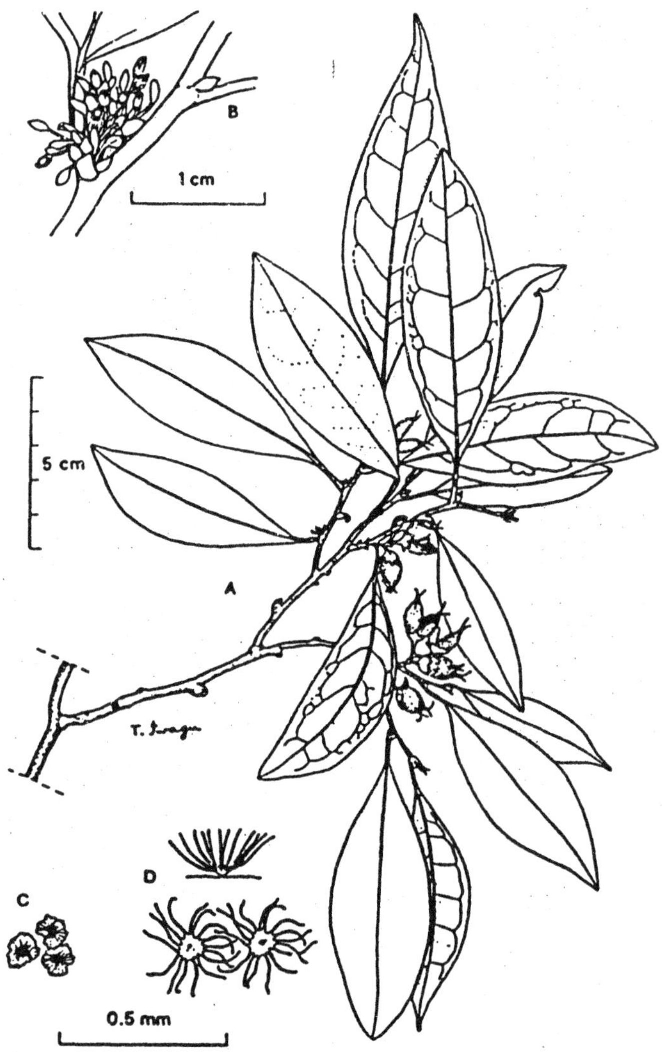

Sycopsis dunnii (A) twig with dehiscing fruit; (B) male inflorescence; (C) (D) scales of glabrescent and tomentose forms.

Notes:

1. Endress (1970) has segregated a new genus *Distyliopsis* from *Sycopsis*. As described, this would include *S. dunnii*. van Steenis (1972) however does not accept *Distyliopsis* as a distinct genus.

KEY TO GENERA IN PAPUASIA

1 widespread genus:

Sycopsis

1. SYCOPSIS

● ● ● ○ ● ○

It is a genus of 7 species distributed from Assam and China to New Guinea and New Britain. The genus *Sycopsis* is represented by a single species, *S. dunnii* Hemsley, in Papuasia. This species, a small tree, occurs from Malaya and the Philippines, throughout mainland New Guinea, to New Britain.

Shrubs or trees to 60 cm d.b.h. Leaves coriaceous, apex acute; tufted hairy domatia present in most vein axils. Petiole to 4-10(-15) mm long, persistently lepidote. Stipules caducous. Inflorescences 3-6 flowered, to 25 mm long, female flowers generally apical, male flowers basal. Subtending bracts densely stellate-tomentose. Receptacle 2-3 mm diameter. Filaments 2.5-4 mm long. Ovary ovoid to conical. fruit ellipsoid to ovoid, 11-13 x 6-8 mm. Persistent style. Opening by 2 valves. seeds 6-8 mm long.

Key to Species in Papuasia

1 widespread species:

Sycopsis dunnii Hemsley in Hook. *Icon. Pl.* 29 (1907) t. 2836.
Distyliopsis dunnii (Hemsley) Endress

Notes:

1. The species occurs from Malaya and the Philippines throughout New Guinea to New Britain. It is locally common in New Guinea from 300 to 2400m. In the Highlands of New Guinea it is a common subcanopy tree in lower montane forest dominated by *Araucaria* and *Castanopsis*, and along *Nothofagus* dominated ridges in the mid montane forests. In Woitape it is common in *Lithocarpus* forest. On New Britain (Fullerborn Harbour) it occurs at 900m in *Castanopsis* forest. It also grows in secondary regrowth.

2. The timber is useful because of its durability. It is used for light construction, panelling and furniture. Traditionally, in the Okapa area, the bark and leaves were used as medicine. Near Minj it has been collected, as an 8 - 9 m tree, in Coffee Gardens. It was possibly planted for use as a medicine.

MYRICALES

MYRICACEAE

● ○ ○ ○ ○ ○

The Myricaceae is a small family of aromatic trees and shrubs which is almost cosmopolitan in distribution. The family is represented by 2 genera (some authorities accept one or three) and some 35 species. One genus from New Caledonia (*Canacomyrica*) differs considerably in floral structure. Apart from this genus, floral structure in the family is relatively consistent.

A group of trees or shrubs, often dotted with yellow glands. Strongly aromatic. A genus of gregarious, heliophilous species, often of volcanic debris. Mainly in mountainous regions.

Diagnostic features

Leaves: alternate, simple or pinnatifid, with or without stipules.

MYRICACEAE

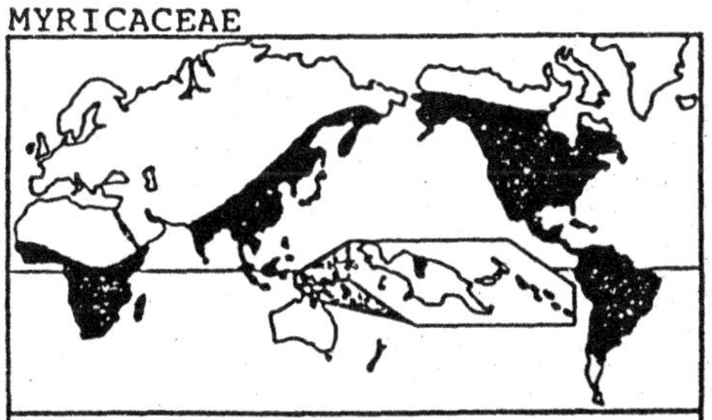

	WORLD	PAPUASIA
Number of genera:	2	1
Number of species:	35	1

Distribution: Almost cosmopolitan.

Economic uses: Fruits used in wax production. Charcoal.

Flowers: unisexual on axillary catkin-like spikes, both on same plant. Male flowers: two bracteoles and disk at base of stamens which are (2-)4(-10), but usually 4. Filaments may be joined at base. Female flowers: 2-4 bracteoles and no disk; ovary superior, style with two short stigma branches.

Fruit: small, rough, waxy drupe with hard endocarp. Seed no endosperm and a straight embryo.

References

Backer, C. A. (1951). Myricaceae. Flora Malesiana. Series I. 4: 276 - 279.

KEY TO GENERA IN PAPUASIA

1 genus:

Myrica

1. MYRICA

● o o o o o

In Papua New Guinea the family is represented by a single species, Myrica javanica, a widely distributed Malaysian species that grows from 900-3300m in altitude.

Small trees or shrubs from 2-10 m tall. Crowns dense. Densely covered with yellow sessile glands when young. Leaves elliptic or ovate, apex rounded. To coarsely serrate, hairless, when young densely covered with yellow glands. Male and female flowers separate, exceptionally a few female flowers among the male ones. Male inflorescences solitary in leaf axils, 4-18 cm long; catkins to 3 cm long; solitary in axils of bracts, bracts sessile, to 2-3 mm long. Stamens 4, rarely 3. Anthers short stalked. Female inflorescences solitary in leaf axils, 3-7 cm long, catkins solitary. Flowers 5 or more, each in axis of a hairy bract. Stigmas 2, sessile. Berries 1 or rarely 2 per catkin, black with bluish violet juice. Acidic, usually not palatable.

Key to Species in Papuasia

1 species in Irian Jaya:

Myrica javanica Blume
Bijdr. (1825) 517.

Myrica javanica (A) flowering twig; (B)-(D) flowers.
(Source: van Steenis (1951)

Notes:

1. It grows in open, sunny, stony localities, often near active volcanoes on ridges and lava streams. Forms a pioneer vegetation, often gregarious or mixed with other shrubs and small trees amongst which it may predominate. The wood produces excellant charcoal. Myrica esculanta Buch-Ham, a second species distriuted from S E Asia to the Sunda Is. and the Philippines, has edible berries.

2. M. javanica is used in Malesia for reforestation.

3. In the National Herbarium at Lae there are 2 collections from Jayapura (Hollandia) at 1200 m altitude. In this area it is a locally common tree to 20 m tall. Collections could be expected in the mountains behind Vanimo.

4. L. J. Brass (Brass 10938, / 10 / 1938), collected a small leaved species of Myrica with young, female flowers. Backer (1951) states it is clearly allied to M. javanica, and may be 'one of the numerous forms of this very polymorphous plant'.

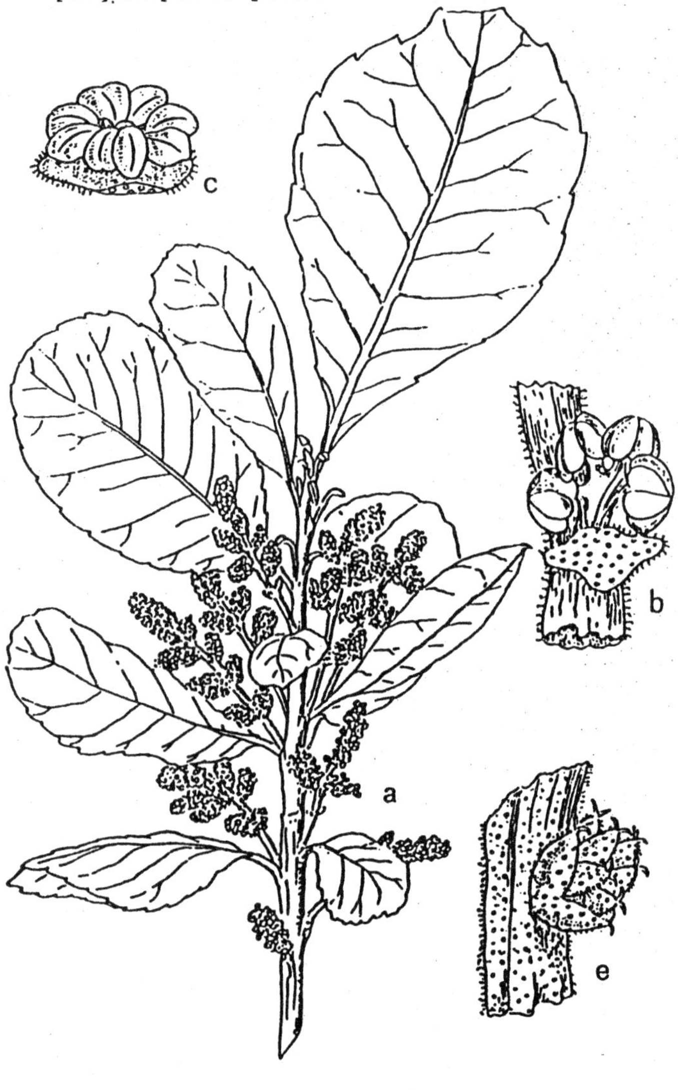

URTICALES

A group of closely related families; woody but also herbaceous; accompanied by a great reduction in floral structure. Leaves mostly simple, usually stipulate. Cronquist (1983) includes six families and about 2200 species. The families are Ulmaceae, Moraceae, Urticaceae, Moraceae, Cannabaceae and Cecropiaceae. The Cannabaceae have often been included in the Moraceae but recently the genera have been treated as a distinct family. Cannabis sativa has become naturalised in some areas in New Guinea.

The relationship between the Moraceae and Urticaceae has also been the subject of much discussion. According to Chew (1963) there is only one distinguishing character between the two families Urticaceae and Moraceae. In the Urticaceae the ovules are always basal and orthotrophus, while in the Moraceae they are apically or laterally fixed and, except for Cecropia never orthrotrophus.

Key to Families in Urticales
(Source: Cronquist (1983)

1. Leaves exstipulate, opposite; nodes unilacunar; gynoecium of 1-3 distinct or partly connate pistils, each with a separate style. .... Barbetaceae*

1. Leaves usually stipulate, variously opposite or alternate; nodes trilacunar or pentalacuar; gynoecium of a single (rarely bilocular) pistil.............. 2

2. Ovary with 2 styles (1 sometimes reduced); ovule apical, pendulous, anatropous..................... 3

2. Ovary with a single style, pseudo-monomerous; ovule basal or nearly so, errect, more or less orthotropous....................... 5

3. Plants without lacticifers, and without milky juice............ 4

3. Plants nearly always with lactici-fers and milky juice. .... Moraceae

4. Woody Plants. Ulmaceae

4. Herbs or herbaceous vines. .... Cannabaceae+

5. Stamens more or less erect in bud, not elastically reflexed in dehiscence; woody plant, cyctoliths wanting....... Cecropiaceae*

5. Stamens inflexed in bud, elastically reflexed in dehiscence; herbs or occasionally half-shrubs, or rarely soft-wooded trees; cystoliths generally present. .... Urticaceae

* Cecropiaceae: represented in Papuasia by Poikilospermum which is treated in The National Herbarium at Lae in the Urticaceae. This is used here.
\+ Introduced to Papuasia.

URTICACEAE

● ● ● ● ● ●

A family of about 52 genera and 1,050 species, mostly distributed in the tropics and temperate (few) regions. Most genera are herbs or small shrubs although the tribe Conocephaleae includes mainly trees in the genera Cecropia and Poikilospermum. The tribe Urticeae is characterised by plants with stinging hairs. It includes Laportea, Urtica, Dendrocnide, and Urera. The stinging hairs inject histimine which causes itching and acetylcholine which gices a burning sensation even from dried plants.

Fibre from some species is used in the textile industry Boehmeria nivea, ramine fibre or China grass. Leaves of nettles are, traditionally, widely used in New Guinea to treat headaches and muscular pains.

URTICACEAE

	WORLD	PAPUASIA
Number of genera:	45	20
Number of species:	1000	244
Distribution:	Tropical and temperate	
Economic uses:	Fiber and weeds	

Diagnostic features

Leaves: alternate or opposite, with stipules; the epidermis is usually marked with prominent cystoliths.

Flowers: inflorescence cymose but often condensed into heads. Flowers greenish, regular, unisexual (flowers on same plants) or rarely bisexual, usually with 4-5 sepal-like lobes. Male flowers with 4 or 5 stamens usually bent downwards and inwards in bud, exploding when ripe. Note that members of Conocephaleae have straight stamens. Female flowers have a superior ovary with one locule containing a single erect basal ovary. There is a single style.

Fruits: usually dry, an achene. Seed rich in oily endosperm, embryo straight.

References

Berg, C. E. (1978). Taxon 27: 39 - 44.

Chew, W.-L. (1963). Flora Malesiana Precursores - XXXIV. A revision of the genus Poikilospermum. Gard. Bull. Sing. 20: 1 - 104.

Chew, W. -L. (1965). Laportea and allied genera. Gard. Bull. Sing. 21(2): 195 - 208.

Chew, W. -L. (1969). Nothocnide in Malesia. Gard. Bull. Sing. 24: 361 - 373.

Chew, W. -L. (1969). A monograph of Dendrocnide. Gard. Bull. Sing. 25: 1 - 104.

Chew, W. -L. (1969). A monograph of Laportea. Gard. Bull. Sing. 25: 111 - 178.

Diels, in White, C. T. (1929). Journ. Arn. Arb.

Engler, A. and Prantl, K. (1889). Die Naturlichen Pflanzenfamilien. 3 (1): 112.

Friis, I. (1981). A synopsis of Girardinia. Kew Bull. 36: 143 - 157.

Hutchinson, J. (1967). The genera of the flowering plants 2: 178 - 195.

Merrill, E. D. (1934). An enumeration of Plants collected in Sumatra by W.N. and C. M.Bangham. Contr. Arn. Arb. 8: 47 - 52.

Perry, L. M. (1951). Plantae Papuanae Archboldianae XX. Journ. Arn. Arb. 32: 369 - 389.

Ridley, (1916). Trans. Linn. Soc. London.-Bot. 2.9: 155 - 157.

van Royen, P. (1982). The Alpine Flora of New Guinea. 3: 2081 - 2156.

Schroter, H. & Winkler, H. (1936). Rep. Sp. Nov. Beih. 83: (1,2).

Schroter, H. (1939). Rep. Sp. Nov. 47: 221 - 222.

Schuman, K. and Lauterbach, C.A.G. (1905). Fl. Deutsch. Schutzgeb. Sudsee Nachtr. 255.

Unruh, H. (1943). Monograph der Gattung Leucosyke. Bot. Jahrb. 43: 190 - 258.

Warburg, O. (1891). Beitrage zur Kenntnis der papuanischen flora. Bot. Jahrb. 13: 230 - 455.

Weddell, H. A. (1856). Monographie de la famille des Urticacees. Paris.

Weddell, H. A. (1869). in A. De Candolle Prodromus systematis naturalis regni vegetabilis 16(1). Paris.

Winkler, H. (1922). Die Urticaceen Papuasiens. Bot. Jahrb. 57: 501 - 608.

Winkler, H. (1924). Nova Guinea Bot. 14: 121 - 128.

Zandee, M. (1969). A new species of Urtica from the Papuan Highlands and an emended distribution of Urtica urens and U. diocia. Blumea 17: 444 - 446.

KEY TO GENERA IN PAPUASIA

1. Hemi - epiphytes with aerial roots. Stamens straight in bud, sometimes inflexed but not bending outwards elastically.
.... 16. Poikilospermum

1. Self supporting plants without aerial roots. Stamens inflexed in bud and bending outwards elastically.
.... 2

2. Herbs, shrubs or rarely trees, armed with stinging hairs. Perianth of female flowers 4-partite or 4-lobed. Leaves alternate or opposite........................ 3

2. Herbs, or rarely shrubs, without stinging hairs.................. 6

3. Achenes straight. Stipules lateral, free. Stigma brush-like. Annual or perennial herbs. Leaves opposite.
.... 19. Urtica

3. Achenes oblique. Stipules ± connate into 1 (but see Laportea). Stigma not brush-like. Herbs, shrubs or trees. Leaves alternate........ 4

4. Perianth segments of female membranous or fleshy, as long as or shorter than fruits............. 5

4. Perianth segments of female united into a ± tube, 2-3-toothed, split to base on 1 side. Tall herbs sometimes prickly.
.... 7. Girardinia*

5. Annual herbs with chartaceous, dentate leaves. Stipules not wholly connate and hence bifid at apex. Female flowers usually borne on winged pedicels.
.... 8. Laportea

5 Perennial shrubs or trees with leathery undulate - margined leaves. Stipules wholly connate. Pedicels not winged.
.... 4. Dendrocnide

6. Stigma brush - like. Perianth of female flower mostly 3-partite. Leaves opposite or alternate, in the latter case very unequal sided at the base.................... 7

6. Stigma not brush-like. Perianth of female flower mostly tubular, rarely absent.................. 10

7. Leaves alternate, unequal - sided, the upper side mostly smaller, often with smaller leaf opposite.
.... 8

7. Leaves opposite, equal or very unequal in pairs, equal-sided. Perianth of female 3-partite with one segment larger.
.... 15. Pilea

8. Female flowers cymose or ± head-like...... 13. Pellionia

8. Female flowers arranged on a fleshy, receptacle................ 9

9. Female flowers on a flattened disc-shaped or lobed receptacle.
.... 5. Elatostema

9. Female flowers head - like on a ± globular, fleshy, receptacle.
.... 18. Procris

10. Perianth of female flowers free or united with ovary, rarely absent. Leaves spirally arranged or opposite.......................... 11

10. Perianth of female flowers always present and free from ovary. Leaves alternate. 12. Parietaria

11. Perianth of female flower well developed........................ 12

11. Perianth of female flower minute or completely reduced.............. 16

12. Perianth of female flower membran-membranous or dry, enclosing the free achene.................... 13

12. Perianth of the female more or less fleshy, enclosing the free achene.
.... 14

12. Perianth of the female closely adnate to achene................ 15

13. Stigma persistent on fruit.
.... 1. Boehmeria

13. Stigma deciduous.
.... 17. Pouzolzia

14. Leaves opposite, often bullate-rugose. Stigma persistent.
.... 2. Cypholophus

14. Leaves alternate or spiral. Small shrubby tree or woody climbers.
.... 15

15. Small shrubby trees.
.... 14. Pipturus

15. Woody climbers.
.... 11. Nothocnide

16. Fruiting perianth thinly fleshy. Stigmas thinly discoid, pilose along margin. Leaves hairy on nerves beneath, not tomentose.
.... 20. Villebrunea

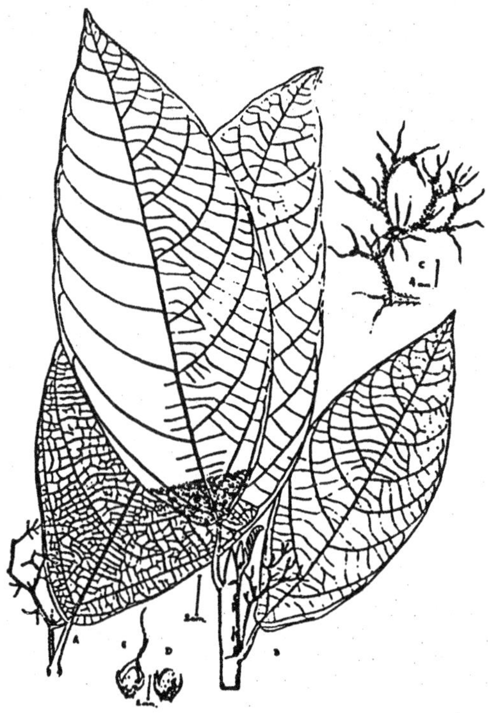

Dendrocnide latifolia A. & B. twigs; C. female inflorescence part; D. female perianth; E. female flower. (Source: Chew (19969).

16. Stigma penicillate. Fruiting perianth succulent. Leaves grey tomentose beneath.
.... 3. Debregeasia

17. Perianth of female flower minute or cup-shaped...................... 18

17. Perianth of female flower completely reduced.
.... 10. Maoutia

18. Stipules 2-lobed or almost comp- completely united. Flowers mostly dioecious, globose - glomerate or capitate.
.... 9. Leucosyke

18. Stipules deeply 2-lobed. Flowers monoecious in small androgynous cymules... 6. Gibbsia

*Notes:

*1. Girardinia has not yet been collected in Papuasia but is known from neighbouring areas in Indonesia. In Girardinia the first formed flower in a cyme is female, followed by bisexual flowers and the last formed flowers are male. However, in the bisexual flowers, the stigma is receptive long before the anthers mature, and by this time the style has often fallen off giving the appearance of male flowers.

1. BOEHMERIA

o ● ● ● ● ●

A genus of some 50 species of tropics and subtropics, without stinging hairs. B. nivea is widely cultivated for its long, silky fibres. However the stems are difficult to decorticate. Of the species that have been described from Papuasia, three species have been transferred to Cyclophorus and one species to Pipturus. Boehmeria platyphylla is represented by two varieties in Papuasia. Traditionally the plant was used for binding and as an ornamental plant.

Small trees or shrubs to herbs, lacking stinging hairs. Leaves, dentate, 3-nerved; stipules free or rarely connate at base within the petiole. Flowers nearly always unisexual, axillary or at nodes or sessile along rachis of spikes or cymes; bracts small; plants monoecious or dioecious. Male flowers: calyx 4-lobed or partite (3- or 5-), lobes valvate; stamens 4 (5 or 3); ovary rudimentary. Female flowers: calyx tubular, in fruit sometimes enlarged into 2 acute angles or wings; ovary sessile or stipitate; stigma elongate, pilose on one side, persistent for some time; ovule basal, erect. Achene enclosed by the calyx.

Key to Species in Papuasia

1 species with 2 varieties
Doehmeria platyphylla D.Don Prodr. Fl. Nepal (1825) 60.
o o ● o o ●
var. celebica Wedd. in D.C. (1869: 211)
o o o o o ●
var. moluccana Wedd. in D.C. (1869: 211)
o o o o o ●

Notes:

1. var. celebica is recorded from Papuasia in K. Sch. (1891: 198), K. Sch. & Holl. (1889: 39) and K. Sch. & Laut. (1900: 239).

2. var. moluccana was recorded from the Solomon Is. in Trop. Woods 81 (1945) 29.

3. The following species described in Boehmeria are synonyms:
Pipturus albidus (Hook. ex Am.) Wedd. B. albida Hook. ex Am.
Cypholophus latifolius (Bl.) Wedd. B. latifolia Bl.
C. melanocarpus (Bl.) Miq. B. melanocarpa Bl.
C. prostratus (Bl.) Miq. B. prostrata Bl.
C. vestitus (Bl.) Miq. B. vestita Bl.

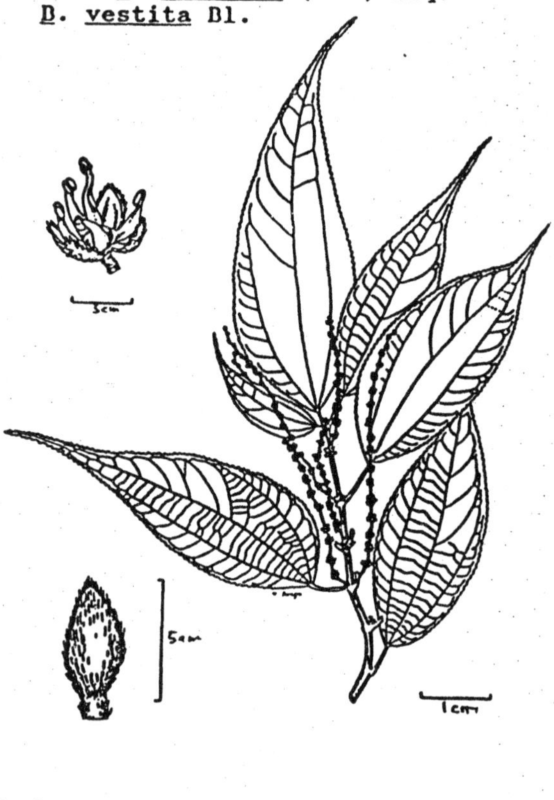

Doehmeria platyphylla var. moluccana
(BSIP 9919)
(Drawn: T. Iwagu)

CONCEPAHALUS

The five species included in this genus by H. Winkler (1922) have all been transfered to Poikilospermum by Chew in 1963.

2. CYPHOLOPHUS

o ● ● ● ● ●

A genus of around 30-35 species from Java, Borneo and the Philippines to Fiji and Samoa. In Papuasia about 27 species of which two occur above 3,000 m altitude (van Royen, 1982). The bark fibres of some species are used to make mats.

Erect shrubs to small trees, sometimes with prostrate branches, no stinging hairs. Stipules 2. Leaves opposite, decussate, often same pair very unequal in size, or pair large leaves alternating with small leaves, margins serrate-dentate. Petioles very unequal in each pair.

Inflorescence small, axillary, many flowered, sometimes clasping the branchlets. Flowers unisexual, plants monoecious or dioecious. Bracts and bracteoles present. Male perianth 4-lobed, valvate; stamens 4, abortive ovary covered with hairs at base. Female perianth tubular, 2- or unequally 4-toothed. No staminodes. Stigma strongly recurved, pilose ir plumose. Nutlets elliptic or obovoid.

List of Species in Papuasia

Cypholophus brassii Diels (1929: 75)

o ● ● o o o

Cypholophus chaemaephyton H. Winkler (1922:582)

● ● ● o o o

Cypholophus decipiens H. Winkler (1922: 573).

var. tenuifolius H. Winkler (1922: 574)

o ● ● ● o ●

Cypholophus englerianus H. Winkler (1922: 578)

● o o o o o

Cypholophus friesianus (K. Sch.) H. Winkler (1922: 548). Pilea friesianus K. Sch. (1905)

o ● o o o o

Cypholophus gjellerupii H. Winkler (1922: 580)

● o o o o o

Cypholophus heterophyllous Wedd. in D.C. Prod. 16.1(185) 235. The New Guinea material is probably C. decipiens

o o o o o o

Cypholophus integer H. Winkler (1922: 584)

o ● o o o o

Cypholophus kerewensis van Royen (1982: 2137)

o o ● o o o

Cypholophus latifolius (Bl.) Wedd. Boehmeria latifolia Bl. Mus. Bot. 2 (1856) 207

o ● ● o o o

Cypholophus ledermannii H. Winkler (1922: 581)

o ● o o o o

Cypholophus melanocarpus (Bl.) Miq. Boehmeria melanocarpa Bl. Mus. Bot. 2 (1856) 210.

● o o o o o

Cypholophus melanocarpoides H. Winkler (1922: 581)

o o o o o o

Cypholophus Bl. H. Winkler lists this as a synonym of C. warburgianus

Cypholophus montanus Ridley (1916: 157)

● o o o o o

Cypholophus nummularis H. Winkler (1922: 582)

o ● o o o o

Cypholophus pachycarpus H. Winkler (1922: 574)

o ● o o o o

Cypholophus patens H. Winkler (1922: 577)

Cypholophus prostratus (Bl.) Miq. Fl. India 1.2 (1859) 243 Boehmeria prostrata Bl. Mus. Bot. 2 (1856) 209.

● o o o o o

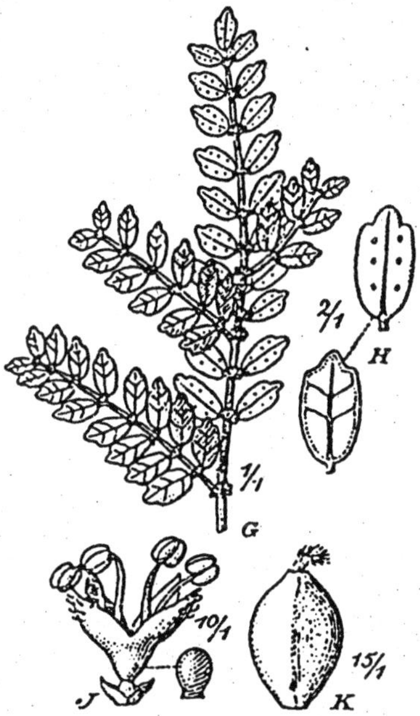

Cypholophus vaccinioides G. stem; H. leaf from above and below; J. male flower with rudimentary ovary; K. fruit. (Source: Winkler (1922)

Cypholophus pulleanus H. Winkler (1922: 584)

● o o o o o o

Cypholophus radicans H. Winkler (1922: 576)

● o o o o o o

Cypholophus reticulatus H. Winkler (1922: 571)

o ● o o o o o

Cypholophus rudis Ridley (1916: 157)

o o o o o o o

Cypholophus trapula H. Winkler (1922: 579)

o ● o o o ●

var. hirsutus (1922: 580)

Cypholophus treubii H. Winkler (1922: 583)

● o o o o o

Cypholophus vaccinioides H. Winkler (1922: 583)

● o o o o o

Cypholophus velutinus H. Winkler (1922: 578)

o ● o o o o

Cypholophus vestitus (Bl.) Miq. Fl. Ind. Bat. 1.2 (1859) 263
Boehmeria vestita Bl. Mus. Bot. 2 (1856) 209

● ● o o o o

Cypholophus warburgianus Laut. in K. Sch. & Laut. (1905: 255)

● ● o o o o

Key to high altitude Species in Papuasia
(Source: van Royen (1982)

1. Larger leaves 0.7-5.0 by 0.6-3.0 cm, limb ovate to elliptic. Petiole to 3-4.5 cm long.
.... C. kerewensis
van Royen (1982: 2137)

1. Larger leaves 2.5-2.8 by 0.7-1.7 cm, limb ovate lanceolate. Petiole to 2.5 mm long.
.... C. montanus
Ridley (1916: 157)

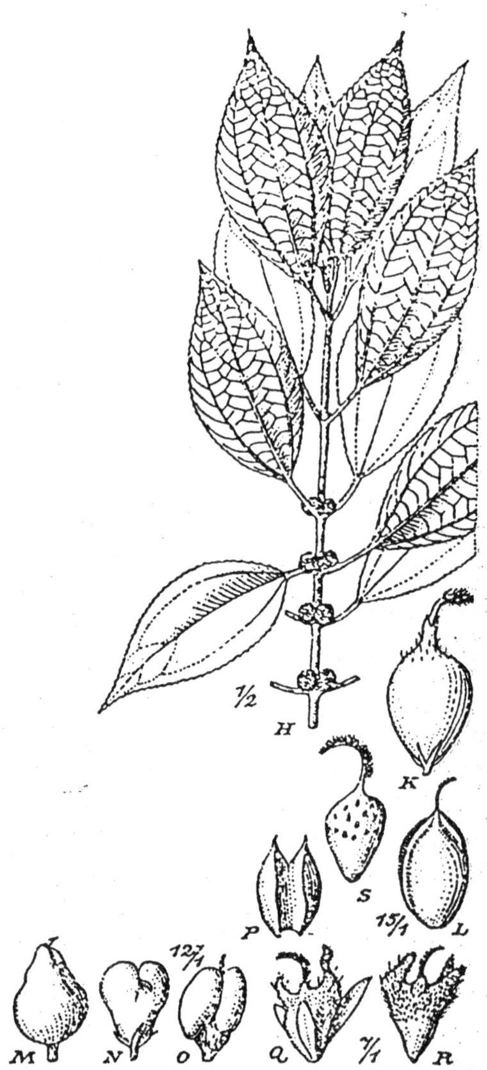

Cypholophus pachycarpus H. stem; J. leaf with cystoliths; K. female flower; L. ovary; M.& N.& O. fruit in frontal, lateral and dorsal view; P. tepal on ripe fruit. (Source: Winkler (1922)

Notes:

1. Both high altitude species are endemic on high mountains in Papuasia.

Key to Species in Papuasia
(Source: Winkler (1922)
(Translated by Dra. C. M. H. Lapre)

1. Leaf venation pinnate. The female perigon with veins the full length.
Subgenus STICHOCEPHALUS
.... 2

1. Leaves with 3-ribs (in C. pulleanus not clearly visible; nearly pinnate); the female perigon without veins.
Subgenus EUCYPHOLOPHUS
.... 3

2. Leaf margins serrate.
..... Cypholophus friesianus

2. Leaf margins only serrate at the top.. Cypholophus integer

3. Halfshrub or shrub with normal leaves. (Section Foliosae)
.... 4

3. High mountain shrubs or shrubby plants with vaccinoide leaves. (Section Vaccinoides).......... 19

4. The pairs of leaves are the same in size and shape................ 5

4. The pairs of leaves clearly different in shape and size. The smallest leaf cordate; longer than the petiole of the longest leaf, but only about 1/3 of the longer leaf.... 14

4. Both leaves of every pair very different, the smallest leaf will not exceed or just slightly exceed the length of the petiole of the larger leaf (but with C. vestitus clearly longer, but then not cordate)........................... 15

5. Leaves large to very large, cordate with a 5 cm long tip at the midrib. 6

5. Leaves small at most 5 cm long, apex obtuse or slightly mucronate. 13

6. Leaves round (orbicular), egg shaped, obovate-oblanceolate or longish but not lanceolate............. 7

6. Leaves lanceolate............... 12

7. Leaves nearly circular (rounded), apex aristate.
..... Cypholophus rotundifolius

7. Leaves not fully rounded........ 8

8. The young stalk, petiole and the underside of the leaves with stiff, flat laying hairs............... 9

8. Young stalk, petiole and underside of the leaves are shaggy........ 11

9. Leaves with equal sided sides except at the top slightly unequally formed; regularly longish.
..... Cypholophus pachycarpus

9. Leaves oblique, egg-shaped, oblong to oval....................... 10

10. Leaves broad egg-shaped, often clearly oblanceolate.
..... Cypholophus latifolius

10. Leaves obovate or eliptic, sometimes slightly falcate.
..... Cypholophus warburgianus

10. Leaves egg-shaped, longish to oblong.
..... Cypholophus decipiens

11. Leaves longish at the base, acute.
..... Cypholophus radicans

11. Leaves egg-shaped at the base, cordate.
..... Cypholophus rudis

12. The width of the leaf is optimal at the middle or at 1/3 from the base; gradually smaller and rounded at the tip, clearly dentate margins.
..... Cypholophus melanocarpus

12. The width of the leaf is optimal at the top 1/3 of the leaf; not clearly dentate.
..... Cypholophus patens

Cypholophus friesianus A. stem; B. leaf with cystoliths; C. fruit. (Source: Winkler (1922)

13. Stalk with rough bristle like hairs; leaves elliptic; apex obtuse; margins denticulate and with a net-like venation. The length of the petiole of the larger leaves 1/2 to 2/3 of the length of the blade of the leaf.
..... Cypholophus reticulatus

13. Stalk slightly bristle like; leaves longish; apex mucronate or mucronulate; dentate; not clearly a net-like venation. The petiole length of the larger leaves at most half of the length of the blade of the leaf. Cypholophus englerianus

13. Stalk with stiff but flat hairs; leaves longish to longish egg-shaped; petiole very short.
..... Cypholophus montanus

14. The whole plant, particularly the young shoots and leaves, soft-haired; leaves with very small wrinkles.
..... Cypholophus velutinus

14. The whole plant, particularly the young shoots and leaves, with laying stiff hairs or the stalks and the petiole rough-haired; leaves rather coarsely wrinkled.
..... Cypholophus trapula

15. Plants prostrate, rooting; stalks thickly soft-haired; leaves egg-shaped or elliptic.
..... Cypholophus prostratus

15. Plants upright; leaves longish to longish-lanceolate............ 16

16. Plants soft-haired; the apexes of the leaves mucronate or mucronulate. Cypholophus nummularis

16. Plants more or less flattly stiff-haired; apex of the leaf cuspidate.
.... 17

17. The young stalk round; leaves longish lanceolate.
..... Cypholophus ledermanni

17. The young stalk clearly with 4 corners; leaves longish, faintly dentate...................... 18

18. Leaves oval-longish; at the base clearly oblique; truncate or slightly rounded; apex apiculate.
..... Cypholophus gjellerupii

18. Leaves longish; at the base slightly oblique, cuneate or truncate; apex caudate.
..... Cypholophus vestitus

19. Small creeping, rooting, shaggy, scrambling plants. Leaves with lumps.
..... Cypholophus chamaephyton

19. Small shrub, not rooting, without hairs; leaves smooth.......... 20

20. Stem and branches clearly with rings; leaves 12-15 mm long, at the top half of the leaf margins dentate.. Cypholophus treubii

20. Stem not clearly with rings or without rings; leaves 6-10 mm long, only at the top at both sides with a single dentation.
..... Cypholophus vaccinioides

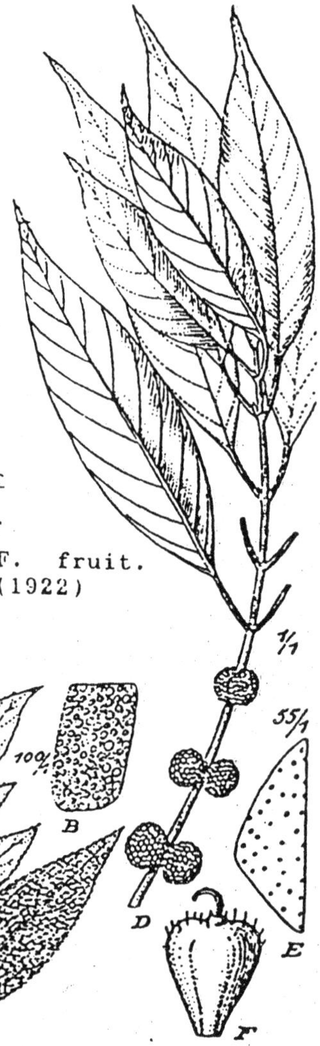

Cypholophus integer

D. stem; E. leaf with cystoliths; F. fruit. (Source: Winkler (1922)

Cypholophus numularis A. stem; B. leaf with cystoliths; C. Inflorescence-below; D.& E. male flower and bud; F. female flower; G. fruit.
(Source: Winkler (1922)

3. DEBREGEASIA

o ● ● ● ● ●

Mabberley (1987) notes that the genus, described by Gaudich., consists of 5 species from T and temperate Asia, and N E Africa. All species of Debregeasia described from Papuasia were transferred to Leucosyke by Winkler (1922). Species are cultivated for fruit (D. edulis in Japan) and fibre.

Monoecious and dioecious trees and shrubs. Leaves alternate, serrate-crenate, 3-nerved, rugose, often snow-white or silvery below; stipules connate into one, 2-lobed, intrapetiolar. Flowers glomerate-capitate, female receptacle slightly thickened; head globose, sessile, axillary or at nodes, or in lax cymes. Bracts in male clusters scarious; male flowers : calyx 4-partite, rarely 3- or 5-partite, segments ovate, valvate or slightly imbricate at the apex, depressed-globose in bud: stamens 4, rarely 3 or 5; rudimentary ovary glabrous or woolly at the base. Female flowers: calyx ovoid or obovoid, mouth contracted, very minutely dentate in fruit succulent; ovary straight, enclosed; stigma sessile or with a distinct style, penicillate with a few hairs, soon disappearing; ovule basal, erect. Achenes closely enclosed in a fleshy calyx and adnate to it (aggregated into spherical syncarps), seed with endosperm, cotyledons small broad.

Debregeasia sp.
NGF 41511 Oksapmin.

List of Species in Papuasia

Debregeasia pulchra Ridley Trans. Linn. Soc. Bot. 9 (1916) 156. Basonym for Leucosyke pulchra (Ridl.) Winkler (1922).
var. alpestris Ridley

Debregeasia rhodopleura Boerl. Basonym for Leucosyke rhodopleura (Boerl.) Winkler (1922)

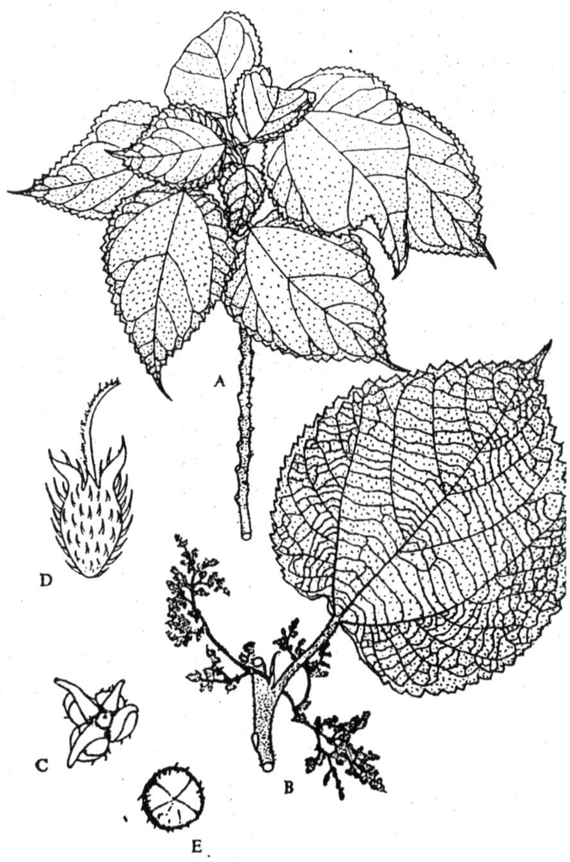

Boehmeria nivea A. leafy shoot; B. inflorescence; C. male flower; D. female flower; E. fruit. This plant is cultivated in China as a source of fibre.
(Source: Purseglove (1977)

4. DENDROCNIDE

o o o o o o

A genus distributed from S and S E Asia, throughout Malesia and Papuasia to N Australia and the Pacific Islands. Chew (1965) recognised 48 species from throughout the range of the genus. From Papuasia Chew (1965) published 12 recombinations for 12 species previously treated as *Laportea*. He also listed 6 species as synonyms.

Perennial shrubs or trees, irritant hairs present, dioecious. Leaves, smooth to undulate, crenulate, coriaceous. Stipules intra-petiolar, wholly connate. Male flowers mainly with 4 tepals and 4 stamens. Female flowers pedicellate or sessile; when sessile arranged distichally or flabellately on swollen extremities of peduncles; pedicellate flowers arranged in fascicles; the pedicels either simple or swollen but never winged. Achenes compressed to bloated, ovate, usually strongly warted, not papery.

Key to Species in Papuasia

(Source: Chew (1969)

1. Shrubby tree with succulent base; lamina ovate; plicate, the lower lateral veins palmate, the upper penninerved; the upper side of the lamina often with 2 fleshy appendages...... *Dendrocnide mirabilis* (Rechinger) Chew (1965: 204)
Laportea mirablilis Rechinger *Denkschr*. *Akad*. *Wiss*. *Math*.-*Natur*. 89 (1914) 543. Tab. 1.

o o ● o o ●

1. Not this combination of characters. 2

2. Lamina peltate, almost round to ovate, minutely dentate. .... *Dendrocnide peltata* (Bl.) Miq. *Jungh*. 1 (1851) 30. Chew (1969: 42)
var. *peltata*
Urtica peltata Bl. *Bijdr*. (1825: 496)

● ● ● o o o

2. Lamina not peltate.............. 3

3. Lamina less than 80 cm long, glabrous to pubescent........... 4

3. Lamina 100 - 150 cm long, ca. 50 cm broad, the upper side glabrous, the lower side pubescent. .... *Dendrocnide kajewskii* Chew (1969: 94)

o o o o o ●

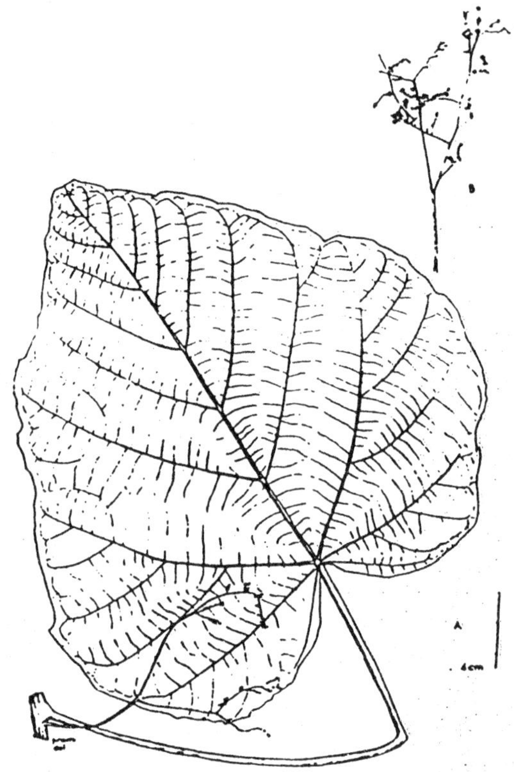

Dendrocnide cordata A. twig; B. male inflorescene.
Source: Chew (1969)

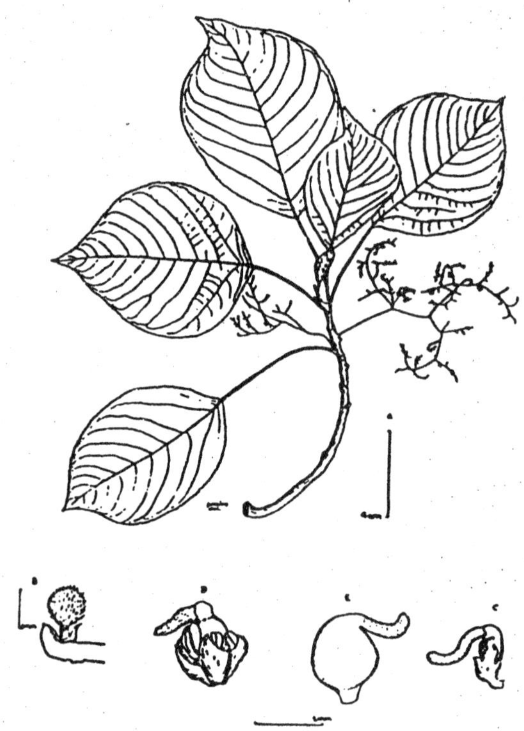

Dendrocnide carriana A. twig; B. male flower; C. female flower; D. bisexual flower; E. achene.
Source: Chew (1969)

4. Female flowers sessile, subsessile at most, sitting in somewhat opposing rows on peduncular extremities. 5

4. Female flowers pedicellate, usually in fascilcles; pedicels often swollen and fleshy at fruit........ 7

5. Pistillode very large, cubical in shape, fleshy, occupying nearly the entire floral space; stamens very small, lodged at sides of avary; infrutescence often fleshy and convulate.... Dendrocnide longifolia (Hemsley) Chew (1965: 203; 1969, 79)
Laportea longifolia Hemsley Hk. Ic. Pl. 26 (1898) tab. 2559, 2560
Dendrocnide salomonensis (Rech.) Chew (1965: 205)
Laportea salomonensis Rech. (1912: 182)
Dendrocnide warburghii (Winkler) Chew (1965: 207)
Laportea warburgii H. Winkler (1922: 509)

● ● o o ● ●

5. Pistillode small, not cubical nor fleshy; stamens larger than pistillode; inflorescence fleshy but not convolute.............. 6

6. Lamina elliptic to oblong, tomentulose on the lower side, base round to cordate; lateral tepals of female flowers often greatly developed..... Dendrocnide rechingeri (Winkler) Chew (1965: 205) Chew (1969: 74)
Laportea rechinger H. Winkler (1922: 506)

● o ● ● o ●

6. Lamina narrowly elliptic to obovate lower side not tomentulose, base cuneate; lateral tepals of female flowers not much longer than the medial ones. .... Dendrocnide schlechteri (Winkler) Chew (1965: 205) Chew (1969: 76 - 79)
Laportea schlechteri H. Winkler (1922: 505)
L. crenulata Wedd. var. nervosa H. Winkler (1922: 507)

● ● ● o ● o

7. Female perianth very shallowly 4-toothed, becoming a loose envelope at the base of the achene at maturity...................... 8

7. Female perianth with well developed tepals enclosing the base of the achene firmly at maturity....... 10

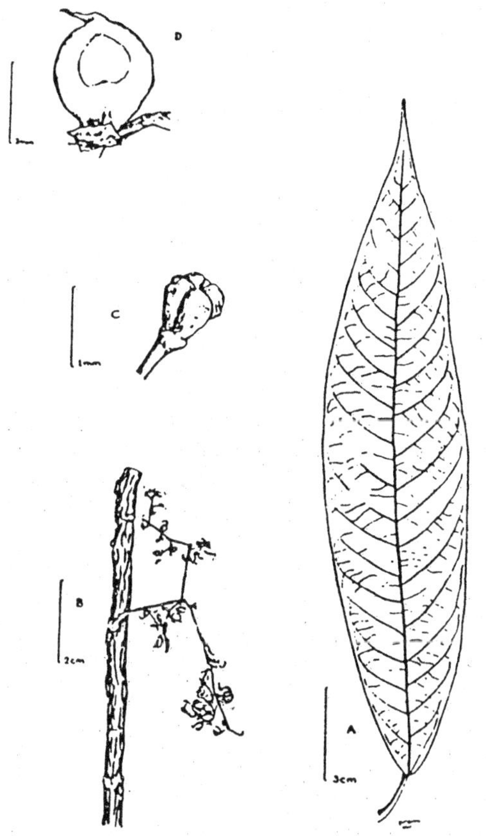

Dendrocnide corallodesme A. leaf; B. cauliflorous female inflorescence; C. male flower; D. achene.
Source: Chew (1969)

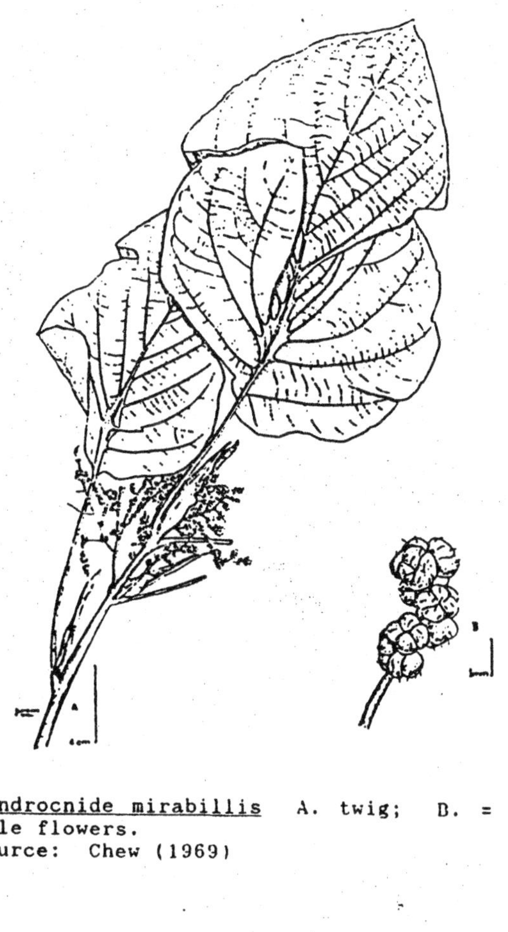

Dendrocnide mirabillis A. twig; B. = male flowers.
Source: Chew (1969)

Dendrocnide morobensis A. twig; B. male flower; C. female flowers; D. achene. Source: Chew (1969)

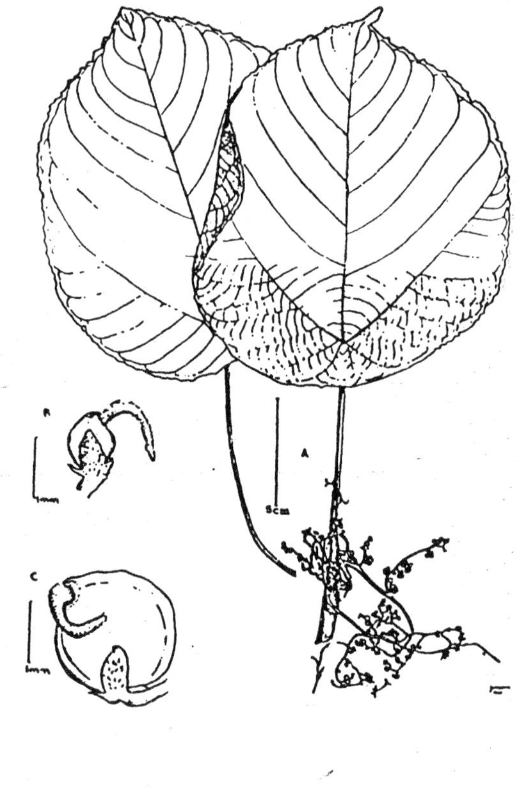

Dendrocnide peltata var. *peltata* A. twig; B. female flower; C. achene. Source: Chew (1969)

Dendrocnide nervosa A. twig; B. male flower; C. achene. Source: Chew (1969)

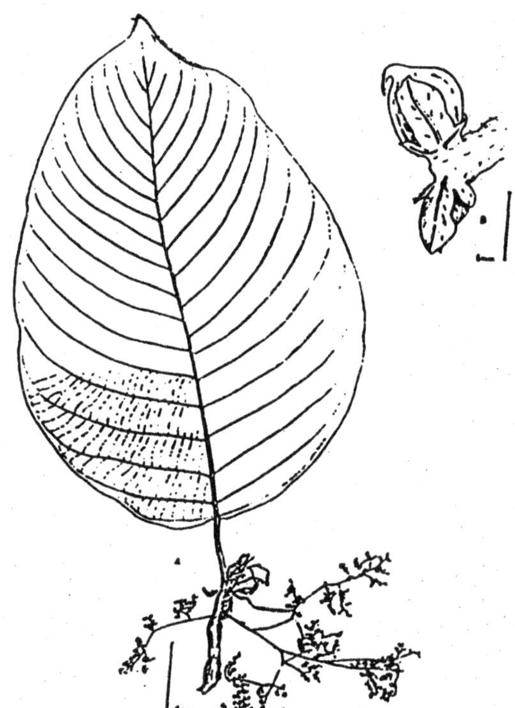

Dendrocnide rechingeri A. twig; B. achene. Source: Chew (1969)

8. Lamina narrowly elliptic, often toothed, very shortlt petiolate, lateral veins short, numerous; male flowers pedicellate, female inflorescences commonly borne on leafless twigs.

.... Dendrocnide corallodesme (Laut.) Chew (1965: 202) Chew (1969: 53 - 55) Laportea corallodesme Laut. in K. Sch. & Laut. (1905: 252)

● ● ● o o o

8. Lamina ovate, often very broadly so, often toothed, fairly long petiolate, lateral veins long, at a low angle to the midrib; male flowers sessile; female inflorescences borne borne on leafy twigs. 9

9. Lamina light villous on the lowerside; male flowers dense villous to wooly.....

Dendrocnide torricellensis (Laut.) Chew (1965: 207) Chew (1969: 51 - 53) Laportea toricellensis Laut in K. Sch. & Laut. (1905: 251)

o ● o o o o

9. Lamina glabrous; male flowers only with irritant hairs, not villous nor wooly.

Dendrocnide ternatensis (Miq.) Chew (1965: 207) Chew (1969: 47 - 51) Laportea ternatensis Miq. Ann. Mus. Bot. Lugd. Bat. 4 (1869) 302 Laportea glabra Ridley (1916: 150) Laportea mammosisetosa H. Winkler (1922: 507) Laportea crenulata Wedd. var. nitida Winkler (1922: 507) Dendrocnide nitida (Winkler) Chew (1965: 204).

● ● ● ● o o

10. Adult lamina entirely glabrous and devoid of irritant hairs........ 11

10. Adult lamina variously glabrous or hairy but always with irritant hairs......................... 12

11. Lamina obovate, the basal lateral veins very short; female pedicels winged, with irritant hairs in two rows...... Dendrocnide nervosa (H. Winkler) Chew (1969: 69) Laportea crenulata var. nervosa H. Winkler (1922: 507).

o ● ● o o ●

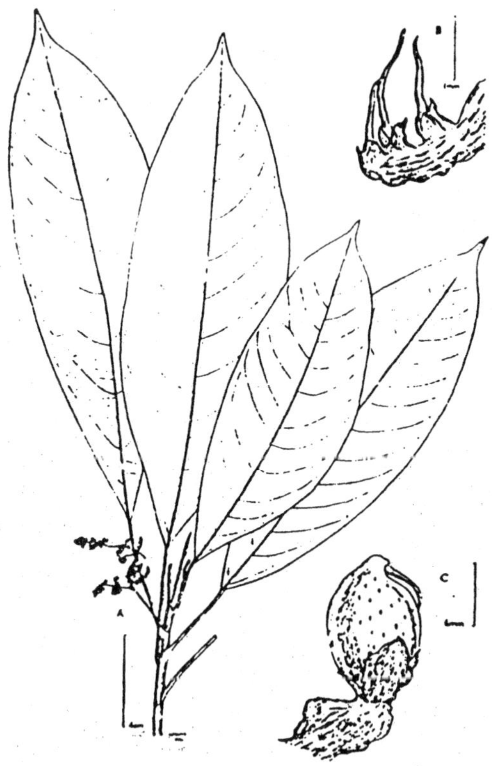

Dendrocnide schlechteri A. twig; B. female flowers; C. achene. Source: Chew (1969)

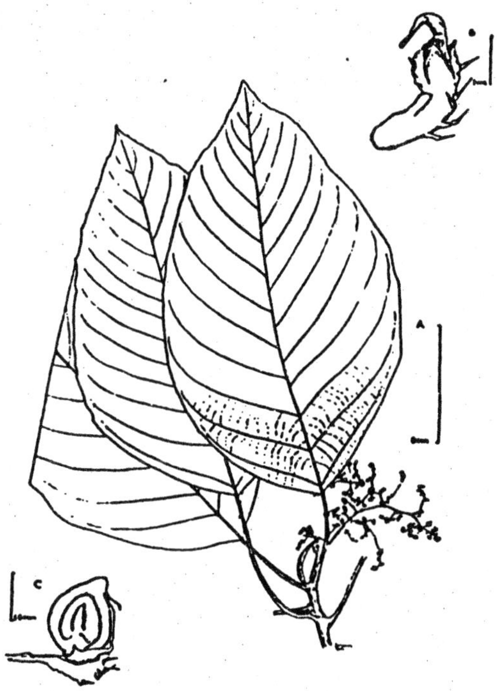

Dendrocnide ternatensis A. twig; B. female flower; C. achene. Source: Chew (1969)

11. Not this combination of characters.
.... Dendrocnide latifolia
(Gaud.) Chew (1965: 203)
Chew (1969: 71 - 74)
Laportea latifolia Gaud.
Bot. Voy. Bonite (1844)
tab. 81.
○ ○ ○ ○ ○ ●

12. Lamina cordate shaped, margin broadly sinuate; petioles usually more than 18 cm long; irritant hairs not on papillate bases.
.... Dendrocnide cordata
(Warb. ex. Winkler) Chew
(1965: 202) Chew (1969:
58 - 60)
Laportea cordata Warb. ex
Winkler (1922: 503)
Laportea gaudichaudiana
Wedd. sensu K. Sch. &
Hollr. (1889: 38)
● ● ● ○ ● ○

12. Lamina ovate, elliptic to pentagonal, margin smooth or dentate, not sinuate; petioles less than 18 cm long; irritant hairs on papillate bases.......................... 13

13. Lamina ovate, with 6 or 7 pairs of lateral veins; the base with one pair very prominent veins.
.... Dendrocnide morobensis
Chew (1969: 67 - 69)
○ ● ○ ○ ○ ○

13. Lamina elliptic; oblong to ovate-pentagonal, with 8-11 pairs of lateral veins; the base with a few pairs of very short veins.
.... Dendrocnide carriana
Chew (1969: 45 - 47)
○ ● ● ○ ○ ○

Notes:

1. In 1922 Winkler pointed out that this species is of 'mixed composition'. The description by Warburg lacks diagnostic characters (Chew 1969: 95). The type was destroyed in Berlin. The stated distribution for Dendrocnide sessiliflora (Warb.) Chew (1965: 206) is New Guinea and the Bismarck Archipelago. Basionym: Laportea sessiliflora Warb. (1891: 292) L. sessilifolia K. Sch. & Laut. (1901: 290)

2. Chew (1965) lists L. crenulata var. nervosa Winkler as a synonym of both D. schlecteri and D. ternatensis.

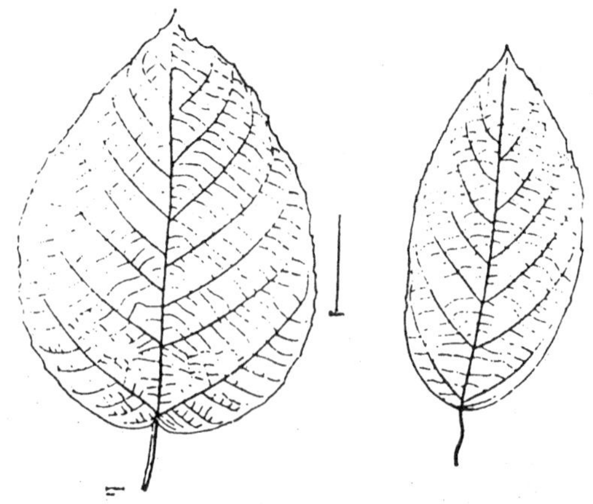

Dendrocnide ternatensis Leaf-forms.
Source: Chew (1969)

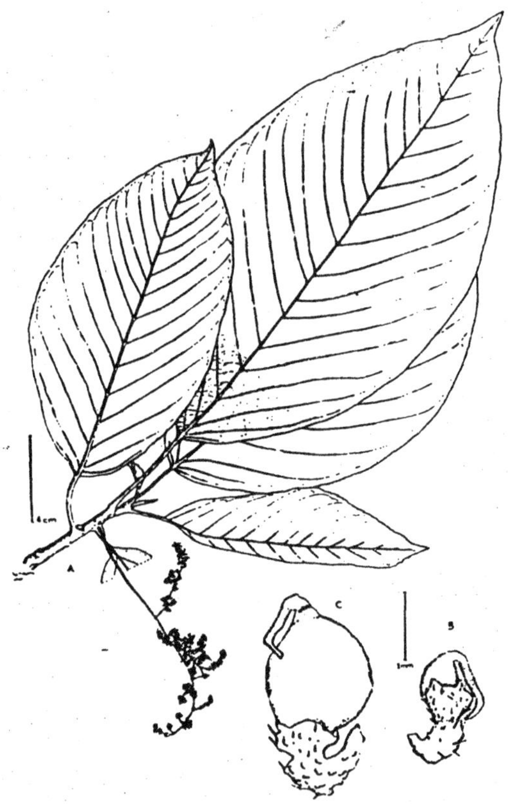

Dendrocnide torricellensis A. twig; B. female flower; C. achene.
Source: Chew (1969)

5. ELATOSTEMA

o o o o o o

A genus of some 200 - 300 species centered in the Old World Tropics, but from Africa and Madagascar to Polynesia. Characterised by the fruit being ejected by staminodes. In New Guinea some 50 species, 4 from high altitudes. E. elegans Winkler from New Guinea is a rheophyte.

An erect or prostrate herb or shrublet, often thickened at nodes, lacking stinging hairs. Stipules axillary, connate. Leaves very unequal in each opposite pair. Smaller caducous, larger often oblique, triplinerved, margin entire or serrate-crenulate in apical regions, bases very unequal; cystoliths usually linear.

Flowers unisexual or bisexual, rarely apogamous. Plants monoecious or dioecious. Male flowers in sessile clusters, sometimes in lax cymes. Perianth 4- to 5-lobed, stamens equal in number to lobes. Abortive ovary present. Inflorescence of female flowers similar to male. Perianth 4- to 5-lobed, lobes subequal. Ovary ellipsoid. Staminodes incurved, scale-like. Stigma shortly penicillate, caducous. Nutlets ovoid or compressed.

The genus is subdivided into several subgenera. An entry on the bottom right hand side under the species gives the subgenus. Note that members of Subgenus Pellionia have been transferred to the genus Pellionia. They are included in the following list for reference but are not printed in Bold Type and are followed by the number (1). Members of Procris are also listed as a separate genus (2).

Check List for Elatostema

(1) Genus Pellionia

(2) Genus Procris

Genus Elatostema

(3) Subgenus: Euelatostema

(4) Subgenus: Elatostematoides

(?) Position unknown.

List of Species in Papuasia
(Source: Winkler (1922)

Elatostema alatocaule H. Winkler (1922: 537; 1924: 123) (3)
o o o o o o

Elatostema angulare H. Winkler (1922: 549 fig 8) (1)

Elatostema annulatum H. Winkler (1922: 528) (3)
o o o o o o

Elatostema articulatum H. Winkler (1922: 525) (3
o o o o o o

Elatostema barbarufa H. Winkler (1922: 546) (3)
o o o o o o

Elatostema beccarii Hilde Schroter Rep. Sp. Nov. Beih. 83 (1936) 125. Pl. 35. (4)
o ● ● ● o ●

Elatostema belense Perry (1951: 374) (3)
o o o o o o

Elatostema biakense H. Winkle (1924: 124) (1)

Elatostema blechnoides Ridley (1916: 154) (3)
o o o o o o

Elatostema calophyllum Rech. (1912: 181, 102) (1)

Elatostema cauliflorum H. Winkler (1922: 556) (1)

Elatostema celaticaule Ridley (1916: 155) (?)
o o o o o o

Elatostema cupulare H. Winkler (1922: 535. Fig. 4) (3)
o o o o o o

var. pedunculatum Perry (1951: 379) (3)
o o o o o o

Elatostema curvitepalum Perry (1951: 386) (1)

Elatostema cyrtandra H. Winkler (1922: 523. Fig. 2) (3)
o o o o o o

Elatostema decurrens H. Winkler (1922: 565) (2)

Elatostema densum H. Winkler (1922: 536) (3)
o ● o o o o

Elatostema doormannianum H. Winkler (1924: 123) (?)
o o o o o o

Elatostema elegans H. Winkler (1922: 526) (3)
● o o o o o

Elatostema eludens Perry (1951: 387) (1)

Elatostema filicinum Ridley (1916: 154) (1)

Elatostema finisterrae Warb. (1892: 13, 19) (3)
o o o o o o

Elatostema fragile H. Winkler (1922: 546) (?3)
o o o o o o

Elatostema frutescens Hassk. (2)

var. caudatum H. Winkler (1922: 565) (2)

var. majusculum K.Sch. (3)
basonym for E. macrophyllum var. majusculum (K. Sch.) H. Winkler

var. novo-guineense (Val.) H. Winkler (1922: 565) Basionym: Procris frutescens var. novo-guineense Val. (1907)
o o o o o o

var. sculptum H.Winkler (1922: 565)

Elatostema fruticulosum K.Sch. (1905: 254) (1)

Elatostema grammicum Perry (1951: 385) (1)

Elatostema grueningii H. Winkler (1922: 563) "gruningii" (2)
var. krausei H. Winkler (1922: 564) (2)

Elatostema hallieri H. Winkler (1922: 556) (1)

Elatostema hoelscherianum H. Winkler (1922: 559. Fig 10: A - E) (1)

Elatostema hoffmannianum H. Winkler (1922: 552) (1)

Elatostema humile Perry (1951: 378) (3)

● o o o o o

Elatostema hymenophyllum H. Winkler (1922: 543. Fig. 7) (3)

o ● o o o o

var. stollei H. Winkler (1922: 544). (3)

o o o o o o

Elatostema inamoenum H.Winkler (1922: 557) (1)

Elatostema inconstans Perry (1951: 384) (1)

Elatostema integrifolium Wedd. in DC. Prod. 16(1): 179 (3)

o ● ● ● ● o

Elatostema jabiense H. Winkler (1922: 561) (1)

Elatostema janowskyi H.Winkler (1922: 559) (1)

Elatostema kietanum Rechinger (1912: 182) (1)

Elatostema kupeiense Perry (1951: 376) (3)

o o o o o o

Elatostema lamii H. Winkler (1924: 125. t. XVB) (1)

Elatostema lanceolatum H. Winkler (1922: 558. fig. 9) (1)

Elatostema lauterbachii H. Winkler (1922: 525) (3)

o o o o o o

Elatostema ledermannii H. Winkler (1922: 531) "ledermanni"(3)

o o o o o o

Elatostema lignescens Hallier f. in Kds. Med. Plantent 19 (1898) 595. nomen. = E. pedunculatum (2)

Elatostema lingelsheimii H. Winkler (1922: 537) (3)

o o o o o o

var. adscendens H. Winkler (1922: 538) (3)

o o o o o o

var. seriatum H.Winkler (1922: 538) (3)

o o o o o o

Elatostema longicollum H. Winkler (1922) 545 (3)

o o o o o o

Elatostema macrophyllum Brongn. (3)

o ● ● o o ●

var. majusculum (K.Sch.) H. Winkler (1922: 526) (3)

o o ● o o ●

Elatostema macropus H. Winkler (1922: 532. Fig. 3) (3)

o o o o o o

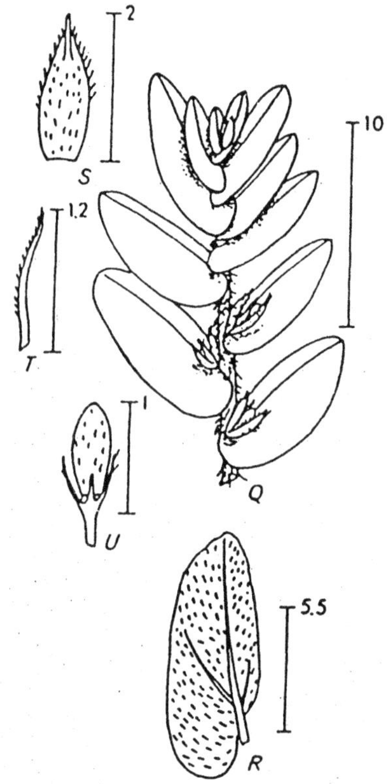

Elatostema celaticaule Q. tip of flowering branchlet, R. leaf underside, S. bract of female inflorescence. T. bracteole of female inflorescence, U. young nutlet with perianth and staminodes. (Kloss s.n.).
(Source: van Royen (1982)

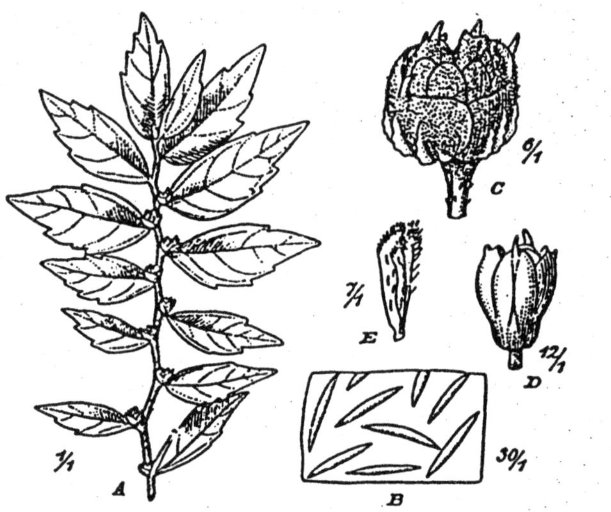

Elatostema lanceolatum A: stem; B. leaf with cystoliths; C. male flower; D. staminodes; E. female flower.
(Source: Winkler (1922)

Elatostema mafuluense Perry (1951: 385) (1)

Elatostema minutiflorum H. Winkler (1924: 126) (1)

Elatostema mongiense Laut. (1914: 239) "mongiensis" (3)

○ ● ● ○ ○ ○

var. gracile H. Winkler (1922: 543) (3)

○ ○ ○ ○ ○ ○

Elatostema morobense Perry (1951: 379) (3)

○ ● ○ ○ ○ ○

Elatostema novae-britanniae Laut. (1905: 253) (1)

Elatostema novo-guineense Warb. (1891: 290) (3)

○ ● ● ● ● ○

var. angustifolium Warb. (1891:291) (3)

○ ○ ○ ○ ○ ○

var. latifolium H. Winkler (1922: 530) (3)

var. stenophyllum Laut. (1905: 255) = var. angustifolium

var. stenopus H. Winkler (1922: 531). (3)

○ ○ ○ ○ ○ ○ ○

var. trapezoideum H. Winkler (1922: 530) (3)

○ ○ ○ ○ ○ ○ ○

var. uber H. Winkler (1922: 531) (3)

○ ○ ○ ○ ○ ○ ○

Elatostema pachypoda Diels (1929: 75, 273) (1)

Elatostema paramelanum. H. Winkler (1922: 556) (1)

Elatostema pauperatum H. Winkler (1922: 551) (1)

Elatostema pedunculatum Forst. (2)

var. angustum H. Winkler (1922: 563) (2)

Elatostema pellionianum Gaud. var. effusum H. Winkler (1922: 549) (1)

var. novae-britanniae (Laut.) H. Winkler (1922: 549) E. novae-britanniae Laut. (1905) (1)

var. pedunculatum H. Winkler (1922: 549) (1)

Elatostema peltifolium H. Winkler (1922: 553) Basionym: Pellionia peltata Ridley (1916) non E. peltata (1)

Elatostema peperomioides H. Winkler (1922: 538. Fig. 5) (3)

○ ● ○ ○ ○ ○

Elatostema perpusillum Perry (1951: 381) (3)

● ○ ○ ○ ○ ○

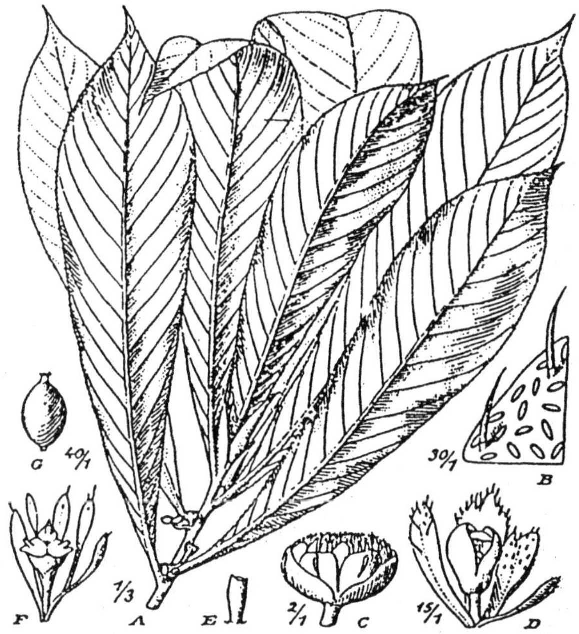

Elatostema cyrtandra A. stem; B. leaf with cystoliths; female inflorescence; D. male flower; E. rudiments of ovary in male flower; F. female flower; G. fruit. (Source: Winkler (1922)

Elatostema sp.
(Source: National Herbarium, Lae).

Elatostema hymenophyllum A. stem; B. leaf with cystoliths; C.& D.& E. male inflorescence, flower, and inner bract. (Source: Winkler (1922)

Elatostema piliferum H.Winkler (1922: 535. Fig. 10 F - J) (3)

o o o o o o

Elatostema platycarpum Perry (1951: 370) (3)

o o o o o o

Elatostema podophyllum Wedd. Ann. Sc. Nat. IV 1 (1854) 189 (3)

o o o o o o

Elatostema polypodioides Ridl. (1916: 154) (3)

o o o o o o

Elatostema poteriifolium Ridl. (1916: 155) (3)

o o o o o o

Elatostema pulleanum H.Winkler (1922: 527) (3)

o o o o o o

Elatostema rectangulare H. Winkler (1922: 566) (2)

Elatostema reiterianum H. Winkler (1922: 543) (3)

● o o o o o

Elatostema reticulatum Wedd. Ann. Sci. Nat. 4:1 (1854) 188 (3)

o o o o o ●

Elatostema retinervium Perry (1951: 371) (3)

● o o o o o

Elatostema rigidum var. laxum Hilde Schroter Rep. Sp. Nov. Beih. 83 (1936) 111 (4)

o o o o o o

Elatostema rudicaule H. Winkler (1922: 540. Fig. 6) (3)

o o o o o o

var. villiferum Perry (1951: 377) (3)

● o ● o o o

Elatostema salomonense Perry (1951: 373) (3)

o o o o o o

Elatostema schroeteri Perry (1951: 375) (3)

● ● o o o o

var. angustilimbum Perry (1951: 375) (3)

o o o o o o

Elatostema serra H. Winkler (1922: 527) (3)

o o o o o o

Elatostema serratifolium H. Schroter Rep. Sp. Nov 47 (1939) 221. non Elmer (1934). = E. schroeteri Perry (3)

o o o o o o

Elatostema sesquifolium Hassk. (3)

o ● o o o o

var. caudatum H. Winkler (1922: 535) (3)

o o o o o o

var. ramosum H. Winkler (1922: 534) (3)

● o o o o o

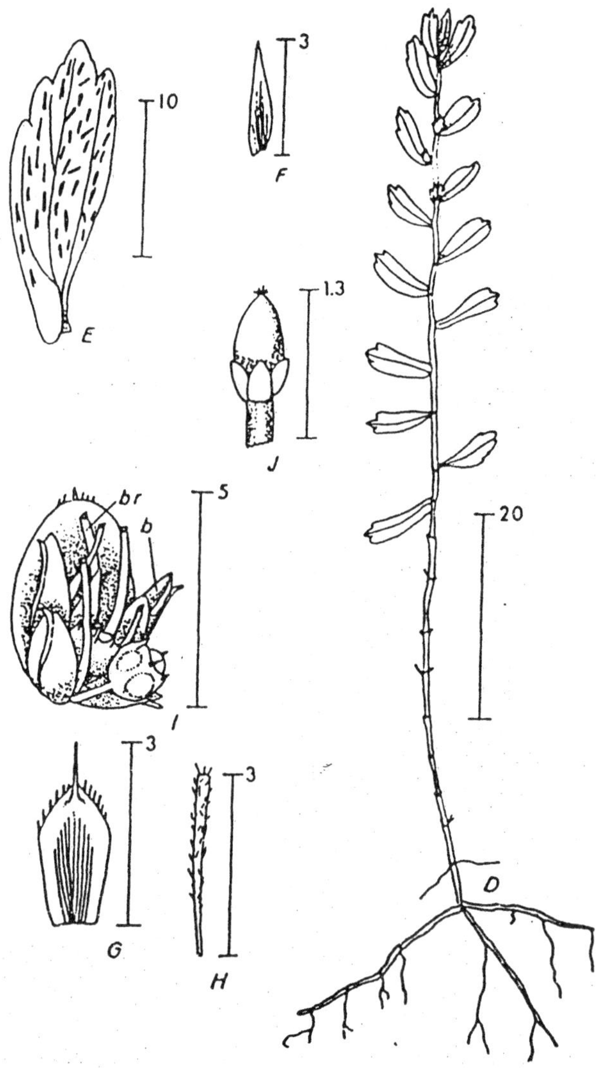

Elatostema morobense D. habit; E. underside of leaf; F. stipule; G. bract of female inflorescence; H. bracteole of same; I. male inflorescence, frontal bract removed (b = bract, br = bracteole, J. young nutlet. (Brass 30750; except G. van Balgooy 579)
(Source: van Royen (1982)

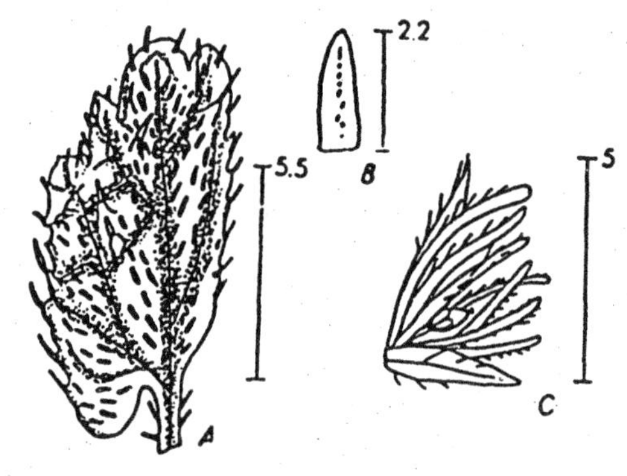

Elatostema poteriifolium A. upper side of leaf; B. stipule; C. female inflorescence. (Kloss s.n.)
(Source: van Royen (1982)

Elatostema sessile Forst et f Char. Gen. (3)

o o o o o o

Elatostema sinuatum (Bl.) Hassk. (1)

var. pedunculatum (H. Winkler) H. Schroter Rep. Sp. Nov. Beih. 83 (1936) 32 (1)

Elatostema stenurum Hallier f. Meded. Rijks. Herb. Leiden 26 (1915) 2 (1)

Elatostema subintegrum H. Winkler (1922: 533) (3)

o o o o o o

Elatostema tenuicaule H. Winkler (1922: 540) (3)

o o o o o o

Elatostema tricanthum Laut. in K. Schum. & Laut. (1905: 255) (3)

o ● o o o o

Elatostema trichomanes H. Winkler (1922: 543) (3)

o o ● o o o

Elatostema tricuspe H. Winkler (1922: 545) (3)

o o o o o o

Elatostema tridens Perry (1951: 382) (3)

o ● o o o o

Elatostema truncicola H. Winkler (1922: 238) (?)

o o o o o o

Elatostema undulatum H.Winkler (1922: 551) (1)

Elatostema urvilleanum Brongn. Bot. Voy. Coq. (1829) 210. Pl. 46. (3)

o o o o o o

Elatostema velutinicaule H. Winkler (1922: 542) (3)

o o o o o o

Elatostema velutinum (Warb.) K. Sch. in K. Sch. & Laut. (1905: 253) Procris velutina Warb. (1893: 189) (1)

var. glabrum K. Sch. (1905: 253) (1)

Elatostema vicinum Perry (1951: 372) (3)

o o o o o o

Elatostema warburgii H.Winkler (1922: 552) (1)

Elatostema weinlandii K. Sch. in K. Sch. & Laut. (1905: 254) (4)

● ● ● o o o

var. dispar H. Winkler (1922: 554) (4)

o o o o o o

var. inamoenum (H. Winkler) H. Schroter in Schroter & Winkler (1936: 118) (4)

o o o o o o

var. kochii (Val.) H. Winkler (1922: 554) Pellionia kochii Val. (1907) (4)

● o o o o o

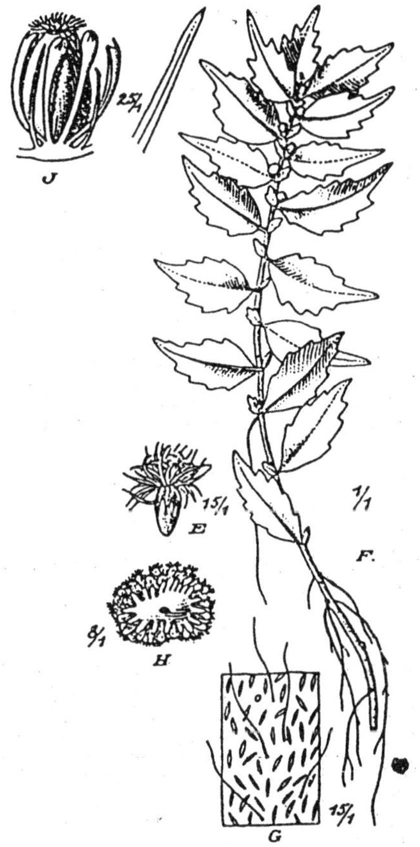

Elatostema piliferum F. stem; G. leaf with cystoliths; E. male flower; J. female flower.
(Source: Winkler (1922)

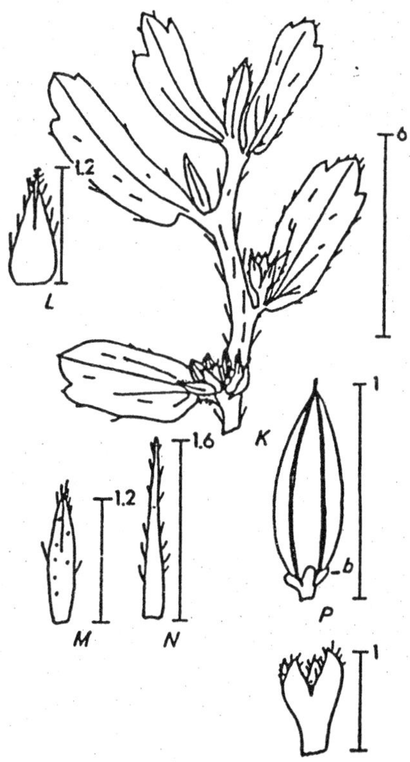

Elatostema peperomioides K. apical part of flowering branch; L. bact of female inflorescence; M.-N. bracteoles of female inflorescence; O. female perianth; P. nutlet with staminodes at base (b. bracteole); (LAE 54213) (Source: van Royen (1982)

var. vestitum H. Winkler (1922: 554) (4)

o o o o o o

var. yulense (Hallier f.) H. Winkler (1922: 555) E. yulense Hallier f. (1915) (4)

o o o o o o

Elatostema whartonense Perry (1951: 381) (3)

o o o o o o

Elatostema yulense Hallier f. Meded. Ryks. Herb. Leiden 26 (1915) 3. Dasionym: E. weinlandii (1)
var. yulense (Hallier f.) H.Winkler

Key to Species of Elatostema

(Source: Winkler (1922)
(Translated by Dra. C. M. H. Lapre)

1. Plants vigorous, at least 40 cm to 2m high. E. novoguineense also smaller but then always more stalks from 1 rhizome, stalks separate, sparsely branched; leaves large, at least 6 cm long; in E.novoguineese also smaller but pointed and often with a drip tip................ 2

1. Plants not vigorous (flabby) parched, or at most 40 cm high or longer and creeping, the one stalk not or sparsely branched; leaves small (not more than 20 cm long) entire, or on top with a few dentations. 14

1. Plants 30 - 60 cm high, the only stalk at the top side numerously branched, densely haired; leaves small, always serrate along whole margin......................... 20

1. Creeping or in clumps; small plants with thin stalks. Leaves unlike the Urticaceae or the nettles....... 23

1. Small creeping plants with nettle like leaves. .... Elatostema longicollum

2. Leaves pinnate, sideveins parallel. 3

2. Leaves pinnate, at least at the base; the nerves re on one side, not parallel................... 7

2. Leaves with 3 prominent nerves (in E. novo-guineense var. not always prominent).................... 10

3. Stalk (at least at the apex), petiole and leaves (at least on the under side) soft-haired......... 4

3. Stalk, petiole and leaves bare, or very thinly hirsute, but not soft haired........................ 5

4. Leaves very long (20 - 30 cm),longish-lanceolate, at the base a little uneven, at both sides gradually merging into the petiole. .... Elatostema cyrtandra

4. Leaves shorter (15 - 20 cm), obovate, at the base very unequal (very dissimilar), at least at the smaller side of the leaves clearly shorter... Elatostema lauterbachii

5. Leaves longish oval, suddenly pointed........................... 6

5. Leaves lanceolate, gradually ending in a drip tip (cuspidate). .... Elatostema elegans

6. Leaves very wrinkled with very prominent nerves underneath; margins coarsely serrate; cystoliths in dried plants coarse and yellow. .... Elatostema finisterrae

6. Leaf nerves indistinct (only in dried plants), wrinkled, unevenly, indistinctly serrated. Cystoliths fine, not yellow. .... Elatostema articulatum

7. Leaf base at the broad side pointed .... Elatostema macrophyllum

7. Leaf base at the broad side cordate 8

8. At most the top 6 - 8 internodes about the same size, the lower ones are longer; stipules scales not prominent at the nodes, not ring-shaped. Leaves not or very slightly sickle shaped (falcate)........ 9

8. The top 10 - 15 internodes about the same length; the rather swollen nodes with prominent, ring-shaped stipule scars. .... Elatostema annulatum

9. Stalk already in the top internode 3 - 4 mm diameter; leaf top suddenly pulled out (aristate); leaf margins entire, crenations at the leaf base less prominent. .... Elatostema pulleanum

9. Stalk at top internodes only 1.5 - 3 mm diameter; leaf top gradually pulled out (cordate); margins dentate....... Elatostema serra

10. Leaf margins on both sides, above the middle or to the base, coarsely dentate........................ 11

10. The leaf margins on the broad side of the leaf to the middle coarsley dentate; on the narrow side dentate only along 1/3 of margin.
.... Elatostema ledermannii

10. Leaves entire or only slightly crenate at apex................ 12

11. Inflorescence without peduncle; stipules rather broad, hyalin.
.... Elatostema sessile

11. Inflorescences with a short or long peduncle; stipules rather longish; sometimes acicular; strong.
...... Elatostema novo-guineense

12. Side veins running, in the broader half of the leaf, at most to the middle: on narrower side of leaf at the most to 2/3 the length of the leaf. Side veins end at leaf margin......................... 13

12. The two side veins ending in the top of the leaf.
.... Elatostema sesquifolium

13. Nodes not swollen, not ringed. Leaves long-lanceolate; male inflorescence (with peduncle) many times longer than the diameter.
.... Elatostema macropus

13. Nodes swollen, with ringed markings (stipule scars); leaves linear-lanceolate; male inflorescences (with peduncle) as long or slightly longer than their diameter
.... Elatostema subintegrum

14. Leaves at the upper and lower side covered loosely with long white hairs..... Elatostema piliferum

14. Leaves not with long white hairs.
.... 15

15. Stalks rough, sometimes nearly chaff-like, hairy.............. 16

15. Stalks without hairs (bare) or nearly bare.
.... Elatostema tenuicaule

16. Leaves with slightly uneven side (oblique), at the base not auriculate.
.... Elatostem cupulare

16. Leaves strongly uneven sided (oblique), at the base of the broad side auriculate................ 17

17. Leaves only slightly auriculate; 1 to 4 dents up to the middle.
.... Elatostema densum

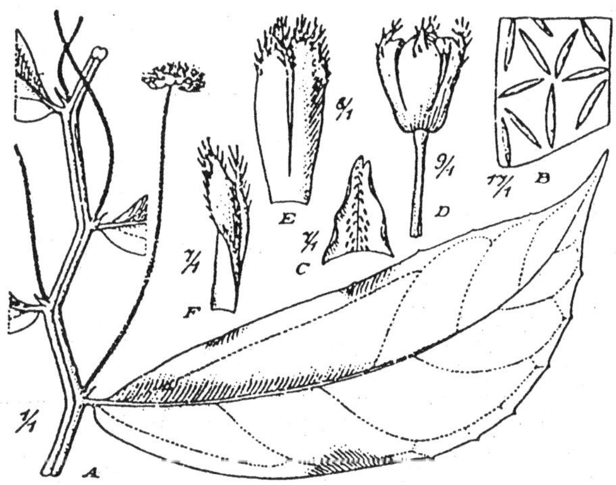

Elatostema macropus A. stem; B. leaf with cystoliths; C. stipule; D. male flower; E. F. inner bract of male inflorescence: dorsal amd lateral view. (Source: Winkler (1922)

Elatostema rudicaule A. stem of female plant; B. leaf with cystoliths; C. stipule; D.& E.& F. male inflorescence inner bract, and flower.
(Source: Winkler (1922)

17. Leaves strongly auriculate, entire or, only near the top, with 1 to 2 dentations..................... 18

18. Stalk unbranched or if forked in more stalks of the same kind.... 19

18. Stalk racemose (clustered branched) the side branches shorter than the main branch.
.... Elatostema peperomioides

19. Leaves entire, or only at the top, slightly dentate 1 to 2 times. The leaves do not overlap each other.
.... Elatostema trichanthum

19. Leaves entire, very densely placed so they touch each other.
.... Elatostema alatocaule

19. Leaves at the tip clearly serrated.
.... Elatostema lingelsheimii

20. Leaves at least 15 cm long, sharply serrated....................... 21

20. Leaves 6 - 12 (-15) mm long; coarsely serrate.
.... Elatostema mongiense

21. Leaves nearly spinose........... 22

21. Leaves dentate-serrate.
.... Elatostema reiterianum

22. Stalks shaggy haired, stipules big, papery (skin-like), double serrated margins... Elatostema rudicaule

22. Stalks velvet-like haired; stipules small, not papery, margins serrate.
.... Elatostema velutinicaule

23. Leaves 2 to 3.5 cm long, pinnately lobed..... Elatostema trichomanes

23. Leaves at most 1.5 cm long, only at tip some dentations............ 24

24. Leaves obovate, oblique, at both sides unevenly dentate; on the small side 1 and, on the broader side, 2 to 3 dentations.
.... Elatostema hymenophyllum

24. Leaves longish-lanceolate, slightly oblique, both leaf sides nearly equal..... Elatostema tricuspe

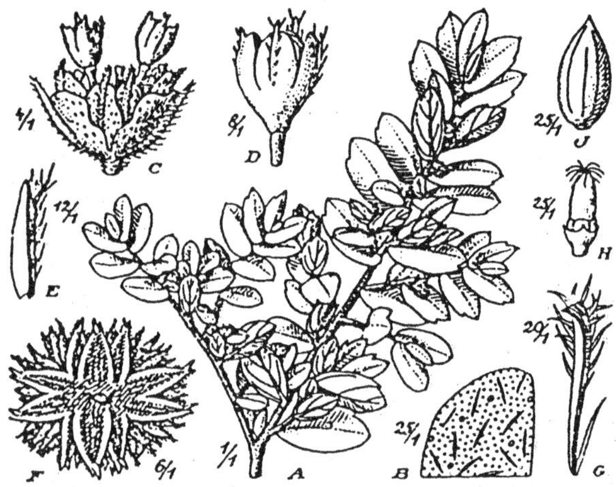

Elatostema peperomioides A. stem; B. leaf with cystoliths; C. inflorescence; D. to G. flower parts.
(Source: Winkler (1922)

ELATOSTEMATOIDES

Brass Journ. Arn. Arb. (1941) 292 discusses this genus (which is unpublished), with reference to Brass 13688. This collection was subsequently identified by Perry (1951) as Elatostema weinlandii var. hochii. which is in subgenus Elatostematoides.

FLEURYA

H. Winkler (1922: 509-510) lists 2 species in Papuasia. Chew (1965) treats all species considered in this genus as a distinct section (Section Fleurya) of Laportea.

List of Species

Fleurya interrupta (L.) Gaud.
Urtica interrupta L. (1753)
=Laportea interrupta (L.) Chew
Fleurya ruderalis (Forst.f) Gaud.
Urtica uderalis Forst f.
=Laportea ruderalis (Forst.f.) Chew

6. GIBBSIA

o o o o o o

This endemic genus Gibbsia is represented by two species in the mountains of Papuasia. The genus was described by Rendle in Gibbs (1917: 129. Fig.1). It is closely related to Leucosyke but there is no trace of a fleshy receptacle in Gibbsia. The fruit is very characteristic: the upper portion of the thinly succulent exocarp separates like a cap and the endocarp enclosing the seed is then readily separable from the cup formed by the union of the somewhat fleshy perianth with the lower part of the exocarp.

Shrubs. Leaves alternate, crenate-serrate, 3-nerved, canescent below; stipules mambranous, connate into one, deeply 2-fid and intrapetiolar, deciduous. Flowers monoecious, in small, axillary androgynous cymules; bracts minute, ovate, scarious. Male flowers: calyx 5-partite, segments valvate, ovate: stamens 5; anthers rounded, dorsifixed, inflexed in bud; rudimentary ovary woolly. Female flowers: calyx broadly cupular, persistent and adnate at the base of the fruit; ovary ovoid, very oblique; stigma apical, sessile, discoid, papillous, margin fimbriate-setulose; ovule erect from the base. Fruit drupaceous, small, very oblique. Seed with fleshy endosperm; cotyledons small, elliptic.

Gibbsia insignis A. Branch, bearing leaves and inflorescences; B. cymes: C. D. flowers; E. male flower with dehiscent anthers; F. stamens; G. dorsal connective of stamen; H. female flowers; I. & J. stigmatic disc with inflexed setae; K. fruit; L. ovule. (Source: Gibbs (1917)

List of Species in Papuasia

1. Petiole 1 - 2.5 cm long; leaves broader, to 2 - 2.5 cm. Bracts larger.... Gibbsia cartenszensis Rendle in Gibbs (1917: 131)

1. Petiole 3 - 7 mm long; leaves only 1.2 - 1.7 cm broad. Bracts smaller. .... Gibbsia insignis Rendle in Gibbs (1917: 130. Fig. 8)

Notes:

1. van Steenis (1934) includes a detailed discussion of the genus in the Bull. Jard. Bot. Buit. Ser. III. 13: 256.

2. G. cartenszensis Rendle has only been collected on Mt Carstensz from 1650 to 2000 m altitude. G. insignis Rendle was collected from Angi Lakes at 2100 m altitude.

7. GIRARDINIA

Friis (1981) revised the genus Girardinia and reduced it to two species which are found in the montane areas of the tropics and warm areas of the Old World. Girardinia diversifolia occurs from West Africa and China, through Korea and Taiwan to Indonesia. This species possibly occurs in Irian Jaya and on this basis the genus was included by Barker (unpubl.) in her keys to the genera of Plants in New Guinea.

The genus belongs to Tribe Urereae which has stinging hairs. It is closely related to Laportea from which it differs in the compact female inflorescences, the 3 fused tepels in the female flowers and the uniformly sculptured achenes.

Tall, herbaceous plants, with long stinging hairs. Monoecious or dioecious by abortion. Leaves alternate, petiolate, variabley divided, in outline elliptic to ovate. Cystoliths dot-like. Intrapetiolar stipules, caducous, fused almost to apex. Flowers in unisexual, axillary inflorescences. Male inflorescence: ± branched, slender, flowers densely clustered. Female inflorescence: as in male but "fiercely armed" with stinging hairs. Male flowers: pedicellate, 4-5 tepals, stamens + tepals in number, rudimentary ovary always present. Female flowers: sessile with 3 tepals fused into a hollow structure, ± enclosing the reflexed ovary; fourth tepal, free, ± rudimentary or lacking. Staminodia absent. Ovary compressed laterally, stigma filiform. Fruit achene, with uniform rugose sculpturing. Embryo in seed broad.

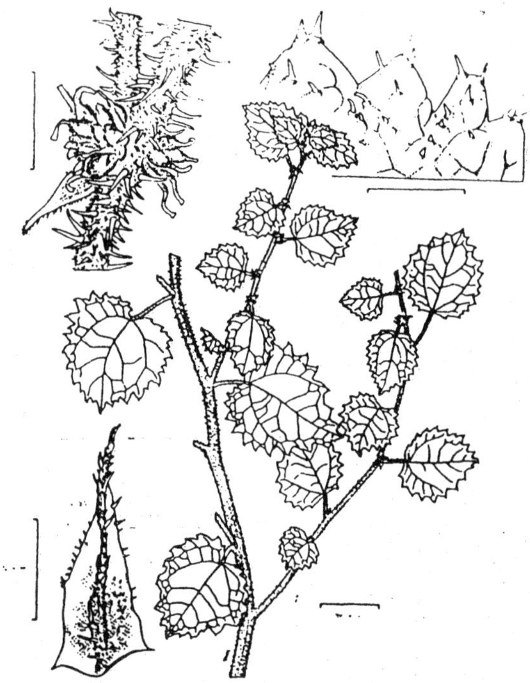

__Girardinia__: represented by a single undescribed species in Papuasia, if the identification from the Rijksherbarium is correct. Drawn from NGF 30152.

8. __LAPORTEA__

o o o o o o

This genus is centered in the Africa - Madagascar region. Of the 22 species 12 are confined to this region, 5 are endemic in Madagascar. Richett & Stafleu __Taxon__ 8 (1959) 262 proposed this genus as a nomen. conserv. Most species are branched herbs less than 2 m high. The stem is weak and woody only at the base. The genus contrasts clearly with __Dendrocnide__ which is a genus of woody trees to 35 m tall.

Leaves simple, petiolate, spirally of alternately arranged with a variously toothed margin. Monoecious or rarely dioecious, usually with irritant hairs. Stipules intra-petiolar, incompletelt connate, bifid at apex. Inflorescence generally paniculate, pedunculate, axillary with unisexual flowers; peduncular bracts very small. Male flowers 4- or 5-tepaled, with an equal number of stamens opposite the tepals; filaments reflexed; pistillode present, small. Female flowers 4 tepalled, the lateral tepals equal sized, the dorsiventral ones greatly unequal sized, much smaller than the laterals, devoid of staminodes; ovary ovoid; style generally linear, rarely very short and brush-like. Achene generally compressed, ovoid to semicircular, sessile or stipitate, 1 to 4 mm long, usually reflexed upon either dorsiventrally or laterally winged pedicels.

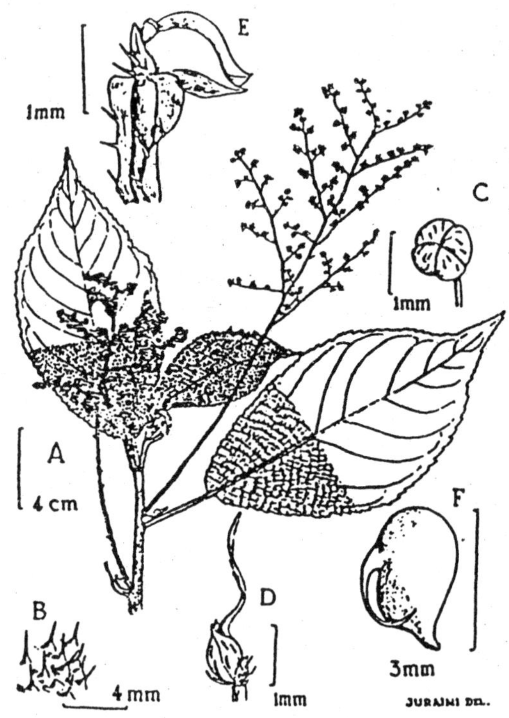

__Laportea decumana__ A. twig; B. leaf surface; C. male flower; D, female flower; E, pedicel; F, achene.

__Key to Sections in Papuasia__

1. Pedicels of female flowers winged laterally and symmetrically. Achenes articulated on pedicels. .... Section __Laportea__

1. Pedicels of female flowers usually winged dorsiventrally and asymmetrically. Achenes not articulated on pedicels. .... Section __Fleurya__

Key to Species in Papuasia
(Source: Chew (1969)

1. Stigma 3-fid, the central branch longer than the lower two; inflorescences appear unbranched due to the great reduction of lateral branches.. Laportea interrupta
○ ● ○ ○ ● ●

1. Stigma simple, without branches; inflorescence variable.......... 2

2. Plants less than 60 cm tall, glabrous and lacking irritant hairs; inflorescence paniculate, well branched, shorter than leaves, male flowers 4-merous, tepals corniculate; stigma almost brush-like 0.2 mm long... L. ruderalis
● ○ ○ ● ● ●

2. Not this combination of characters. .... L. decumana
● ● ● ○ ○ ○

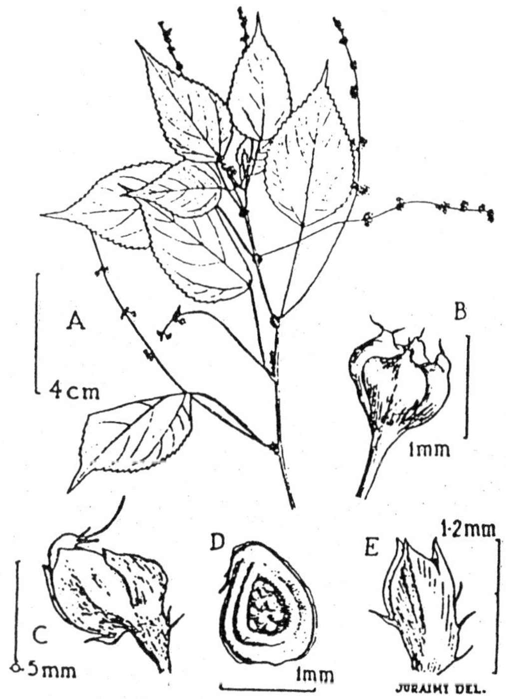

Laportea interrupta A. twig; B. male flower; D. achene; E. female perianth. (Source: Chew (1969)

List of Species with Synonomy

Section Laportea

Laportea decumana (Roxb.) Wedd. Ann. Sc. Nat. (1854) 181
Urtica decumana Roxb. Fl. Ind. 3 (1832) 587.
Laportea armata Warb. (1891: 293).
Laportea humilis Laut. in K. Sch. & Laut. (1905: 251)
var. pedunculata (Laut. & K. Sch.) Winkler (1922: 504)
Laportea pedunculata Laut. & K.Sch. in K. Sch. & Laut. (1900: 290)

Section Fleurya

Laportea interrupta (L.) Chew (1965: 200)
Urtica interrupta L. Sp. Pl. (1753)
Laportea ruderalis (Forst. f.) Chew (1965: 201)
Urtica ruderalis Forst. f. Fl. Ins. Austr. Prodr. (1786) 66

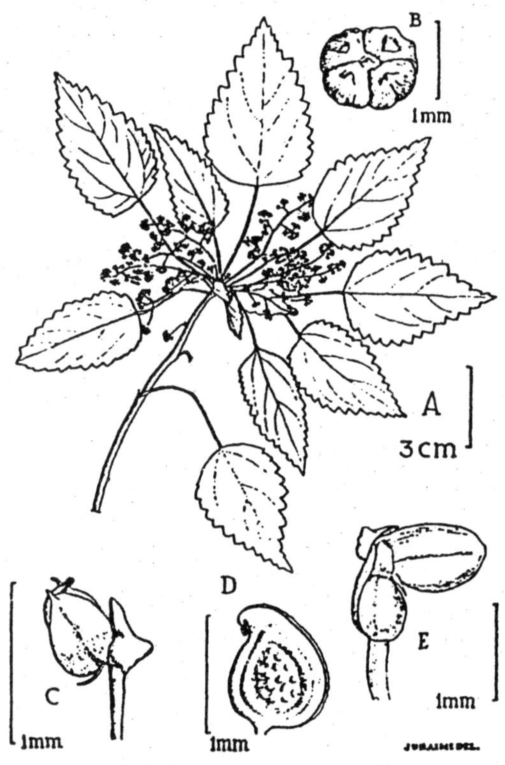

Laportea ruderalis A. twig; B, male flower; C, female flower; D. achene; E, female pedicel and perianth.
(Source: Chew (1969)

Notes

1. L. decumana is distributed from South Borneo to New Guinea. L. interrupta is distributed from Africa to the Pacific, the most widely distributed species in the genus. It is essentially a weed species (gardens, plantations, roadsides) but is constant in taxonomic characters.

2. L. ruderalis occurs from Christmas Is. to Polynesia. It is a small herb devoid of irritant hairs.

LECANTHUS

See Procris

Lecanthus wightii Weddell. Ann. Sc. Nat IV Bot. 1 (1854) 187. This species was recorded by Ridley Trans. Linn. Soc. II. Bot. 9 (1916) 152. SW

9. LEUCOSYKE

o o o o o o

A genus of 36 species from Malaysia to Polynesia. Leucosyke was recorded from Papuasia by H. Winkler (1922: 603 - 606) who recognised four species. Urunh revised the genus for Malesia and the Pacific (1942). Ten species were recognised from Papuasia. L. capitellata is distributed from Taiwan and Java to New Guinea. It is used locally for medicines and the fibre for ropes.

Small trees or shrubs. Leaves long petioled, alternate, crenate-serrate or nearly entire, 3-nerved, haory below; stipules connate into 1, 2-lobed or almost entire, intrapetiolar; flowers often dioecious, globose-glomerate or capitate, fruiting receptacle fleshy. Clusters axillary, sessile or pedunculate, solitary or 2-4 together, sometimes shortly racemose. Bracts in the male c;usters scarious: male Flowers: calyx 5- to 4-partite, segments valvate; stamens 5-4; rudimentary ovary glabrous or woolly. Female flowers: calyx cupular, 5-4 toothed or lobed; ovary obliquely ovoid; stigma sessile, penicillate or long-papillous; ovules basal, erect. Achene ovate, more or less compressed, calyx not enlarged. Seed with endosperm; cotyledons elliptic.

List of Species from Papuasia

Leucosyke alba (Bl.) Zoll. et Mor. (1854) 100. Urtica alba Bl. (1825) 495.

o o o o o o

Leucosyke australis M. Unruh (1943: 220)

o ● o o ● ●

var. eu-australis M. Unruh (1943: 221)

o o o o o ●

Leucosyke capitellata (Poir.) Wedd. in D.C. Prod. Urtica capitellata Poir (1804) 227

o ● ● o ● ●

var. eucapitellata M. Unruh (1943: 233)

o o o o o o

Leucosyke clemensii M. Unruh (1943: 230)

o o o o o o

var. eu-clemensii M. Unruh (1943: 230)

o o o o o o

var. longe-pedicellata M. Unruh (1943: 231)

o o o o o o

var. pilosula (1943: 231)

o o o o o o

Leucosyke corymbulosa

o o o o o o

Leucosyke montana H. Winkler (1922: 605)

● o o o o o

Leucosyke ochroneura (Bl.) Wedd. (186() 235. Missiessya ochroneura Bl. (1856) 164. L. sumatrana var. ochroneura Miq.

o o o o o o

var. eu-ochroneura M. Unruh (1943: 231)

o o o o o o

Leucosyke pulchra (Ridley) H. Winkler (1922: 606)
Debregeasia pulchra Ridley (1916)

● o o o o o

Leucosyke rhodopleura (Bl.) H. Winkler (1922: 604) Villebrunea rhodopleura Bl. 131 (1856) 168. Oreocnide rhodopleura (Bl.) Miq. (1859) 21. Debregeasia rhodopleura (Bl.) Boerl. (1900) 380.

● o o o o o

Leucosyke salomonensis M. Unruh (1943: 217)

o ● ● o ● ●

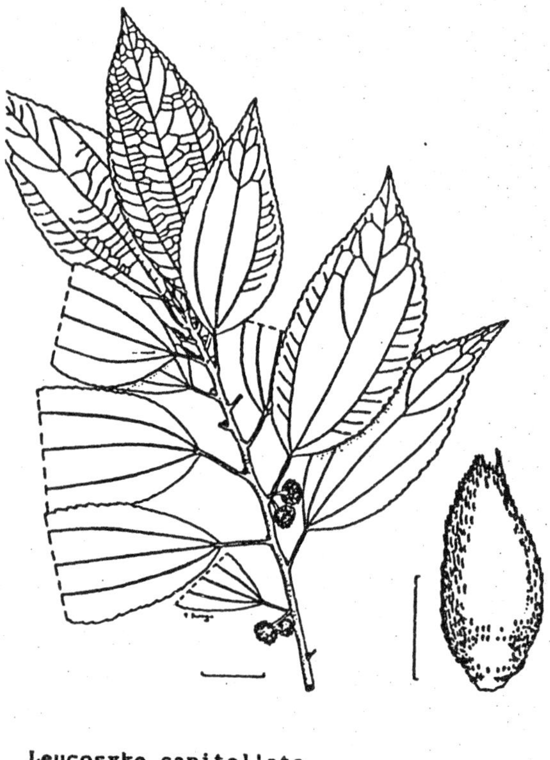

Leucosyke capitellata
NGF 42033.

10. MAOUTIA

o o o o o ●

A genus of 15 species from India and Malaysia to Polynesia. Female flowers lacking a perianth. The fibre of M. puya, distributed from Himalaya to Burma, is used for cloth and sails.

Shrubs. Leaves alternate, crenate to serrate, 3-nerved, white below; stipules connate, intrapetiolar, 3-lobed. Monoecious or dioecious. Male flowers: calyx 5-partite, segments acute, valvate or inflexed imbricate; stamens 4; ovary rudimentary, densely woolly. Female flowers: calyx minute or absent, ovary straight; stigma small; ovule basal, erect. Achene ovate, compressed. Seed with little endosperm.

List of Species in Papuasia

Maoutia cf. M. ambigua Wedd.

● ● ● o o o

Maoutia gracilis H. Winkler (1924: 133)

o o o o o o

Maoutia grandifolia (Ridley) H. Winkler (1922: 607)
Pipturus grandifolius Ridley (1916)

● o o o o o

Maoutia lanceolata Ridley (1916: 158)

● o o o o o

Maoutia rugosa Warburg (1891: 290)

o o o o o o

Maoutia salomonensis Rechinger (1912: 181)

o o o o o o

Notes

1. The species Maoutia ambigua is the basonym for Robinsoniodendron ambiguum (Wedd.) Merr. (1917). Three varieties were described by H. Winkler (1922: 607 - 608). These are listed, with authorities, under the genus Robinsiodendron

Key to Species in Papuasia
(Source: Winkler (1922)
(Translated by Dra. C. M. H. Lapre)

1. Leaves to 25 cm long and 16 cm broad; petioles 18 to 20 cm long. ..... Maoutia grandifolia

1 Leaves smaller and petioles shorter............................ 2

2. Leaves egg-shaped, to longish. ..... Maoutia ambigua

2. Leaves lanceolate. ..... Maoutia lanceolata

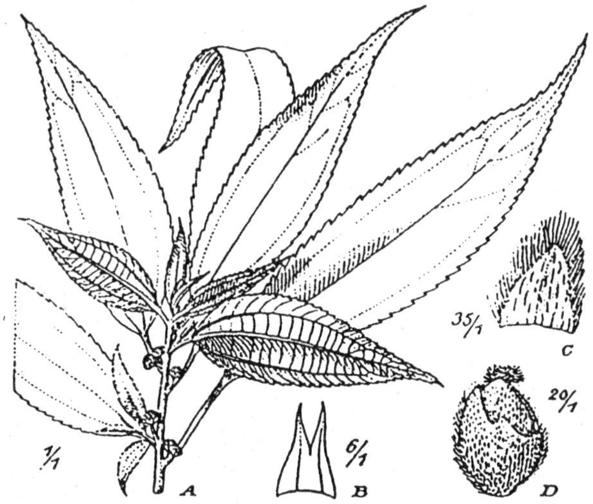

Leucosyhe rhodopleura A. Oberer Teil der Pflanze, B Nebenblatt, C Braktee, D Frucht in Perigon.
(Source: Winkler (1922)

Maoutia cf. M. ambigua Hartley 10811. Musgrave R.W.P.

MEMORIALIS

Memorials hirta (Bl.) Wedd. in D.C. Prod. 16.1 (18) 235.

Recorded in India, Java and New Hanover ? by Engler (Bot. Jahrb 7 (1886) 450) and K. Sch. (Bot. Jahrb. 9 (1887) 199.

MISSIESSYA

Missiessya ochroneura Bl is the basionym for Leucosyke achroneura (Bl.) Wedd. (Unruch 1943).

11 NOTHOCNIDE

o o o o o o

The genus Pseudopipturus was described by Skottsberg in 1933 to include climbing genera typified by Urtica repanda described by Blume. The 'lost' generic name Nothocnide was taken by Backer and Bakhuizen in 1965 in the Flora of Java as an older name for Pseudopipturus. Chew (1969) published a revision of the genus which includes 5 species centered in Malesia. All four Malesian species occur in Papuasia. The genus differs from Pipturus in the following characters: (a) climbers or scramblers; (b) lamina margin smooth not toothed; (c) underside of lamina without the white felt (except for N. discolor; (d) the female perianth fleshy and adnate to the ovary.

Dioecious soft-woody climbers devoid of irritant hairs. Leaves spirally arranged, stipulate, petiolate, simple. Lamina elliptic to obovate, prominently 3-veined, with fine intercostals almost perpendicular to the midrib; margin smooth; the upper side often dense with punctiform cystoliths, usually strigose. Stipules connate, intrpetiolar, bifid at apex, the two ridges on the outside strigose. Inflorescence spicate, axillary, with flowers in glomerules at regular intervals all the unbranched peduncle. Male flowers sessile to subsessile, ca. 1 mm long and broad, 4-merous; the rudimentary ovary surrounded by long thin white hairs. Female flowers sessile; perianth minutely toothed at the apex, ca. 1 mm. long, adnate to the ovary; stigma long ligulate. Achenes 1 - 1.5 mm. long and broad, obovoid often spherical.

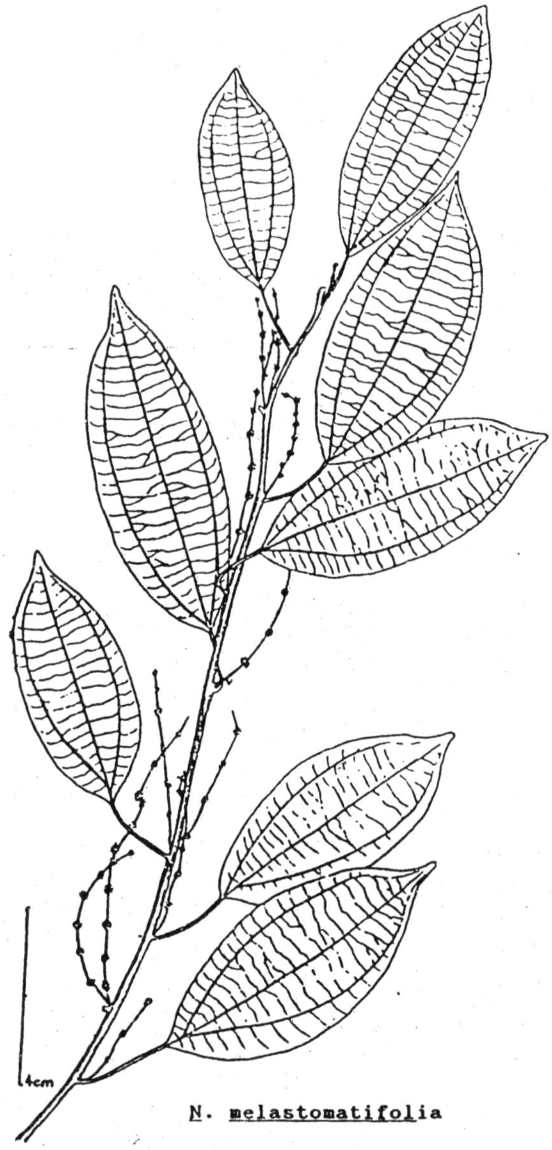

N. melastomatifolia

Key to Species in Papuasia

1. Underside of lamina covered with dense white felt. .... Nothocnide discolor (C. B. Robb) Chew (1969: 368)
   Pipturus discolor C. B. Robb Philip. J. Sc. Bot. 6: 15 (1911)
   Pseudopipturus discolor (C. B. Robb) Skottsb. Acta. Bot. Gothob. 8: (1913) 117

o ● o o o o

1. Underside of lamina glabrous to villose but without the dense white felt.......................... 2

2. Undersides of lamina densely pubescent to villose; twigs, petioles and peduncles densely strigose. .... Nothocnide mollissima (Bl.) Chew (1969: 365 - 367)
   Urtica mollissima Bl. Bijdr. (1825: 501)
   Pipturus mollissimus (Bl.) Wedd.
   Pipturus repandus var. rufescens Winkler (1922: 593)
   Pseudopipturus mollissimus (Bl.) Merr. Papers Mich. Acad. Sc. 23 (1938) 177

o ● ● o o ●

2. Not this combination of characters. 3

3. Lamina with only one pair of main lateral veins arising from the base and reaching to the apex. .... Nothocnide melastomatifolia (K. Sch.) Chew (1969: Fl. Kais. Wilhelm. Land (1889: 37) Winkler (1922: 593).
Pseudopipturs melastomatifolius (K. Sch.) Skottsb. Acta Hort. Gothob. 8 (1933) 117.

o ● o o o o

3. Lamina with 2 to 3 pairs of main lateral veins; one pair arising from the base but the others arising from the mid-rib near the apex...... Nothocnide repanda (Bl.) Bl. Mus. Bot. Lugd. Bat. 2 (1856). Tab. 14
Urtica repanda Bl. Bijdr. (1825: 501).

c ● ● ● ● ●

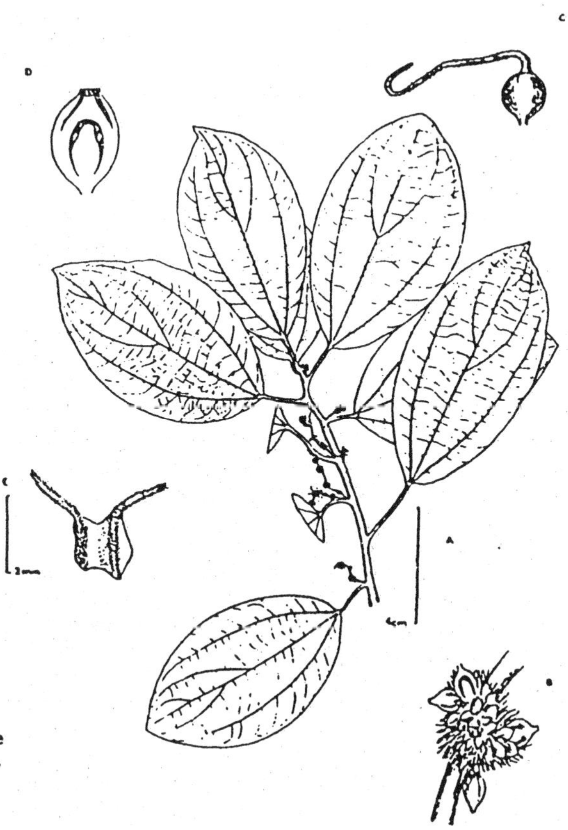

N repanda A. twig; B. & C. female flowers; D. female flower (v.s.); E. stipule.

OREOCNIDE

Species of Oreocnide are treated in the genera Villebrunea and Leucosyke.

12. PARIETARIA

o o o o o o

A subcosmopolitan genus of some 10-20 species. In New Guinea represented by a single widespread species occuring from Africa, throughout Malesia and Australasia to the Galapogos Is.

Small much branched, flaccid herb to 40 cm tall. Stems glabrescent to hairy, angular to subterete. Stipules 2 or none, to 0.2 mm long. Leaves 6-40 mm long, finely ciliate along margin or glabrous; (3-)10-30 by (2-)13-20 mm, base rounded or cordate, obtuse to acuminate at tip; single or triplinerved when lateral nerves suprabasal; cystoliths punctiform. Petioles densely pilose. Inflorescences 2-8 flowered, to 6 mm long. Bracts present or absent. Pedicels 0-0.2 mm long. Involucre 3-lobed.

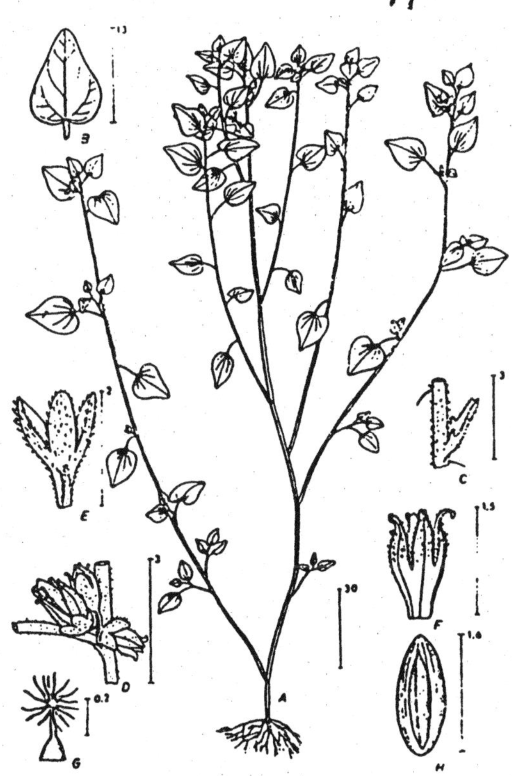

Parietaria debilis A. habit, B. leaf, C. part of stem with base of petiole and stipules, D. inflorescence, most of the pubescence left out, E. bracteoles, F. corolla, G. tip of ovary, style and stigma, H. nutlet. (Brass and Meyer-Dress 10345).
(Source: van Royen (1982)

Key to Species in Papuasia

1 species:
.... Parietaria debilis Forster Prodr. Fl. Aust. (1786) 73. Winkler (1922) 608.

● ● ● o o o

Notes:

1. Collected on Mt Wilhelmina by Versteeg (Type: 2520) and Brass and Meyer-Drees 10345. This species is of circumpacific distribution.

2. Growing in damp, shady areas from 3800 to 3850 m altitude.

3. Parietaria indica L. and Parietaria zeylanica L. are both treated as species of Pouzolzia.

13. PELLIONIA

o o o o o o

A genus of some 50 species from T E Asia and the Pacific. A group of monoecious or dioecious herbs and shrubs. Some ornamentals cultivated for colured leaves and stems. Treated by Winkler (1922: 546 to 562) as a subgenus of Elatostema but here recognised as a distinct genus. Elatostema subgenus Pellionia was revised by Schroter and Winkler in 1939.

Herbs or rarely shrublets. Leaves alternate, very unequal at base, the narrower side facing the branch, entire or dentate, often with a small leaf alongside: stipules intrapetiolar, undivided, small or obsolete. Flowers dioecious or monoecious, densely cymose or subcapitate, axillary. Male flowers: 5- (rarely 4-) partite, segments imbricate, with a mucro or horn at the apex: stamens 5-4; rudimentary ovary small. Female flowers: calyx-segments 5 -4, equal or unequal, nude or dorsally appendaged in the sterile flowers; staminodes scale-like, inflexed, or minute or absent. Ovary straight; stigma sessile; penicillate. Ovule basal, erect. Achene compressed or subterete; seed with very little endosperm; cotyledons ovate.

List of Species in Papuasia

Pellionia angulare (H. Winkler) comb. nov. Elatostema angulare H. Winkler (1922: 549. Fig. 8)

o ● o o o o

Pellionia biakense (H. Winkler) comb. nov. Elatostema biakense H. Winkler (1924: 124)

o o o o o o

Pellionia calophyllum Rech. Elatostema calophyllum Rech. (1912: 181, 102)

o o o o o ●

Pellionia cauliflorum (H. Winkler) comb. nov. Elatostema cauliflorum H. Winkler (1922: 556)

o ● o o o o

Pellionia curvipetalum (Perry) comb. nov. Elatostema curvitepalum Perry (1951: 386)

o o ● o o o

Pellionia eludens (Perry) comb. nov. Elatostema eludens Perry (1951: 387)

o o o o o o

Pellionia filicinum (Ridley) comb. nov. Elatostema filicinum Ridley (1916: 154)

● o o o o o

Pellionia fruticulosum (K. Sch. in K. Sch. & Laut.) comb. nov. Elatostema fruticulosum K.Sch. (1905: 254)

o ● o o o o

Pellionia grammicum (Perry) comb. nov. Elatostema grammicum Perry (1951: 385)

o o o o o o

Pellionia hallieri (H. Winkler) comb. nov. Elatostema hallieri H. Winkler (1922: 556)

o ● o o o o

Pellionia hoelscherianum (H. Winkler) comb. nov. Elatostema hoelscherianum H. Winkler (1922: 559. Fig 10: A - E)

● o o o o o

Pellionia hoffmannianum (H. Winkler) comb. nov. Elatostema hoffmannianum H. Winkler (1922: 552)

● o o o o o

Pellionia inamoenum (H. Winkler) comb. nov. Elatostema inamoenum H. Winkler (1922: 557)

● o o o o o

Pellionia inconstans (Perry) comb. nov. Elatostema inconstans Perry (1951: 384)

o o o o o o

Pellionia jabiense (H. Winkler) comb. nov. Elatostema jabiense H. Winkler (1922: 561)

● o o o o o

Pellionia janowskyi (H. Winkler) comb. nov. Elatostema janowskyi H.Winkler (1922: 559)

● o o o o o

Pellionia kietanum (Rechinger) comb. nov. Elatostema kietanum Rechinger (1912: 182)

o o o o o ●

Pellionia lamii (H. Winkler) comb. nov. Elatostema lamii H. Winkler (1924: 125. t. XVD)

o o o o o o

Pellionia lanceolatum (H. Winkler) comb. nov. Elatostema lanceolatum H. Winkler (1922: 558. fig. 9)

o ● o o o ●

Pellionis mafuluense (Perry) comb. nov. Elatostema mafuluense Perry (1951: 385)

o o o o o o

Pellionia minutiflorum (H. Winkler) comb. nov. Elatostema minutiflorum H. Winkler (1924: 126)

o o o o o o

Pellionia novae-britanniae (Laut.) comb. nov. Elatostema novae-britanniae Laut. (1905: 253)

o ● o o o ●

Pellionia pachypoda (Diels) comb. nov. Elatostema pachypoda Diels (1929: 75, 273)

o ● o o o o

Pellionia paramel(an)um (H. Winkler) comb. nov. Elatostema paramelanum H. Winkler (1922: 556)

o ● o o o o

Pellionia pauperatum (H. Winkler) comb. nov. Elatostema pauperatum H. Winkler (1922: 551)

● o ● o o o

Pellionia pellionianum (Gaud.) comb. nov. Elatostema pellionianum Gaud.

● ● ● o ● o

var. effusum H. Winkler (1922: 549)

o ● o o o o

var. novae-britanniae (Laut.) comb. nov. E. novae-britanniae Laut. (1905)

o o o o ● ●

var. pedunculatum (H. Winkler 1922: 549) comb. nov.

o ● o o o o

Pellionia peltata Ridley (1916) Elatostema peltifolium H. Winkler (1922: 553). (1916) non E. peltata

● o o o o o

Pellionia sinuatum (Bl.) comb. nov. Elatostema sinuatum (Bl.) Hassk. Catalog. Pl. Bogor (1844) 79

o o o o o o

var. pedunculatum (H. Winkler) comb. nov. H. Schroter Rep. Sp. Nov. Beih. 83 (1936) 32

o ● ● ● o o

Pellionia stenurum (Hallier f.) comb. nov. Elatostema stenurum Hallier f. Meded. Rijks. Herb. Leiden 26 (1915) 2

o o o o o o

Pellionium undulatum (H. Winkler) comb. nov. Elatostema undulatum H. Winkler (1922: 551)

o ● o o o o

Pellionium velutinum (Warb.) comb. nov. Elatostema velutinum (Warb.) K. Sch. in K. Sch. & Laut. (1905: 253) Procris velutina Warb. (1893: 189)

o ● o o o o

var. glabrum (K. Sch. in K. Sch. & Laut. 1905: 253) comb. nov.

o o o o o o

Pellionia angulare A. stem with flowers; B. cross section of stem; C. leaf with cystoliths; D. bud; E. malr flowers.
(Source: Winkler (1922)

Pellionia warburgii (H. Winkler) comb. nov. Elatostema warburgii H. Winkler (1922: 552)

o ● o o o o

Pellionium yulense (Hallier f.) comb. nov. Elatostema yulense Meded. Ryks. Herb. Leiden. 26 (1915) 3. Basionym: E. weinlandii

o o o o o o

var. yulense (Hallier f.) comb. nov.

o o o o o o

Additional Species

P. elatostemoides Gaud. Freyc. Voy. Bot.

P. nigrescens Warb. (1891) Elatostema warburgii Winkler (1922) 552.

P. weinlandii K. Schum. comb. nov. Elatostema weinlandii K. Schum. in Schum. & Laut. (1905) 254 Elatostema weinlandii Hall. f. (1915) 2
var. vestitum Winkler (1922) 554
var. dispar Winkler (1922) 554
var. kochii (Valeton)
Pellionia kochii Valeton (1907) 6. Elatostema weinlandii var. kochii (Valet.) Winkler (1922) 554.

P. lanceolatum

Key to Species in Papuasia

(Source: Winkler (1922)
(Translated by Dra. C. M. H. Lapre)

1. Plants with herblike stalks or with stalk becoming woody, in the last mentioned case tall, at least 40-50 cm high, stems at least 3-4 mm diameter. Leaves at least 8 cm long and 2.5 cm broad, longish looking. 2

1. Plants frail, with herblike or woody stalks. If woody with entire leaf margins.................... 17

2. Plants not or slightly branched. Leaves 8 cm or more long, at least 2.5 cm broad. Longish looking... 3

2. Plants abundantly branched. Leaves at most 8-9 cm long at most 2 cm broad. Lanceolate or longish lanceolate.......................... 12

3. Venation pinnate or seemingly palmate because the side nerves at the base are running towards the tip. 4

3. Leaves with three nerves........ 10

4. Stalk already at the top internodes 3-5 mm diameter, towards the ground up to 10 mm diameter. Leaves 24-30 cm long, 10-16 cm broad (compare also P. angulare) with, at the top 2mm thick and at the bottom 6 mm thick needles (points). ..... Pellionia velutinum

4. Stalk thinner, leaves at utmost 20 cm long and 8 cm broad (in a special case bigger); apex cuspidate........................... 5

5. Stalk not becoming woody. ..... Pellionia pellionianum

5. Stalk, at least at the base, becoming woody...................... 6

6. Inflorescences with long peduncles. ..... Pellionia angulare

6. Inflorescences sessile.......... 7

7. The flower tufts of 1-6 flowers. ..... Pellionia pauperata

7. The flower tufts with many flowers. 8

8. Leaf margins entire. ..... Pellionia hoffmannianum

8. Leaf margins crenate or dentate. 9

9. Margins crenate. Leaves short, cordate. Pellionia undulatum

9. Margins dentate. Leaves short, deltoid. Pellionia warburgii

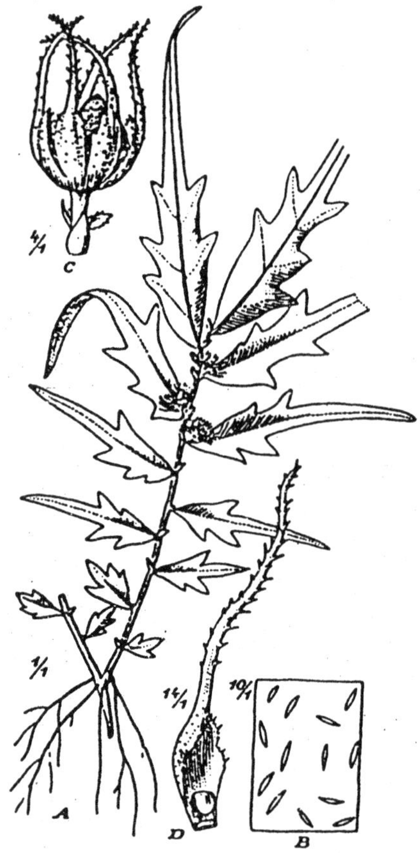

Pellionia hoelscherianum A. stem; B. leaf with cystoliths; C. flower; D. petals. (Source: Winkler (1922)

10. Leaves peltate (sometimes only slightly). ..... Pellionia peltifolium

10. Leaves not peltate............. 11

11. Leaves longish, egg-shaped. Male inflorescences much longer than the petiole. ..... Pellionia weinlandii

11. Leaves longish-lanceolate. Male inflorescences shorter or slightly longer than the petiole. ..... Pellionia kietanum

12. Venation of the leaves pinnate. 13

12. Leaves with three nerves (ribs). 16

13. Margins serrate. ..... Pellionia calophyllum

13. Margins dentate................ 14

14. Margins irregularly denticulate. ..... Pellionia cauliflorum

14. Margins regularly dentate...... 15

15. Leaves 1.5-4 cm long, apex cuspidate or mucronate; at the base of leaves oblique. Venation pinnate, by an arch connecting each vein.
..... Pellionia hallieri

15. Leaves 4-9 cm long, cordate, apex caudate; at the base of the leaf not or very slightly oblique. Pinnate venation by an arch connecting the veins.
..... Pellionia paramelanum

16. Leaves with three indistinct ner-nerves. Lower pair of side nervesonly at a small angle towards the top of the leaf, as with the next pair of side nerves. Apex mucronulate or mucronate.
..... Pellionia fructiculosum

16. Leaves clearly with three nerves, leaves longer and apex caudate.
..... Pellionia inamomum

17. Leaf margins entire.
..... Pellionia lanceolatum

17. Leaf margins dentate or lobed-dentate........................ 18

18. Apex cuspidate.
..... Pellionia hoelscherianum

18. Apex acuminate or acute........ 19

19. Leaves at the base slightly oblique and broader, at the narrow side obtuse and at the broad side rounded at the base.
..... Pellionia filicinum

19. Leaves at the base narrower and clearly oblique; on the narrow side a sharply wedged shape, on the broad side acute or cordate.... 20

20. Leaves lanceolate to linear-lanceolate; at both sides with 3-4 big dents.
..... Pellionia jaukowski

20. Leaves longish lanceolate, the larger leaves mostly with 6-8 dents.
..... Pellionia jabiense

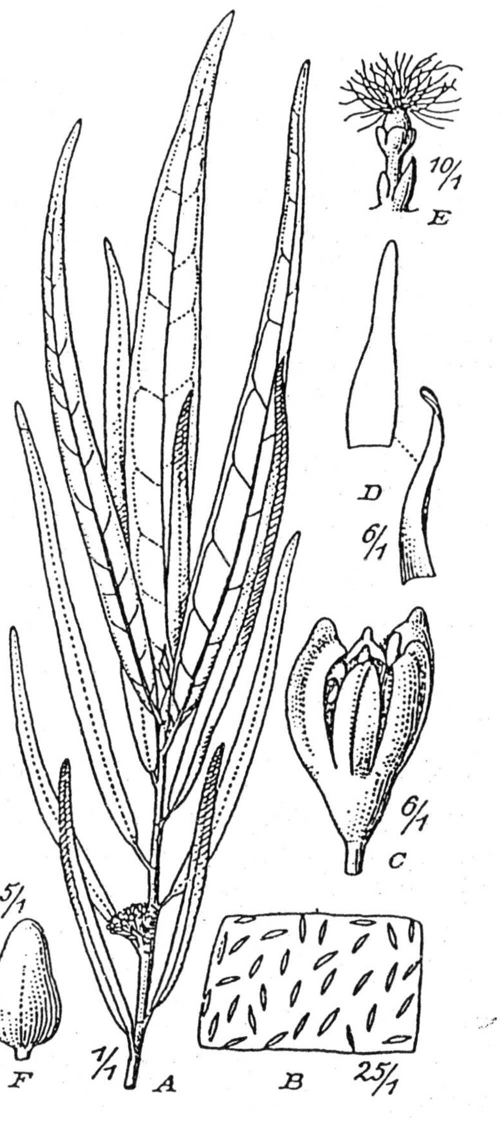

Pellionia lanceolatum A. stem; B. leaf with cystoliths; C. flower; D. petals; E male flower; F. fruit.
(Source: Winkler (1922)

14. PILEA

o o o o o o

A genus of 250 (to 600) tropical and warm temperate species. Some 45 species are known from Papuasia. P. cadierei, the aluminium plant, has leaves with silvery stripes.

Annual or perennial herbs, sometimes creeping or diffuse; leaves opposite, equal or very unequal pairs, entire or dentate; stipules connate into 1, intrapetiolar. Plants monoecious or dioecious. Male flowers: calyx 4-partite (rarely 3-2), segments concave, sometimes connate into a truncate cap; stamens 4 (3-2); ovary rudimentary. Female flowers: calyx 3-partite, segments unequal; staminodes opposite the segments, scale-like, minute; ovary straight; stigma sessile; ovule basal, erect. Achene ovate or orbicular, compressed, not or only partly enclosed by the calyx. Seed with scarcely any endosperm.

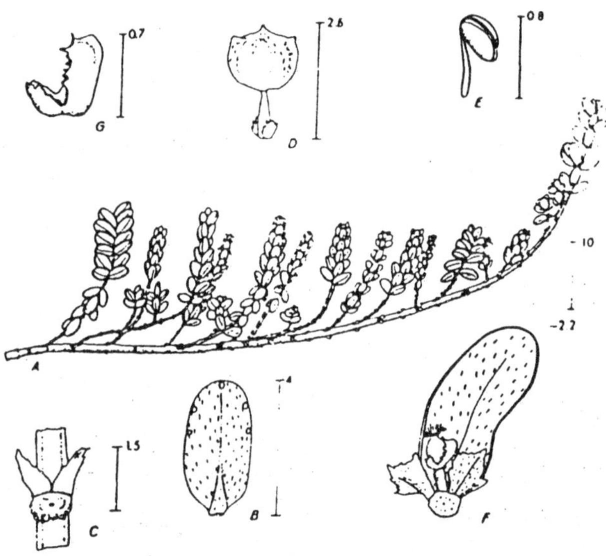

Pilea alpestris A. flowering branch; B. leaf, underside, C. stipules, D. male flower, E. stamen, F. female flower in axil of leaf, G. corolla of female flower. (Kloss s.n.)
(Source: van Royen (1982)

Key to High Altitude Species in Papuasia
(Source: van Royen (1982)

1. Leaves sparsely adpressed bristly below, somewhat denser below than above..... Pilea hederacea

1. Leaves glabrous on either side.. 2

2. Cystoliths punctiform. Perianth lobes of female flowers free.... 3

2. Cystoliths linear or none. Perianth lobes of female flowers connate at base.......................... 4

3. Creeping herbs. .... Pilea lacorum

3. Erect bush. Pilea zaranensis

4. Stipules 2- to 6-laciniate...... 5

4. Stipules entire................ 6

5. Leaves oblong-lanceolate, oblong-ovate, oblong, or rhombate-obovate; plicate-dentate along margin, often in apical 1/2 only, underside with scattered red glands. .... Pilea rubiacea var. latifolia

5. Leaves rhombate or ovate-rhombate, dentate along entire margin, underside without red glands. .... Pilea michaelensis

6. Margin of leaves dentate or serrate, sometimes minutely so.. 17

6. Margin entire or slightly wavy.. 7

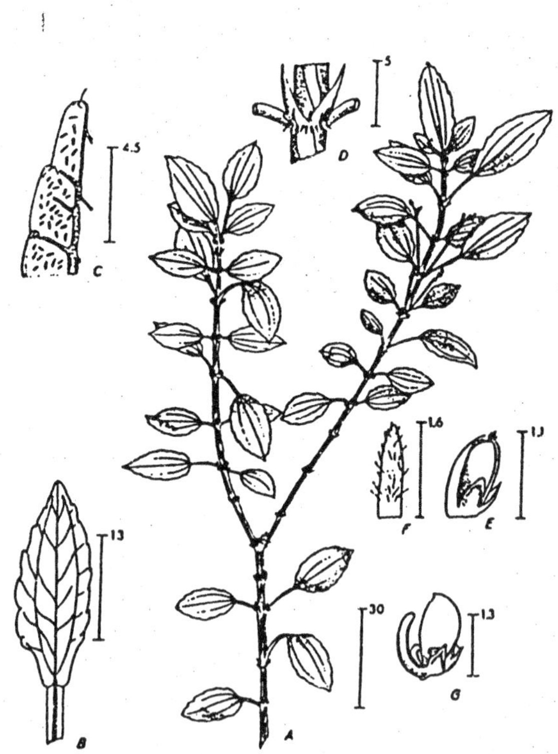

Pilea brassii A. flowering branch, B. leaf, upper side, C. margin of part of limb of leaf, upper side, D. stipules, E. female flower, F. staminode, G. nutlet. (McVean and Wade ANU 7183).
(Source: van Royen (1982)

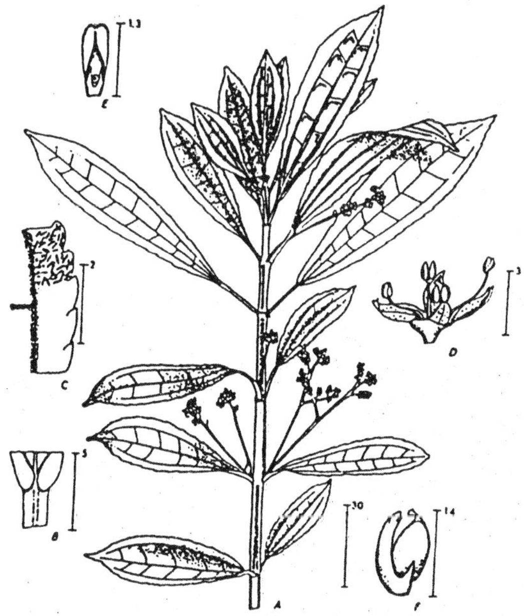

Pilea cuneata A. fruiting branchlet, B. base of leaf, upper side, C. part of margin of limb of leaf, D. male flower, E. abaxial view of female flower, ovary removed, F. female flower. (Hoogland 9553, except E - F. after van Royen 11297).

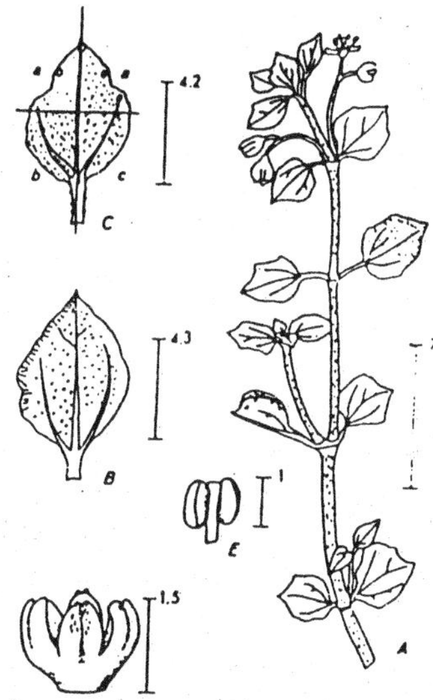

Pilea hederacea A. flowering branchlet; B. leaf, upper side, pubescence left out, at left two types of cystoliths, at right the linear type; C. leaf, underside, at a only showing one type of cystoliths, at b only the pubescence, at c only the second type of cystoliths; D. male flower; E. anther.

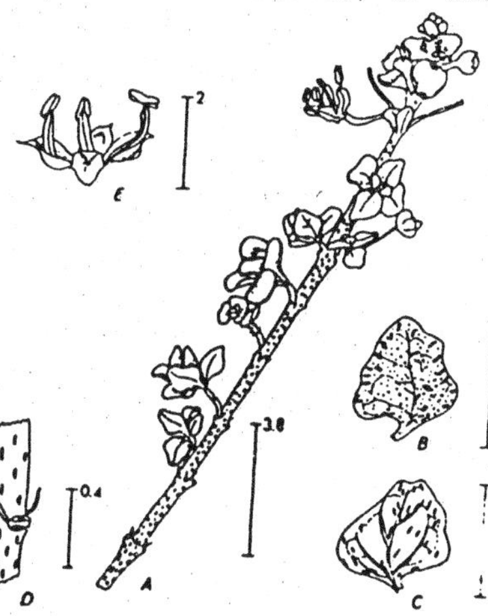

Pilea helxinoides A. flowering branchlet; B. leaf, upper side; C. leaf, underside; D. stipules; E. male flower. (Kloss s.n.).
(Source: van Royen (1982)

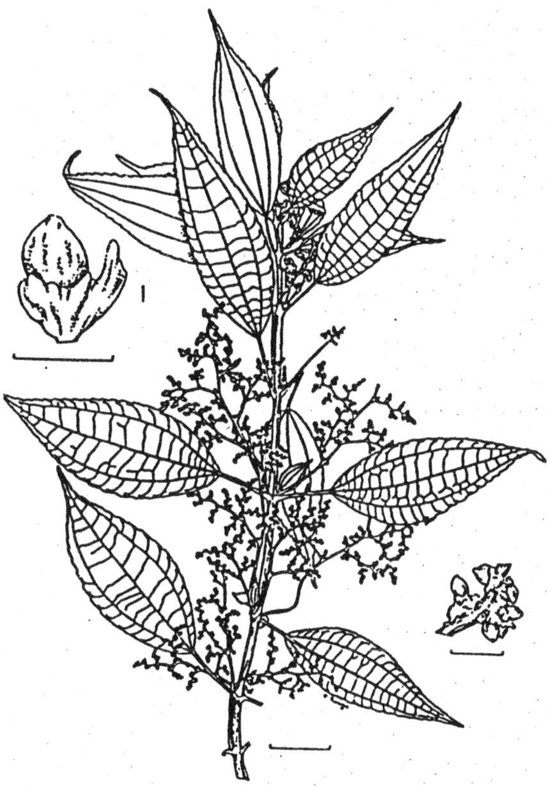

Pilea effusa A. twig; B. fruit; C. individual fruit in cup. (Brass 29650)
(Drawn: T. Iwagu)

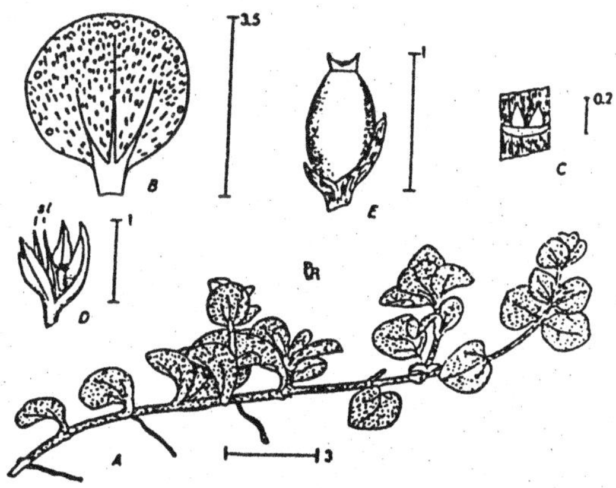

Pilea johniana A. habit; B. leaf, underside; C. stipules; D. female flower (st = staminodes); E. nutlet. (NGF 20023).
(Source: van Royen (1982)

7. Leaves minute, 1 - 1.5 by 1 - 1.5 mm........ Pilea minutissima

7. Leaves larger.................. 8

8. Leaves under 10 mm long......... 9

8. Leaves well over 10 mm long, basal nerves ascending at less than 30o from midrib.
.... Pilea victoriensis

9. Leaves pinnatinerved, without basal nerves........................ 10

9. Leaves triplinerved or 1-nerved, or seemingly nerveless, sometimes with several suprabasal nerves as well.
.... 11

10. Leaves fleshy, not brown punctate below..... Pilea helxinoides

10. Leaves thin, brown punctate below.
.... Pilea stellarioides

11. Cystoliths on upper side of leaves parallel to each other and near midrib at almost right angles to the latter. Leaves 1- to 3-nerved, ovate, indistinctly 3- to 5-lobed.
.... 12

11. Cystoliths on upper side of leaves sometimes parallel to each other, at a sharp angle to midrib or even parallel to latter, or scattered. Leaves triplinerved, 1-nerved, or lacking nerves................ 13

12. Inflorescences long-stalked, flowers in subumbelliform cymes. Leaves without round glands on underside along margins.
.... Pilea stellarioides

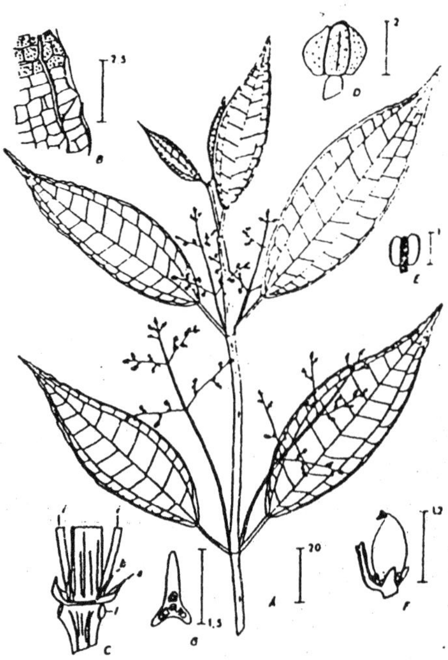

Pilea melastomoides A. flowering branch; B. part of margin of limb of leaf; C. stipules (s), bracts (b), base of inflorescences (i); D. male flower; E. anther; F. nutlet; G. female flower from above, showing three tepals and staminodes, ovary removed. (ANU 7217, D. E. Jacobs 8657).
(Source: van Royen (1982)

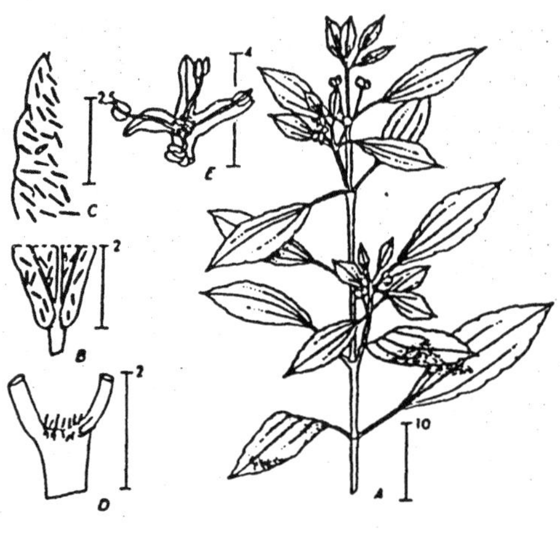

Pilea michaelensis A. flowering branch; B. base of limb of leaf; C. cystoliths in limb of leaf; D. part of stem with stipules. (Brass 31286).
(Source: van Royen (1982)

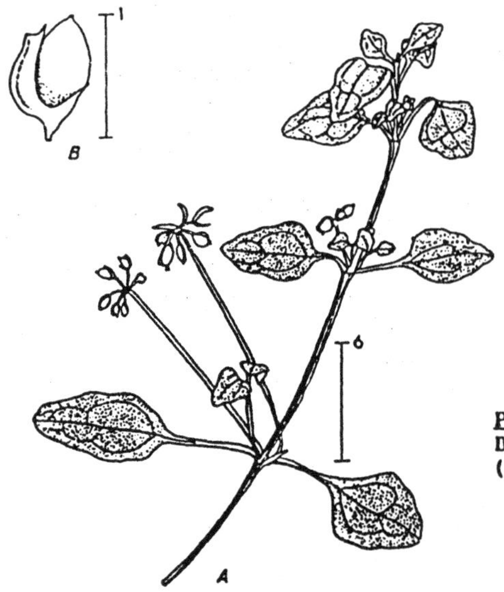

Pilea lacorum A. flowering branchlet; B. female flower. (NGF 14651).
(Source: van Royen (1982)

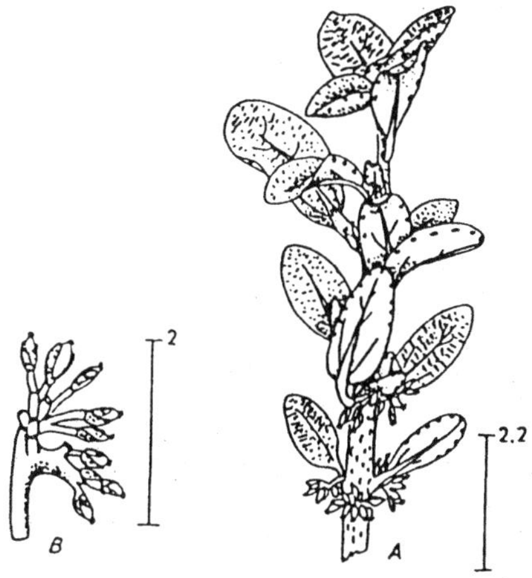

Pilea montis-wilhelmi A. flowering branchlet; B. female flowers. (ANU 7257). (Source: van Royen (1982)

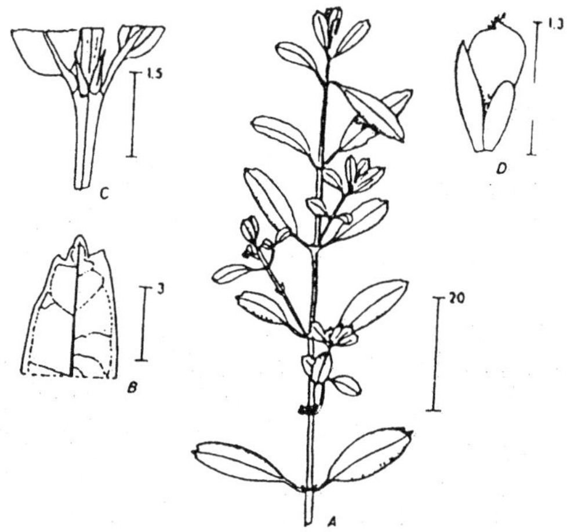

Pilea zaranensis A. flowering branch; B. tip of leaf; C. part of stem with stipules and bases of leaves; D. female flower. (NGF 16169).
(Source: van Royen (1982)

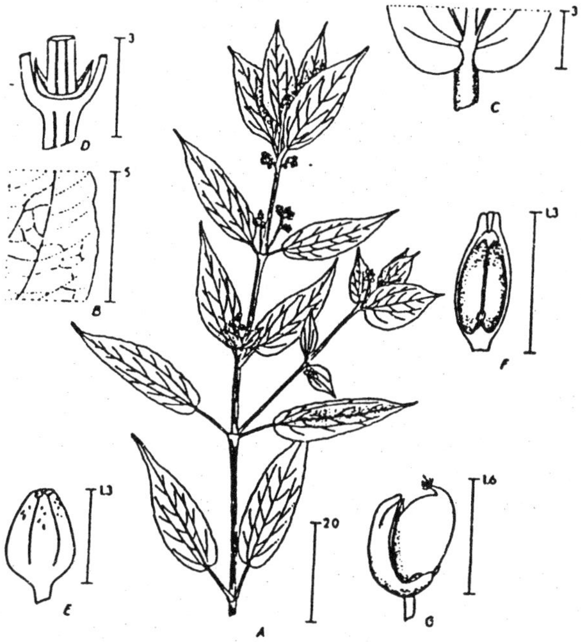

Pilea victoriensis A. flowering branch; B. part of limb of leaf; C. base of limb of leaf; D. stipules; E. male flower bud; F. male flower, one tepal removed; G. female flower. (LAE 61709, E-F Craven 2966).
(Source: van Royen (1982)

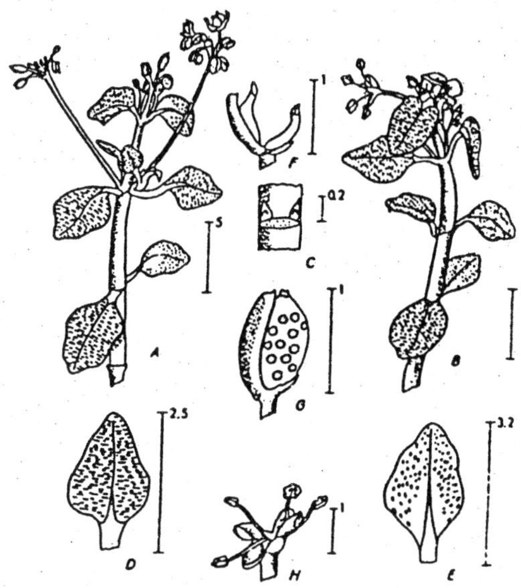

Pilea stellarioides A. Branchlet with female flowers; B. branchlet with male flowers; C. stipules; D. leaf, upper side; E. leaf, underside; F. sterile female flower; G. nutlet; H. male flower. (Vink 16265).
(Source: van Royen (1982)

12. Inflorescence sessile or sub-sessile, flowers seemingly scorpoid arranged. Leaves with 3 - 6 round glands on either side of underside along margin.
.... Pilea montis-wilhelmi

13. Leaves pandurate, obovate, or rhombate, widest above the middle... 14

13. Leaves widest below or in middle. 15

14. Stipules linear. Leaves pandurate, apiculate. Pilea pandurata

14. Stipules ovate, dentate. Leaves obovate, obtuse.
.... Pilea versteegii

15. Leaves oblong, elliptic, or rhombate...... Pilea alpestris

16. Leaves 1-nerved, with 1 large pellucid gland at tip.
.... Pilea thymifolia

16. Leaves triplinerved, with several large pellucid glands along margin.
.... Pilea johniana

17. Leaves triplinerved, without any nerves in between connecting the three nerves or only a few in apical parts.
.... Pilea rubiacea var. latifolia

17. Leaves triplinerved with few to many subtransverse nerves between the three nerves............... 18

18. Inflorescence unbranched, flowers in clusters, rarely branched in apical parts with the flower-clusters on the first branchlets, or sometimes on secondary branchlets...... Pilea cuneata

18. Inflorescence branched from basal 1/3 upwards, flowers scattered. 19

19. Leaves ovate-elliptic to obovate-elliptic, 3 - 20 by 1.6 - 10 cm. Stamens conoid, glabrous.
.... Pilea melastomoides

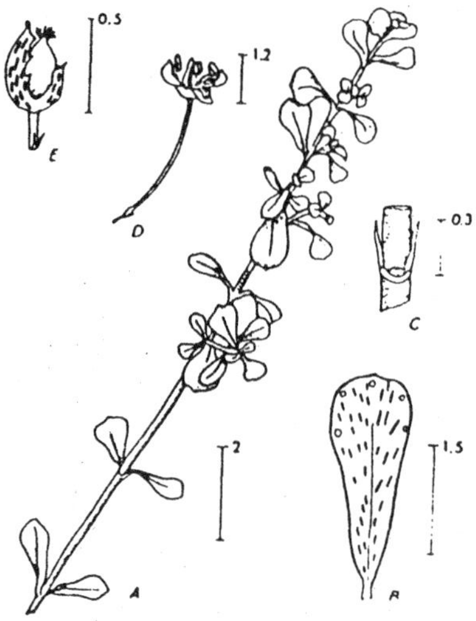

Pilea pandurata A. Flowering branchlet; B. leaf, underside; C. stipules; D. male flower; E. female flower. (van Balgooy 303, except D after Pullen 5154). (Source: van Royen (1982)

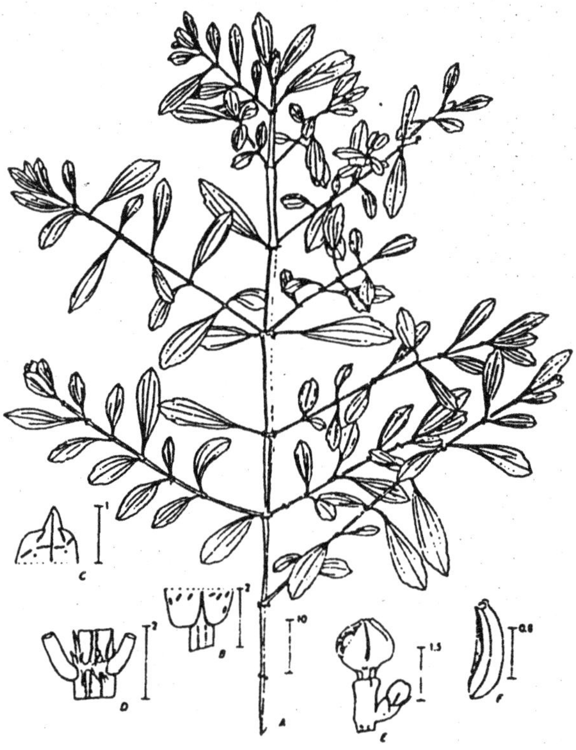

Pilea rubiacea var. latifolia A. flowering branch; B. base of limb of leaf; C. tip of limb of leaf; D. part of stem with stipules and bases of leaves, E. male inflorescence; F. tepal of male flower. (Brass 9292). (Source: van Royen (1982)

19. Leaves broadly elliptic, ovate, or lanceolate-elliptic, rarely deltoid orbicular, 0.9 - 2.8 by 0.9 - 1.7 cm. Staminodes oblong, sparsely pilose on outside.
.... Pilea brassii

List of Species in Papuasia

Pilea alpestris Ridley (1916: 152) van Royen (1982: 2095)

● o o o o o

Pilea brassii Chew in van Royen (1982: 2114)

o ● o o o o

Pilea caespitosa H. Winkler (1919: 419)

o o o o o o

Pilea cordata H. Winkler (1922: 514)

● ● o o o o

Pilea cuneata H. Winkler (1922: 513) van Royen (1982: 2117)

● ● ● o ● o

Pilea effusa H. Winkler (1922: 515)

o ● ● o o o

Pilea friesianus K. Sch. Basionym for Cyclophorus friesianus

Pilea hederacea van Royen (1982: 2097)

o o o o o o

Pilea helxinoides Ridley (1916: 151) van Royen (1982: 2099)

● o o o o o

Pilea johniana Stapf. Trans. Linn. Soc. Bot. 4 (1894) 227. van Royen (1982: 2093)

o ● o o o o

Pilea lacorum van Royen (1982: 2112

o o o o o o

Pilea lamii H. Winkler (1924: 120)

o o o o o o

Pilea ledermannii H. Winkler (1922: 517)

o ● o o o o

Pilea melastomoides (Poiret) Blume Mus. Bot. Lugd. Bat. 2 (1852: 54)

o ● ● o ● o

Pilea michaelensis van Royen (1982: 2131)

o o o o o o

Pilea minutissima H. Winkler (1922: 519)

● o o o o o

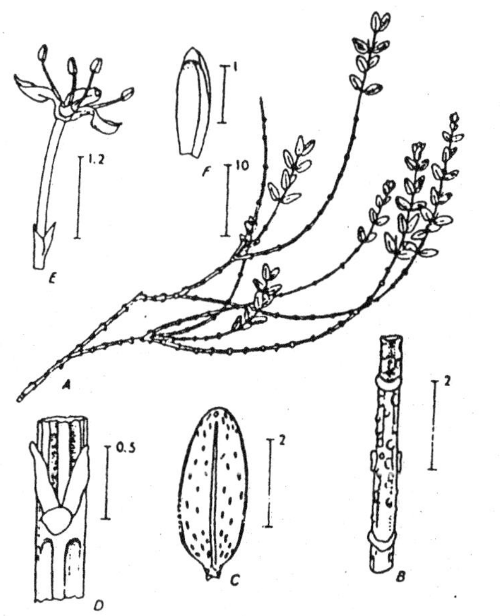

Pilea thymifolia A. flowering branch; B. part of branchlet; C. leaf, underside; D. stipules; E. male flower; F. free part of corolla of male flower. (Kloes s.n.).
(Source: van Royen (1982)

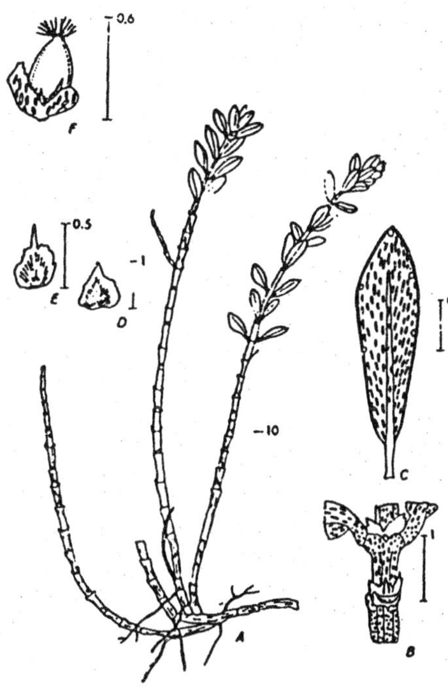

Pilea versteegii A. habit; B. stipules, in lower parts frontal, in higher parts lateral; C. leaf, underside; D. bract, E. bracteole; F. female flower. (van Royen NGF 16155)
(Source: van Royen (1982)

Pilea montis-wilhelmi van Royen (1982: 2110)

o o o o o o

Pilea pandurata van Royen (1982: 2101)

o o o o o o

Pilea papuana H. Winkler (1922: 519)

• • • • o o

Pilea pellis-crocidili H. Winkler (1922: 512)

o • o o o o

Pilea plicatidentata H. Winkler (1922: 513)

• • o o o o

Pilea rubiacea Ridley (1916: 150). Pilea roemeri H. Winkler (1919: 418) van Royen (1982: 2123) var. rubiacea

• • o o o o

var. latifolia (Winkler) Winkler (1922: 518) P. roemeri var. latifolia Winkler (1919)
Pilea trinervia Ridley (1916: 151) van Royen (1982: 2125)

• o o o o o

Pilea schlechteri H. Winkler (1922: 515)

o • • o o o

Pilea smilacifolia var. serrata Ridley (1916: 151)

o o o o o o

Pilea stellarioides H. Winkler (1922: 518) van Royen (1982: 2105)

o • • o o o

Pilea stenoneura H. Winkler (1922: 512)

o o o o o o

Pilea thymifolia Ridley (1916: 152) van Royen (1982: 2107)

• o o o o o

Pilea thymoidea H. Winkler (1919: 419) van Royen (1982: 2109)

• • o o o o

Pilea versteegii H. Winkler (1919: 418) van Royen (1982: 2103)

o o o o o o

Pilea victoriensis van Royen (1982: 2128)

o o o o o o

Pilea zaranensis van Royen (1982: 2126)

o o o o o o

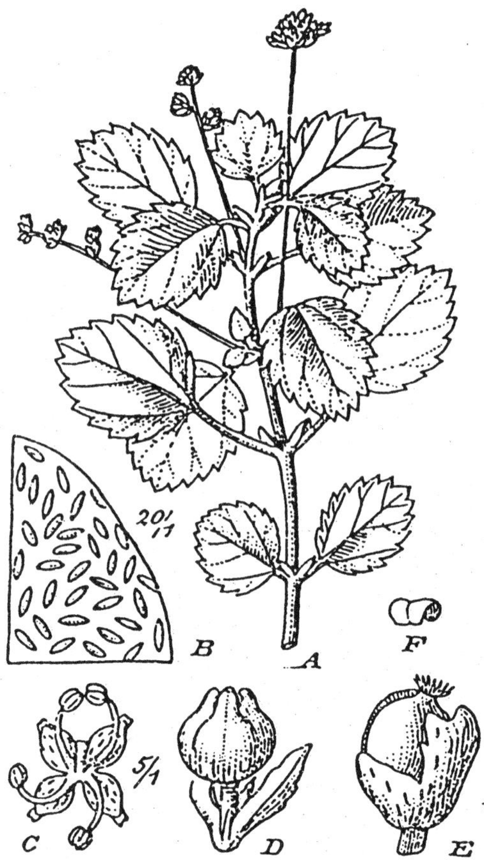

Pilea schlechteri A. stem; B. leaf with cystoliths; C. male flower; D. female flower bud; D. female flower; E. staminodes.
(Source: Winkler (1922)

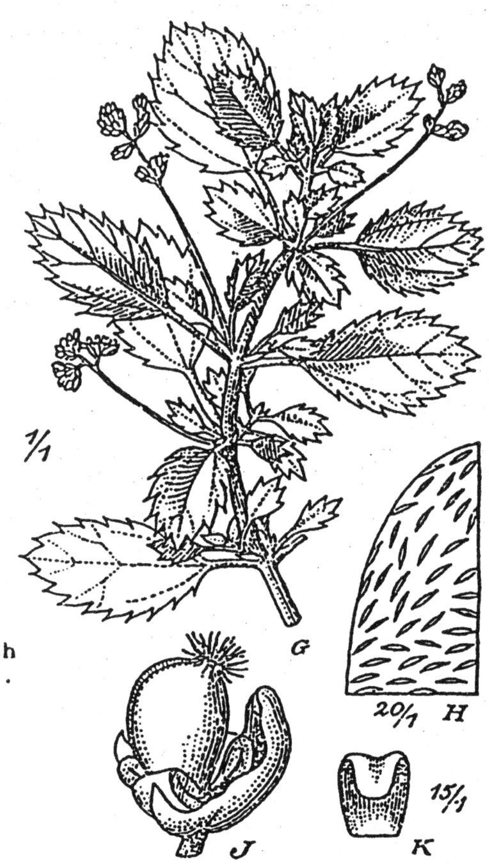

Pilea ledermanni G. stem; H. leaf with cystoliths; J. female flower; K. staminodes.
(Source: Winkler (1922)

Key to Species of Pilea in Papuasia
(Source: Winkler (1922)
(Translated by Dra. C. M. M. Lapre)

1. Plants strong, upright, or laying down and rooting at the nodes but not growing in clumps........... 2

1. Plants delicate, with very small leaves to 5 mm long, laying down, growing in clumps or very small bushes......................... 10

2. Herb-like plants or shrub-like; leaves large, at least 5 cm long. 3

2. Plants herblike or shrublike with a light foliage (without many leaves) leaves small (up to 2 cm long); elliptic eggshaped to orbiculate (round)................. 8

2. Plants bushy or shrubby, with a close foliage. Leaves at most 1 cm. long, lanceolate (var. latifolia obovate).. Pilea rubiacea

3. Leaves nearly leather-like; both side veins of the middle vein close to the margin of at most in middle of lamina. Pilea stenoreura

3. Leaves papery: the side veins closer to the midrib........... 4

4. Stipules attached (like a wornout collar) round the base of the petiole, leaving a scar after shedding...................... 5

4. Stipules not attached around the base of the petiole............ 6

5. Leaves lanceolate, 12-18 cm. long, with sideveins coming perpendicular from the midrib. .... Pilea pellis-crocodili

5. Leaves sagitate - lanceolate, 5-12 cm. long, with side veins coming in a narrower angle towards the tip. .... Pilea cuneata

6. Leaves longish-lanceolate, 8-20 cm. long, cystoliths obvious on the upperside of the leaves......... 7

6. Leaves longish, 6-12 cm., at least at the upperside without cystoliths. .... Pilea effusa

7. Leaves broadening towards the base, at the base truncate, very rarely attenuate. Pilea caudata

7. Leaves broader in the middle or towards the top; cuneate at the base...... Pilea plicatidentata

8. Plants bare (lacking hairs)..... 9

8. Plants with thick rough hairs, leaves crenulate, the notches ending in hairs. .... Pilea ledermanni

9. Laeves round shaped, mostly obtuse, rarely acute. .... Pilea schlechteri

9. Leaves deltoide (egg-shaped), with a short tip. .... Pilea papuana

10. Leaves with three ribs (nerves), at the underside without glandular spots at the margins; stipules very small linear, free............. 11

10. Leaves with one rib, underneath on the margins with sunken glandular spots, one spot at the tip of the mid-vein; stipules small or large, not linear........... 13

11. Stalk not becoming woody; leaves flimsy, thin; underneath with brown specks.... Pilea stellarioides

11. Stalk becoming slightly woody; leaves thick, underneath without brown specks.................... 12

12. Stalk nobby because of the many cystoliths; leaves orbicular. .... Pilea helxinoides

12. Stalk clearly without cystoliths; leaves rhomboid (oblanceolate). .... Pilea versteegii

13. Leaves up to 5 mm. long, stipules clearly brown, growing into the petiole, margins frayed. .... Pilea alpestris

13. Leaves smaller; stipules small, not grown into the petiole (free), dentate......................... 14

14. Leaves oblanceolate, underneath, besides the spot at the tip, also 2 other glandular spots at the margin; stipules strongly pointed. .... Pilea minutissima

14. Leaves oblanceolate or oval, underneath next to the spot at the tip, many other glandular spots at the margin, stipules not pointed.... 15

15. Leaves closely placed, at most 2 mm. long, oval, underneath with 2-3 glandular spots on each side of the middle nerve. .... Pilea thymifolia

15. Leaves more scarsely spaced, to 3mm. long, oblanceolate, underneath with 5-6 glandular spots on each side of the middle nerve. .... Pilea thymoidea

o o o o o o

A genus of about 40 species from Mascarenes and the Malay Archipelago to the Pacific Is. and Australia. 14 species in Papuasia. Some bark is used for fibre and bark cloth in the Pacific.

Trees or tall shrubs, rarely scandent. Leaves alternate, entire or crenate-serrate, 3-5-nerved, often wooly below. Stipules unitedinto one, 2-fid, intrapetiolar, caducous. Flowers sessile axillary or spicate; bracts small. Flowers dioecious (rarely monoecious). Male flowers: calyx 5-4-fid, lobes acute, valvate; stamens 5-4; ovary rudimentary, woolly. Female flowers: calyx ovoid, narrowed at apex, mouth minutely dentate, thinly fleshy in fruit; ovary enclosed, adherent to calyx; stigma linear, exserted, villous on 1 side, deciduous. Achene closely enclosed by slightly enlarged calyx.

List of Species in Papuasia

Pipturus albidus H. Winkler (1922: 586-594)

o o ● o o o

Pipturus argenteus (Forst.f.) Wedd. in D.C. 16.1 (1869) 235
Urtica argentea Forst.f. (1786)
var. incanus (Bl.) H. Winkler (1924: 131) P. incanus (Bl.) Wedd.
Pipturus velutinus H. Winkler (1919: 584)

● ● ● ● ● ●

Pipturus cinnamomeus Ridley (1916: 156)

o o o o o o

Pipturus grandifolius Ridley (1916: 156)

o o o o o o

Pipturus ledermannii H. Winkler (1919: 176)

o ● o o o o

Pipturus lithospermum H. Winkler (1922: 512)

o ● o o o o

Pipturus melastomatifolius K. Sch. in K. Sch. & Hollr. (1889: 37)

o ● o o o o

Pipturus papuanus Gibbs (1917: 128)

● o o o o o

Pipturus propinquus Wedd. Fl. Vit. (1868) 244

o o o o o o

Pipturus pullei H. Winkler (1919: 175)

● ● ● o o o

Pipturus repandus (Bl.) Wedd. Arch. Mus. Nat. Hist. 9 (1857) 448
Urtica repanda Bl. (1825)

● ● o o o o

var. rufescens H. Winkler (1922: 593)

o ● o o o o

Pipturus sp. A. habit; B.C. fruits. (Drawn: T. Iwagu)

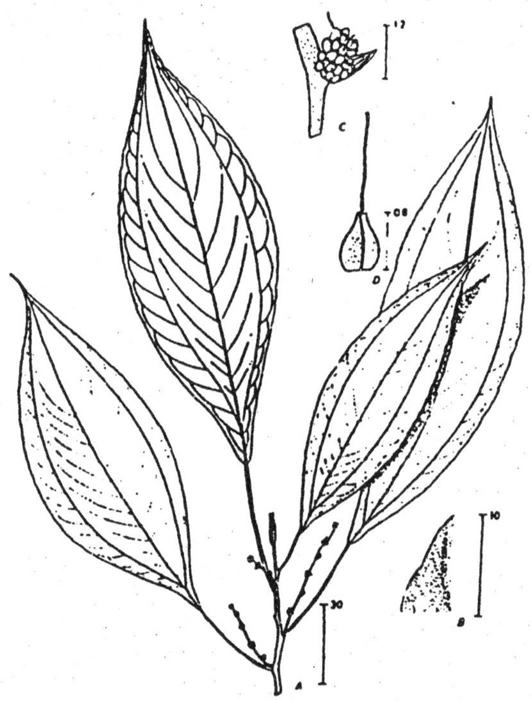

Pipturus cinnamoneus A. flowering branch; B. part of margin of limb of leaf: upper side; C. female inflorescence; D. female flower. (Kloss s.n.). (Source:

Pipturus subinteger H. Winkler (1922: 584)

● ○ ○ ○ ○ ○

Pipturus verticillatus H. Winkler (1922: 589)

○ ● ○ ○ ○ ○

Key to Species in Papuasia

(Source: Winkler (1922)
(Translated by Dra. C. M. H. Lapre)

1. Plants climbing, with few hairs. The underside of the leaves not velvety........................ 2

1. Plants shrub-like, shrubs or trees; strongly haired. Leaves, on the underside, grey- white- or brown-felty.......................... 3

2. Leaves with two side veins at the base and additional short side veins at the top.
..... Pipturus repandus

2. Lacking the side veins at the top of the leaf or very rarely one or two side veins visible.
..... Pipturus melastomatifolius

3. Flower heads simple spikes or branched panicles.............. 4

3. Flower heads in spikes or panicles and individually situated at the leaf node, or all the flower heads are situated individually in the leaf node...................... 6

4. Leaves evenly and finely wrinkled; inflorescences (only females are known) multiple whorl-like branched... Pipturus verticilatus

4. Leaves not evenly, finely wrinkled. Inflorescence a spike or panicle.
.... 5

5. Twigs with flat stiff hairs; leaves clearly serrate.
..... Pipturus argenteus

5. Twigs with short, thick, velvety hairs; leaves not clearly serrate.
..... Pipturus subinteger

6. Flower heads in spikes or panicles with a few individual flowers situated at the leaf nodes. Leaf margins entire.
..... Pipturus ledermanni

6. Flower heads individual situated at the leaf node.................. 7

7. Leaves with entire margins, lanceolate. Pipturus papuanus

7. Leaf margins serrate, egg-shaped.
.... 8

Pipturus ledermanni A stem with male flowers; B. male flower; C. bract.
(Source: Winkler (1922)

Pipturus montanus A. flowering branch; B. leaf margin; C. stipule; D. female inflorescence.
(Source: van Royen (1982)

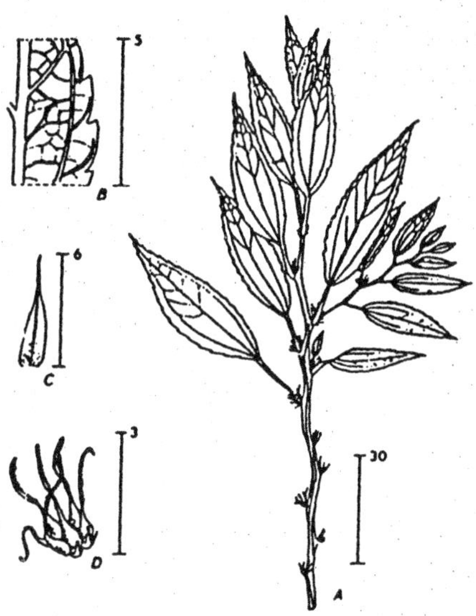

8. Plants in general with more upright soft hairs. Leaves swollen or thick.
..... Pipturus pullei

8. Plants with laying stiff hairs. leaves thin.................... 9

9. leaves dentate-serrate, on the underside with short, rough hairs.
..... Pipturus lithospermum

9. Leaves spinose, on the underside with short, soft hairs.
..... Pipturus albidus

16. POIKILOSPERMUM

o o o o o o

A genus of some 20 species which is confined entirely to the Rain Forests of Indo-Malaysia. It is divided into two subgenera: Pikilospermum mainly found in Philippines and E Malaysia and Ligulistigma, which is mainly W Malaysian. There are four species in Papuasia: the genus has however not been collected on Manus, New Ireland or the Solomons. Merrill (1943) described 36 species but only 20 were maintained by Chew (1963). Following Chew (1963) the genus Concephalus Bl. is included in Poikilospermum.

The genus occupies an intermediate position between the Moraceae: the vegetative structures are moraceous, the reproductive parts urticaceous.

Dioecious woody scramblers. Lamina variable; with petiole, usually coriaceous, cystoliths arranged in circular groups on adaxial surface, on the upper surface arranged along the veins. Petioles canalised on upper surface. Stipules connate, intrapetiol-ar, very coriaceous; strongly crescentic (rarely) to straight and long; caducous to persistent.

Inflorescence axillary, cymose, often solitary, rarely paired to numerous. Flowers generally small to minute. Male flowers: sessile to subsessile; perianth glabrous to thick pubescent; tepals 2 to 4, free, usually strongly incurved at apex; stamens 2 to 4, opposite the tepals; pistillode always present, small to large. Female flowers: perianth glabrous to pubescent, apex 4-toothed or 4-lobed, lobes 2 + 2. Ovary superior, basal and orthotropous; stigma subsessile, even rudimentary stamens absent. Fruits dry, small brownish achenes, pericarp slightly warty, at least half, often more, exserted from perianth (in Ligulistigma totally enclosed by remnant perianth) it. Embryo straight.

Key to Species in Papuasia

(Source: Chew (1963)

1. Filaments as broad as tepals, stamens 4; flowers subsessile, lamina vary large, at least 40 - 70 cm long by 18 - 25 cm broad; obovate, lateral veins very prominent, closely set, more than 18 pairs; intercostals very straight and prominent, very numerous and closely set.
.... P. nobile (Ridl.) Merril (1934: 50) Fig. 18.
Concepalus nobilis Ridley Trans. Linn. Soc. Bot. 9 (1916) 150.

o o o o o o

Poikilosperum amboinense A. NGF 19267
(Drawn: T. Iwagu)

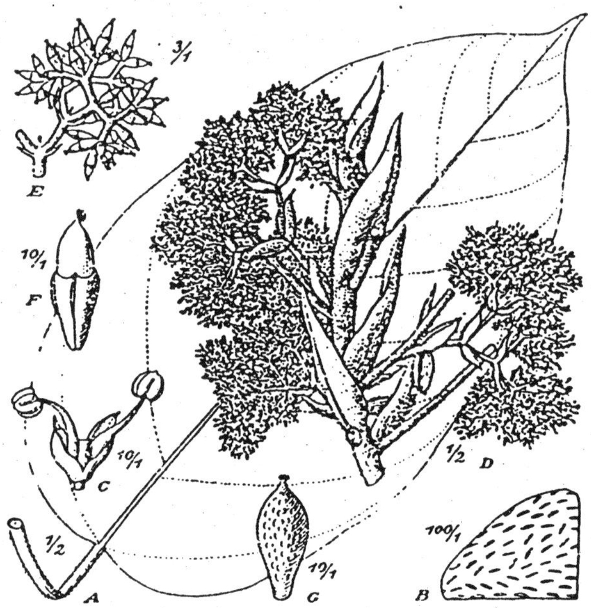

Poikilospermum hirsutus A. stem; B. leaf with cystoliths; C. male flower; D. fe,ale inflorescence; E.F. female flowers; G. fruit.
(Source: Winkler (1922)

1. Filaments much narrower than tepals, stamens 2 - 4; flowers subsessile to pedicellate; lamina small, 2 cm long by 3 cm broad, to as large as those of the above species, elliptic or ovoid ie. broadest at or below the middle; lateral veins 6 - 16 pairs rarely 18; intercostals almost not evident, to very numerous and prominent.......................... 2

2. Male and female flowers distinctly pedicellate; pedicel as long as, sometimes longer than, perianth; peduncles distinctly canescent; lamina large elliptic; lateral veins 13 to 18 pairs, closely set, intercostals prominent.
.... P. paxianum (H. Winkler) Merr. (1934: 51)
Concephalus paxianus H. Winkler (1922: 602)
C. amboineensis var. longifolius Winkler (1922: 601).

o o o o o o

2. Male and female flowers sessile or subsessile; pedicel shorter than perianth; only the solitary terminal flowers distinctly pedicellate; peduncles glabrous to thickly pubescent, very rarely canescent; lamina variable; lateral veins less than 14 pairs........ 3

3. Perianth of male flowers splitting irregularly into 2 to 5 unequal parts; secondary peduncles of male inflorescences of unequal lenths, very thin and flimsy; flowers free, or at most 5 in each agglomeration; usually more than 2 inflorescences at each axil; plants very frail; twigs thin and weak; lamina eliptic; intercostals very faint.
.... P. inaequale Chew (1963: 53) Fig. 17
Type: NGF 6850.

o ● ● o ● o

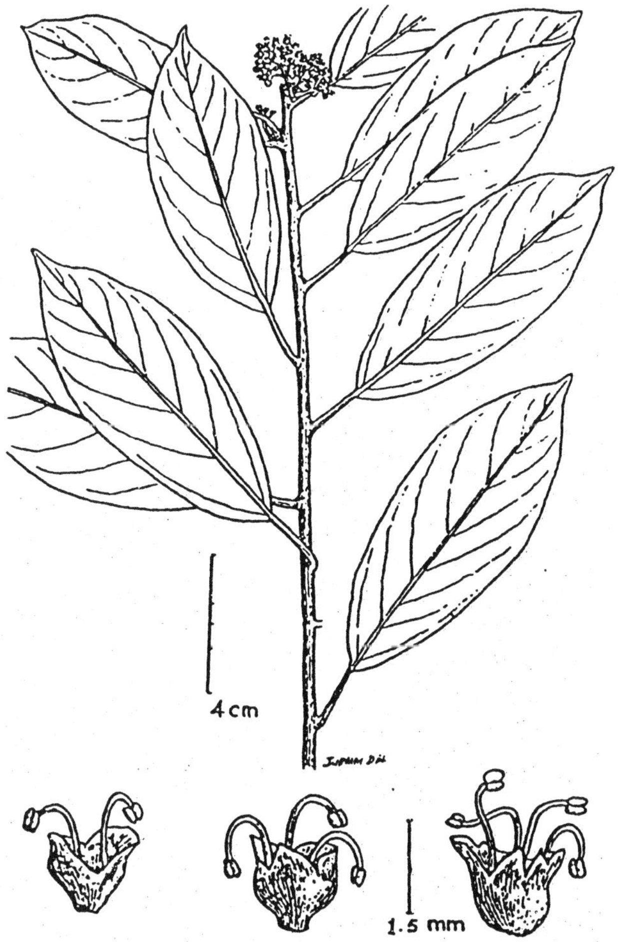

Poikilospermum inaequale A. Male twig. Perianth with 2-5 irregular tepals. (Source: Chew (1963)

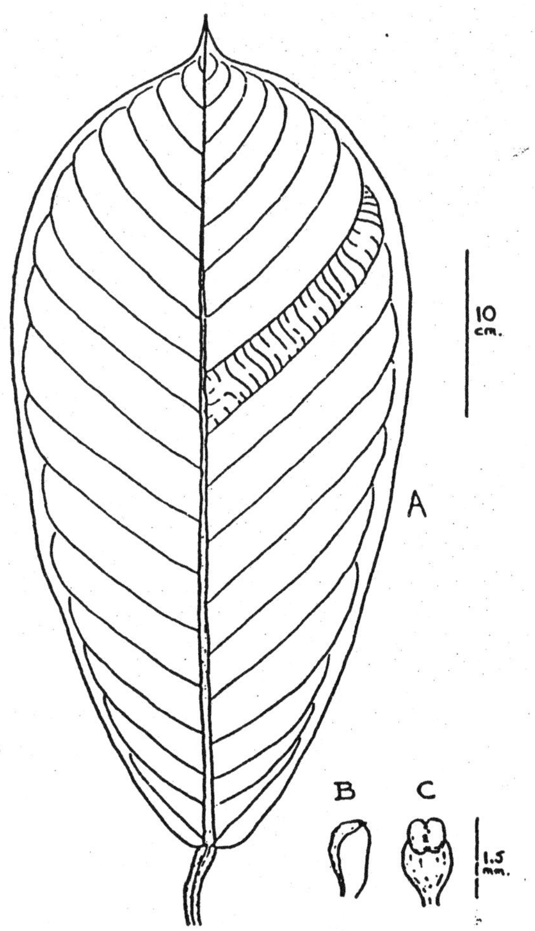

Poikilospermum nobile A. lamina; B. tepal; C. stamen. (Source: Chew (1963)

4. Perianth of male flowers splitting regularly into 2 to 4 equal parts; peduncles stouter; secondary peduncles usually equal in length; more than 5 flowers in each male floral agglomeration; generally less than 2 inflorescences at each axil; plants stout; twigs thick to very robust; lamina variable; intercostals prominent.
.... P. amboinense Miq. Ann. Mus. Bot. Lugd. Bat. 1 (1864) 203.
Concephalus amboinensis (Miq.) Warb.
C. forbesii Moore J. Bot. London 63 (1925) suppl. 112.
C. hirsutus Winkler (1922) 598. t. 14
C. subscaber Winkler (1922) 599
C. gjellerupii Winkler (1922) 602
C. papuanus Barg.-Petr. Nuovo G. Bot. Ital. n. ser. 9 (1902) 228 t. 13.

o ● ● ● ● o

Notes:

1. P. amboinense is a common species in riverine forests from sea-level to 550 m altitude.

2. A common climber in swampy forests, P. inaequale occurs mostly above 1000m. It is the only species with the male perianth splitting irregularly into unequal parts.

3. P. nobile is known from only two specimens. It is distinct is shape and venation of lamina, and is only species in Papuasia to have the filaments as broad as the tepals.

4 The presence of distinct pedicels distinguishes P. paxianum.

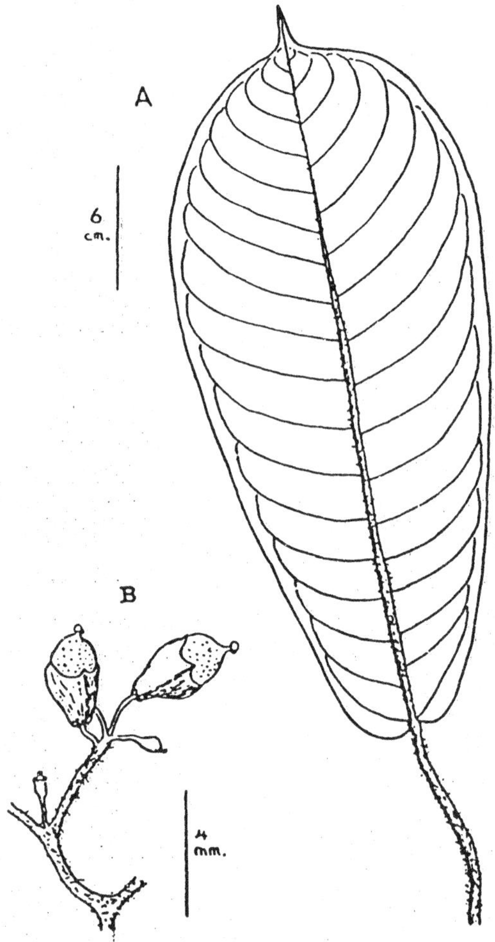

Poikilospermum paxianum A. lamina; B. young and matured female flowers. (Source: Chew (1963))

17. POUZOLZIA

o o o o o o

A genus of 50 tropical species. In Papuasia some 4 species. Fibres used for fishing nets. Hutchinson (1967) lists several genera as synonyms for Pouzolzia.

Herbs or shrubs. Leaves alternate or the lower very rarely all opposite, mostly entire, 3-nerved, upper gradually smaller, upper most sometimes passing into bracts; stipules free, often persistent. Flowers monoecious, rarely dioecious, clustered in the leaf-axils; bracts small. Male flowers: calyx 5- or 4- lobed or partite, rarely 3-partite, segments valvate, either concave at the apex or abruptly inflexed and transversely plicate on the back; stamens 5-4, rarely 3; rudimentary ovary clavate or oblong. Female flowers: calyx tunbular, often beaked, mouth contracted, 2-4-toothed; ovary included; stigma filiform, villous on one side; ovule basal, erect. Achene enclosed by the marcescent or scarsely changed calyx, often free from it; seed with little or no endosperm; cotyledons ovate.

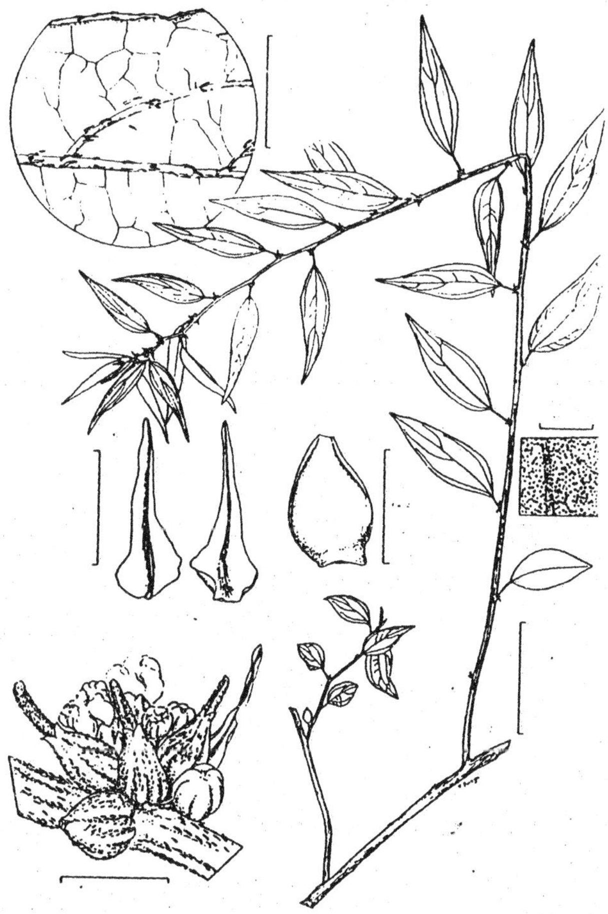

List of Species in Papuasia

Pouzolzia papuana Laut. & K. Sch. (1900: 294).

o ● o o o o

Pouzolzia pentandra

● ● o o ● o

Pouzolzia quiquenervis Oenn. in Horsf. Pl. Java. 66

o o o o o o

Pouzolzia zeylanica (L.) Benn. Pl. Java (1838) 67
Pouzolzia hirta

o ● ● o o ●

Key to Species in Papuasia

(Source: Winkler (1922)
(Translated by Dra. C. M. H. Lapre)

1. The side nerves not going to the tip of the leaf, but going to the margins...................... 2

1. The side nerves going to the tip of the leaves..................... 3

2. Leaves entire, lanceolate to egg-shaped.
..... Pouzolzia indica

2. Leaves closely serrate, broad egg-shaped.
..... Pouzolzia papuana

3. Plants scarcely branched; bracts of the inflorescences, at least the upper bracts very small, sheathing.
..... Pouzolzia pentandra

3. Plants with many branches; bracts of the inflorescence like the other leaves or slightly smaller.
..... Pouzolzia hirta

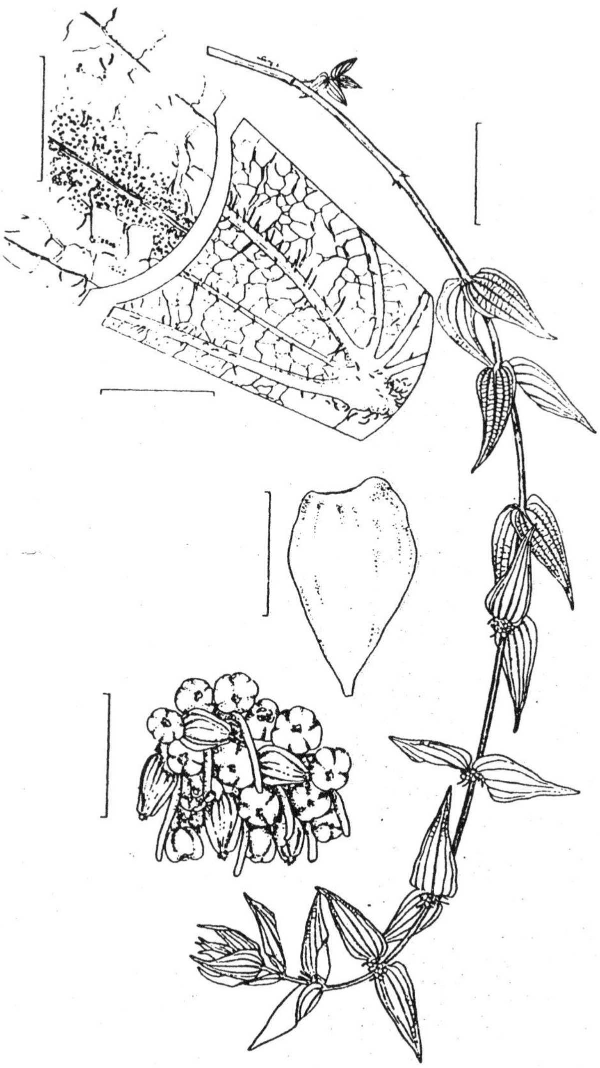

Pouzolzia hirta A. habit; B. flower; C. seed; D.E. lower and upper surface of leaf. (Drawn: T. Iwagu).

Captions for Figures on Page 53

Above:
Pouzolzia zeylanica A. habit; B. lower surface of leaf; C. upper surface of leaf; D. flower; E. fruit; F.G. upper and lower surface of stipules.
(Drawn: T. Iwagu).

Below:
Pouzolzia hirta A. habit; B. fruit; C. seed; D. flower; E.F. lower and upper surface of leaves.
(Drawn: T. Iwagu)

18. PROCRIS

o o o o o o

A genus of 20 species in the Old World Tropics. It is represented by a single species in Papuasia. van Royen (1982) has included this species in the genus Lecanthus which is represented by a one (van Royen) or 3 (Mabberley) species. Winkler (1922, 1924) treated Procris as a distinct subgenus in Elatostema. Schroter (1939) treated Procris as a separate genus.

Small herb to 30 cm tall, stems finely ribbed, nodes swollen. Stipules axillary, with a few marginal hairs. Leaves to 36 mm long; base cuneate, tip acute, with 2-4 basal nerves; margin coarsely dentate-serrate, with 2-15 teeth on either side. Petioles to 25 mm long, pillose along margin, grooved above.

Inflorescences 1-2 per axil. Heads to 15 mm across borne on peduncles to 5 cm long. Male flowers: pedicels to 2.5 mm long, tepals glabrous, to 1.5-2.5 mm long, stamens 1-1.5 mm long. Female flowers: fertile short stalked, sterile long stalked; pedicels at first 0.2 mm long elongating up to 2.5 mm in fruit; tepals to 1.5 mm long, boatshaped. Staminodes to 0.2 mm long, if present. Ovary glabrous, stigma penicillate with long hairs to 0.5 mm long. Nutlets fusiform-ovoid, slightly longer than ovary.

List of Species in Papuasia

Procris decurrens (H. Winkler) comb. nov. Elatostema decurrens H. Winkler (1922: 565)

● o o o o o

Procris rectangulare (H. Winkler) comb. nov. Elatostema rectangulare H. Winkler (1922: 566)

o ● o o o o

Procris frutescens (Hassk.) comb. nov. Elatostema frutescens Hassk.

o ● ● ● ● ●

var. caudatum (H. Winkler) comb. nov. (1922: 565)

o ● o o o o

var. sculptum (H.Winkler) comb. nov. (1922: 565)

Procris grueningii (H. Winkler) comb. nov. Elatostema grueningii H. Winkler (1922: 563) "gruningii"

o ● o o o o

var. krausei (H. Winkler) comb. nov. (1922: 564)

o o o o o o

Elatostema lignescens Hallier f. in Kds. Med. Plantent 19 (1898) 595. nomen. = Procris pedunculatum

Procris pedunculatum (Forst.) comb. nov. Elatostema pedunculatum Forst.

● ● ● ● ● ●

var. angustum (H. Winkler) comb. nov. (1922: 563)

o ● o o o o

Procris peduncularis Wallich ex Royale Himal. Mts. (1839) t.82, f.2.

Lecanthus peduncularis (Wallich ex Royale) Weddell, in DC Prodr. 16 (1869) 164.

o ● ● ● o ●

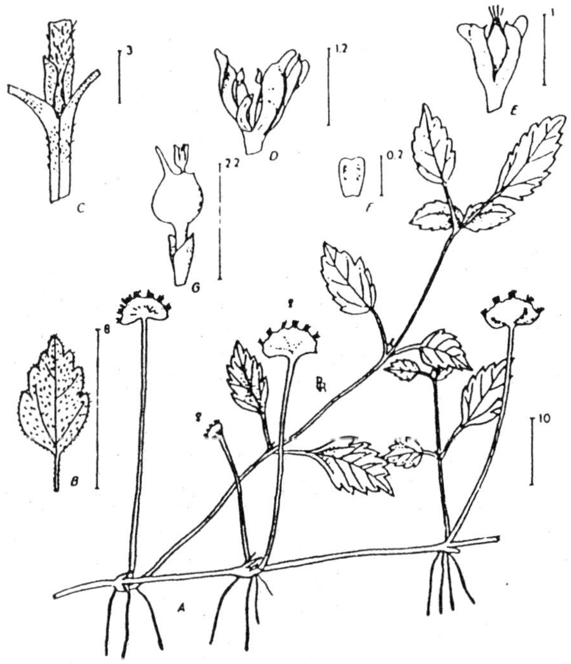

Procris peduncularis A. habit; B. leaf; C. stipules; D. sterile female flower; E. fertile female flower; F. staminode of fertile female flower; G. sterile flower. (Hoogland 9184) (Source: van Royen (1982)

Key to Species in Papuasia
(Source: Winkler (1922)
(Translated by Dra. C. M. H. Lapre)

1. Leaves succulent; side nerves and other venation not clearly observable.......................... 2

1. Leaves not succulent; side nerves and other venation clearly observable.......................... 3

2. Leaves mostly entire; apex acute, at the tip rounded.
..... Procris pedunculatum

2. Leaves irregularly dentate; apex emarginate or cleft.
..... Procris gruningii

3. Side nerves pointed in the direction of the tip (apex)......... 4

3. Side nerves perpendicular to the midrib.
..... Procris rectangulare

4. Leaves longish, suddenly pointed (cuspidate), not decurrent.
..... Procris fruetescens

4. Leaves lanceolate, caudate, decurrent.. Procris decurrens

Notes:

1. Procris peduncularis is appar-ently rare but this is probably an artifact of collecting. A montane plant, widespread throughout mainland New Guinea between 1370 and 3560 m altitude.

2. It grows in damp, shady places, on riverbanks, terrestrial in wet forests, and on gravel banks.

3. Procris frutescens Blume is the basonym for Elatostema frutescens (Bl.) Hassak. var. novoguineensis described by Valeton (1907: 7) was tansferred as a variety of this species by H. Winkler (1922). Elatostemma pedunculata was originally, together with var. eu-pedunculata, described as Procris pedunculata. Procris grandis is a synonym. Elatostema velutina is based on a species of Procris described by Warburg.

PSEUDOPIPTURUS

This name used by Brass in Journ. Arn. Arb. 22 (1941) 292, is a latter name for Nothocnide. The specimen referred to is Brass 13078. It is not listed in the revision of that genus by Chew (1969).

ROBINSONIODENDRON

Merrill in Inter. Rumph Herb. Amb. (1917) 204, listed Robinsoniodendron ambiguum (Wedd.) Merr. as the only species of a new genus. It was based on Maoutia ambigua Wedd. (1855/6). The genus is no longer accepted as being distinct and is now treated as part of Maoutia.

18. URTICA

o o o o o o

A genus of 45 species, subcosmopolitan in distribution. Usually with alkaloids and stinging hairs. The stinging hairs are bulb-like projections which break off along a line of weakness leaving a sharp edge which breaks the skin and injects histamine causing itching and acetylcholine causing burning. The shoots of some species eaten as spinach.

Herbs with stinging hairs. Monoecious or dioecious. Stipules lateral to base of petioles, free or fused (not in Papuasia). Leaves opposite and decussate, equal in each pair. Flowers unisexual, pedicel short 0-0.5 mm. Flowers with 4-tepals. Male flower with stamens to c. 1 mm long; abortive ovary cup-shaped. Female flowers: ovary glabrous, style very short, stigma penicillate with long hairs. Nutlets pale yellowish brown, ovoid; 1.5-2.0 mm long.

Key to Species in Papuasia

1 endemic species:

.... Urtica papuana Zandee (1969) 444.

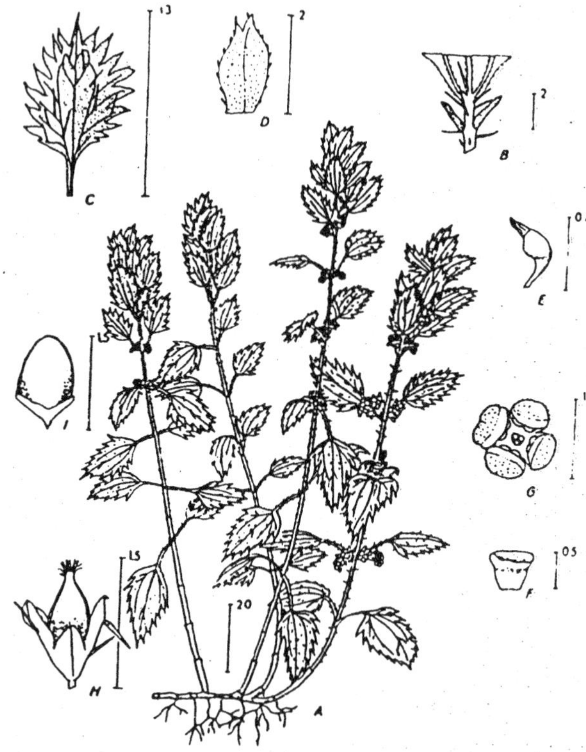

Urtica papuana A. habit; B. stipules; C. leaf; D. single lobe of limb of leaf; E. tepal of male flower; F. abortive ovary in male flower; G. male flower from above, with cuplike abortive ovary in centre; H. female flower; I. nutlet. (Brass 91929)
(Source: van Royen (1982)

Notes;

1. Known from a single collection (Brass 91929 NO) from Lake Habbema in Irian Jaya. Collected in a thicket of Rubus at 3225 m.

2. Some ten species of Urtica have been described or recorded from Papuasia. These have been subsequently transferred to several genera including: Fleurya (interrupta); Laportea (decumana); Leucosyke (capitellata, rhodopleura); Pipturus (argenteus, repanda); Pouzozlzia (hirta, pentandra);

20. VILLEBRUNEA

o o o o o o

A genus of 8 erect or scandent shrubs, or trees, distributed from Sri Lanka to Japan and Papuasia. V. integrifolia from T Asia is used as a source of fibre. Three species in Papuasia.

Trees or shrubs, leaves alternate, 3-nerved or pinnately nerved, entire or crenulate; stipules intrapetiolar, connate into 1. Flowers dioecious, densely clustered, solitary or in lax cymes. Male flowers: calyx 4-partite, valvate, subglobose in bud, stamens 4, ovary rudimentary. Female flowers: calyx tubular, adnate to ovary, minutely toothed; stigma sessile, small or discoid, long ciliate on margin; ovule basal erect. Achene adnate to thinly fleshy calyx. Seed with endosperm.

List of Species in Papuasia

Villebrunea fasciculata Warb. (1891) 289

Villebrunea rufescens Blume Mus. Bot. Lugd. Bat. (1856) 166 var. pubescens Blume Mus. Bot. 2 (1856) 166

Villebrunea trinervis Wedd. Ann. Sc. Nat. 4 (1854) 1869

Key to Species in Papuasia

(Source: Winkler (1922)
(Translated by Dra. C. M. H. Lapre)

1. Leaves with pinnate venation. ..... Villebrunea rufescens

1. Leaves with three ribs.......... 2

2. Inflorescences not too compact. Male buds round, with few greyish hairs. Villebrunes trinervis

2. Inflorescence very compact. Male bud acuminate, with grey - white hairs. ..... Villebrunea fasciculata

Notes:

1. Several species described by Blume in the genus Villebrunea have been transferred to other genera:
Oreocnide murina (Bl.) Miq.
Leucosyke rhodophura (Bl.) Winkler
Oreocnide rufescens (Bl.) Miq.

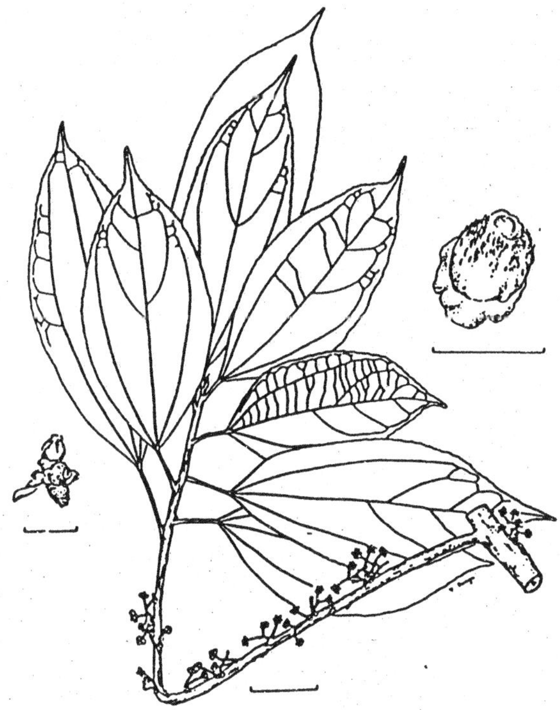

Villebrunea trinervis A.
NGF 48589 Madang.
(Drawn: T. Iwagu)

CANNABIDACEAE

o ● ● o o o

A family of 2 genera and 3 species. A small but economically important family including the genus Humulus. Hops are much cultivated in temperate regions for flavouring beer and for medicinal purposes. Hemp, Cannabis sativa L. is grown for its fibre which is used for making ropes, and as a source of 'hashish' which is exuded by the leaves and stem.

Erect or scandent herbs. Leaves alternate or opposite, simple, undivided or palmately lobed; stipules present. Flowers dioecious, axillary, male paniculate, female sessile, crowded or strobilate, the female with large persistent imbricate bracts. Male flowers: calyx 5-partite, segments imbricate, petals absent; stamens 5, filaments short: anthers erect in bud, 2-locular, opening by short slits lengthwise; rudimentary ovary absent. Female flowers: calyx very thinly membranous, closely applied to the ovary, entire; ovary sessile: style central, 2-partite, branches caducous; ovary pendulous. Achene more or less enclosed by the membranous calyx; seed with fleshy unilateral endosperm, curved embryo and thick fleshy endosperm.

List of Species in Papuasia

1 Introduced Species

.... Cannabis sativa

Notes:

1. This plant occurs widely in the Highlands of Papua New Guinea, in areas it has reportedly become naturalised as an escapee from cultivation.

Cannabis sativa A. habit.
(Source: Purseglove (1977)

MORACEAE

o o o o o o

An economically important family of mainly tropical and subtropical trees and shrubs. The family is widely distributed in both hemispheres. It includes some 48 genera and circa. 1,200 species of trees, shrubs or climbers. The most important commercial plants are the fig (Ficus), the mulberry (Morus) and the breadfruit (Artocarpus). Mulberries are of less importance with the decline in the silk industry. Several species, including species of Ficus and Castilloa elastica (Panama rubber), were used as a source of latex in rubber manufacture. Several species were used as sources of timber. Throughout the highlands of New Guinea, Ficus damaropsis was cultivated for its edible young leaves and fruit.

In many tropical rainforests the seeds of Ficus are an important food source for birds. Flowers are usually wind pollinated or, in figs, pollinated by insects. The figs are particularly interesting in their pollination mechanisms. This will be discussed under Ficus. Fig wasps also provide an important food source for insect eating birds.

The character that distinguishes the family most clearly is the presence of a milky sap containing latex. The sap is produced in the embryo and is produced throughout the life of the plant. Its function is not known.

Classification systems within the family are variable. Corner (1962) recognises 6 tribes whilst Hutchinson (1967) recognises 8. The delimitations are often radically different.

Note:
The family Cannabidaceae includes Cannabis sativa and Humulus lupulus. It is sometimes included in the Moraceae.

Diagnostic features

Leaves: simple, alternate, spiral or rarely opposite, entire or lobed. Stipules are usually present but often deciduous. Leaf margins are entire or toothed.

Flowers: all flowers are unisexual, bourne either on the same , mixed male and female flowers, or on separate plants. In heads on flattened or hollowed recetacles or in catkins. The male flower has 4 more or less fused tepals and 4 stamens (rarely 1, 2 or 5) opposite the tepals. The female flowers has a four member perianth (often much reduced or fused to the ovary) and 2 carpels, one abortive. The ovary contains a single, pendulous, ovule. Within the family there is a transition from a superior to an inferior ovary, some genera being intermediate.

MORACEAE

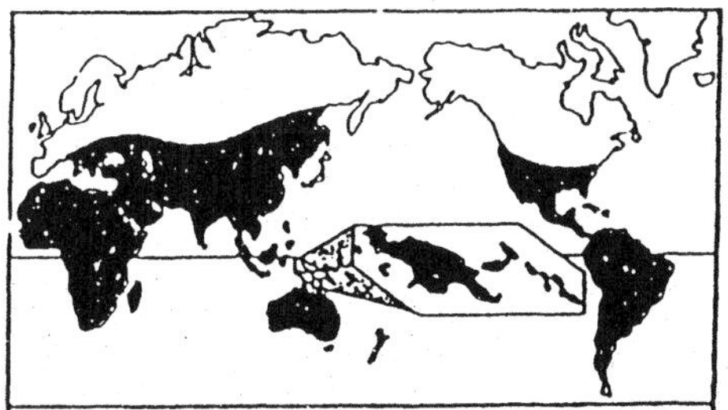

	WORLD	PAPUASIA
Number of genera:	75	14
Number of species:	3,000	188

Distribution: Tropics and subtropics to temperate.

Economic uses: Fruits (mulberries, breadfruit) and fiber (hemp)

Fruit: the ripe fruit is often fleshy and edible. The fleshy parts are produced from the swollen receptacle in which the seeds are embedded, not by the ovary. The style is persistent. The seed is with or without an endosperm.

References

Berg, C. C. (1977). The Castilleae: a tribe of the Moraceae. Acta Bot. Neerl. 26: 73 - 82.

Corner, E. J. H. (1939). A revision of Ficus, subgen. Synoecia. Gard. Bull. Sing. 10: 82 - 161.

Corner, E. J. H. (1958). An introduction to the distribution of Ficus. Reinwardtia 4: 15 - 45.

Corner, E. J. H. (1960). Taxonomic notes on Ficus, Asia and Australia. Sections 1 - 4. Gard. Bull. Sing. 17: 368 - 485.

Corner, E. J. H. (1960). Taxonomic notes on Ficus, Asia and Australia. Sections 5. Gard. Bull. Sing. 18: 1-35.

Corner, E. J. H. (1960). Taxonomic notes on Ficus, Asia and Australia. Sections 6. Gard. Bull. Sing. 18: 35-69.

Corner, E. J. H. (1961). Taxonomic notes on Ficus, Asia and Australia. Addendum. Gard. Bull. Sing. 18: 83 - 97.

Corner, E. J. H. (1962). Taxonomic notes on Ficus, Asia and Australia. Addendum II. Gard. Bull. Sing. 19: 385 - 415.

Corner, E. J. H. (1962). The classification of the Moraceae. Gard. Bull. Sing. 19: 187 - 252.

Corner, E. J. H. (1965). A check list of Ficus in Asia and Australia. Gard. Bull. Sing. 21: 1 - 186.

Corner, E. J. H. (1967). Ficus in the Solomon Islands. Phil. Trans. Roy. Soc. B 253: 23 - 159.

Corner, E. J. H. 1969). Ficus section Adenosperma. Phil. Trans. Roy. Soc. B 256: 319 - 355.

Corner, E. J. H. (1970). New species of Streblus and Ficus. Blumea 18: 393 - 411.

Corner, E. J. H.(1972). New taxa of Ficus. Blumea 20: 427 - 432.

Corner, E. J. H.(1975). New taxa of Ficus. II. Blumea 22: 299 - 309.

Corner, E. J. H. (1975). The evolution of Streblus with a new species of Sect. Bleekrodea. Phytomorphology 25: 1 - 12.

Corner, E. J. H. (1978a). Ficus glaberrima and the pedunculate species of Ficus subgenus Urostigma in Asia and Australia. Phil. Trans. Roy. Soc. B 281: 373 - 406.

Diels, L. (1935). Die Moraceen von Papuasien. Bot. Jahrb. 67: 170 - 235.

Hutchinson, J. (1967). The genera of the Flowering Plants. 2: 151 - 175. Clarendon Press.

Jarrett, F. M. (1959). Studies in Artocarpus and allied genera. I. General considerations. J. Arn. Arb. 40: 1 - 29.

Jarrett, F. M. (1959). Studies in Artocarpus and allied genera. II. A revision of the genus Prainea. J. Arn. Arb. 40: 30 - 37.

Jarrett, F. M. (1959). Studies in Artocarpus and allied genera. III. Artocarpus subgenus Artocarpus. J. Arn. Arb. 40: 113 - 155.

Jarrett, F. M. (1959). Studies in Artocarpus and allied genera. IV. Artocarpus subgenus Pseudojaca. Arn. Arb. 40: 298 - 368.

Jarrett, F. M. (1959). Studies in Artocarpus and allied genera. V. A revision of Paratocarpus and Hulletlia. J. Arn. Arb. 41: 320 - 340.

Jarrett, F. M. (1960). Journ. Arn. Arb. 41: 320.

Lauterbach et Schumann, K. (1901). Fl. Deutsch. Schutzgeb. Suds. 267.

Summerhays, V. S. (1929). Ficus species collected for the Arnold Arboretum by L. J. Brass. J. Arn. Arb. 10: 142 - 154.

Summerhays, V. S. (1941). Additiona to our knowledge of the figs of New Guinea. J. Arn. Arb. 22: 81 - 109.

Yasue, Y. (1978). Occurence of mulberry trees Morus sp. in New Guinea. Nogaku Kenkyu 57(2): 87 - 92.

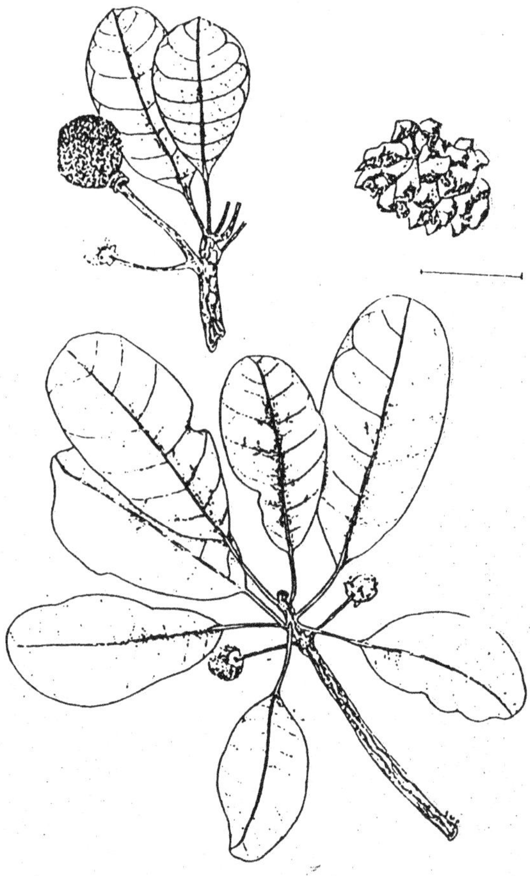

Parartocarpus venosus A. fruiting branch; B. immature fruit. (Drawn: R. Bialostocki)

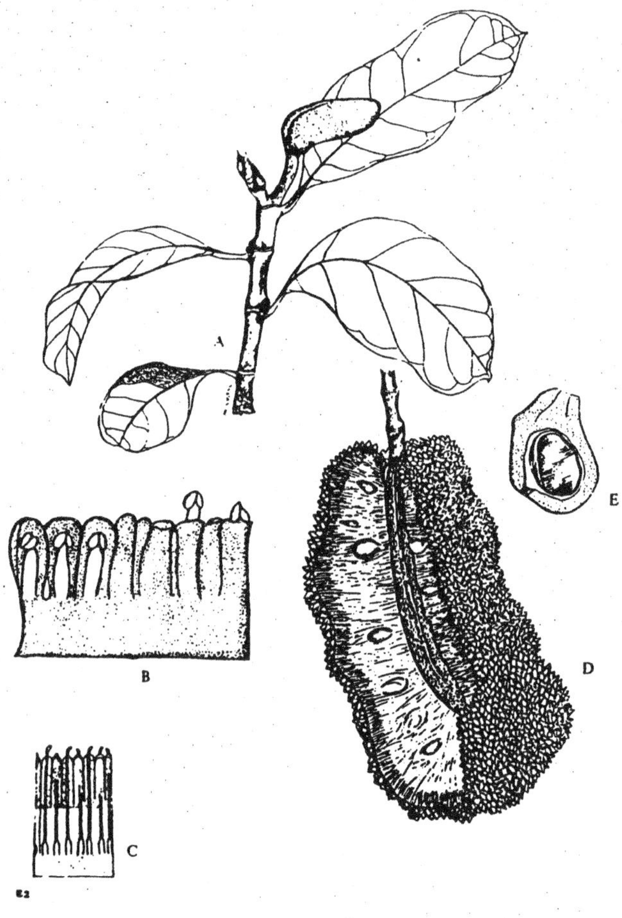

Artocarpus heterophyllus A. shoot with male inflorescence; B. LS male inflorescence; C. female inflorescence; D. seed. (Source: Purseglove (1977)

Key to Genera in Papuasia
(Source: Corner (1962)

Flowers inside urn-shaped recept receptacle, numerous. Styles not extruded from receptacles. Female flowers insect inhabited (blastophagous). Stamens, when 2 or more, with inward facing anthers
7. Ficus

Flowers not enclosed, or if so, styles exserted from receptacles, not insect inhabited. Stamens mostly outward facing. Female flowers mostly sessile.................. 2

2. Female inflorescences racemose or spicate with slender axis or 1-flowered, or the inflorescence bisexual; cymose or spicate, neither capitate or discoid. Ovaries mostly free, if inferior then not in sockets. Male inflorescence paniculate, racemose, spicate or head like, often with a sterile groove. Stamens 5 or 1, filaments mostly inflexed in bud, in somr cases straight; rudimentary ovary usually present. Spined or not.. 3

2 Female inflorescences head-like or thickly spicate, mostly syncarpous (carpels joined). Ovaries often immersed in sockets of the receptacle or joined with it. Male flowers mostly lacking rudimentary ovary.......................... 6

3 Seeds small, 1-2 mm wide, somewhat compressed; endocarp hard, thin and brittle. Cotyledons flat. Male flowers with rudimentary ovary, filaments 4, bent inwards towards bud... 4

3 Seeds 4 12 mm wide, rounded. Cotyledons often folded or much thickened, often unequal, leaves distichous. Spinous or not...... 5

4 Herb. Inflorescence, bisexual or unisexual. 6. Fatoua

4 Trees. Inflorescence racemose or spicate, unbranched, unisexual. Dioecious or monoecious.
.... 10. Morus

5 Ovary inferior. Not spinous.
.... 14. Trophis

5 Ovary superior. Spinous or not.
. ..... 13. Streblus

6 Male and female inflorescence discoid, obconic or urn-shaped with an involucre of several bracts. Female inflorescence sometimes 1-flowered. Carpels joined or not. Stamens 8-1, the filaments straight in the bud. Not spinous.................... 7

6 Male and female inflorescence of different shape, involucrate or not. Female inflorescence thickly spicate to globular-head-like, never 1 flowered; carpels united. Male inflorescence paniculate, racemose, spicate or head-like. Stamens 4 1, the filaments straight or inflexed in bud. Spinous or not.
... . 9

7 Receptacle of both sexes with several or numerous flowers..... 8

7 Receptacle of male flowers many flowered, of female 1-4 flowered. (Perianth absent in female flowers)
.... 2. Antiaris

8 Receptacle disc-like or concave. Perianth parts of female remaining free. Stipules joined into 1 large, deciduous ligule?.
.... 5. Castilla

8 Receptacle globose. Perianth of female flowers becoming thick and fleshy but remaining free in upper part...... 1. Antiaropsis

9. Female perianth parts 4, free (at least in distal half), well-developed, decussate, overlapping fleshy in fruit. Bracts and/or tepals with 2-7 immersed yellow glands. Seeds compressed with woody endocarp. Dioecious, spiny.
.... 8. Maclura

10. Male inflorescence racemose-spicate, unbranched, with a sterile groove; filaments bent inwards in bud; rudimentary oveary present. Female inflorescence utricular (small bladder); ovaries not in cells of receptacles; style long. Embryo curved. Dioecious unarmed.
.... 12

10. Male inflorescence club-shaped or head-like; if spicate without a sterile groove and with filaments straight in bud................. 13

11. Syncarp globose, thickly set with slender, stalked bracts of various shapes ± covering drupes. Drupes exserted from perianth. Seeds 2-3 mm long, endocarp crustaceous to woody. Cotyledons equal, folded together lengthwise. Stipules membranous... 4. Broussonetia

11. Syncarp with few strongly projecting drupes each enclosed by utricular perianth, bracts short. Seed 6-7 mm long; endocarp membranous. Cotyledons very unequal, the larger thickly fleshy and folded. Climber.
.... 9. Malaisia

12. Dioecious. Male and female heads similar, globose. Perianth utricular. Ovary on surface? Stamen 1. Drupes projecting strongly from syncarp Embryo straight, cotyledons thick, equal.
.... 12. Prainea

12. Monoecious. Drupes embedded in the more or less fleshy syncarp..... 13

13. Male head usually of different shape to female, varying spicate to club-shaped to head-like. Flowers not embedded. Stamens 1. Female perianth utricular, much reduced or none. Ovaries sunk in receptacular sockets or hidden beneath the layer of joimed utricular perianths. Embryo straight or slightly curved.
.... 3. Artocarpus

13 Male and female heads similar, globose, distinctly involucrate with bracts. Stamens 2-3, sunk like ovaries in receptacular sockets, the filaments and style projecting between the free conic or truncate sepals. Embryo curved.
.... 11. Parartocarpus

Antiaropsis decipiens A. branch with fruit; V. flowers enlarged.
(Drawn: R. Bialostocki)

1. ANTIAROPSIS

o o o o o o

Endemic in New Guinea, the genus was described by Schumnann. Corner (1962: 250) notes 1-2 species. A leptocaul tree or shrub; leaves alternate, shortly petiolate, entire, long acuminate, pinnately nerved: nervesprominently lobed within the margin; stipulea lateral, very caducous.

Flowers dioecious, densely capitate, receptacle densely involucrate with imbricate lanceolate interfloral bracts. Male flowers: calyx-segments 4, spathulate, imbricate; stamens 4, filaments short, erect, free; anthers linear oblong; rudimentary ovary filiform. Female flowers: bracts scale-like, free: ovary free: style 2-lobed, stigma filiform; ovules pendulous from the apex. Fruit dry, exocarp crustaceous, style persistent; endosperm absent; cotyledons involute, equal.

Key to Species in Papuasia

A single species (see Notes below):

.... Antiaropsis decipiens K. Schum. (1889: 40) Laut. et K. Schum: (1901) 267. Diels (1936: 174)

o o o o o o

Notes:

1. Corner (1962: 250) notes that there may be a second species described as A. decipiens K. Schum. var. parvifolia Diels (1936: 174). The type species was collected by Romer (Nr. 867) in Hellwig Mts, 11/1909 at 750 m altitude. McAdam s.n. apparently collected similar material at Wau in 1939. His cspecimen has a small single female inflorescence 4-5 mm wide with a single female flower. There are insufficient collections to decide if this is a new species, but it shows how the one flowered state of Antiaris may have arisen.

2. ANTIARIS

o o o o o o

Corner (1962) reduced the seventeen described species to four. The most widespread species A. toxicaria is the most variable. Variations in this species are merely in size and shape of the leaves and fruit. This species occurs in lowland tropical rainforest from W Africa to S China and Fiji. It reaches 1,500 m altitude in Borneo and New Guinea. Many taxonomic complications arise with the small tree or shrub, the form cultivated in the Solomon Islands, Fiji and Tonga. It agrees in essential characters with var. macrophylla from New Guinea and Vanuatu where it shows all gradations to a large tree.

Trees or shrubs. Leaves alternate, pinnately nerved, entire or serrate: stipules lateral or connate within the petiole, small, very caducous: lamina ovate-elliptic or elliptic, attenuate to the acute apex, base cordate to widely cuneate, scabridulous to smooth.

Flowers: monoecious. Male flowers: densely capitate, females solitary; receptacle of both sexes with many seriate, imbricate bracts; receptacles axillary, shortly pedunculate, males often paired or few, flat or convex. Male flowers: calyx 4- to 3-partite, segments spathulate, apex concave, imbricate; stamrns 4 or 3, filaments short, erect, free; anthers long; rudimentary ovary lacking. Female flowers: calyx none; ovary enclosed in the involucre and adnate yo the receptacle: style 2-partite, branches subulate, exserted: ovule pendulous. Fruit a fleshy drupe; 20-45 by 17-26 mm; pericarp fused with the enlarged receptacle. Seed lacking endosperm.

Key to Species in Papuasia

A single species in Papuasia:

.... Antiaris toxicaria Lesch. Ann. Mus. Hist. Nat. Paris. (1810: 478).

var. macrophylla (R.Br.) Corner (1962: 248). Antiaris macrophylla R.Br. Flind. Voy. 2 (1814) 602, t.5.

o o o o o o

Notes:

1. var. macrophylla is distributed from the Philippines to Fiji, Tonga and N Australia. The shrubby plant in Melanesia is probably a cultivar.

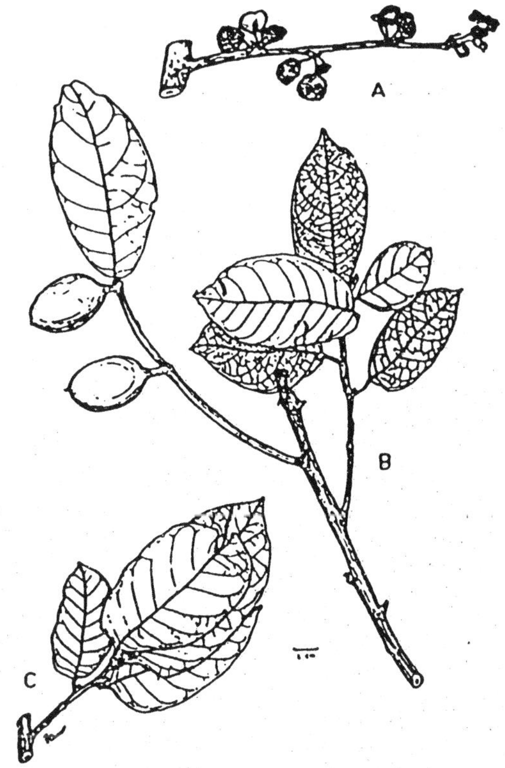

Antiaris toxicaria A. branch with male flowers; B. branch with fruits; C. branch with female flowers.
(Source: National Herbarium, Lae)

Artocarpus vriesianus var. refractus A. stem with mature fruits.
(Source: National Herbarium, Lae)

3. ARTOCARPUS

o o o o o o

The genus shows the extreme reduction in the male inflorescence to a spike or head of minute unistaminate flowers with a tubular perianth and a minute anther (Corner 1962). In Papuasia 5 species, with 1 species, the jackfruit introduced. Two species A. communis the breadfruit, and A. heterophyllus the Jackfruit are cultivated throughout the tropics. Jarrett (1959) considers that, in all probability, the Island of New Guinea is the origin of the Breadfruit.

Small to large trees. Leaves spirally arranged (subg. Artocarpus) or alternate and distichous (subg. Pseudojaca), simple and entire to pinnatifid; penninerved; thinly to thickly coriaceous, glabrous to pubescent or hispid; hairs/glands of epidermis deeply immersed to superficial with globose heads, to flattened and peltate. Stipules paired, large, intrapetiolar, leaving distinct scars.

Inflorescence unisexual, capitate, pedunculate. Solitary or paired in leaf-axils, raely borne on short shoots on older wood.; numerous flowers completely covering surface, the perianths enclosing a single ovary or stamen, mixed with numerous bracts.

Male flowers: calyx 4- to 2-lobed or partite; stamen 1; filament erect, often flattened; rudimentary ovary lacking. Female flowers: calyx tubular, immersed at the base in the receptacle, with a minute 3- to 4-toothed pore at the apex; ovary enclosed; style central or more or less lateral, apex stigmatose spathulate or subpeltate, rarely 2- to 3-lobed, ovary pendulous. Fruit a syncarp, sometimes large and edible.

Key to Species in Papuasia
(Source: Jarrett 1959)

1. Inflorescences narrowly cylindric, the surface entirely or partly covered by the peltate heads of numerous bracts, persistent in both sexes.
..... Artocarpus sepicanus
Diels (1935: 176)
as "sepicana".

o o o o o o

1. Inflorescences not as above..... 2

2. Twigs (4-)6-22 mm. thick, villous, varying glabrous or with patent acicular hairs; adult leaves entire or pinnatifid: male head 1-30 by 0.7-3(-5.5) cm.: syncarp processes rough from the apices of deflexed, inflated hairs (except in A. communis), styles bifid or simple.
..... Artocarpus communis
J. R. & G. Forster Char. Gen. (1776) 101. T. 51.
Artocarpus incisus Linn. F. Suppl. Pl. (1781: 411)
Artocarpus incisa Linn. F. var. muricata For. Borneo 628 (1902)
Artocarpus leeuwenii Diels (1935: 175).
Artocarpus papuana Diels (1935: 175)

o o o o o o

2. Twigs 2-6 mm. thick, glabrous or pilose from patent, rufous hairs; adult leaves entire; male head 2-7.5 by 0.5-1.2 (-3) cm.; syncarp processes puberulent or minutely hispid, styles simple.......... 3

3. Cauliflorous or ramiflorous; leaves with the lateral veins curved, 6-1 pairs; syncarp 15-100 by 10-50 cm., fruiting perianths markedly fleshy. 4

3. Not cauliflorous or ramiflorous; leaves with the lateral veins straight; syncarps 6.5-8.5 by 2-3.5 cm., fruiting perianths not fleshy.
..... Artocarpus teysmannii
Miq. Fl. Ind. Bat. Suppl. 418, (1861)

o o o o o o

4. Twigs and peduncles usually pilose from patent, rufous hairs, c. 3 mm. long; base of leaf abrupt intercostals to c. 10; inflorescences without a basal annulus.
..... Artocarpus integer
(Thunb.) Merr. Interpr. Rumph. Herb. Amb. (1917) 190.
Corner (1939: 56. t. 1,2.)
Rademachia integra Thunb. (1776: 254)
var. integer

o o o o o o

4. Twigs and peduncles glabrous; base of leaf decurrent, intercostals c. 10-14; inflorescences with a basal annulus formed by the enlargement of the top of the peduncle into a narrow flange.
..... Artocarpus heterophyllus
Lamarck, Encycl. Meth. (1789: 210).

o o o o o o

Key to specimens with male inflorescences (Source: Jarrett 1959)

1. Head with the surface largely covered by the peltate heads of numerous bracts.
..... - Artocarpus sepicanus

1. Heads with bracts few or entirely lacking, cylindric, varying ellipsoid or clavate................ 2

2. Numerous filiform or clavate processes present between flowers and projecting 1-2 mm from the surface. Artocarpus teysmannii

2. Projecting processes absent or a few scarcely emergent.......... 3

3. Lateral veins of leaf 6-10 pairs, curved......................... 4

3. Lateral veins of leaf 10-25 pairs, straight.
..... Artocarpus communis

4. Peduncle and twigs glabrous; head with a basal annulus formed by the enlargement of the peduncle into a flange to 3 mm. wide; base of leaf decurrent onto petiole.
..... Artocarpus heterophyllous

4. Peduncles and twigs pilose with patent, flexuous, rufous hairs c. 3 mm. wide; base of leaf decurrent onto petiole.
..... Artocarpus integer

Key to Specimens with Female Inflorescences (Source: Jarrett (1959)

1. Processes covering the syncarp all alike, perforate, with styles exserted at anthesis.
..... Artocarpus teysmannii

1. Processes covering the syncarp markedly dimorphic, the longer ones solid, flexuous, the shorter perforate with styles exserted at anthesis...................... 2

2. Syncarp 15-100 by 10-50 cm., fruiting perianths thick-walled, fleshy; cauliferous; adult leaves entire.
.... 3

2. Syncarp smaller, or fruiting perianths thin-walled or lacking (seeds not developed); not cauliferous. 4

3. Syncarp with a basal annulus formed by the enlargement of the top of the peduncle into a flange to 5 mm. wide (sometimes obscure)_; peduncle and twigs glabrous; base of leaf decurrent onto petiole.
..... Artocarpus heterophyllous

3. Syncarp without a basal annulus; peduncle and twigs pilose with patent, rufose hairs c. 3 mm. long, varying glabrous; base of leaf abrupt.
...... Artocarpus integer

4. Syncarp more than 10 cm. across, covered by attenuate or conical processes to 15 mm. long, varying to low facets or areolae, cultivars often seedless.
..... Artocarous communis

4. Syncarp less than 10 cm. across. 5

5. Syncarp with numerous peltate bracts, the heads to 1 mm. across, the surface areolate with very short, truncate, velutinous processes, (?) becoming nearly smooth.
..... Artocarpus sepicanus

5. Syncarp without peltate bracts. Processes conical at maturity.
..... Artocarpus teysmannii

Notes:

1. The breadfruit (A. communis) is of world wide distribution but possibly originates from New Guinea where some, at least, of the few seeded or seedless varieties were developed (Jarrett 1959: 311-312).

2. A. integer occurs only in the Vogelkop Peninsula in Papuasia but is more widely distributed in Malesia.

3. A. heterophyllus, the Jackfruit, originates from India (probably the western Ghats). It is locally cultivated throughout Papuasia.

4. The species A. teysmannii occuts from Malaya to Irian Jaya.

5. A. sepicanus Diels is an endemic species growing in the lowlands up to 180 m altitude. It is often common in swampy sites.

4. BROUSSONETIA

o o o o o o

A genus of some 7 species divided into 2 rather similar sections (Corner, 1962). In Papuasia there are species.

A group of dioecious trees, shrubs or climbers, usually deciduous,unarmed. Leaves spirally arranged to opposite or distichous, palmately lobed to undivided Base cordate to rounded or widely cuneate; dentate to entire, membranous; intercostals numerous; stipules membranous.

Male inflorescence racemose to spicate, with a sterile groove, interfloral bracts generally stalked, those bordering the sterile groove sessile; flowers shortly pedicellate to sessile, tepals 4, valvate; filaments 4, inflexed in bud; pistillode spicate, minute. Female inflorescence capitate, without a sterile groove; bracts as in male, becoming stipitate in fruit; perianth utricular with 2-4 teeth or small lobes; ovary sessile; style 1(-2), long. Syncarp globose, 10-25 mm wide, flowers not connate, ripening orange red, more or less pulpy. Drupes thinly pulpy, invested by a thin pulpy, yellow to red, perianth. Seeds 1.7-2.5 mm long, faintly keeled; embryo curved; cotyledons thinly fleshy, equal.

Broussonetia papyrifera A. stem with fruit; B. inflorescence; C. stamen. (Drawn: R. Bialostocki)

List of Species in Papuasia

CALPIDOCHLAMYS

Described as an endemic genus with 2 species by Diels (1935: 172). The genus, as described by Diels, included trees with entire, elliptic to oblanceolate-oblong leaves which were pinnately nerved; stipules narrowly lanceolate, deciduous. Male flowers unknown. Female inflorescence spicate, with dense minutely peltate bracts. Female flowers: calyx urceolate-tubular, adnate to the ovary, limb free, 4-toothed; style 2-armed, arms thick; ovule 1, pendulous. Fruit drupaceous, subglobose, pericarp juicy-fleshy; endocarp bony-crustaceous. It is here treated as part of the genus Trophis.

Synonymy of Species from Papuasia

Calpidochlamys drupacea Diels (1935: 173)

Calpidochlamys branderhorstii Diels (1935: 173)

5. CASTILLA

o o o o o o

A genus of 3 tropical American rainforest species. Formerly important as a source of rubber (Castilla elastica), now replaced by Hevea (Euphorbiaceae).

Lactiferous tree. Leaves alternate, often large, entire or dentate, pinnately nerved; stipules connate into one, large and deciduous, intrapetiolar. Flowers monoecious, densely capitate, in flator concave receptacles, involucrate with several series of bracts. Male receptacles often several in the axils or defoliate nodes, female solitary, sessile or nearly so. Male flowers: calyx none; stamens (single flowers) numerous on the receptacle, crowded, intermixed with bracteoles, filamemts straight; anthers ovate. Female flowers: calyx ovoid, 4-lobed, lobes thick, closely contiguous; ovary enclosed, adnate to the calyx; style short, branches filiform; ovary pendulous. Calyx enlarged in fruit, succulent, enclosing the achene; seed withouy endosperm; cotyledons thick, equal; radicle small, inferior.

Key to Species in Papuasia

1 introduced species:
..... Castilla elastica Cerv.

o o o o o o

Notes:

1. Used for rubber production before the use of Hevea.

CUDRANIA

Mabberley (1987) states that this is a genus of 5 species distributed from China to Australia. Diels (1936: 174) records a single species in Papuasia.

..... Cudrania javensis Trec.

This is here treated in the genus Fatoua.

DAMMAROPSIS

Described as an endemic genus in Papuasia by Warburg (1891: 296). Diels (1936: 204) gives the following combination:

..... Ficus dammaropsis Diels (1936: 204)

Basionym: Dammaropsis kingiana Warburg (1891: 296)

6. FATOUA

o o o o o o

A monotypic genus of distributed from China and Japan to N. Australia and the Pacific Islands. In Papuasia the single species.

Herbs, leaves alternate, dentate, pinnately nerved and subtrinerved; stipules lateral, deciduous. Flowers monoecious, in axillary androgynous subcapitate cymes. Male flowers: calyx deeply 4-lobed, lobes valvate; stamens 4, filaments reflexed in bud; rudimentary ovary small, conical. Female flowers: calyx sub 6-partite, segments narrow, valvate; ovary oblique; style sublateral, filiform, undivided or with a short tooth; ovule pendulate. Achene enclosed by the persistent calyx, obliquely globose; seed without endosperm; embryo incurved.

List of Species in Papuasia

A single species:

..... Fatoua japonica (Thunb.) Bl.

Laut. & Schumann (1901: 265)
Diels (1936: 171)

o o o o o o

Castilla elastica A. stem with fruit, B. fruit enlarged
(Drawn: R. Bialostocki)

Fatoua pilosa A. branch with flowers; B. inflorescence.
(Drawn: R. Bialostocki)

7. FICUS

o o o o o o

A genus with over 800 species in tropics and warmtemperate regions, especially in Indomalesia. Papuasia, with some 300 species, is an important centre of diversity for the genus. Many species are used for bark cloth, sources of rubber, fibres, paper, timber, medicine and ornamentals and food. In the highlands plants of Ficus dammaropsis are cultivated in gardens and the young leaves are eaten. Several have religous significance, sacred tree of Hindus.

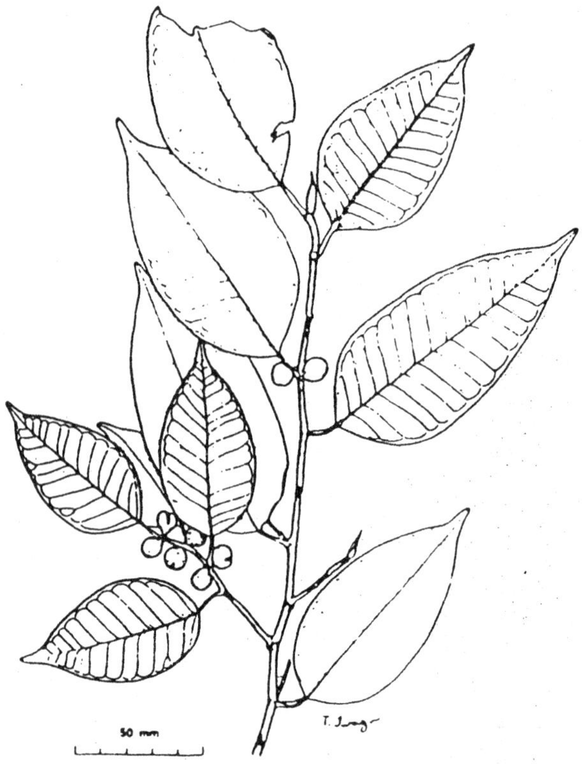

Ficus benjamina A. habit. (Drawn: T. Iwagu)

Pollination and Dispersal

Dispersal is a major problem in Ficus as the pollinating insect must accompany the host. There is much evidence that the insects are host specific (Corner 1958).

Classification of Ficus

Corner (1959) divides the family into three subgenera:

subgen. Pharmacosycea Miq.
The monoecious figs.

subgen. Urostigma (Gasp.) Miq.
The monoecious banyans.

subgen. Ficus
The dioecious figs.

Ficus polyantha A. stem with fruit; B. leaf; C.D.E. flowers; F. TS fruit; G.H.I. flower parts.
(Source: National Herbarium, Lae)

Distribution of Ficus
(Source: Corner 1958)
Endemics in brackets.

Section	NG	SI
Pharmacosycea	12 (8)	3
Urostigma	24 (9)	6
Sycomorus	4 (2)	1
Adenosperma	16 (9)	4 (1)
Sycidium	29 (13)	15 (4)
Ficus	31 (23)	4
Synoecia	4 (4)	-
Totals	138 (79)	43 (11)

KEY TO THE SUBGENERA OF FICUS

(Source: Corner (1965)

1. Dioecious. No interfloral bracts. .... D. FICUS

1. Monoecious. Figs often with interfloral bracts, without lateral bracts. Gall-and female flowers often similar. Tepals red or with white edges. Seeds smooth...... 2

2. Banyans, strangling figs, or creeping with adventitious roots. Lamina often coriaceous, entire, not plicate, with a gland at the back of the petiole-apex or none. Fig often with an outer and an inner layer of sclerotic cells in the wall, or with a single inner layer: interfloral bracts present. Stamen 1. Style mostly simple. Rarely cauliflorous. Pantropical. .... A. UROSTIGMA

2. Trees, rarely shrubs, not epiphytic. Leaf-glands in the axils of the main basal nerves or none. Scleritic cells in the fig-wall diffuse or none. Stamens 2(-3), or 1, often with a pistillode...... 3

3. Male flowers disperse or ostiolar, often pedicellate; stamens 1-3: filaments free or slightly joined. Ovary white or reddish at the base on the stylar side; stigma generally bifid. Tepals entire. Interfloral bracts often present. Leaves mostly entire (dentate in some saplings), not plicate. Rarely cauliflorous. .... B. PHARMACOSYCEA

3. Male flowers ostiolar, sessile, compressed; stamens 2 (1-3): filaments joined in the lower part. Gall- and female (sessile) flowers distinct, ovary dark red; stigma simple.... C. SYCOMORUS

Note:
Separate keys are given to the series in the genus Ficus as an aid to identification.

A. FICUS Subgenus UROSTIGMA

Key to the Sections of Ficus subgenus Urostigma

1. Fig-orifice as a pore or bilabiate or triadiate slit closed by 2 (or 3) series of inflexed (not interlocking) apical and internal bracts. Male flowers disperse. Cystoliths amphigenous, hypergenous or none. Anthers with 2 pollen-sacs, dehiscing with one crescentic slit. Stigma bifid or simple. Ovary with a red mark at the base or with red apex and then partly immersed in the fig wall. .... A5. Malvanthera

1. Fig orifice more or less closed by interlocking apical and internal bracts. Anthers normal. Stigma simple......................... 2

2. Ovary wholly red-brown, or in upper half. Male flowers ostiolar or disperse. Tepals generally narrow, acute. Basal bracts 3, or as a cupule. Cystoliths hypogenous or none. Venation mostly with intercostals... A1. Urostigma

2. Ovaries white or with a red mark at the base. Male flowers disperse. Cystoliths amphigenous, hypogenous, or none........................ 3

3. Ovary white. Venation with or without intercostals. Cystoliths hypogenous or amphigenous. .... A2. Leucogyne

3. Ovary with a red mark at base.. 4

4. Venation with the secondary lateral nerves as prominent as the primary. Stipule long. Fig-peduncle short, thick. Basal bracts early caducous. Cystoliths hypergenous. .... A4. Stilpnophyllum

4. Without such character. Cystoliths mostly amphigenous. .... A3. Conosycea

A1. Section Urostigma

Key to the series of section Urostigma

1. Fig with abundant internal chaffy-vesicular bristles. Basal bracts persistent. .... A.1.3 ser. Caulobotryeae

1. Fig without internal bristles or few and minute................. 2

2. Basal bracts caducous. Pedunculate fig....... A.1.2 ser. Superbae

2. Basal bracts persistent. .... A.1.1 ser. Religiosae

A.1.1 Series Religiosae

Key to species in series Religiosae

A single species: .... F. saxophila Bl. (1753) Indochina to Irian Jaya.

o o o o o o

A.1.2 Series Superbae

Key to species in series Superbae

A single species:
.... *F. prasinicarpa* Elmer (1937)
F. glabella Bl. v. *papuana* King (1887). Philippines to Solomon Is.

o o o o o o

A.1.3 Series Caulobotryae

Key to species in series Caulobotryae

1. Lamina ovate to elliptic, obovate, or sublanceolate; lateral nerves 5-11 pairs.. *F. virens* Ait. (1789)

v. *virens*
v. *sublanceolata* (Miq.) Corner (1960).
Both varieties are distributed from Ceylon and India to the Solomon Is.

o o o o o o

1 Lamina oblong, lateral nerves 9-16 pairs; figs in groups of 2-8 in leaf axils and on the twigs below the leaves.
.... *F. caulocarpa* Miq. (1867) v. *caulocarpa*. Distributed from Sri Lanka to New Guinea.

o o o o o o

A3. Section Conosycea

Key to subsections of Section Conosycea

1. Venation with intercostals. Figs pedunculate, sessile or pedicellate....... subsection Consycea

1. Venation without intercostals, or 1-2 vague intercostals. Figs mostly sessile........................ 2

2. Secondary lateral nerves almost or quite as prominent as the primary and subparallel.
.... subsection Benjamina

3. Secondary lateral nerves not so prominent. subsection Dictyoneuron

Subsection Consycea

Key to series of subsection Conosycea

1. Fig pedunculate ripening green, brown or brownish orange.
...... A.3.1 series Validae

1. Fig sessile or the body pedicellate, ripening yellow to red.
.... A.3.2 series Drupaceae

A.3.1
Key to species in series Validae

A single species:
.... *F. novoguineensis* Corner (1961) Eastern New Guinea, New Britain.

o o o o o o

A.3.2
Key to species in series Drupaceae

1. Fig oblong, ellipsoid, or pyriform -pedicellate; basal bracts often small or concealed; apical bracts mostly umbonate (subser. Drupaceae)
A single species:
.... *F. drupaceae* Thunb. (1786).

v. *drupaceae*
Burma to Papuasia and Queensland.

o o o o o o

v. *glabrata* Corner (1960)
Indonesia to Solomon Is.

o o o o o o

v. *pedicellata* Corner (1960)
Eastern Highlands of New Guinea.

o o o o o o

1. Fig subglobose to ellipsoid; basal bracts mostly well developed; apical bracts in a disc (subseries Crassirameae)

A single species:
.... *F. crassiramea* Miq. (1867) v. *patellifera* (Warb.) Corner (1960). *F. patellifera* Warb. (1905)

o o o o o o

Subsection Dictyoneuron

A.3.3
Key to species in series Glaberrimae

A single species:
.... *F. glaberrima* Bl. (1825) v. *bracteata* Corner (1960) India to New Guinea.

o o o o o o

A.3
Subsection Benjamineae

Key to the series in subsec. Benjamineae

1. Leaves drying pale green to pale brown, thinly coriaceous, the nerves raised on both sides; basal nerves not elongate.
...... A.3.4 ser. Benjamineae

1. Leaves drying dark brown, thinly to thickly coriaceous; petiole often black; basal nerves elongate or not... A.3.5 ser. Callophylleae

A.3.4
Key to the species in series Benjamineae

1. Fig 20-50 by 14-25 mm, ellipsoid to oblong. Lamina 8-20 by 3-9 cm. .... F. subcordata Bl. (1825) v. subcordata Tonkin and Java to Papuasia and Vanuatu.

o o o o o o

1. Fig smaller or, if as wide, then pyriform-pedicellate. .... F. benjamina L. (1767)

v. benjamina India to Solomon Is. and north Australia.

o o o o o o

v. bracteata Corner (1960) Celebes anf Philippines to New Guinea and New Britain.

o o o o o o

v. nuda (Miq.) Barrett (1951) NE India and S China to New Guinea.

o o o o o o

A.3.5
Key to species in series Callophylleae

1 Lamina obtuse or obtusely acumin-acuminate, elliptic to obovate, often thickly coriaceous; primary lateral nerves 6-11 pairs, not raised above; basal nerve elongate. .... F. microcarpa L. f. (1781)

v. microcarpa
Sri Lanka, India and S China to New Britain and Australia.

o o o o o o

v. hillii (F.M. Dailey) Corner (1960). F. hillii F. M. Dailey (1891).
Sunda Is., New Guinea to Australia and New Caledonia.

o o o o o o

v. latifolia (Miq.) Corner (1960). Urostigma accedens Miq. v. latifolia Miq. (1859)
Celebes and Moluccas to New Guinea and Queensland.

o o o o o o

v. naumannii (Engl.) Corner (1960) F. naumannii Engl. (1886) F. dahlii K. Schum (1898) F. regnans Diels (1935)
E New Guinea, New Britain, New Ireland to Solomon Is.

o o o o o o

1. **Lamina acutely acuminate, mostly thinly coriaceous, nerves finely raised on both sides, elliptic (5-12 cm long); primary lateral nerves 7-16 pairs; basal nerves not or little elongate................ 2**

2. Fig 13-18 by 10-14 mm. Primary lateral nerves 10-14 pairs. Basal bracts free. .... F. archboldiana Summerh. (1941) F. retusa L. v. papuana Diels (1935). E New Guinea.

o o o o o o

2. Fig smaller not subglobose. Fig-body not knobbly. Petiole black. 3

3. Basal bracts entirely connate as a thick saucer; body 5-7 mm wide. Primary lateral nerves 13-16 pairs. .... F. patellata Corner (1960) Irian Jaya.

o o o o o o

3. Basal bracts free............... 4

4. Basal bracts 2.5-4 by 3-6 mm, wider than long; body 5-7 mm wide. Primary lateral nerves 7-13 pairs. .... F. benjaminoides Corner (1960) Irian Jaya to Papua.

o o o o o o

4. Basal bracts 1.5-3.5 mm long and wide; body 6-11 mm wide, mostly with a few short internal bristles. Primary lateral nerves 6-9 pairs; basal nerves often elongate. .... F. microcarpa

A.4
Section Stilpnophyllum

Key to species in section Stilpnophyllum

A single species:
.... F. elastica Roxb. ex Hornem. (1819) Cultivated: India-rubber.

o o o o o o

A.5
Section Malvanthera

Key to series of section Malvanthera

1. **Anthers reniform, dehiscing longitudinally and cresentically. ..... A.5.1 series Malvanthereae**

1. **Anthers depressed-subglobose, dehiscing transversely. Seed-ovaries immersed in the fig-wall; stigma bifid. Fig 7-11 by 4-5 cm, woody. Stipule and venation as in F. elastica. ..... A.5.2 series Cyclanthereae**

A.4.1
Key to the species in ser. Malvanthereae

1. **Basal bracts 3, 0.5-3 mm long, inconspicuous, persistent or, if caducous, narrowly crescentic. Secondary lateral nerves almost as conspicuous as primary (more than 10 pairs). Seed-ovaries partly immersed in the fig-wall. Stigma bifid or simple................ 2 (subseries Hesperidiiformes)**

1 Basal bracts 2 or 3, relatively large, very early caducous...... 6

2. Fig-body 15-40 by 7-23 mm, ellipsoid to cylindric, apex somewhat umbonate to shortly rostrate; peduncle 3-15 by 1.5-4 mm, rather slender, not dilated at apex. .... F. xylosycia Diels (1935)

v. xylosycia
Eastern New Guinea.

o o o o o o

v. cylindrocarpa (Diels) Corner
F. cylindrocarpa Diels (1935).
Eastern New Guinea to Solomon Is.

o o o o o o

2. Fig-body larger and/or the peduncle stouter and, often, dilated upwards as a cupule adnate to the base of the fig-body.................. 3

3. Lamina ovate - cordate to broadly elliptic, very coriaceous; basal nerves 1/4 to 1/2 lamina, conspicuous. Peduncle not dilated at the apex...... F. augusta Corner (1960) Eastern Highlands.

o o o o o o

3 Lamina elliptic; basal nerves short, inconspicuous. Peduncle generally dilated at apex....... 4

4. Peduncle 10-50 by 3-5mm (-10 mm living), distinct, dilated or not at apex; fig-apex umbonate. .... F. hesperidiiformis King (1887) F. sclerotiara Diels (1935)

v. hesperidiiformis
Irian Jaya, New Guinea and Papua.

o o o o o o

v. myrmekiocarpa
Papua.

o o o o o o

4. Peduncle short, thick, 6-8 mm long and wide, dilated 10-15 mm wide at the apex...................... 5

5. Fig-apex strongly umbonate with 2(-4) gibberous apical bracts. Primary lateral nerves 15-23 pairs. .... F. mafuluensis Summerh (1941) Papua.

o o o o o o

5 Fig apex not umbonate, the orifice as a triradiate groove. Primary lateral nerves 24-27 pairs. ..... F. heteromeka Corner (1961) Irian Jaya.

o o o o o o

6. Seed-ovaries not immersed in the fig-wall; stigma simple. Stipule 1-5 cm long. Lamina 2-16 by 1.5-7 cm; primary lateral nerves mostly 8-13 pairs. Fig 6-14 mm wide, subglobose. (subseries Platypodeae) .... F. obliqua Forst. f. (1786) v. obliqua Celebes through New Guines to New Caledonia, Vanuatu and Australia.

o o o o o o

6. Seed-ovaries partly immersed in the fig-wall: stigma simple or shortly bifid. Figs, stipules and lamina generally larger, with more numerous primary lateral nerves...... 7 (subseries Malvanthereae)

7. Fig-body 7 - 10 by 6-9 mm, the orifice as a minute slit: peduncle 2-4 by 2 mm, the apex slightly dilated. Lamina 4-14 by 1.5-5.5 cm. .... F. rhizophoriphylla King (1887) New Guinea.

o o o o o o

7. Fig 10-15 by 10-18 mm, the orifice as a slit. Lamina larger. .... F. glandifera Summerh. (1932)

v. glandifera
Papua to New Britain, Solomon Is and Vanuatu.

o o o o o o

v. brachysyce Corner (1960)
Celebes to New Guinea.

o o o o o o

A.5.2
Key to species in series Cyclantherae

A single species:
.... F. sterrocarpa Diels (1935)

v. sterrocarpa
New Guinea and Papua.

o o o o o o

v. pubigemma Diels (1935)
Sepik.

o o o o o o

C. Subgenus PHARMACOSYCEA

Key to series of subgenus Pharmacosycea

1. Leaves drying grey-green to pale brownish; reticulations often raised on one or both sides; basal glands raely present. Fig-peduncles well developed; sclerotic cells throughout the fig-wall. Stamens generally 2 in the ostiolar flowers. Medium to large trees.
.... C.1.1 series Vasculosae

1. Leaf generally drying brown; basal glands 2, generally distinct.... 2

2. Leaf more or less coriaceous; stipules often elongate. Fig-peduncles often short. Stamen generaly 1. Large trees.
.... C.1.2 series Nervosae

2. Leaf membranous to coriaceous; stipule not elongate. Stamen 1,2, or 3. Shrubs to medium sized trees.
.. C.1.3 series Austrocaledonieae

C.1.1
Key to species in series Vasculosae

1. Leaves not nitid, membranous to chartaceous-subcoriaceous: cystoliths amphigenous to hypogenous. Deciduous.....(subseries Albipilae)
.... F. albipila (Miq.) King (1888) Covellia albipila Miq. (1860)

v. albipila
From Thailand to Irian Jaya and the mainland of Papua New Guinea.
o o o o o o

1. Leaves nitid, coriaceous; cystoliths hypogenous................ 2
(subseries Vasculosae)
None

C.1.2
Key to species in series Nervosae

1. Fig sessile, sclerotic cells abundant in the wall. Tepals 3-4, free or shortly joined. Stipule often prominent. Glabrous or thinly appressedly hairy.................. 2

1. Fig stalked..................... 4

2. Basal bracts 3.5-11 mm long. Fig often ellipsoid. Lamina lanceolate-elliptic, attenuate acute to obtuse; lateral nerves 11-19 pairs, at a wide angle. Stipules 2-5 cm long. Male flowers disperse, abundant; stamens 1.
.... F. pachystemon Warb. (1905) F. mangiferifolia Laut. et K. Schum (1901) F. brassii Summerh. (1929) F. saxicola Summerh. (1929); F. aechmophylla Summerh. (1933). Mainland Papua New Guinea to New Britain and New Ireland.
o o o o o o

2. Basal bracts 1-3.5 mm long. Fig subglobose..................... 3

3. Male flowers ostiolar. Lamina elliptic to lanceolate; lateral nerves 9-16 pairs. Fig 7-10 mm wide.
.... F. subtrinervia Laut. et K. Schum (1901)

v. subtrinervia
Eastern New Guinea to New Britain.
o o o o o o

v. doormaniana (Diels) Corner (1960) F. doormaniana Diels (1935) F. behrmanniana Diels (1935). Throughout New Guinea.
o o o o o o

3. Male flower disperse. Fig 6-8 or 10-15 mm wide. Disperse male flowers pedicellate. Stipule 1-4 cm long. Thinly appressedly hairy.
.... F. pubinervis Bl. (1825) v. pubinervis Sumatra to Aru Islands (Irian Jaya).
o o o o o o

4. Male flowers ostiolar. Fig 12-20 mm wide, basal bracts caducous: lamina small: lateral nerves 4-6 pairs, impressed above.
.... F. microdictya Diels (1935). New Guinea, Papua. Note: Corner (1965) keys out this species here. It is in subsect. Papuasyce.

4. Male flowers disperse (in some cases sparse)................... 5

5. Fig 20-40 mm wide. Stamens 2-3. Stipule not elongate. Fig-wall 5-10 mm thick, wholly sclerotic; cavity compressed. Lamina thickly coriaceous, elliptic, obtuse, base cuneate; lateral nerves 7-9 pairs. Tepals free.
.... F. pachysycia Diels ex Corner (1960). New Guinea: Morobe Province.
o o o o o o

5. Fig smaller or, if 20-30 mm wide, then the stipule elongate. Stamen 1 or 2........................ 6

6. Stipule prominent, often curved, 20-75 mm long. Stamen 1......... 7

6. Stipule not so prominent........ 11

7. Lateral nerves 5-9 pairs, 0-3 vague intercostals. Leaves more or less distichous...................... 8

7. Lateral nerves 9-14 pairs or more. Intercostals often numerous. Perianth often gamophyllous........ 9

8. Twigs, stipules, underside of midrib thinly appressedly pale brown hairy. Fig-wall with abundant sclerotic cells. Tepals free or shortly joined.. F. pubinervis Bl. (1825) v. pubinervis Sumatra to Aru Is.

o o o o o o

8. Glabrous. Fig-wall with few or no scleritic cells. Perianth gamophyllous, 3-4 short lobes. .... F. subnervosa Corner (1960) New Guinea (Key Is.?)

o o o o o o

9. Lamina obtuse to subacute, rarely acute, coriaceous; lateral nerves at a wide angle, scarcely raised below. Fig 9-12 mm wide, wall with abundant sclerotic cells. Glabrous or minutely appressedly hairy. ... F. hombroniana Corner (1960) Moluccas throughout New Guinea to Bougainville.

o o o o o o

9. Lamina acute to acuminate; lateral nerves curved-ascending, raised below. Fig larger............... 10

10. Glabrous. Stipule 2-5 cm long. Fig 18-30 mm wide, with or without sclerotic cells; peduncle 2-15 mm. Leaves distichous. .... F. polyantha Warb. (1905) Philippines to Solomon Is.

o o o o o o

10. Thinly appressedly pale brown puberulous. Stipule 15-30 mm long. Fig 14-20 mm wide, wall with abundant sclerotic cells; peduncle to 3 mm long. Fig without pedical, apical bracts projecting 1-2 mm. .... F. edelfeltii King (1887) F. edelfeltii v. glyptoneura Diels (1935)

v. edelfeltii New Guinea

o o o o o o

v. bougainvillei Corner (1960) Bougainville Is.

o o o o o o

11. Leaves spirally arranged. Lateral nerves strongly raised below. Not cauliflorous. Not cauliflorous. .... F. edelfeltii King

11. Lateral nerves scarsely raised below. Lamina often small or obtuse. 12

12. Lamina broadly elliptic, lateral nerves curved-ascending. Fig 10-17 mm wide, peduncle to 6 mm wide, wall with few or no sclerotic cells. Perianth gamophyllous, 3-4 lobed. Stamen 1. .... F. subnervosa

12. Lamina small, lateral nerves at a wide angle. Fig 6-12 mm wide, peduncle none, wall with abundant sclerotic cells. Fig pyriform. Tepals more or less free. Stamens 2......... F. ihuensis Summerh. (1929) Papua.

o o o o o o

Species Incertae Sedis

Twigs brown velutinate, hairs 0.2 mm long. Lamina 20-33 by 11-17 cm; lateral nerves 15-22 pairs; petiole 15-20 by 4-5 mm. Fig stalked.

.... F. hadroneura Diels (1935) New Guinea to New Britain.

o o o o o o

Twigs brown villous, hairs 0.5-1 mm long. Lamina 10-26 by 5-14 cm. narrowly auricled at the base; lateral nerves 10-15 pairs; petiole 6-20 by 3 mm. Fig unknown.

.... F. pseudojaca Corner (1960) Irian Jaya to New Guinea.

o o o o o o

C.1.3
Key to species in ser. Austrocaledoniae

A single species:

.... F. smithii Horne ex Baker (1883) v. smithii Solomon Is. to Fiji.

o o o o o o

C. FICUS Subgenus SYCOMORUS

A single species;

.... F. racemosa L. (1753) v. racemosa Sri Lanka to Australia and New Guinea.

o o o o o o

D. FICUS Subgenus FICUS

Keys to the Sections of Ficus subgenus Ficus

1. Root climbers. Leaves distichous, entire mostly coriaceous: appressed bathyphylls usually present. Fig with a collar of 3 basal bracts, no lateral bracts. Female stigma generally bifid. Seeds oblong, generally compressed and keeled. Hairs generally closely septate. Microscopic gland-hairs peltate...... 2

1. Trees or shrubs, in some cases epiphytic, strangling or scrambling (rarely root-climbing). Seed not oblong-compressed. Hairs not septate. Microscopic gland-hairs various, mostly not peltate..... 3

2. Male and neuter flowers disperse, abundant, pedicellate. Stamens 1, not mucronate. Tepals mostly 3, red or white. Fig usually large, often cauliflorous.
.... D3. Kalosyce

2. Male and neuter flowers ostiolar or, if disperse, mostly sessile. Stamens 2, mucronate or not. Tepals 3-5, red. Fig with the internal bracts of the orifice generally strongly inflexed and not interlocking.
.... D2. Rhizocladus

3. Perianth gamophyllous (saccate, cupular, annular), entire (or split by the enlarging ovary), membranous or none. Male flowers ostiolar; stamen 1, less often 2. Gall-stigma widely infundibuliform: female stigma simple. Lamina mostly without basal glands: often with a subnodal gland on the twig. Figs with or without lateral bracts, mostly with a collar of 3 basal bracts, often cauliflorous.
.... D8. Sycocarpus

3. Perianth of separate tepals or, if joined, then lobate. Gall-stigma narrowly infundibuliform to subclavate. Lamina often with glands in the axils of the main basal nerves; subnodal glands in some cases.......................... 4

4. Stamens 2 or more, in a few species 1. Fig with a collar of 3 basal bracts, no lateral bracts. Leaves symmetric...................... 5

4. Stamen 1 or, if 2, then the fig with lateral bracts or without a collar of basal bracts. Seeds smooth or minutely reticulate... 6

5. Large, often buttressed, trees, mostly cauliflorous. Male flowers ostiolar, often sessile and compressed: filaments shortly joined. Tepals pallid or red, membranous, entire or laciniate. Stigma thick, conical, simple.
.... D7. Neomorphe

5. Shrubs to medium sized trees, rarely creepers or cauliflorous. Male flowers mostly pedicellate, ostiolar or disperse, not compressed; filaments mostly free. Tepals dark red or pink and entire, or colourless and setose (then with copious chaffy internal bristles). Style slender; female stigma often bifid.
.... D1. Ficus

6. Male flowers disperse and ostiolar; stamens generally 2. Fig ramiflorous, pedicellate without a collar of basal bracts, with lateral bracts on the body. Ovary white.
.... D4. Sinosycidium

6. Male flower ostiolar. Stigma simple.......................... 7

7. Seed strongly compressed, generally with a double keel at the base, ovate-oblong to auriculiform. Flowers mostly sessile. Ovaries dark red or red-spotted, compressed; style strongly lateral to gynobasic, glabrous. Leaves entire, usually symmetric. Trees or shrubs, never epiphytic.
.... D6. Adenosperma

7. Seeds lenticular to shortly oblong, rarely with a double keel. Flowers often pedicellate. Ovary white: style subterminal, rarely strongly lateral. Leaves often toothed or asymmetric. Trees, shrubs, banyans, epiphytes or scrambling climbers.
.... D5. Sycidium

D1. Section Ficus

Keys to the subsections and series of section Ficus

1. Seeds smooth, not or slightly keeled. Tepals mostly glabrous. Leaves with hypogenous cystoliths.
(subsection Ficus)
Male flowers disperse. Gall-flowers both sessile and pedicellate, the tepals nearly or quite as long as the sessile or stipitate ovary. Female tepals long or short.
.... D.1.1 series Podosyceae

1. Seed tuberculate or echinate, with a double keel at the base, (if smooth or faintly keeled, then the tepals white and densely setose). Male flowers ostiolar. Cystoliths none.......................... 2
(subsection Eriosycea)
Tepals red, glabrous or ciliolate on the edges or the tip.
.... D.1.2 series Eriosyceae

D.1.1
Key to the species in series Podosyceae

A single species:
.... F. pedunculosa Miq. (1848)

v. pedunculosa
Formosa to Irian Jaya (Japan Is.)
o o o o o o

v. segaarensis
Ceram to New Guinea.

o o o o o o

D.i.2
Key to the species in ser. Eriosyceae (Subseries Dehiscentes)

1. Twigs 5-10 mm thick; lamina 7-18 cm wide, ovate-cordate; fig 15-18 mm wide...... F. ruficaulis Merr. (1904) Formosa to Philippines, Celebes and New Guinea (?).

o o o o o o

1 Twigs 3-6 mm thick; lamina and figs smaller... F. glandulifera Corner (1960) v. villosa Corner (1960) Borneo to New Guinea.

o o o o o o

D.2
Section Rhizocladus

Key to the series of Section Rhizocladus

1. Male and neuter flowers disperse, mostly sessile: leaves distichous. ..... D.2.3 series Trichocarpeae

1. Male and neuter flowers ostiolar in one to several rows, the male mostly stalked (except those next to the orifice)................ 2

2. Anthers not mucronate, short: filament free (?slightly joined): internal bristles absent: fig wall with sclerotic cells: figs ripening red, generally small, orifice sunken, without pedicel. Leaves distichous, often small and blunt, with the veins not or scarsely raised below.
.... D.2.2 series Disticoideae

2. Anthers mucronate. Filaments joined. Fig ripening red, often pedicellate, the orifice often sunken, with or without internal bristles and sclerotic cells: basal bracts often caducous. Leaves distichous, the veins impressed above or invisible. D.2.1 series Ramentaceae

D.2.1
Key to the species in series Ramentacaeae

1 Apical bracts concealed in sunken fig-orifice.................... 2

1 Apical bracts exposed, the fig-orifice not or scarsely sunken. (subseries Dalanotae).............. 8

2. Hairs irritant, especially on the figs. Fig-wall with sclerotic cells (except the small F. convexa). New Guinea (subseries Irritantes).. 11

2. Hairs not irritant............. 3

3. Fig-wall with sclerotic cells. (sub series Pantonianae)............. 4

3. Fig-wall without sclerotic cells. (subseries Ramentaceae)........ 15

4. Young parts floccoso-tomentose with undulate hairs. Lamina 10-15 cm. long; basal nerves short. .... F. floccifera Diels (1935) New Guinea and Papua.

o o o o o o

4. Glabrous or with straight hairs. Lamina often rather small; basal nerves often elongate.......... 5

5. Fig sessile, 7 mm wide. .... F. amblisyce Corner (1960) Irian Jaya.

o o o o o o

5. Fig stalked.................... 6

6. Fig-apex contracted into a papilla 1mm high. Lamina very coriaceous. .... F. ampulliformis Corner (1962) Irian Jaya.

o o o o o o

6. Fig-apex not so contracted...... 7

7. Fig 7 mm wide; internal bristles abundant. Lamina to 5.5 by 2.5 cm. .... F. hypobrunnea Corner (1962) Irian Jaya.

o o o o o o

7. Fig larger; internal bristles none or few, minute. Lamina often larger..... F. pantoniana King (1887)

v. pantoniana
Moluccas to New Guinea, New Britain and Queensland.

o o o o o o

v. colobocarpa
New Guinea.

o o o o o o

v. rhytidophloea
Papua.

o o o o o o

8. Fig stalked; basal bracts caducous. 9

8. Figs sessile, not umbonate (unless from projecting apical bracts). Internal bristles sparse or none. 10

9. Fig strongly umbonate, villous as the twigs, petioles, and underside of the lamina. Internal bristles abundant.
.... F. balanota Diels (1935) New Guinea.

o o o o o o

9. Figs not umbonate, puberulous, soon glabrous as the leaves and twigs. Internal bristles none.
.... F. oxymitroides Corner (1961)

v. oxymitroides
New Guinea, Eastern Highlands.

o o o o o o

v. brevipes
Mainland New Guinea.

o o o o o o

10. Apical bracts prominent; basal bracts caducous. Lamina 4-7 cm long, lateral nerves 5-6 pairs. Hairs to 1 mm long.
.... F. sageretina Diels (1935) New Guinea: Finisterre Mts

o o o o o o

10. Apical bracts not prominent; basal bracts persistent. Lamina 1-3 cm long, very coriaceous, hard, bullate; lateral nerves 3-4 pairs. Hairs to 3 mm long.
.... F. ceanothifolia Corner (1960) Irian Jaya: Wissel Lakes

o o o o o o

11. Basal bracts 3-6 mm long. Fig 15-35 mm wide, peduncle 1-10 mm long. 12

11. Basal bracts 1.5-3 mm long. Fig 7-12 mm wide, peduncle short. Lamina 5-28 cm long.................. 13

12. Lamina thinly coriaceous, the nerves not or slightly impresseed above. Basal bracts caducous. Internal bristles few, small or none. .. F. odoardii King (1887) F. conspicabilis King (1888) F. trichosphaeridia Diels (1935)

v. odoardi
New Guinea.

o o o o o o

v. glabrata Corner (1960)
F. cinnabarina S. Moore (1923)
New Guinea.

o o o o o o

12. Lamina coriaceous-brittle, the nerves impressed above. Basal bracts subpersistent. Internal bristles few, small or none.
.... F. sphaerocarpa Corner (1960) Papua: Boridi.

o o o o o o

13. Fig stalked. Hairs 1-2.5 mm long.
.. .. 14

13. Fig subsessile, 7-10 mm wide. Hairs 2-8 mm long on twigs and petioles.
.... F. insculpta Summerh. (1941) New Guinea.

o o o o o o

14. Figs 10-12 mm wide; internal bristles often abundant. Lamina 5-28 cm long. F. irritans Summerh. (1941) New Guinea.

o o o o o o

14. Fig 7-9 mm wide; internal bristles none or few, minute. Lamina to 8 cm long, stiffly coriaceous, convex.
.... F. convexa Corner (1962) New Guinea: Eastern Highlands.

o o o o o o

15. Figs lanuginous or villosulous, orifice slightly sunken. Lamina persistently hairy beneath; venation rather lax.
.... F. baeuerleni King (1887) F. mespiloides King (1888) F. hollrungii Laut. et K. Schum (1901) F. laurentina Diels (1935) F. baeuerleni v. glabrata Diels (1935)

v. baeuerleni
New Guinea to Solomon Is.

o o o o o o

v. vulcaniformis Corner (1960)
New Guinea to New Britain and Solomon Is.

o o o o o o

15. Fig soon glabrous, long-pedicellate, orifice slightly sunken. Lamina soon glabrous; nerves and intercostals crowded.
.... F. camptandra Diels (1935) Aru Is to New Guinea.

o o o o o o

D.2.2
Key to species in series Distichoideae

1. Figs 8-16 mm wide.............. 2

1. Figs 4-8 mm wide............... 3

2. Lamina obtuse to succate. Fig pedunculate. Twigs glabrous. .... F. distichoidea Diels (1935)

v. platyphylla Diels (1935)
F. pernitida Diels (1935)

i. Fig 8-12 mm wide; basal bracts persistent.
v. distichoidea
New Guinea and Papua.

o o o o o o

ii. Fig 14-16 mm wide; basal bract caducous.
v. megacarpa Corner (1960)
New Guinea: Eastern Higfhlands.

o o o o o o

2. Lamina ovate-acute to acuminate. Fig subsessile. Twigs appressedly hairy..... F. ovatacuta Corner (1962) New Guinea: E. Highlands.

o o o o o o

3. Lamina acute to subacuminate: basal nerves short. .... F. phatnophylla Diels (1935)

i. Fig sessile; basal bracts slowly caducous; lamina thinly coriaceous.
v. glochidioides Corner (1960)
Irian Jaya: Rouffaer River.

o o o o o o

ii. Fig peduncle to 2.5 mm long.

ii.i Lamina thinly coriaceous, subacute, lanceolate elliptic; Basal bracts caducous.
v. meiocarpa (Diels) Corner (1960). F. meiocarpa Diels (1935).
New Guinea: Sepik.

o o o o o o

ii.ii Lamina stiffly coriaceous, hard elliptic or ovate-elliptic, acuminate; basal bracts persistent.
v. phatnophylla
New Guinea: Sepik.

o o o o o o

4. Lamina obtuse or retuse; basal nerves often elongate; basal bracts caducous...................... 5

4. Lamina suborbicular to broadly elliptic; basal nerves 1/3 to 1/2 of lamina. Fig peduncle 1-4 mm long..... F. calodictya Summerh. (1929)

i. Fig 4-5 mm wide. Tepals free or shortly joined.
v. calodictya
Halmahera and Ternate to New Guinea.

o o o o o o

ii. Fig 6-8 mm wide. Tepals gamophyllous with 2-4 short lobes.
v. gamophylla Corner (1960)
Irian Jaya.

o o o o o o

5. Lamina small, elliptic to obovate, base cuneate, basal nerves rarely conspicuous. .... F. agapetoides Diels (1935)

i. Fig sessile or nearly so. Leaves stiffly coriaceous; stomata superficial.
v. agapetoides
New Guinea, Papua.

o o o o o o

ii. Fig-peduncles 1-3 mm long. Leaves thinly coriaceous; stomata deeply sunken.
v. solomonensis Corner (1960)
Solomon Is.: Bougainville and Guadalcanal.

o o o o o o

D.2.3
Key to species of series Trichocarpeae

1. Lamina not thickly lanuginous beneath with undulate hairs. 2

1. Lamina thickly lanuginous beneath with undulate hairs concealing the reticulations. Fig lanuginous... 8

2. Fig glabrous or thinly hairy and soon glabrescent............. 3

2. Fig lanuginous or villous....... 6

3. Lamina thinly appressedly hairy beneath along the nerves or glabrous. Twigs minutely brown puberulous to glabrous.......................... 4

3. Lamina closely villous or hispid beneath with erect brown hairs. Basal bracts 1-1.5 mm long. Robust climbers.......................... 5

4. Robust climbers. Lamina more than 9 by 5 cm. Fig 15-25 mm wide; basal bracts 2-4 mm long. .... F. nasuta Summerh. (1933)

v. nasuta
Solomon Is. Santa Cruz.

o o o o o o

v. glabrata Corner (1960)
New Guinea and Papua.

o o o o o o

Slender climbers Lamina smaller. Fig 5 8 mm wide; basal bracts 1-1.5 mm long. Gall flowers with the lower half of the tepals yellow and indurate Gall figs with sclerotic cells in the walls, seed-figs without. . . . *F. pleiadenia* Diels (1935) New Guinea.

o o o o o o

Fig body thinly appressedly puberulous, 8 10 mm wide; peduncle 10-15 mm long; sclerotic cells abundant in the wall (gall fig). Leaf not bullate; cystoliths amphigenous. .... *F. alococarpa* Diels (1935) New Guinea: Sepik.

o o o o o o

5 Fig-body soon glabrous, 10-12 mm wide; peduncle 5-9 mm long; no sclerotic cells in the fig-wall (gall-fig) Leaf not bullate; cystoliths amphigenous. . . . . *F. phaeobullata* Corner (1960) Papua: Rouna

o o o o o o

6 Twigs and petioles with dull brown spreading hairs 2 3 mm long, underside of nerves finely tomentose with undulate hairs. Lamina 9 17 by 4-12 cm. Fig 15 18 mm wide, basal bracts caducous . *F. fuscata* Summerh (1941) Papua

o o o o o o

6. Without such spreading hairs. Figs smaller; basal bracts persistent. . 7

7 Lamina 7-14 by 5.5-10 cm, glabrous. Fig 10-12 mm wide; basal bracts 3-6 mm long. *F. supfiana* Schlcht ex Diels (1935) New Guinea.

o o o o o o

7. Lamina 4-8 by 2-3.5 cm, nerves thinly hairy beneath. Fig 6-8 mm wide; basal bracts 1-2 mm long. . . . . . . *F. semilanata* Corner (1960) Papua.

o o o o o o

8 Robust climber. Twigs 4-5 mm thick. Lamina more than 9 by 5 cm. Fig body 15-20 mm wide; basal bracts 3 4 mm long.. *F. hypophaea* Schltr ex Diels (1935) Papua New Guinea

o o o o o o

8 Slender climbers. Twigs 2 mm thick Lamina smaller Fig-body 7 10 mm wide, basal bracts 3-4 mm long. . 9

9. Indumentum rusty cinnamon. Lamina acuminate (without cystoliths) Fig 7-8 mm wide, the seed-fig without sclerotic cells in the wall . . *F. cinnamomea* Corner (1960) Papua.

o o o o o o

9. Indumentum dull btown. Lamina attenuato-acute. Fig-body 10 mm wide, the seed-fig with sclerotic cells in the wall. .... *F. hypophaeola* Corner (1960) Irian Jaya.

o o o o o o

D.3 Section *Kalosyce*

Only series *Punctatae* is represented in Papuasia.

Key to species in series *Punctatae*

1 Veins not impressed above, invisible or as fine raised lines; lamina thinly coriaceous, often drying greenish; the veins not so coarse below (subseries *Punctatae*) .... *F. scratchleyana* King (1887)

v. *scratchleyana*
New Guinea to New Britain.

o o o o o o

v *aurantiola* Corner (1960)
New Guinea.

o o o o o o

v. *rhopalosycia* (Diels) Corner (1960) *F. rhopalosycia* Diels (1935) Irian Jaya: Hellwig Mts.

o o o o o o

1 Veins impressed on the upperside of the stiffly coriaceous lamina, drying brown; veins coarse and prominent beneath (subseries *Ruginerviae*) .... *F. gymnorygma* Summerh (1941) *F. scratchleyana* King v *pleiotricha* Diels (1935)

o o o o o o

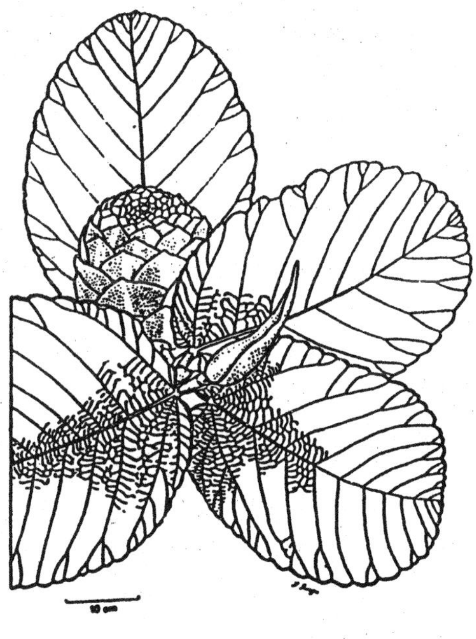

Ficus dammaropsis A. stem with fruit
(Drawn: T. Iwagu)

D.5
Section Sycidium

Key to subsections and series of Section Sycidium

1. Seeds lenticular (as long as wide), slightly keeled all round or in the upper half or not at all. Male flowers mostly without a normal gall-ovary, often with a rudiment. Leaves variously arranged. Figs often pedunculate with a collar of basal bracts. Cystoliths often papillate. Trees, shrubs, not epiphytic (subsection Sycidium). 2

1. Seed shortly oblong, generally keeled or gibbous at the apex. Ovary white. Male flowers generally with a normal gall-ovary. Leaves distichous, short-petiolate. Figs not pedunculate. Cystoliths not papillate. Trees, shrubs or climbers, often epiphytic (subsection Palaeomorphe)............. 5

2. Figs on long-hanging rami-cauliferous twigs. Fig-body 4-8 mm wide, without lateral bracts. Tepals red. Ovary red. Style (female) hairy. Leaves long-petiolate to distichous and short-petiolate. Hairs dark brown, wiry.
..... D.5.1 Series Pungentes

2. Figs axillary or on short leafless twigs from trunk and branches, rarely geocarpic. Ovary white.. 3

3. Leaves short-petiolate, variously arranged, commonly distichous, often asymmetric. Figs often pedunculate with a collar of basal bracts. Tepals red, pink, or white. Hairs not purple or dark brown.
..... D.5.4 Series Scabrae

3. Leaves spirally arranged or decussate, mostly long-petiolate, symmetric. Figs pedicellate or nearly sessile, without a collar of basal bracts........................ 4

4. Hairs dark purple or brown. Fig-body often with lateral bracts. Tepals dark red.
..... D.5.2 Ser. Phaeopilosae

4. Hairs white, or merely scabridulous -glabrous. Tepals purple, reddish or white.
.... D.5.3 Series Copiosae

D.5.1
Key to species of series Pungentes

1. Leaves spirally arranged, ovate-cordate, long-petiolate. Hairs 2-10 mm long on twigs and petioles.
.... F. pungens Reinw. ex Bl. (1825) Philippines to New Guinea and New Britain.

o o o o o o

1. Leaves distichous, elliptic-obovate, short petiolate. Hairs 1-2 mm long on twigs and petioles.
.... F. petrotica Diels (1935) New Guinea.

o o o o o o

D.5.2
Key to species of series Phaeopilosae

1. Perianth cupular, 3-4-toothed. Fig 6 mm wide, brown strigose, cauliflorous (?geocarpic). Lamina narrowly elliptic-obovate, dentate. Stipules caducous. Cystoliths hypogenous.... F. badiopurpurea Diels (1935) New Guinea: Sepik River.

o o o o o o

1. Tepals free, 4-6............... 2

2. Fig 5-12 mm wide. Internal bristles abundant. Twigs 1.5-3.5 mm thick. Stipules small, caducous........ 3

2. Figs mostly larger. Twigs thicker. Stipules often large and persistent............................ 4

3. Fig stalk 3-18 mm long. Petiole 1-8 cm long or more. Fulvous to dark brown hairy. Cystoliths amphigenous....... F. gul Laut. et K. Schum (1901)

v. gul
Borneo to New Guinea and New Britain.

o o o o o o

v. eubracteata Corner (1961)
New Guinea.

o o o o o o

v. lasiocarpa Corner (1960)
Halmahera to Irian Jaya.

o o o o o o

v. solomonensis Corner (1960)
Bougainville to San Cristoval.

o o o o o o

3. Fig stalk 0-2 mm long. Petiole to 1.5 cm long. Dark purple brown or black hairy. Cystoliths hypogenous.
.... F. phaeosyce Laut. et K. Schum. (1961).
New Guinea, Papua.

o o o o o o

4. Fig 8-12 mm wide, purple-strigose, glabrescent; lateral bracts 1-2.5 mm long. Irritantly purple-brown, bristly. Cystoliths amphigenous, varying hypogenous.
.... F. porphyrochaete
Corner (1960) Papua, Solomon Is.
o o o o o o

4. Figs and lateral bracts larger.. 5

5. Fig 20-25 mm wide, globose to ellipsoid; lateral bracts 3-6 mm long; internal bristles none. Gall- and female flowers with long pedicels (-4 mm). Lamina obovate, nerves thinly apressedly brown hairy below. Cystoliths amphigenous.. F. conocephalifolia
Ridley (1917) New Guinea.
o o o o o o

5. Fig 10-17 mm wide (excluding hairs and bracts); lateral bracts 5-16 mm long. Cystoliths hypogenous. Densely strigose hairy............. 6

6. Leaf ovate or elliptic, base cordate. Stipules persistent. Internal bristles few, minute. Flowers subsessile, the fig hollow.
.... F. complex Corner
(1960) New Guinea, Papua.
o o o o o o

6. Leaf-base cuneate; stipules caducous. Internal bristles abundant. Flowers pedicellate. Lamina elliptic....... F. eustephana Diels
(1935) New Guinea.
o o o o o o

D.5.3
Key to species of series Copiosae

1. Cystoliths hypogenous. Figs cauliferous with projecting apical bracts and, often, conspicuous lateral bracts. Leaves not opposite..... 2

1. Cystoliths amphigenous. Figs axillary to cauliflorous, the apical bracts not projecting, lateral bracts small. Leaves often opposite........................... 3

2. Lamina ovate, lateral nerves 5-8 pairs. Fig small, with a few latlateral bracts. Tepals dark red. Internal bristles copious, long.
.... F. griseifolia
Corner (1960) Papua.
o o o o o o

2. Lamina obovate, lateral nerves 9-13 pairs. Fig large, pyriform, stalks to 70 mm long, no lateral bracts.
.... F. primaria Corner
(1960) New Guinea to New Britain.
o o o o o o

3. Twigs 3-8 mm thick. Leaves mostly opposite; lamina 6-20 cm wide, the base cordate to widely cuneate. Figs cauliferous, 12-20 mm wide, the stalks 12-55 mm long. Tepals dark red or purple, fading pink to white. Male flowers rarely with a gall-ovary. F. copiosa Steud.
(1841)

v. copiosa
Celebes to Solomon Is. and Queensland.
o o o o o o

v. pubescens Corner (1960)
Moluccas to Solomon Is. and Queensland.
o o o o o o

3. Twigs 1.5 - 3 mm thick. Leaves usually spirally arranged; lamina 2-12 cm wide, the base cuneate or subcordate in saplings. Figs axillary to cauliferous, 9-13 mm wide, the stalks to 35 mm long. Tepals dark red or white. Male flowers often with an ovary. Shrubs or small trees................ 4

4. Figs minutely scabridulous or nearly smooth.. F. wassa Roxb.
(1832)

v. wassa
Moluccas to Solomon Is and Vanuatu.
o o o o o o

v. nubigens (Diels) Corner (1960). F. nubigena Diels (1935)
F. caroli Diels (1935)
New Guinea.
o o o o o o

4. Figs densely echinate-muriculate, axillary, the stalk to 7 mm long. Tepals white.
.... F. hystricicarpa
Warb. (1905) New Guinea and Papua
o o o o o o

D.5.4
Key to species of series Scrabrae

1. Leaves opposite or spirally arranged, not or slightly asymmetric. Cystoliths amphigenous or as microscopic hairs on the underside of the lamina.................. 2

1. Leaves distichous. Cystoliths amphigenous or hypogenous.......... 4

2. Leaf-base rounded to cordate, apex blunt to subacute. Harshly scabrid, hispid villous, the leaves even with sculeate hairs. Apical bracts of fig obtuse, not or scarcely projecting.
.... F. opposita Miq. (1848)

v. opposita
Java, New Guinea nd Papua to Australia.
o o o o o o

v. micrantha (Miq.) Corner (1960).
Irian Jaya and Papua to Australia.
o o o o o o

2. Leaf-base cuneate or, if subcordate, then the apex acute to acuminate. Not (white to brown) villous or aculeolate.................. 3

3. Tepals dark red, fading pink. Fig becoming cauliflorous, pedicels to 20 mm long, no collar of basal bracts.... F. wassa var. nubigena

3. Tepals white. Fig not cauliflorous, peduncle to 10 mm long, with a collar of basal bracts.
.... F. cumingii Miq. (1848)

v. androbrota (Summerh.) Corner (1960)
F. androbrota Summerh. (1929)
F. dichroa Summerh. (1929)
New Guinea.
o o o o o o

4. Leaf-base cuneate, often narrowly, not or little asymmetric........ 5

4. Leaf-base broadly cuneate to cordate, asymmetric................ 10

5. Hairy. Lamina caudate, not scabrid. Fig 10-12 mm wide, peduncle 1-2 mm long. Internal bristles copious, long. Tepals hairy, red, fading white. Cystoliths hypogenous.
.... F. macrorrhyncha Laut. ex K. Schum.

v. leptodictya (Diels) Corner (1960) F. ledtodicty Diels (1935)
New Guinea.
o o o o o o

5. Without such characters......... 6

6. Tepals dark red. Cystoliths amphigenous. Glabrous-scabridulous.. 7

6. Tepals whit or pale pink........ 8

7. Fig axillary, solitary. Lateral nerves 7-11 pairs: intercostals 2-5.
.... F. imbricata Corner (1960)

v. imbricata
Bougainville.
o o o o o o

v. subcordata Corner (1961)
F. tonsa Miq. v. subcordata Corner (1960).
Bougainville.
o o o o o o

7. Figs cauliflorous. Lateral nerves 11-16 pairs: intercostals 5-11.
.... F. arawensis Corner (1961) Bougainville.
o o o o o o

8. Fig pedunculate with a collar of basal bracts, 10-20 mm wide. Fig with abundant internal bristles. Tepals white. petiole with a single gland an the back at the junction with the midrib.
.... F. schumanniana Warb. (1905) New Guinea and New Britain.
o o o o o o

8. Figs mostly pedicellate with scattered lateral bracts on the stalk. Leaves mostly scabrid.
.... F. ampelas Burm. f. (1768)

v. ampelas
Sumatra to Irian Jaya.
o o o o o o

v. soronensis (King) Corner (1960). F. blepharosepala Warb. (1905).
Amboina to New Guinea and New Britain.
o o o o o o

9. Leaf-base strongly asymmetric, the broader side generally concealing the petiole.................. 10

9. Leaf-base not so strongly asymmetric, the petiole not concealed.
.... 11

10. Apical bracts strongly projecting 2-3 mm. Densley hispid. Twigs 5-6 mm thick. Stipules 20-30 mm long. Fig ellipsoid, 15 by 12 mm, with lateral bracts; solitary.
.... F. gryllus Corner (1960) Solomon Is. Ysabel.
o o o o o o

10. Apical bracts not or scarcely projecting. Twigs 1.5-3.5 mm.
.... F. erinobotrya Corner (1960)

v. erinobotrya
New Britain.

o o o o o o

v. solomonensis Corner (1960)
Solomon Is. Bougainville to South Georgia.

o o o o o o

f. glabrior Corner (1960)
Solomon Is. Bougainville, Guadalcanal, San Cristoval.

o o o o o o

11. Lamina subacute to rather obtusely subacuminate, intercostals lax. Basal bracts to 1 mm long. Apical bracts not projecting. Tepals white or pinkish.
.... F. erinobotrya Corner (1960)

11. Lamina acute, acutely acuminate, cuspidate, or caudate. Intercostals generally numerous and regular. 12

12 Figs axillary or on the twigs behind the leaves, not ramiflorous or cauliflorous, usually with a collar of basal bracts. Cystoliths amphigenous.................. 13

12. Figs becoming ramiflorous in clusters or cauliflorous........... 16

13. Petiole more than 10 mm long. Lamina ovate- to obovate-elliptic, hairy to globose. Fig 10-15 mm wide, basal bracts to 1 mm long. Tepals reddish pink fading white, glabrous or sparsely hairy. Seed minutely reticulate. Large tree; latex serous.
.... F. melinocarpa Bl. (1825)

v. melinocarpa
Sumatra to Solomon Is.

o o o o o o

v. hololampra (Diels) Corner (1960). F. hololampra Diels (1935). F. collinsii Elm. (1937).
Sumatra to Solomon Is.

o o o o o o

13. Petiole usually less than 10 mm long. Tepals often very hairy. Basal bracts 1-2 mm long. Latex white......................... 14

14. Tepals white. Leaf-base cordate: lateral nerves 6-10 pairs. Fig 8-14 mm wide: internal bristles to 1 mm long, copious. Twigs and petioles white or yellow hairy.
.... F. chrysochaete Corner (1960) Bougaiville to Ysabel.

o o o o o o

14 Tepals red.................... 15

15. Fig with slender pedicel 3-18 mm long, and slender peduncle. No sclerotic cells. Lamina with a small basal auricle on the acroscopic side.. F. imbricata

15. Fig with a short peduncle to 5 mm long, no pedicel, with sclerotic cells. Lamina asymmetric, not auricled..... F. quercetorum Corner (1961) F. tonsa Miq. var. aspera Corner (1960) New Guinea, Papua.

o o o o o o

16. Lamina oblong-elliptic, large.
.... F. oleracea Corner (1960)

v. oleracea Corner (1960)
Bougainville to New Georgia and San Cristoval.

o o o o o o

v. pungens Corner (1960)
Dougainville.

o o o o o o

16. Lamina elliptic to ovate-elliptic, small to medium size. Figs 6-12 mm wide...... F. trachypison K.

Schum (1901) F. lima K. Schum (1901) F. pteleaephylla S. Moore (1923) F. xanthosyce Summerh. (1929)

v. trachypison
Amboina to Solomon Is.

o o o o o o

v. pallida Corner (1960)
New Guinea, Papua, Bougainville.

o o o o o o

D.5.5
Key to species of subsection Palaeomorphe series Pallidae

1. Lamina more or less asymmetric, often subrhombic or angled, obtuse to acute or shortly acuminate, basr rounded or subcordate to cuneate, often drying brown areolate beneath. Lateral nerves 3-9 pairs. Fig 10-17 mm wide. Tepals white. Stomata superficial or sunken.
.... F. tinctoria Forst. f.

ssp. tinctoria
Formosa through Philippines and New Guinea to australia and Polynesia.

o o o o o o

1. Lamina symmetric or nearly so, acuminate, base cuneate, not angular, drying grey-green. Lateral nerves 7-13 pairs, (4-6 in small leaves). Fig 7-10 mm wide. Tepals white or red. Stomata superficial.
..... *F. virgata* Reinw. ex Bl (1825)

v. *virgata*
Formosa, Philippines, throughout Papuasia to Australia and New Caledonia.

o o o o o o

v. *sessilis* (Dur.) Corner (1960) *F. philippinensis* v. *sessilis* Dur. (1872)
Luzon, New Guinea, New Britain, Solomon Is. to Vanuatu and New Caledonia.

o o o o o o

D.5.6
Key to Species in series *Subulatae*

1 Glabrous or the young parts thinly appressedly puberulous. Lamina acuminate to caudate, base usually cuneate, not scabrid. Seed slightly keeled. Tepals mostly puberulous...... *F. subulata* Bl. (1825)

v. *subulata*
China to New Britain and Solomons.

o o o o o o

v. *gracillima* (Diels) Corner (1960) *F. gracillima* Diels (1935) *F. otariophylla* Diels (1935).

o o o o o o

1. Hispid with light fulvous or brown hairs, glabrescent. Lamina acute to acuminate, base subcordate to asymmetric and subauricled on one side, often scabrid. Seed compressed, strongly keeled. Male flower mostly without an ovary. Tepals glabrous...... *F. armiti* King (1887) *F. fuscipes* Warb. (1905)
Central and Eastern New Guinea.

o o o o o o

D.5.7
Key to species of series *Minutuliflorae*

A single species:
..... *F. aurita* Bl. (1825)

v. *aurita*
Borneo and Philippines to Irian Jaya.

o o o o o o

D.6 Section *Adenosperma*

Key to Series of Section *Adenosperma*

1. Cystoliths (as minute pustules in dried leaves) on both sides of the lamina.
..... D.6.1 series *Amphigenae*

1. Cystoliths only on the lower side of the leaf (not minutely pustulate above.
..... D.6.2 series *Hypogenae*

D.6.1
Key to species of series *Amphigenae*

1. Fig sessile; lateral bracts conspicuous, 4-10 mm long; basal bracts 3-5, 5-13 mm long. Internal bristles abundant.............. 2

1. Fig pedunculate or, if sessile, with smaller basal bracts. Stipules not conspicuous................ 4

2. Stipules to 2 cm long, caducous. Twigs 3-4 mm thick. Lamina to 20 cm long, base narrowly cuneate, costas 6-8 pairs. Fig 15 mm wide, basal bracts 5-9 mm long. Twigs, petioles, and figs densely fulvous hairy..... *F. ochrochlora* Ridley (1916) New Guinea.

o o o o o o

2. Stipules 2-5 cm long, persistent. Twigs 5-10 mm thick. Lamina 25-45 cm long, costas 10-14 pairs. Petioles short, stout.......... 3

3. Stipules ovate-saccate, apiculate. Lamina obovate, gradually attenuate to the narrowly cordate base. Fig 20 mm wide, basal bracts 5-7 mm long. Tepals 4-6, oblong spathulate, free. *F. saccata* Corner (1960) New Guinea to Papua.

o o o o o o

3. Stipules broadly lanceolate. Lamina broadly elliptic to obovate, rounded subcordate to widely cuneate at the base. Fig 20-35 mm wide, basal bracts 9-13 mm long. Tepals 3(-4), ovate-lanceolate, gamophyllous round the stalked ovaries.
.... *F. megalophylla* Diels (1935) New Guinea.

o o o o o o

4. Leaves distichous.............. 5

4. Leaves spirally arranged. Trees or shrubs with *Terminalia*-branching.
.... 6

5. Brown hairy. Leaves lanceolate, pendulous. Fig-peduncles 10-12 mm long. Internal bristles abundant. .... F. endochaete Summerh. (1941) New Guinea.

o o o o o o

5. Thinly white puberulous to glabrous. Leaves elliptic, subacute, scabrid. Fig-peduncle 2-6 mm. Internal bristles none. Stomata sunken.... F. umbonata Reinw. ex Bl. (1825) Covellia umbonata (Reinw.) Miq. (1859). Moluccas to Irian Jaya.

o o o o o o

6. Leaves with regular intercostals. 7

6. Leaves with few, lax or irregular intercostals. Figs mostly on the twigs......................... 9

7. Closely fulvous hairy. Leaf-base widely cuneate to cordate (lanceolate in riparian forms). Fig 12-16 mm wide, becoming cauliferous. .... F. mollior Benth. (1873) F. gazellae Engl. (1886) F. chrysolaena Laut. et K. Schum. (1901) F. dielsii Warb. (1905) F. stelechosycia Diels (1935).

v. mollior
Amboina to Bougainville and Queensland.

o o o o o o

v. pseudocovellia Corner (1960) Irian Jaya: Mamberamo.

o o o o o o

v. sessilis Corner (1960)
Papua and Papuan Islands.

o o o o o o

f. riparia Corner (1960)
Irian Jaya: Sorong.

o o o o o o

7. Glabrous or thinly appressedly hairy. Leaf-base cuneate........ 8

8. Fig 6-8 mm wide, ramiflorous. .... F. comitis King (1888) F. nuruensis Warb. (1905) New Guinea, Papua (?).

o o o o o o

8. Fig 10-18 mm wide, ribbed. Perianth gamophyllous, at least in the lower part. Style not strongly lateral. Seed without double based keel. Lamina often somewhat auricled at the base.. F. indigofera Rechinger (1912) F. bukaensis Rechinger (1912) F. kietana Rechinger (1912)

o o o o o o

9. Riverside shrub to 2 m high. Leaves small, narrow, often greyish -white beneath. Fig 9-22 mm wide, lateral bracts often equatorial. .... F. arbuscula Laut. et K. Schum. (1901) F. charadrophila Summerh. (1929) Halmahera to New Guinea.

o o o o o o

9. Trees. Leaves not whitish beneath. Lateral bracts scattered or none. 10

10. Leaves drying brittle-chartaceous, venation obscure, smooth or scabrid. Glabrous or minutely puberulous. Fig-stalks 10 mm or more long, slender. Internal bristles few or none...... F. verticillaria Corner (1960)

v. verticillaris
Solomon Is.

o o o o o o

v. robusta Corner (1960)
San Cristoval, Guadalcanal.

o o o o o o

10. Leaves membranous to subcoriaceous, not brittle or scabrid, venation distinct. Mostly hairy. Fig-stalks mostly less than 10 mm long, rather thick..... F. adenosperma Miq. (1867) F. pauper King (1888) F. trichoneura Summerh. (1932).

v. adenosperma
Celebes, throughout Papuasia (exclu-ding Solomons?) to New Hebrides.

o o o o o o

v. chaetophora (Warb.) Corner (1960). F. novae-hannoverae Engl. (1886) F. chaetophora Warb. (1905) F. thelostoma Warb. (1905).
New Guinea, New Britain.

o o o o o o

v. glabra Corner (1960)
F. depressa Benth. (1873) F. frutescens F. M. Dailey (1914) F. turbinata Ridley (1916)
Throughout Papuasia to Queensland.

o o o o o o

v. microlepia Corner (1960)
New Guinea.

o o o o o o

D.6.2
Key to the species of series Hypogenae

1. Glabrous, or the young parts minutely puberulous. Lamina often obtuse. Internal brsitles none or few and minute..................... 2

1. Hairy. Stipule not prominent. Twigs generally grey to dark brown. Lamina acuminate: intercostals numerous, regular. Internal bristles mostly conspicuous. Often cauliferous........................... 3

2. Stipules 25-45 mm long, lanceolate-subulate, very prominent. Twigs brown. Lamina large obovate, 6-15 cm wide. Fig 15 mm wide or more, orifice umbonate.
.... F. austrina Corner (1960) New Guinea to Solomon Is. and Vanuatu.

o o o o o o

2. Stipules shorter. Twigs pale cream-ochraceous. Lamina 2-7 cm wide, obtuse to shortly acuminate. Fig 10-12 mm wide, the orifice plane or sunken.... F. casearioides King (1887) F. kingii F. Muell. (1890) F. hylobia Diels (1935).

v. casearioides
Morotai, New Guinea

o o o o o o

v. gamosepala Corner (1960)

Ternate, New Guinea.

o o o o o o

3. Strigose-villous with brown hairs 1-3 mm long, without an undercoat of tomentose undulate hairs. Fig 20-30 mm wide, without lateral bracts: basal bracts 6-8 by 4-5 mm, early caducous.
.... F. subcuneata Miq. (1867) F. stoechotricha Diels (1935) F. trichoneura Diels v. latifolia Diels (1935) F. formosa Summerh. (1941). Celebes to New Guinea and New Britain.

o o o o o o

3. Thinly appressedly hairy or tomentose with an undercoat of undulate hairs. Basal bracts smaller, persistent. Lateral bracts usually present, small................. 4

4. Cauliferous on leafless twigs to 150 cm. Fig 7-8 mm wide, peduncle 3-5 mm long. Fulvous villous without an undercoat of undulate hairs.
.... F. pleioclada Diels (1935) New Guinea: Sepik.

o o o o o o

4. Figs mostly larger, axillary, or, if cauliforous, on much shorter twigs.......................... 5

5. Stipules 10-30 mm long, lanceolate, persistent. Lamina lanceolate-elliptic, petiole 4-14 mm long. Fig axillary 6-8 mm wide. Thinly appressedly brown hairy or glabrous. Slender riverside tree.
.... F. tenella Corner (1960) Papua.

o o o o o o

5. Stipules caducous. Petiole longer. Fig larger. Large forest trees, becoming cauliflorous.......... 6

6. Fig 20-30 mm wide, densely villous-tomentose. Leaf-base cordate to wi widely cuneate, basal nerves prominent..... F. funiculosa Corner (1960) F. trichoneura Diels v. lachnocarpa Diels (1935). New Guinea, New Britain.

o o o o o o

6. Fig 10-15 mm wide. Leaf-base cuneate or narrowly cuneate-subcordate: basal nerves short........ 7

7. Fig persistently and densely tomentose with undulate hairs, as the twigs and stipules; peduncle short, to 6 mm long: basal bracts 1-2 mm long...... F. trichosera Diels (1935) F. incompta Diels (1935).

v. trichosera
New Guinea.

o o o o o o

v. glabristipula Corner (1960)
Papua.

o o o o o o

7. Fig glabrescent, thinly appressedly hairy with or without a coat of undulate hairs (as the twigs): peduncle or pedicel longer: basal bracts 1-2 mm long.
.... F. erythrosperma Miq. (1867) F. pycnoneura Laut. et K. Schum. (1901) F. lachnocarpa Warb. (1905) F. xanthoxyla Summerh. (1929) Celebes to Solomon Is.

o o o o o o

D.7
Key to species of section Neomorphe series Variegatae

1. Tepals entire. Fig-wall without inward processes. Basal bracts 1-2 mm long. Lamina often acuminate. Twigs slender (subser. Variegatae Bl. (1825)
.... F. variegata Bl. (1825).

v. variegata
Burma to Solomon Is. and Queensland

o o o o o o

1. Tepals dentate laciniate. Fig-wall thick with inward processes bearing the flowers. Basal bracts 2-9 mm long. Lamina ovate-cordate, blunt to acute. Twigs stout (subser. Laciniatae Corner)............. 2

2. Glabrous or thinly hairy. Figs without internal bristles; basal bracts persistent. Cauliflorous. .... F. nodosa Teysm. et Binn. (1866) F. du Laut. et K. Schum. (1901) F. suringari ? Carr (1866) Moluccas to Solomon Is.

o o o o o o

2. Hispid-tomentose or villous. Figs with abundant internal bristles; basal bracts caducous........... 3

3. Cauliflorous. Leaves scabrid. .... F. robusta Corner (1960) New Guinea

o o o o o o

3 Figs solitary on the twigs. Leaves smooth above, villous beneath. .... F. semivestita Corner (1960) New Guinea.

o o o o o o

D.8
Key to subsections of Section Sycocarpus

1. Seed 1.4-2 mm long, more or less auriculiform, subcompressed, obtusely keeled, smooth, hilum not prominent. Stamens 1 or 2. ... D.8.1 Subsect. Auriculisperma

1. Seed 0.6-1.5 mm long, differently shaped........................... 2

2. Fig very large, 5-13 cm wide, axillary, covered by very large lateral bracts. Stamens 2. Ovary white. Seed shortly oblong, hilum not prominent. .. D.8.2 Subsect. Dammaropsis

2. Fig smaller or not covered with lateral bracts.................. 3

3. Ovary white. Seed 1-1.5 mm long, lenticular, scarsely keeled, smooth, hilum not prominent. Figs pedunculate; internal bristles none. ... D.8.3 Subsect. Papuasyce

3. Ovary red. Seed 0.6-1.3 mm long, more or less keeled, smooth or rough, the hilum prominent. Figs various: internal bristles present or not. Stamens 1(-2) ... D.8.4 Subsect. Sycocarpus

D.8.1
Key to Species of subsection Auriculisperma

1. Twigs and leaves with stiff brown hairs 3-7 mm long. Fig sessile: basal bracts 7-15 mm long. Cystoliths hypogenous (series Cynaroides) 2

1. Glabrous or with short pale hairs. Figs pedunculate. Cystoliths mostly amphigenous.................. 3

2. Stipules to 3 cm long, caducous. Lateral nerves 7-8 pairs. Fig-body with lanceolate lateral bracts. .... F. lancibracteata Corner (1960) Bougainville Is.

o o o o o o

2. Stipules 9-13 cm long, sericeous, persistent, concealing the figs. Lateral nerves 9-13 pairs. Figs without lateral bracts. .... F. cynaroides Corner (1960) Bougainville Is.

o o o o o o

3. Freely branched trees. Twigs and leaves smaller. Figs axillary and on the twigs, slightly ridged towards the apex (series Vitiensis) .... F. indigofera Rechinger (1912) F. bukaensis Rechinger (1912) F. kietana Rechinger (1912). Solomon Is.

o o o o o o

3. Sparingly branched, pachycaulous trees. Twigs 5-20 mm thick. Lamina over 30 cm long, cordate-auricled. Figs axillary or cauliflorous. (series Theophrastoides).... 4

4. Twigs 12-20 mm thick. Lamina 80-170 cm long, scabrid. Figs cauliflorous in large bunches, verrucose, with basal and lateral bracts 9-16 mm long: peduncle 35-80 mm long. .... F. salomonensis Rechinger (1912) Bougainville, Isabel Is.

o o o o o o

4. Twigs 5-10 mm thick. Lamina 30-90 cm long, smooth. Figs axillary, concealed, becoming cauliflorous, not verrucose, without lateral bracts: basal bracts 1.5-9 mm long: peduncle 3-9 mm. .... F. theophrastoides Seem. (1868) Solomon Is. Fiji.

o o o o o o

D.8.2
Key to Species of subsect. Dammaropsis

A single species:
.... F. dammaropsis Diels (1933) Dammaropsis kingii Warb. (1891)

v. dammaropsis
Irian Jaya, New Guinea.
o o o o o o

v. obtusa Corner (1960)
New Guinea, Papua.
o o o o o o

D.8.3
Key to Species of subsection Papuasycae

1. Dioecious. Fig 20-40 mm wide, stalk 15-100 mm long. Lamina obovate; lateral nerves 6-9 pairs. Cauliflorous... F. itoana Diels (1935) New Guinea, New Britain.
o o o o o o

1. Monoecious. Fig often axillary, 12-20 mm wide; stalk to 10 mm long; basal bracts caducous. Lamina elliptic, small; lateral nerves 4-6 pairs, impressed above in the dried leaf...... F. microdictya Diels (1935) New Gunea, Papua.
o o o o o o

D.8.4
Key to Series and subseries of subsect. Sycocarpus

1. Upperside of lamina closely punctate from cystoliths.
... D.8.4.1 Ser. Longetuberculatae

1. Upperside of lamina not so punctate ser. Tuberculifasciculatae...... 2

2. Figs with large lateral bracts obscuring the body. Brown Hairy. Long-petiolate. No internal bristles. D.8.4.2 subser. Praestantes

2. Figs without lateral bracts or these not obscuring the body.... 3

3 Figs axillary. Basal bracts 4-16 mm long.
... D.8.4.6 Subser. Axillares

3. Figs cauliflorous or basal bracts shorter........................ 4

4. Intercostals few, lax. Glabrous or white hairy, or thinly appressedly brown hairy. Figs without lateral bracts.... D.8.4.7 subser. Tuberculifasciculatae

4. Intercostals numerous, regular.. 5

5. Internal bristles abundant.
... D.8.4.2 subser. Calopilinae

5. No internal bristles, or few. White to dark brown hairy............ 6

6. Figs generally with lateral bracts. Leaves generally scabrid and opposite. D.8.4.5 subser. Hispidae

6. Figs without lateral bracts. Leaves not or slightly scabrid, variously arranged.
... D.8.4.4 subser. Congestae

D.8.4.1
Key to species of ser. Longetuberculatae

A single species:
.... F. botryocarpa Miq. (1867)

v. scabrida Corner (1960)
New Guinea, New Britain.
o o o o o o

D.8.4.2-7
Following are keys to the species in the subseries of series Tuberculifasciculatae

D.8.4.2
Key to species of subseries Praestantes

1. Lateral bracts obtuse. Cauliflorous fruiting twigs very stout, with out internodes. Basal nerves of lamina short; lateral nerves 10-12 pairs..... F. praestans Corner (1960) New Britain
o o o o o o

1. Lateral bracts acute, lanceolate. Cauliforous fruiting twigs 4-5 mm wide, with internodes, branching. Basal nerves 1/4 to 1/2 lamina; lateral nerves 6-10 pairs.
.... F. longibracteata Corner (1960)

v. longibracteata
Solomon Is.: Bougainville, Guadalcanal, Florida.
o o o o o o

v. lebetoides
Guadalcanal.
o o o o o o

D.8.4.3
Key to species of subseries Calopilinae

1. Basal bracts 5-10 mm long. Apical bracts projecting or gibbous.... 2

1. Basal bracts shorter: apical bracts often appressed................ 4

2. Figs 15-17 mm wide on stolon-like twigs 1-3 mm wide from the branches and trunk. Twigs 3-5 mm thick. Appressedly brown hairy.
.... F. profusa Corner (1960) Bougainville.
o o o o o o

2. Figs larger on shorter stout twigs with short internodes. Twigs 5-10 mm thick...................... 3

3. Stipules persistent. Leaf-base cuneate; basal nerves 1/4 to 1/2 lamina: lateral nerves 6-1- pairs. Fig 25-30 mm wide.
.... F. pachyrachis Laut. et K. Schum. (1901) F. hypoglauca Laut. et K. Schum. (1901) F. pachythyrsa Diels (1935).

v. pachyrachis
New Guinea.
o o o o o o

v. porrecta Corner (1960)
Irian Jaya.
o o o o o o

3. Stipules caducous. Leaf-base more or less cordate: basal nerves short: lateral nerves 9-13 pairs. Fig 17-22 mm wide.
.... F. neobritannica Corner (1961) New Britain.
o o o o o o

4. Fruiting on geocarpic stolons; figs usually ripening red............ 5

4. Not fruiting with geocarpic stolons
.... 6

5. Basal bracts 2-4 mm long; fig-body 10-25 mm wide, pyriform, ellipsoid, or subglobose. Lamina often asymmetric and the leaves inclining to distichous; lateral nerves 6-11 pairs, crowded towards the short basal nerves.
.... F. vrieseana Miq. (1867)

v. vrieseana
Java to New Guinea.
o o o o o o

f. obliqua Corner (1960)
Malaya, Sumatra, New Guinea.
o o o o o o

5. Basal bracts 4-6 mm long: fig-body 20-35 mm wide, depressed globose with a rosette of apical bracts. Lamina more or less symmetric; lateral nerves 5-7 pairs, not approximated to the elongate basal nerves 1/4 to 1/3 lamina. Twigs and leaves persistently ferruginous villous... F. iodotricha Diels (1935) New Guinea.
o o o o o o

6. Leaves distichous, strongly asymmetric at the base, short-petiolate.
.... 7

6. Leaves spirally arranged or opposite; if distichous, then not strongly asymmetric at the base...... 9

7. Figs 20-30 mm wide, on short stout twigs. Lamina oblong, elliptic.
.... F. papuana Corner (1960) New Guinea, Papua.
o o o o o o

7. Fig 10-17 mm wide. Lamina obovate.
.... 8

8. Figs in large bunches at the base of the trunk; peduncles 30-80 mm long...... F. bernaysii King (1887) New Guinea, New Britain, Solomon Is.(?).
o o o o o o

8. Figs in small clusters on the trunk and main branches; peduncles 7-15 mm long. Lamina often narrow.
.... F. adelpha Laut. et K. Schum. (1901) F. platysycia Diels (1935) New Guinea.
o o o o o o

9. Figs more than 20 mm wide...... 10

9. Figs less than 20 mm wide...... 11

10. Hairs stiff, spreading, bright brown. Figs 18-25 mm wide, subglobose to pyriform. Leaf-base shallowly cordate to rounded. Stipules caducous.. F. calopilina Diels (1935) F. setistyla Warb. apud K. Schum. (1905, Nachtr. Fl. Schutzgeb non Fedde Rep. 1: 77).
New Guinea, New Britain.
o o o o o o

10. Hairs appressed, white to brown. Figs 18-25 mm wide, depressed globose. Leaf-base more or less cuneate. Stipules persistent, conspicuous.. F. sublimata Corner (1960) Irian Jaya, Papua (southern)
o o o o o o

11. Basal bracts early caducous. Figs on hanging twigs to 1 m long. Style nearly or quite glabrous. Hairs 2-4 mm long on twigs and petioles.. F. baccaureoides Corner (1960) Bougainville.

o o o o o o

11 Basal bracts persistent. Style hairy (female)................ 12

12. Hairs 2-4 mm long. Petiole to 15 mm long. Stipules persistent. Figs on short twigs.
.... F. hahliana Diels (1935) New Guinea.

o o o o o o

12. Hairs to 1.5 mm long. Petiole longer. Stipules caducous. Figs on simple or sparingly branched, stout, spiciform twigs without internodes.
.... F. d'albertisii King (1888) Papua.

o o o o o o

D 8.4.4
Key to species of subseries Congestae

1 Fig-orifice closed by a rosette of many apical bracts, often sunken: figs 15-40 mm wide, in large clusters on the trunk and branches.
.... F. congesta Roxb. (1832)

v. congesta
Celebes to Solomon Is. and Queensland.

o o o o o o

v. chlamersii (King) Corner (1960) F. chlamersii King (1887) F. caulothyrsa Laut. et K. Schum. (1901) Amboina to Solomon Is.

o o o o o o

1 Fig-orifice closed by 5 apical bracts, not sunken or in a rosette: body 12-20 mm wide. Figs axillary, cauliflorous, or on stolons.
.... F. subcongesta Corner (1961)

v. subcongesta
New Britain.

o o o o o o

v. symmetrica Corner (1961) Bougainville Is.

o o o o o o

D.8.4.5
Key to species of subseries Hispidae

1. Figs with a few conspicuous lateral bracts, at least round the orifice.
.... F. hispida Linn. f. (1781)

v. hispida
Ceylon and India to Queensland and Papua.

o o o o o o

1. Figs without lateral bracts.
.... F. hispidioides S. Moore (1923)

v. hispidioides
Papua, ? to N Australia

o o o o o o

v. flavescens Corner (1961)
Papua, New Britain.

o o o o o o

v. succosa Corner (1961)
New Guinea, Papua, New Britain.

o o o o o o

D.8.4.6
Key to Species of subseries Axillares

A single species:
.... F. saurauioides Diels (1935) New Guinea.

o o o o o o

D.8.4.7
Key to species of subseries Tuberculifasciculatae

1. Figs not ridged. Female style glarous to hairy.
.... F. arfakensis King (1888) F. hylophila Laut. et K. Schum. (1901) F. stenothyrsus Laut. et K. Schum (1901) F. palustris Laut. et K. Schum. (1901) F. tristipula Warb. (1905) Aru Is through New Guinea to Solomon Is.

o o o o o o

1. Figs ridged towards the apex. Female style hairy............. 2

2. Leaves scabrid.
.... F. bougainvillei Rechinger (1912) Bougainville.

o o o o o o

2. Leaves smooth.................. 3

3. Lamina not acuminate. Galbrous or villous with pale spreading hairs. Figs depressed globose, axillary or cauliflorous.
.... F. septica Burm. f. (1768)

v. septica
Sumatra to Solomon Is, Vanuatu and Queensland.

o o o o o o

v. cauliflora Corner (1960)
Timor to Solomon Is, Vanuatu and Queensland.

o o o o o o

3 Lamina acuminate. Figs generally cauliflorous. Thinly appressedly, white to brown hairy........... 4

4. Lamina 3-6 cm wide. Peduncles 2-6 mm long: basal bracts persistent: fig-body strongly ridged, not pedicelate.
.... F. ternatana Miq. (1867) Covellia ternatana Miq (1859) F. flagellaria Diels (1935). Ternate to New Guinea.

o o o o o o

6. Lamina 5-10 cm wide. Peduncle 5-20 mm long: basal bracts caducous: body slightly ridged, distinctly pedicelate.
.... F. macrothyrsa Corner (1961) Bougainville.

o o o o o o

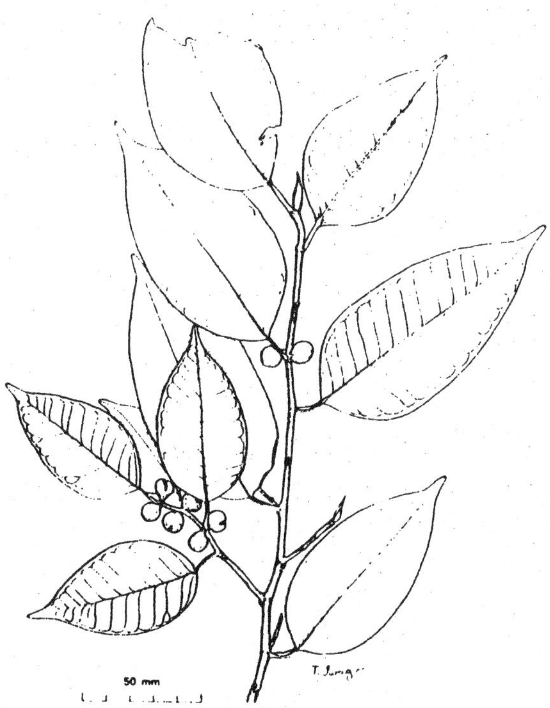

Ficus benjamina A. habit.
(Drawn: T Iwagu)

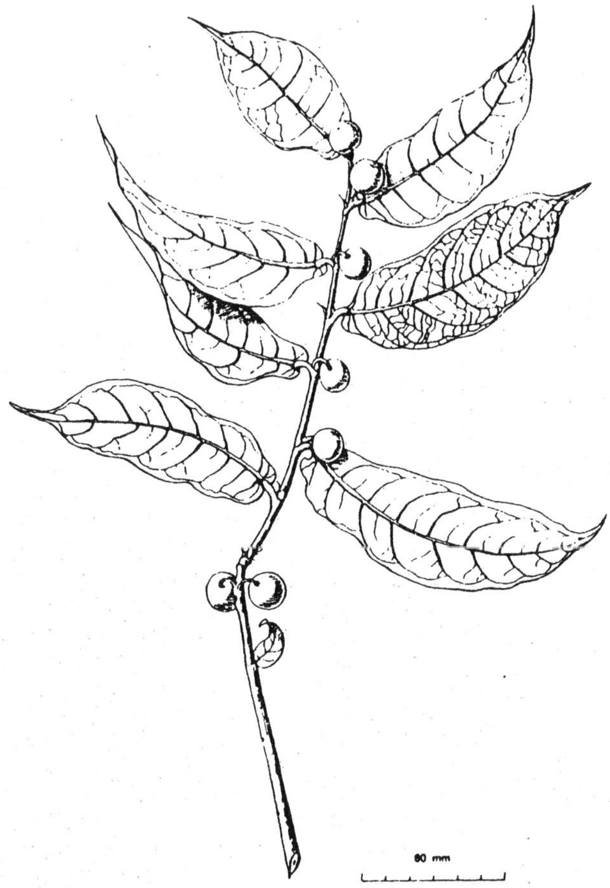

Ficus subulata A. habit with fruit.
(Source: National Herbarium, Lae)

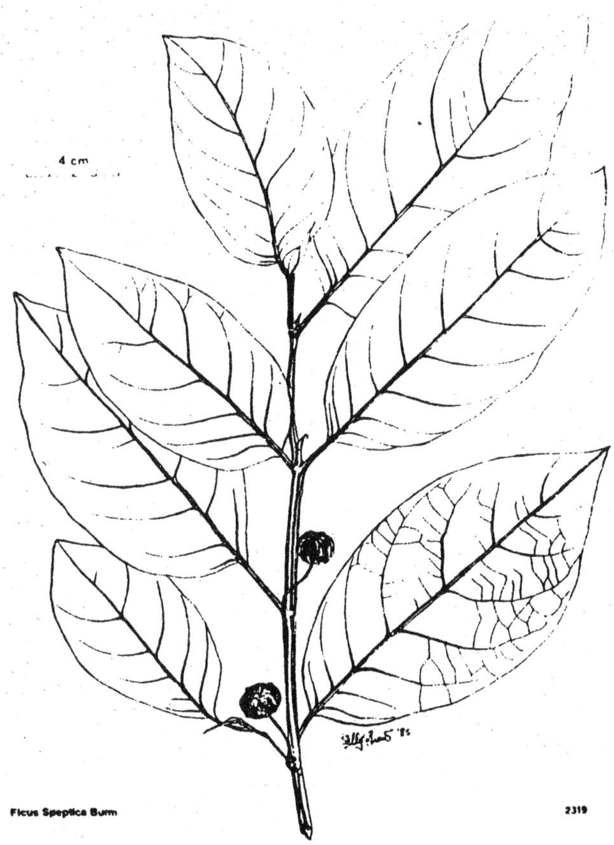

Ficus septica A. habit.
(Source: National Herbarium, Lae)

8. MACLURA

o o o o o o

A genus of 12 species in America, Africa and Asia. Timber traditionally used for bows and war-clubs in USA. Also used for sleepers, fence posts. Root used as a source of yellow dye and leaves used to feed silkworms. All Papuasian species are in Section Cudrania. Both species in Papuasia are climbers.

Dioecious, spiny, tree, shrub or climber. Leaves spirally arranged to distich-ous, simple, lacking glands. Male flowers in racemes and female flowers in heads. Male flowers 4-merous; tepals free; anthers small. Female flowers sessile; tepals 4, decussate, fleshy in fruit. Fruit a yellow syncarp of achenes enclosed in a fleshy perianth on a fleshy receptacle. Seed compressed.

Cudrania sp. A. leaves; B. branch; C. stipular spines; D. flowerbuds; E. inflorescence.
(Drawn: R. Bialostocki)

Key to Species in Papuasia
(Source: Corner (1962)

1 Lateral nerves in arching strongly, 6-9(14) pairs; intercostals none or indistinct; petiole 13-15 mm. Tepals closely villosulous, all with 2 immersed glands. Filaments very short.
.... M. cochinchinensis (Lour.) Corner (1962: 239). Vanieria cochinchonensis Lour. Fl. Cochinch. (1790 : 564.

o o o o o o

1 Lateral nerves ascending, slightly inarching; intercostals 2-5, distinct. Filaments 1-1.5 times as long as the anthers. Petiole often longer.... M. amboinensis Blume Mus. Bot. Lugd. Bat. 2 (1856) 84.

o o o o o o o

Lateral nerves 6-9 pairs.
.... var. amboinensis

o o o o o o o

Lateral nerves 4-6 pairs.
.... var. paucinervia

o o o o o o o

Notes:

1 M. amboinensis var. amboinensis is distributed from Thailand to New Guinea. In contrast var. paucinervia occurs only in the Philippines and New Guinea.

2 M conchchinensis is distributed from Sri Lanka and India to New Britain, Australia and New Caledonia.

Malaisia scandens A. branch with flowers; B. inflorescence; C. fruit.
(Drawn: R. Bialostocki)

9. MALAISIA

o o o o o o

A single species from S E and E Asia to E Australia and Polynesia.

Shrubs, sometimes scandent. Leaves alternate, entire or obscurely dentate, prominently pinnately nerved and reticulate below; stipules lateral, small, deciduous. Flowers dioecious, male in cylindric catkin-like spikes, female globose-capitate; receptacle small; male spikes shortly pedunculate; female heads small, pedunculate, hoary tomentose with closely contiguous fleshy bracteoles; female flowers immersed along the bracteoles, 1-2 often only fertile, remainder sterile. Male flowers calyx 3-(rarely 4-)partite, segments valvate; stamens 3 (rarely 4), filaments inflexed in bud; rudimentary ovary small. Female flowers: calyx urceolate, mouth contracted, scarcely denticulate; ovary straight, included; style central, deeply 2-lobed, branches filiform; ovary pendulous from below the apex. Fruiting head slightly enlarged; fruit enclosed by the thinly fleshy calyx; seeds without endosperm or rarely with thin endosperm intruded between the cotyledons; embryo globose, one cotyledon large, split on one side, the other small and folded around the radicle.

Key to Species in Papuasia

A single species:
...... Malaisia scandens (Lour.) Bl.

Laut. & Schumann (1901: 171)
Diels (1936: 171)

o o o o o o

10. MORUS

o o o o o o

A genus with 7 subtropical, deciduous, monoecious, trees. Common name: mulberry. Male flowers in catkins, females in pseudospikes. Wind pollinated. M. nigra the black mulberry, has been cultivated for millenia, used for juicy fruit. The timber is also used for furniture, inlays etc. It is easily propagated using cuttings to 2.5 m long. Introduced into New Guinea. Widely cultivated in the Wau Bulolo area and suitable for Highlands.

Trees or shrubs; leaves alternate, dentate, undivided or 3-lobed, base 3-nerved; stipules lateral, small, caducous. Flowers monoecious or dioecious, spicate; spikes axillary, solitary, shortly pedunculate, male long, cylindric, catkin-like, female long or short.

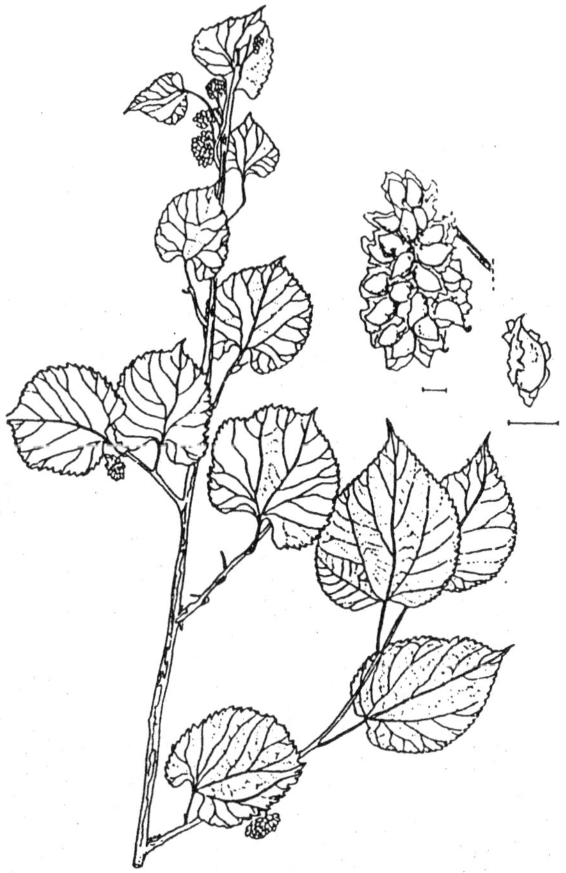

Morus sp. A. Fruiting branch; B. fruit; C. enlarged fruit.
(Drawn: R. Bialostocki)

Male flowers: calyx 4-partite, segments imbricate; stamens 4, filaments inflexed in bud, soon exserted; rudimentary ovary turbinate. Female flowers: calyx segments 4, imbricate, succulent in fruit; ovary enclosed, ovoid or subglobose; style central, 2-partite, branches linear, equal; ovule pendulous. Fruits enclosed by the enlarged succulent calyces; seed subglobose, with fleshy endosperm; embryo incurved, cotyledons oblong, equal.

Key to Species in Papuasia

1 introduced species:
..... Morus alba L.

o o o o o o

Notes:

1. Commonly cultivated as a source of fruits at mid altitudes in New Guinea.

11. PARARTOCARPUS

o o o o o o

A genus with 4 species, from Indo-Malaya to New Guinea and the Solomon Islands (Jarrett 1960). Some with edible seeds.

Tall trees. Leaves alternate, petiolate, pinnately nerved; stipules covering the terminal bud, deciduous; flowers dioecious of both sexes globose-capitate heads, axillary, with crowded peltate bracts mixed with flowers. Male flowers: calyx absent; stamens singly crowded among the bracts; filaments erect, short thick; anthers ovate,

Key to Species in Papuasia

..... Parartocarpus involucrata (K. Schum.) Warburg ex Laut. & Schumann (1901: 267)

o o o o o o

..... Parartocarpus papuana Moore J. of Dot. 61 (1923) Suppl. 52.

o o o o o o

Parartocarpus venosus A. fruiting branch; B. immature fruit. (Drawn: R. Bialostocki)

12. PRAINEA

o o o o o o

A genus with 4 species in Malesia revised by Jarrett in 1959. From Malaya to New Guinea. Jarrett discusses the reasons for treating Prainea as a distinct genus rather than as a section of Artocarpus as proposed by Renner in 1907. A single species is described from Papuasia.

Trees to 35 m tall (or climbers, not in N.G.). Leaves alternate and distichous, simple, entire, penninerved, glabrous. Gland hairs sunken in epidermis. Stipules paired, small to 5 mm long, lateral or intrapetiolarm scars non-amplexicaul.

Inflorescences unisexual, capitate, heads globose, or in male short-obovoid, pedunculate, solitary or paired in leaf axils. Plants dioecious, perianths tubular, with many clusters of yellow resin canals in the mesophyll. the single stamen with straight filament and small anther. Stigmatic arms short. Endocarp membranous.

Key to Species in Papuasia

A single species:

.... Prainea papuana Beccari For. Borneo (1902 635).
Artocarpus papuanus Renner (1907: 367)
Prainea microcephala J. J. Smith
Parartocarpus papuanus S. Moore J. Bot. 61 (1923) Diels (1936: 175)

o o o o o o

Prainea papuana A. habit; B. fruit. (Drawn: R. Bialostocki)

Notes:

1. Locally common in lowland evergreen forest up to an altitude of 1050 m.

2. Jarrett (1959) points out that there is a major geographical variation in the diameter of the head of the female at anthesis. In eastern New Guinea the head has a diameter of 10-15 mm. and the surface remains compact at maturity. In the Vogelkop and Moluccas the diameter is only abou 5 mm. at anthesis and following elongation of the flowers the surface becomes loose. A parallel variation occurs in the size of the male head. However collections from northern Irian Jaya are intermediate (8-10 mm.). In consequence no subspecies were recognised by Jarrett.

PSEUDOMORUS

A single species recorded by Diels (1936: 171). This genus is no longer recognised.

.. Pseudomorus brunoniana (Endl.) Bur. var. australiana Bur.
Laut. & Schumann (1901: 266).
Diels (1936: 171)

13. STREBLUS

o o o o o o

A genus of 22 species in Madagascar, Asia and Australasia. S. asper is used locally in Thailand for paper making. S. (Bleekrodea) tonkinensis is a possible source of rubber.

Trees, unarmed or spinous, monoecious or dioecious. Leaves distichous, denticulate or entire, mostly with short-petiole, without glands. Inflorescences axillary, bisexual or unisexual, pedunculate. Male flowers with pistillode, (3-)4(-5)-merous, ; tepals more or less valvate, free or shortly joined; filaments inflexed in bud; anthers small, reniform, extrose. Female flowers 4-merous; tepal decussate, imbricate, free or shortly joined; ovary superior. Drupe large, 5-14 mm wide. Seeds large, round.

Key to Species in Papuasia
(Source: Corner (1962)

1 Bracts reniform or peltate. Male inflorescences often long, female often many-flowered, never 1-flowered. Lamina more or less membranous, usually without hypodermis........................ 2

1. Bracts ovate - lanceolate, sessile. Male and female spikes 10-25mm long, often few flowered, or the female flowers solitary. Lamina narrowly elliptic to lanceolate, mostly thinly coriaceous and with a hypodermis (1-cell thick); petiole 2-12 mm; intercostals none...... 3

2. Cotyledons flat, rather small; radicle incumbent. Lamina often scabrid, basal nerves often as long as the laterals.
.... S. pendulinus (Endl.) F. Muell. Fragm. Phyt. Austral. 6 (1868) 192.
Morus pendulina Endl. Prodr. Fl. Norf. (1833) 40.

o o o o o o

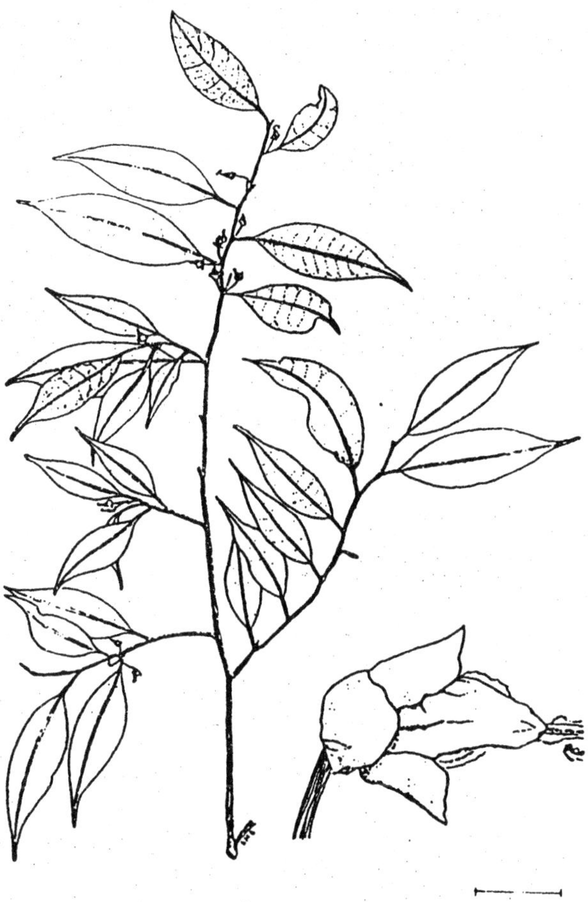

Streblus glaber A. Branch with fruit; B. immature fruit, enlarged.
(Drawn: R. Bialostocki)

2 Cotyledons foliaceous, conduplicate, the radicle incumbent in the sinus of the cotyledons. Lamina smooth, basal nerves shorter than the lateral

S. solomonensis
Corner (1962: 224).

o o o o o o

3 Lamina dentate, shortly acuminate, often stiffly coriaceous; lateral nerves 6-13 pairs. Male flowers 3-7 per spike, 2.5-3.5 mm wide in bud, subglomerate.

.... S. urophyllous
Diels (1936: 172)
var. urophyllous
var. salicifolius
Corner (1962: 225)

o o o o o o

3. Lamina entire or crenulate, rather long attenuate or acuminate or caudate; lateral nerves 9-18 pairs. Male flowers often numerous, smaller, not glomerate.

. . . S. glaber
(Merr.) Corner (1962:221)
Gironniera glabra Merr.
Phil. J. Sci. 1 (1906) suppl. 42.

o o o o o o

Notes:

1. Two varieties of S. glaber are recognised by Corner. One variety is widespread from Malay and the Philippines to New Guinea, in montane rainforest from 700-2500 m altitude. The second variety, var. australianus is restricted to Queensland.

2 S. pendulinus is a widepsread and variable plant. Young trees have large leaves with 10-16 pairs of lateral nerves; large trees have smaller leaves with 7-12 pairs of lateral nerves. The species occurs in Papuasia, New Caledonia and Queensland.

3. S. solomonensis is a common rain forest tree; the young leaves are eaten, the timber used in house construction. The species is characterised by an extremely short style and stigmatic arms. It is distributed on Bougainville and Guadalcanal.

4. Corner (1962) emphasises that S. urophyllous is closely related to S. glaber and perhaps only a high mountain state of it. Corner loc. cit. recognises 2 varieites, the type which is widely distributed in the montane zone of New Guinea between 2400 and 3000 m altitude, and var. salicifolius which was described from a collection by Brass (30220) at 3000 m on Mt Wilhelm.

14. TROPHIS

o o o o o o

A genus of 11 species in T America, Madagascar and Malesia. In Papuasia three species, 2 local endemics in the Sepik and Irian Jaya and 1 widespread species. All three species in New Guinea are in section Calpidochlamys.

Trees or shrubs, unarmed, dioecious. Leaves distichous. Stipules paired, small, caudaceous. Inflorescences axillary, solitary or paired, bracts very small. Flowers 4-merous. Male tepals valvate, joined below; filaments inflexed in bud. Female tepals small, free; ovary inferior, the ovule immersed in the rece[tac;e; styles short with 2 long stigmatic arms. Drupe inferior, tepals persistent, indehiscent. Seeds round to oblong, cotyledons fleshy.

Key to Species in Papuasia
(Source: (Corner 1962)

1. Female inflorescence 15-40 mm long, many-flowered. Endocarp not angled. Lamina acuminate, base cuneate; lateral nerves 9-16(-20) pairs.

.... T. philippinensis
(Bur.) Corner (1962: 231)
Urommorus philippinensis Bur.
DC Prodr. 17 (1873) 237.

o o o o o o

1. Female inflorescence 30-70 mm long, with few fetile flowers. Endocarp 3-angled, pyramidal............ 2

2. Lamina acuminate, base rounded; lateral naerves 15 pairs.

.... T. drupacea
(Diels) Corner (1962:231).
Calpidochlamys drupacea
Diels (1935: 173).

o o o o o o

2. Lamina attenuate or subacuminate, base cuneate; lateral nerves 8-10 pairs.... T. branderhorstii
(Diels) Corner (1962:231).
Calpidochlamys branderhorstii
Diels (1935: 173).

o o o o o o

Notes:

1. Corner (1962) considers that the two species endemic in Papuasia could be synonymous. These are braderhorstii described from Irian Jaya and drupacea described from the Sepik.

2. Trophis philippinensis is widespread from the Philippines and Celebes to Papuasia.

ULMACEAE

o o o o o o

A family of 15 genera and c. 200 species occuring in the tropics, subtropics and temperate areas. Divided into two tribes: the Ulmeae which is N Temperate in distribution with Ulmus the most common genus; and the Celtideae which is tropical and subtropical in distribution and characterised by Celtis and Trema. In Papuasia 5 genera and 16 species. Found from sea-level to 2000 m altitude. Parasponia has recently been intensively studied following the discovery that its roots contain N fixing bacteria. Timber is harvested from Ulmus and Celtis.

ULMACEAE

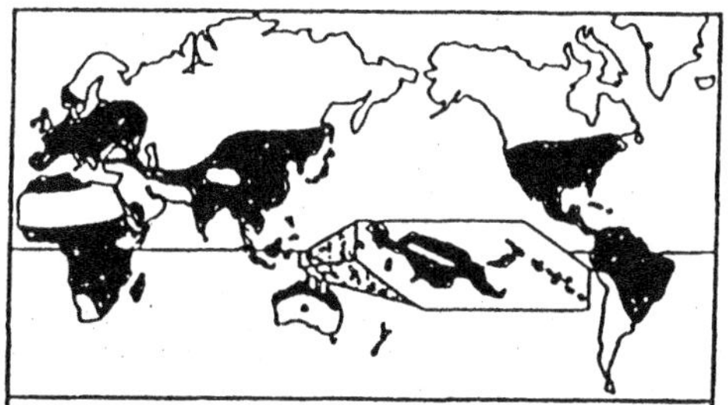

	WORLD	PAPUASIA
Number of genera:	16	5
Number of species:	2,000	15

Distribution: Two groups. Tropical, subtropical and N Temperate.

Economic uses: Hardwood timbers and ornamentals

Diagnostic features

Leaves: alternate and simple, with stipules that are shed as the leaves unfold.

Flowers: bisexual or unisexual, inconspicuous. Stamens arise at base of the calyx; usually with one stamen opposite each clayx-lobe. Petals absent. Ovary superior, styles two, with stigmatic surface on inner face.

Fruit: a nut, drupe or samara, containinga a single seed with little or no endosperm.

References

Elias, (1970). J. Arn. Arb. 51: 37. f.2.

Hutchinson, J. (1967). Genera of the Flowering Plants 2: 149

Merrill, E.D. & Perry, L.M. (1939). Journ. Arn. Arb. 20: 325 - 328.

Soepadmo, E. (1977). Ulmaceae. Flora Malesiana Ser. I. 8: 31 - 76.

KEY TO GENERA IN PAPUASIA

(Source: Soepadmo (1977)

1 Leaves triplinerved at base, or if pinnately nerved the stipules do not leave a circular scar around the node; lateral nerves less than 5 pairs 2

1. Leaves pinnately nerved; lateral nerves more than 5 pairs; stipules free but overlapping each other, im falling leaving a circular scar around the node. .... 3. Gironniera

2. Stipules intrapetiolar, connate. .... 4. Parasponia

2. Stipules extrapetiolar, free.... 3

3. Female flowers borne in condensed, multiflowered raceme. Perianth of male flower induplicate-valvate. Fruit compressed, elliptic lens-shaped in cross-section. .... 5. Trema

3. Female flowers solitary in the axils of leaves or borne in a cymoid cluster of 2-3. Perianth lobes of male flowers imbricate. Fruit faintly 3-5 angular in cross-section........................ 4

4. Male flowers borne in a 2-3 flowered cymoid inflorescence, or in a much branched paniculate, subterminal inflorescence. Female flowers borne in a racemose cluster of 2-10, inflorescence much branched, staminodes mostly present. Cotyledons broad, variously folded or curved. 2. Celtis

4. Male flowers organised in a 2- to 3-flowered raceme. Female flowers always solitary or raely borne in a 2- to 3-flowered racemose, bisexual inflorescence; staminodes always absent. Cotyledons narrow, incurved....... 1. Aphananthe

1. APHANANTHE

o o o o o o

A genus of world wide distribution in tropical and temperate Asia, Mexico, Madagascar and Australia, represented by about 4-5 species. In Papuasia with one species in seasonally dry areas up to 750 m altitude. Often in mixed-Eucalyptus forest, or occasionally forming dense thickets.

Shrub to medium sized tree to 28 m tall. Tree often fluted, occasionally with suckers. Stipules subulate. Leaves glabrous; margin dentate, serrate or rarely subequal; apex acute; midrib andnerves strongly raised beneath. Male inflorescence 1-3 cm diameter, 10 - 50 flowered; male flowers to 1-1.5 mm across. Female flowers always solitary. Fruit ovoid-globose, 6-8 by 4-6 mm. Pedicel to 5-1o mm long.

Key to Species in Papuasia

1 species:
Aphananthe philippinensis Planch. Ann. Sc. Nat. III 10 (1848) 337.

o o o o o o

Notes;

1. Mainly occuring in areas with a seasonal climate, on various soil types. Grows in savannah mixed Eucalyptus forest), hill and gallery forest, from sea-level to 750 m altitude. Often abundant in dense stands on hillsides.

2. This species grows from Luzon throughout Papuasia to Queensland and New South Wales.

2. CELTIS

o o o o o o

A genus of 50 to 60 species, distributed in tropical and temperate areas. Plants are apparently pollinated both by wind and insects. In Papuasia there are 6 species (none are endemic). Several species form a distinct part of the understorey of lowland forests. Other species are confined to areas with a seasonal climate., or on well drained soils such as limestone.

Small to large monoecious or polygamo-monoecious trees. Bark often conspicuously warty. Stipules thick and tough. Leaves entire or not, 3-nerved at base. Inflorescences branched, few-to many-flowered. Staminate inflorescences on lower and leafless parts or in axils of new shoots. Other flowers at distal ends of axis. Perianth lobes 4-5, imbricate in bud. Drupe fleshy; exocarp thick and firm, mesocarp thin and fleshy, endocarp hard and persistent. Seed: endosperm scanty, cotyledons broad and foliaceous. Germination epigeal.

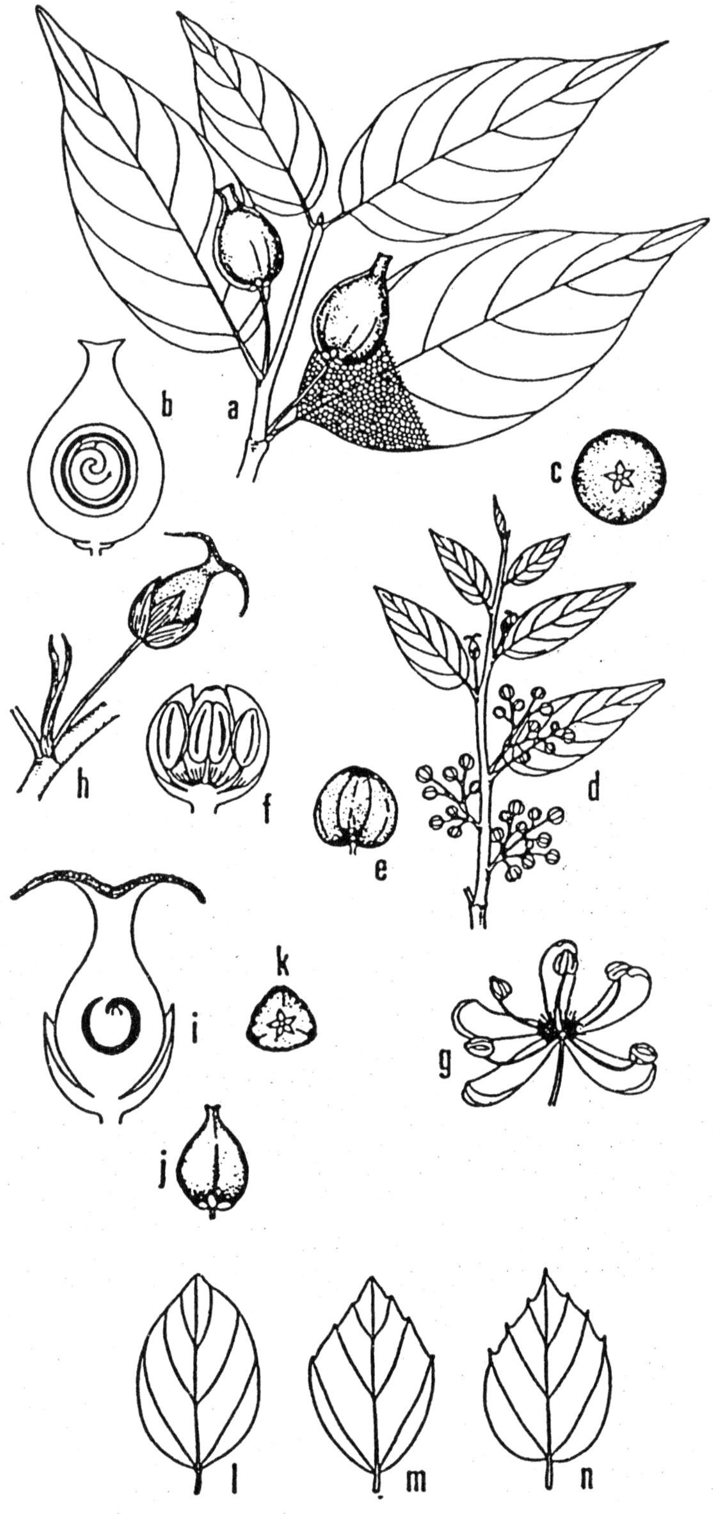

Aphananthe cuspidata A. habit; B.C. fruit; D. young shoot, flowering; E.F.G. male flowers; H. I. female flowers.
A. philippinensis J.K. fruit; L.M.N. variation in leaf form.
(Source: Soepadmo (1977)

Key to Species in Papuasia
(Source: Soepadmo (1977)

1. Leaves serrulate to denticulate at least in the upper half. Nerves 1-3 (-4) pairs. Male inflorescences racemose, 5-20 flowered, male flowers 5-merous. Pistillode reduced in size or absent. Stigmatic arms entire. Fruit strongly beaked. .... C. rubrovenia
Elmer. Leafl. Philip. Bot. 2 (1908) 464.
C. similis Merr. & Perry J. Arn. Arb. 22 (1941) 253.

o o o o o o

1. Leaves entire or nearly so. Male infloresecnece a much-branched, many-flowered panicle with up to 150 flowers. Pistillode rather well developed, c. 1-1.5 by 0.5-1 mm. Stigmatic arms bilobed or bifid at the tip........................ 2

2. Leaves rugose, brittle, sparsely pubescent beneath; midrib and lateral nerves strongly raised beneath. Stipules not peltately attached, free from one another. Fruit densely appressed-hairy. .... C. rigescens
(Miq.) Planch. in DC. Prod.
C. nymanii K. Sch. in K. Sch. & Laut. (1905) 240. Bot. Jahrb. 50 (1913) 311.

o o o o o o

2. Leaves not rugose, not brittle, glabrous; midrib and lateral nerves only slightly raised beneath. Stipules peltately attached, overlapping. Fruit glabrous......... 3

3. Leaves with (2-)3-5 pairs of nerves 4

3. Leaves with 1 pair of nerves.... 5

4. Nerves 3-5 pairs; lowest pair running to 1/3 to half the length of leaf. Inflorescence mostly male or female. Stamens of female flowers rudimentary, non-functional. Fruit ovoid, feebly 4-angular in CS; endocarp with reticulate ridges.... C. paniculata
(Endl.) Planch. Ann. Soc. Nat. III 10 (1848) 305.

o o o o o o

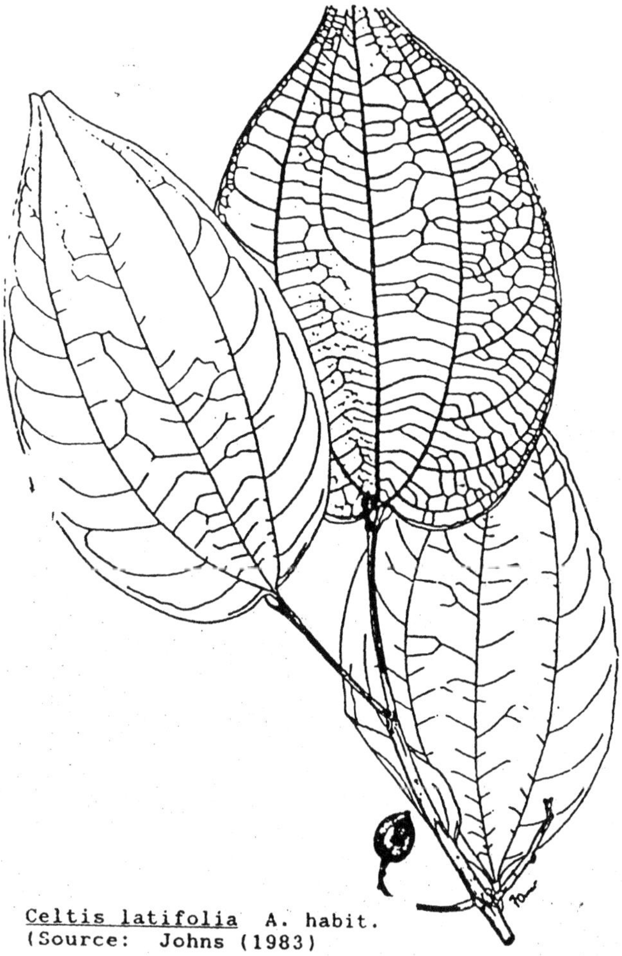

Celtis latifolia A. habit.
(Source: Johns (1983)

Celtis paniculata A. habit.
(Source: Johns (1983)

4. Nerves 1-2(-3) pairs; lowest pair running up to 2/3 the length of leaf. Inflorescence male or mixed with female. Stamens of bisexual flowers well developed and functional. Fruit ellipsoid, ± terete; endocarp smooth.
.... C. philippinensis
Blanco Fl.Philip. (1837) 197.
var. wightii (Planch.) Soepadmo

o o o o o o

5. Leaf symmetrical. Male or bisexual inflorescence. Male inflorescence 15-40 flowered. Cotyledons curved but not folded.................. 6

5. Leaf oblique. Inflorescence male and bisexual; 60-150 flowered. Cotyledons curved and many times folded.... C. hilderbrandii
Soepadmo Fl. Mal8 (1977) 63

o o o o o o

6. Leaves (4-)8-14(-18) by (2-)3-6(-8) cm; reticulations fine, dense. Ovary glabrous. Fruit 8-15 by 7-12 mm; endocarp smooth.
.... C. philippinensis

var. philippinensis

o o o o o o

6. Leaves (8-)15-18(-25) cm by (6-)8-12(-18) cm; reticulations coarse, wide spaced. Ovary densely appressed-pubescent. Fruit 15-25 by 10-18 mm; endocarp pitted.
.... C. latifolia
(Bl.) Planch. in DC. Prod. 17 (1873) 186.
C. kajewskii Merr. & Perry J. Arn. Arb. 22 (1941) 254.

o o o o o o

Notes:

1. Some species are common locally in lowland rainforest.

2. C. philippinensis includes two varieties. var. philippinensis has larger leaves 9-12 cm with one pair of nerves usually running the full length of the leaf. Fruit large 10-15 by 8-12 mm. var. wightii has smaller leaves 5-7 cm with 1-3 pairs of nerves. Fruit smaller, 8-12 by 6-10 mm. Recognition of these varieties must be based on a combination of characters. Individual characters cannot alone be used to describe the varieties.

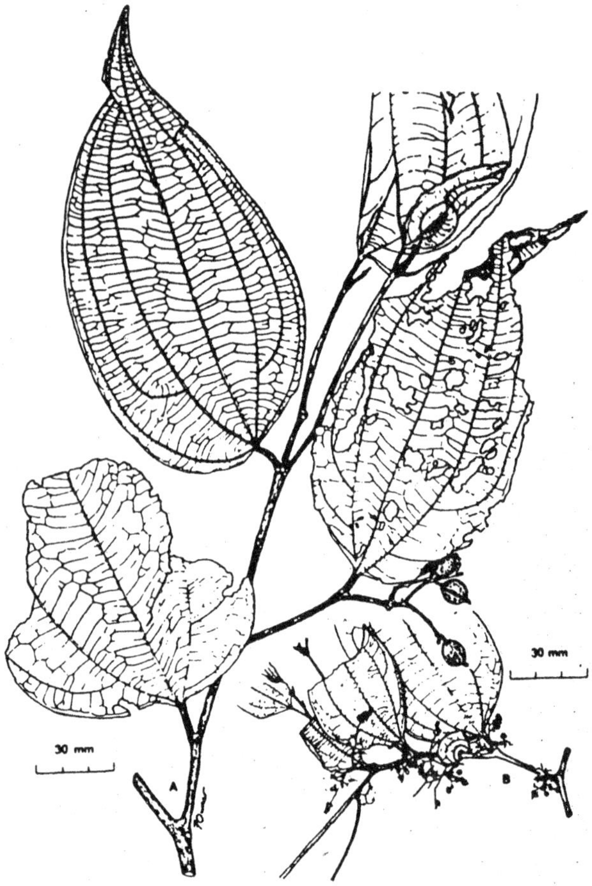

Celtis philipensis A. habit.
(Source: Johns (1983)

Celtis philippensis A. habit; B.C. bisexual flowers; D.E. male flowers; F.G. fruits.
(Source: Soepadmo (1977)

3 GIRONNIERA

o o o o o o

A genus of 6 species from Sri Lanka and S E Asia throughout Melanesia and Polynesia to Fiji and Samoa. Four species in Papuasia. These are understorey trees of primary and secondary forests up to 1300 m. Mainly in rainforest especially along streams.

Shrubs to large trees, rarely buttressed. Stipules extrapetiolar, free but overlappingeach other, completely enclosing the bud, on falling leaving a circular scar around the node. Leaves pinnately nerved, nerves parallel.

Inflorescence axillary or on older leaf less branches. 1 to many flowered. Male flowers globular, perianth lobes 5, pistillode present, rudimentary or well developed. Female flowers compressed, 4 5 lobed, usually very unequal in size, persistent, staminodes absent; stigmatic arms to 1.5 cm long, rather long persistent. Drupe: exocarp thin, strongly adnate the hard and persistent endocarp Seed lacking endosperm; cotyledons fleshy, narrow, equal.

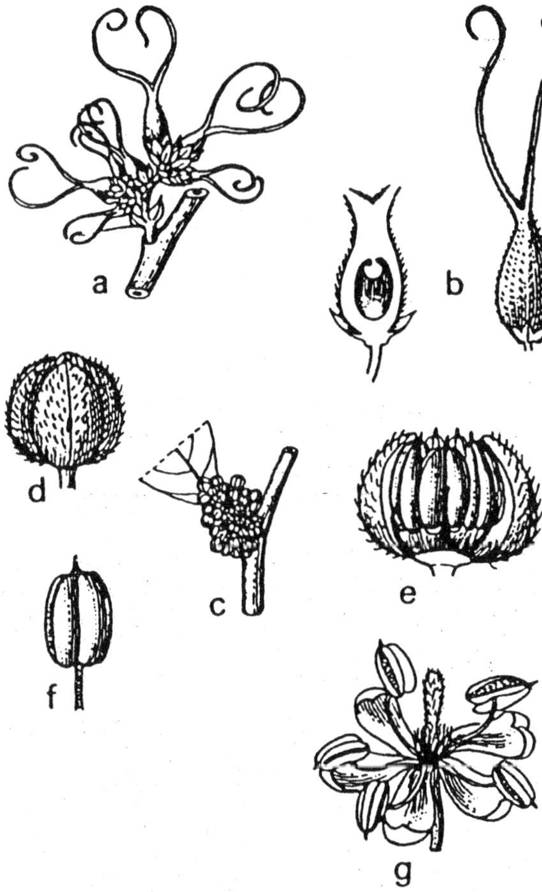

Gironniera celtidifolia A. B. female flowers; C. G. male flowers. (Source: Soepadmo (1977)

Key to Species in Papuasia
(Source: Soepadmo (1977)

1 Male and female inflorescence and infutescens a condensed, capitate thyrse or panicle, axillary or borne on older leafless branchlets Male flowers with a rather well developed pistillode Leaves thick-leathery, distantly serrate or subentire. Terminal bud (stipules) up to 4.5 by 0.5 cm.
..... G. celtidifolia
Gaudich. Voy. Donite
(1844) t. 85

o o o o o o

1 Male and female inflorescence and infutescens a lax, branched or unbranched panicle, thyrse or raceme, always axillary. Male flowers with a strongly rudimentary pistillode. Leaves chartaceous to thick leathery, mostly entire. Terminal bud (stipules) less than 3 cm by less than 0.3 cm.......... 2

2 Mature leaf densely pubescent beneath Male inflorescence a much branched thyrse; male flowers arranged in condensed cymoid clusters of 5 10 along the axes.
G. hirta Ridley J. Str. Br. R. As. Soc. 82 (1920) 194

o o o o o o

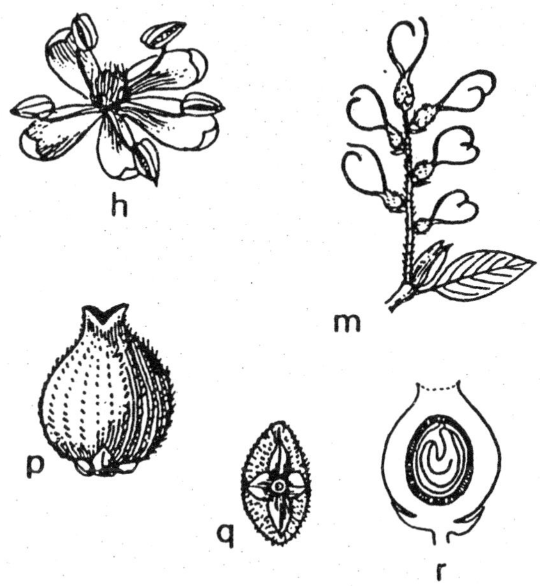

Gironniera subaequalis H. male; M. young infrutescence; P. R. fruits. (Source: Soepadmo (1977)

2. Mature leaf glabrous. Male inflorescence unbranched or a branched raceme or panicle; male flowers solitary along the axes......... 3

3. Mature leaf thick coriaceous, broad ovate, elliptic, elliptic oblong, broadest below or at the middle, index 2 2.5, base unequal sided. Terminal bud 2 3 by 0.2-0.4 cm. Well developed male inflorescence a much branched panicle carrying 40-100 flowers. Well developed female inflorescence carrying (2-) 5 - 10 (15) flowers (fruits). Mature fruits 10-12 by 9 5 mm.
.... G. subaequalis
Planch. Ann. Sc. Nat.
III, 10 (1848) 339

o o o o o o

3. Mature leaf chartaceous to thin coriaceous, elliptic lanceolate or elliptic obovate, broadest at or slightly above the middle, index (2.5)3(3.5), base base more or less equal sided. Terminal bud 2 by 0.5 cm. Well developed male inflorescence a panicle, unbranched, carrying 15 30 flowers. Well developed female inflorescence carrying 1 3(5) flowers (fruits). Mature fruits 7 8 by 5-6 mm.
.... G. rhamnifolia Bl.
Mus. Bot. 2 (1856) 74.
t. 25.

o o o o o o

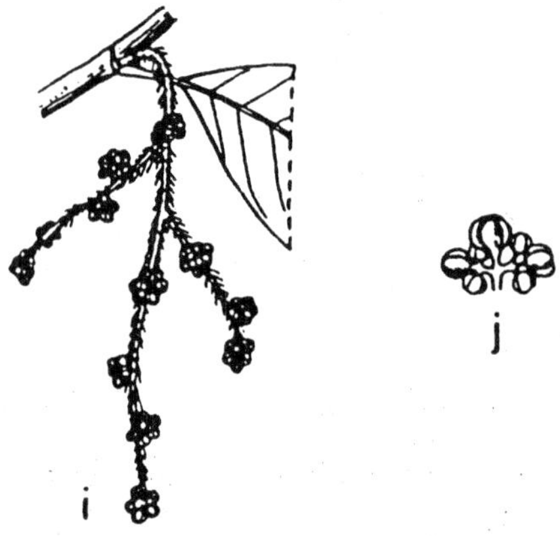

Gironniera nervosa I. J. male flowers; (Source: Soepadmo (1977))

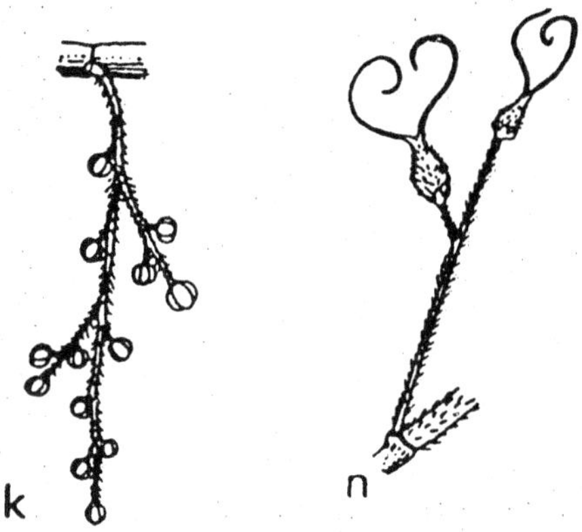

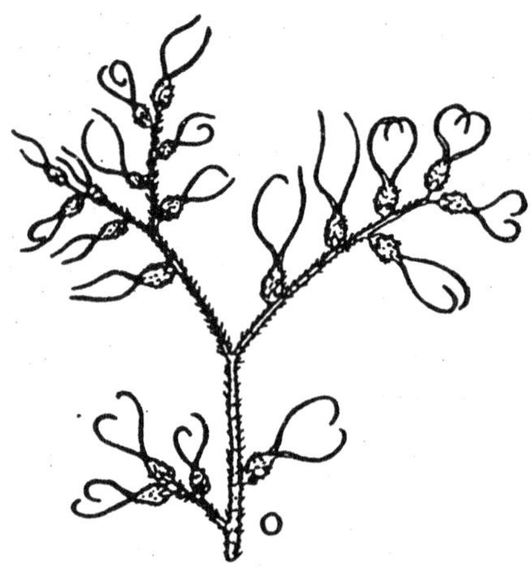

Notes:

1 The mode of germination is unknown for Girronniera.

2. G celtididifolia grows as an understorey shrub or tree in both primary and secondary forest from sea level to 1200 m altitude. G. hirta has a more restricted distribution, mainly being found in lowland forest below 700 m altitude. G. parvifolis is a shrub or small sized tree occuring mostly in soil derived from limestone below 1300 m altitude. G. rhamnifolia commonly grows in rocky and stoney habitats below 1000 m altitude.

Gironniera parvifolia K. male inflorescence; N. young infrutescence. (Source: Soepadmo (1977)

Gironniera rhamnifolia L. male inflorescence

o o o o o o

A genus of 5 species distributed from Malesia, throughout Papuasia to Polynesia. In New Guinea the genus occurs as a pioneer from sea level to 2000 m altitude. Parasponia is morphologically very similar to Trema but can easily distinguished from the latter by its imbricate perianth lobes of the male flower and intrapetiolar, connate stipules enclosing the terminal bud.

Shrubs to medium-sized trees. Bark grey brown. Stipules intrapetiolar, connate enclosing the terminal bud, caducous. Leaves triplinerved at base mostly glabrous except for the midrib and lateral nerves. Inflorescence axillary, much branched, many flowered, densely appressed pubescent. Flowers 5 merous. Male flowers: globose, perianth lobes imbricate in bud. Female flowers: ovoid conical, staminodes absent; stigmatic arms short, simple. Drupe ovoid, pericarp fleshy and fibrous, endocarp hard and stony. Seed: endosperm scanty or copious; cotyledons equal.

Key to Species in Papuasia
(Source: Soepadmo (1977)

1 Basal nerves running up throughout the length of the leaf or nearly so 2

1 Basal nerves running up to 0.5-0.7 the length of the leaf.......... 3

2 Leaf thick coriaceous, lower surface densely set with soft erect hairs; margin distinctly serrate; reticulations prominent beneath. Inflorescence male or female at anthesis condensed and shorter than the petiole
P. rigida Merr. & Perry. J. Arn. Arb 22 (1941) 254.

o o o o o o

2 Leaf chartaceous to thin-coriaceous, lower surface glabrous or sparsely appressed hairy; margin finely serrulate to subentire; reticulations obscure. Inflorescence male or bisexual at anthesis lax and longer than the petiole.
P. melastomatifolia J. J. Smith. Nova Guinea 8 (1914) 891. t. 158
P. simulans Merr. & Perry J. Arn. Arb. 22 (1941) 255

o o o o o o

Captions for Figure on Pages 103 and 104

Parasponia rigida A. twig; B.C. stipules; D. leaf undersurface; E.F. male flower; G.H.I. female flowers; J.K. fruit. (Source: Soepadmo (1977)

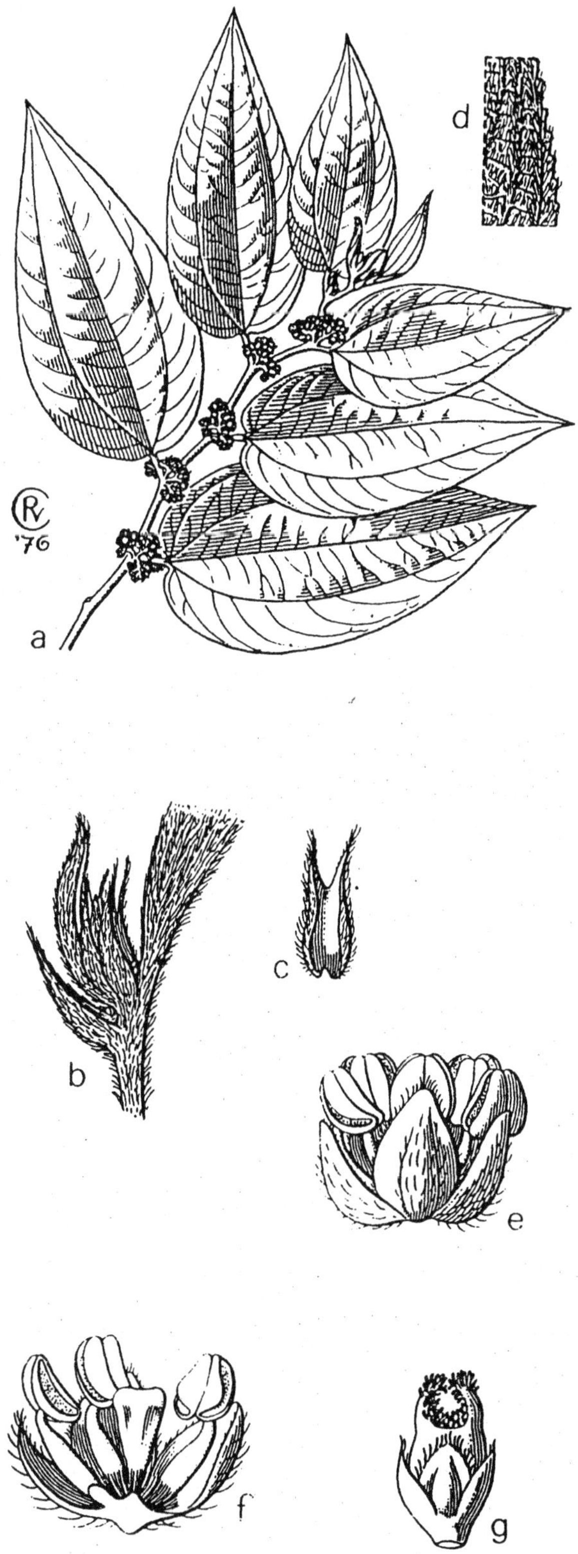

Leaf thick coriaceous, upper surface strongly rugose and scabrous, lower surface densely pubescent, midrib, nerves, and reticulations prominent beneath. .. . P rugosa Bl. Mus. Bot 2 (1856) 66

o o o o o o

3 Leaf chartaceous to thin coriaceous, upper surface not or hardly rugose nor scabrous, lower surface sparsely appressed pubescent or glabrous; midrib, nerves, and reticulations only slightly raised beneath... P. andersonii (Planch) Planch in DC Prod. 17 (1873) 193
P. paucinervia Merr. & Perry J. Arn. Arb. 20 (1939) 324

o o o o o o

Notes

1 Common and/or dominant in seral communities from sea level to 2000 m.

2 Several species have traditional uses: the inner bark of P. rugosa was used as ropes in both house or fence building; P. andersonii was used as medicine in Solomon Is.

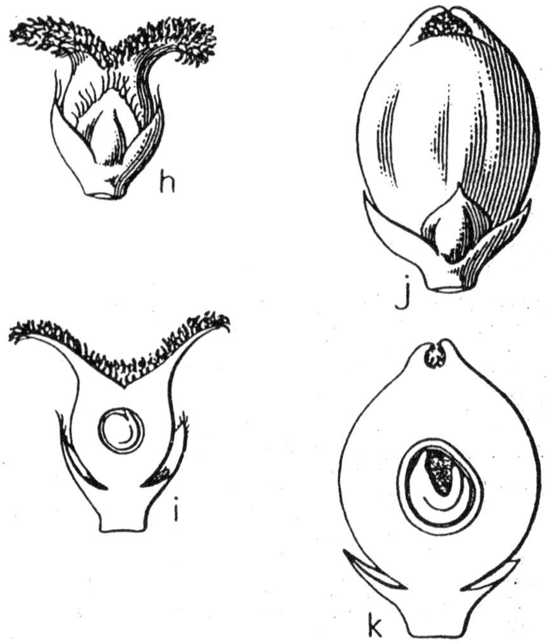

5 TREMA

o o o o o o

A genus of 10 15 species widely distributed in tropica and subtropics. In Papuasia 3 species The genus is homogenous and closely related to Celtis and Parasponia. Throughout the range of the genus it seems to grow well and often gregariously in newly opened secondary habitats It occupies a wide variety of sites from limestone to recent volcanic ash and alluvium.

Trees or shrubs, butressed. Plants monoecious Inovations densely set with simple bulbous hairs and/or multicellular capitate glandular hairs. Terminal bud enclosed overlapping, free, extra petiolar, caducous stipules. Leaves: + glabrous above; penninerved; base tri plinerved, cordate to acute, often unequal sided, margin variously serrate or denticulate Apex acuminate to acute or caudate, petiole sulcate.

Inflorescence axillary, paniculate or thyrsoid, many flowered, condensed or lax Bracts minute, ovate acute, caducous Male flowers: globular; perianth 4 5 lobed lobes induplicate valvate in bud, boat shaped: stamens glabrous, introse: filaments subulate, glabrous, incurved in bud, anthers subglobose to reniform. dorsifixed near the base, pistillodes present Female flowers ovoid; perianth 4 5 lobed; staminodes absent or rarely present; ovary glabrous, ovoid, sessile, slightly compressed; style short Drupe ovoid or subglobose, slightly compressed, glabrous; exocarp fleshy and fibrous, endocarp stony and very hard. Seeds with copious or scanty endosperm.

Key to Species in Papuasia
(Source: Soepadmo (1977)

1. Leaves 5 10 by 2-4 cm, glabrous or rarely sparsely pubescent beneath, chartaceous to thin-coriaceous; base rounded, rounded-attenuate, truncate, or very rarely subcordate, mostly symmetrical; nerves 2-4(-5) pairs. Inflorescence always with a slender axis. Mature fruits c. 3 by 2 mm, orange or red in colour.... T. cannabina Lour. Fl. Coch. 2 (1790) 563

o o o o o o

1 Leaves 10 15 by 3 7 cm, densely and variously hairy beneath. thin to thick-coriaceous; base cordate, subcordate or rounded. mostly asymmetrical; nerves 4 8 pairs In florescence with a stout or slender axis. Mature fruits black, 3 5 by 3 4 mm.. 2

2. Leaves beneath completely covered with matted glaucous to silvery straight appressed or curly, erect hairs and short. multicellular capitate-glandular hairs (epidermis invisible even under high magnification) very often distinctly

discolourous with the upper surface darker in colour and weekly scabrate. Male inflorescence up to 2.5 cm. Fruit ± globular.
.... *T. orientalis* (L.) Blume *Mus. Bot.* 2 (1856) 62
Laut. (1913: 320)

o o o o o o

2. Leaves beneath (densely or sparsely, not completely) set with velvety, greyish-brown erect hairs only (epidermis clearly visible between the hairs even under low magnifications), mostly concolourous, dark-brown to blackish in dry specimens; upper surface scabrate. Male inflorescence up to 5 cm. Fruit ovoid, compressed.
.... *Trema tomentosa* (Roxb.) Hara. *Fl. E. Himal.* 2 (1971) 19.
K. Sch. & Laut. (1900: 264)

o o o o o o

Notes:

1. *Trema cannabina* is common as a pioneer along roadsides, forest margins and in young secondary communities. It grows from sealevel to 1200 m. The species occurs from China and Burma to New Caledonia and W Polynesia.

2. Soepadmo (1977) notes that 3 rather (but not completely) distinct varieties can be recognised. Only 1 occurs in Papuasia where it is common in the lowlands. The seeds are bird dispersed. In its distribution in the lowlands and its method of seed dispersal this species is similar to *T. tomentosa*. Soepadmo (1977) even suggests that the two may represnt a single species.

Trema orientalis A. branch with flowers.
(Source: Johns (1982)

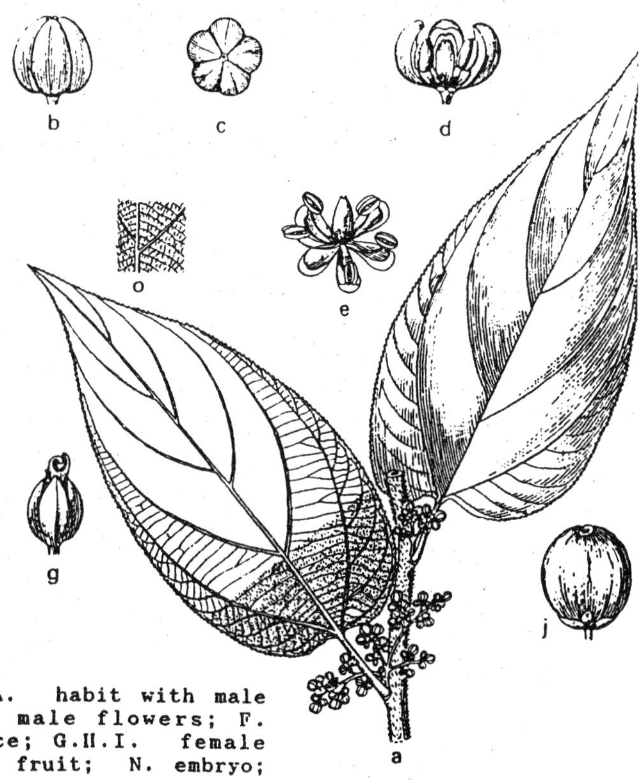

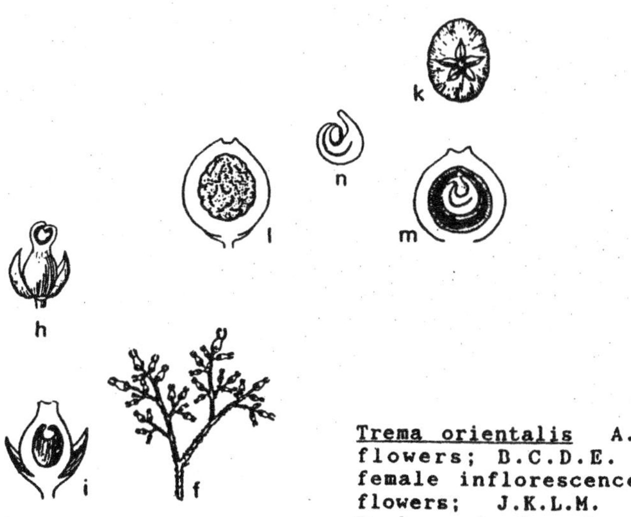

Trema orientalis A. habit with male flowers; B.C.D.E. male flowers; F. female inflorescence; G.H.I. female flowers; J.K.L.M. fruit; N. embryo; O. lower leaf surface.
(Source: Soepadmo (1977)

JUGLANDALES

The order includes 2 families: the Juglandaceae with 60 species and the monotypis family, Rhoipeteleaceae. A group of aromatic trees or rarely shrubs, with pinnately compound leaves.

JUGLANDACEAE

JUGLANDACEAE

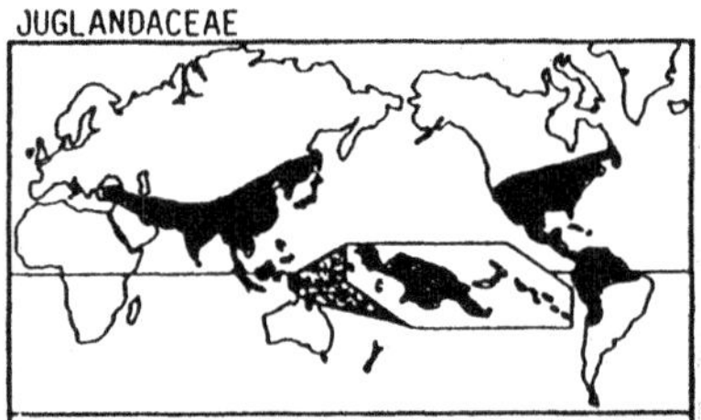

	WORLD	PAPUASIA
Number of genera:	7	1
Number of species:	59	2
Distribution:	Temperate and warm areas.	
Economic uses:	Timber from Juglans	

o o o o o o

A family of 7 genera and 59 species in temperate and warm areas in N Hemisphere, S America and Malesia. Most are wind pollinated trees or shrubs. The main timber producing genus is Juglans. The family is best known for its edible nuts, walnuts, produced mainly from Juglans regia. Engelhardia is sawn for timber in Papuasia. Carya is the source of the pecan nut. Some are grown for their ornamental value. Trees, bole with deep buttresses.

Diagnostic features

Leaves: alternate, rarely opposite, pinnate, without stipules.

Flowers: bracteate, unisexual with both flowers on same plant, monoecious. Perianth 4-lobed or reduced or absent. Male flowers in catkins; 3-40 free stamens in 2 or more series; filaments short. Wind Pollinated. Female flowers in small erect spikes, ovary inferior, consisting of 2 fused carpels. Style short with 2 stigmas.

Fruits: a drupe or nut. Seeds lack endosperm.

References

Barker, W. R. (1981). Juglandaceae. Hanbooks of the Flora of Papua New Guinea. 2: 202 - 205.

Jacobs, M. (1960). Juglandaceae. Flora Malesiana ser. 1. 6: 143 - 154.

Manning, W. E. (1966). New combinations and notes on Engelhardia of the Old World. Bull. Torrey Bot. Club 93: 34 - 52.

Manning, W. E. (1978). The classification within the Juglandaceae. Ann. Missouri Bot. Gard. 65: 1058 - 1087.

Key to Genera in Papuasia

1 genus:
.... 1. Engelhardia

1. ENGELHARDIA

● ● ● o o o

A genus of 5 species distributed from S India and S China, throughout Malesia to the New Guinea mainland. In Papuasia it is represented by 2 species, one of which is widespread, the other known from a single collection.

Trees dioecious. Leaves deciduous, pinnate, stipules absent; leaflets ± asymmetrical. Flowers small unisexual, arranged singly in catkins, subtended by a bract; perianth absent or with 4-lobes in 2 series; stamens 3-8 in Papuasia to many, subsessile. Ovary inferior, ovule 1; style terminal, erect. Fruit 1-seeded; seed without endosperm.

Key to Species in Papuasia

1. Leaflets entire. Petiole and rachis combined (2-)3-8.5(-10) cm. long.
.... Englehardia rigida
Bl. Bijdr. Fl. Ned. Ind. 10 (1825) 528.
Engelhardia lepidota Schlechter (1913:)

1-2(-3) pairs of leaflets; fruiting catkins 10-15 cm. long; fruit 2.5 - 2.7 cm long.
var. subsimplicifolia (Merr.) Manning.
Engelhardia subsimplicifolia Merrill

3-5 pairs of leaflets and longer catkins and fruits.
var. rigida

1. Leaflets conspicuously serrate. Petiole and leaf rachis combined (3-)5-18(-21) cm. long.
.... Engelhardia serrata
Bl. Fl. Jav. Jugl. 14 (1829) t. 4, 5c.

Engelhardtia rigida A. stem with female flowers; B. stem with male flower spikes. (Source: Johns (1983)

Notes:

1. E. rigida is widespread throughout the mainland of New Guinea. The species also occurs throughout Malesia but is rare. The timber is used for interior work. In the Western Province it is used for canoes and buildings. The species occurs from sea-level to 2600 m.

2. Subglabrous forms of E. rigida have been attributed to E. spicata Lesch. ex Dl. from Java westwards. They have flat greyish scales. The specimens from Papuasia bear occasional golden-yellow scales and have the catkins typical of Engelhardia rigida (Barker, 1981).

3. A single sterile collection of Engelhardia serrata was made near the Wissel Lakes in Irian Jaya at an altitude of 1800 m.

DAPHNIPHYLLALES

The order consists of a single family and genus with c. 10 species native to E Asia and Malesia. The family is an isolated endline comparable to Casuarina. The species have a unique alkaloid.

DAPHNIPHYLLACEAE

● ● ● o ● o

A monotypic family confined to Eastern Asia and Malaysia. In Papuasia it is represented by a single species which can form a shrub to 8 m tall, or a much branched slender tree to 12(-23) m tall and 45 cm diameter. Dioecious or monoecious tree or shrub.

Diagnostic features

Leaves: simple, spirally arranged, entire.

Flowers: small, unisexual, axillary (in Papuasia); racemes subtended by 1-several whorls of deciduous bracts,perianth a single whorl of 3-6 tepals. Male flowers: stamens in single whor; of 5-12; fused at apex. Female flowers, with or without a single whorl of staminaodia; ovary superior; styles erect with 1-2(-4) recurved lobes.

Fruit: a drupe with 1(-2) seeds with fleshy endosperm.

References

Barker, W. R. (1981). Handbooks of the Flora of Papua New Guinea II. Daphniphyllaceae. pp. 34 - 37.

Huang, T. C. (1965). Monograph of Daphniphyllum. Taiwania 11: 57 - 98.

Huang, T. C. (1966). Monograph of Daphniphyllum. Taiwania 12: 132 - 234.

Rosenthal, K. (1919). in Engler. Pfl. R. 68: 1 - 16. Daphniphyllaceae.

KEY TO GENERA IN PAPUASIA

1 genus: 1. Daphniphyllum

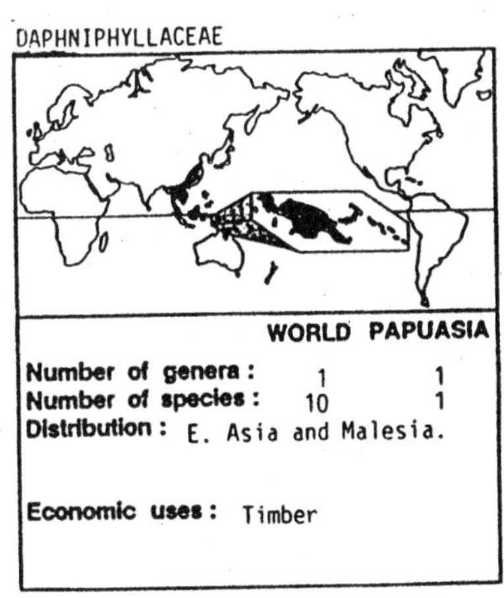

	WORLD	PAPUASIA
Number of genera:	1	1
Number of species:	10	1
Distribution:	E. Asia and Malesia.	
Economic uses:	Timber	

1. Daphniphyllum

● ● ● ○ ● ○

Shrubs or trees to 23 m tall. Leaves usually densely packed at end of branches and in spaced clusters along them; petioles long slender, deep pink to red, channelled above. Blade leathery lanceolate to broad elliptic, glabrous. Inflorescemces unisexual racemeswith 20-30 flowers. Male flowers: stamens 5-9, free. Female flowers: staminodes 6-9; gynoecium 2-3.5 mm long tapering into short, stout, erect style. ; stigmas 2, stout. Fruit globular to broad ellipsoid, 8-15 mm long excluding persistent style; pericarp fleshy. Ripe fruit blue, purple or black.

Key to species in Papuasia

1 species;

Daphniphyllum gracile Gage (1918) 480
D. papuanum Hallier (1918)
non D. gracile Rosenthal (1919)

Key to varieties in Papuasia
(Source: Barker 1981)

1. Mature fruit smooth (at least in dried specimens). Leaves with short acumen to 1-(2-5) mm long, obtuse to rounded or subtruncate behind, usually glaucous, rarely green beneath.... var. gracile

1. Mature fruits tuberculate (at least dried specimens). Leaves with long, often slender acumen (1.5-)3-25 mm long, acute to obtuse behind, usually green, sometimes glaucous beneath.... var. tuberculatum Huang (1966) 159

Notes:

1. The species occurs thoughout mainland New Guinea and also from New Britain and New Ireland. It has been collected in the Celebes.

2. Both varieties occur throughout the range of the species. They occur from (850-)1000 to 3500 m in montane and subalpine forest and also in exposed summit shrubberies. Locally common.

3. var. tuberculum occurs at lower altitudes than var. gracile. The former is normally restricted to midmontane forests.

4. Barker (1981) states that it is doubtful if D. gracile warrant distinction from D. glaucescens Bl. which is widespread to India, S China and Japan. The collection BW 14296 by Sleumer and Vink at 2100 m altitude from the Anggi Gigi Lakes in Irian Jaya has been identified as D. glaucescens Bl. but the material at Lae is not adequate to see if the calyx is longer than the anthers, or not.

4. Hemsley described Daphniphyllum conglutinosum (Kew Bull. 1895: 137)but this has been shown to be Pimeleodendron (Huang 1966).

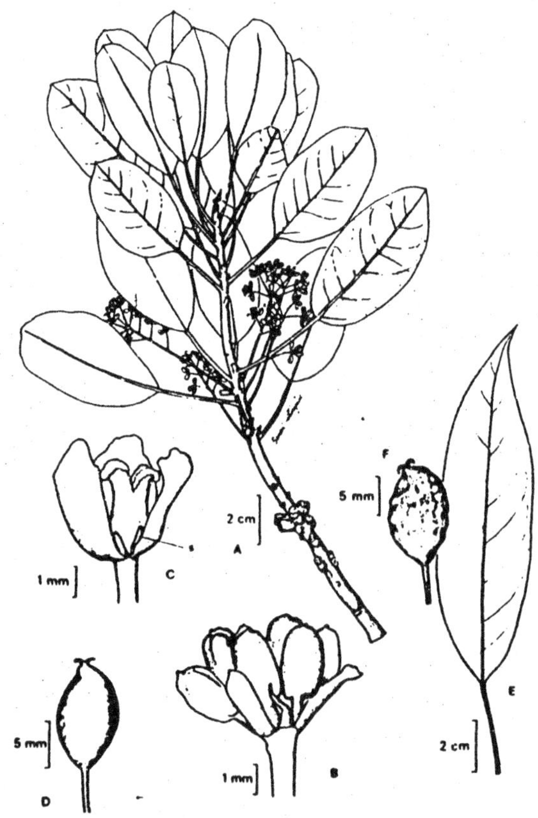

Daphniphyllum gracile var gracile A. flowering twig; B. male flower; C. female flower with staminode; D. fruit; var. tuberculatum E. leaf; F. fruit. (Source: Barker (1981)

FAGALES

The order Fagales includes three families and more than 900 species. The Fagaceae is the only family represented in Papuasia. The other families are the Betulaceae, which consists of 6 genera and about 170 species and the Balanopaceae. The Betulaceae is found predominantly in the north temperate zone. Both Betula and Alnus are valuable timber trees, and are known in the timber trade as birch and alder. The family Balanopaceae has a single genus Balanops and twelve species.

The Fagaceae is distributed in the northern hemisphere where it is represented by the beeches (Fagus), the oaks (Quercus) and the chesnuts (Castanea). In the southern hemisphere the southern beech (Nothofagus) dominates the comparable forests. Tropical genera are Castanopsis and Lithocarpus.

FAGACEAE

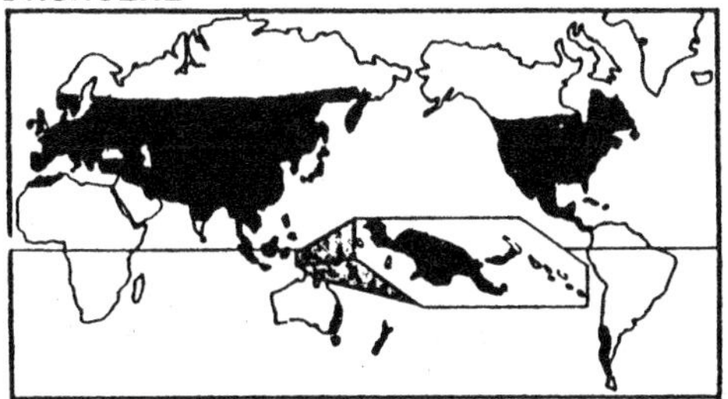

	WORLD	PAPUASIA
Number of genera:	8	3
Number of species:	900	24

Distribution: Temperate and tropical forests.

Economic uses: Hardwood timber and edible fruits.

FAGACEAE

● ● ● ● ● ○

In South East Asia, Malaysia and Papuasia, the mixed mountain forest is largely dominated by the evergreen members of this family. Three genera occur in these forests in Papuuasia. Both Castanopsis and Lithocarpus also are abundant in the montane forests of Indonesia and the Philippines where they are represented by 186 and 110 species respectively. Nothofagus, the southern beech, required continuous land for migration. It is consequently not very efficient in dispersal and has not migrated further into Malesia than New Guinea.

Castanopsis and Lithocarpus have obviously migrated down the Indonesian Archipelago in recent times, probably post Miocene. In Papuasia Castanopsis is represented by a single species and Lithocarpus by 10 species. Nothofagus is represented in New Guinea by 13 species. Of the 24 species of Fagaceae occuring in Papuasia, 22 are endemic.

Most of the tree species are important for timber production. Commercial cork is derived from Quercus suber. Many species have edible nuts. Lithocarpus nuts were eaten by early man in New Guinea Several species are grown as ornamental (shade) trees.

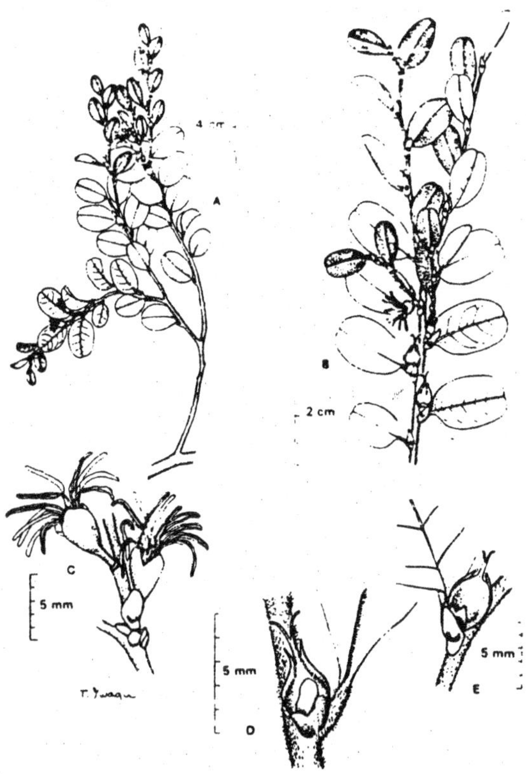

Nothofagus pullei (A) twig; (B) tw wth male flowers; (C) male flower (D) female flowers; (E) female flow with peltate stipules.
(Source: R. J. Johns (1984)

Diagnostic features

Leaves: Deciduous (not in Papuasia) or evergreen, leaves alternate, simple, entire with usually deciduous stipules.

Flowers: unisexual, in catkins or small spikes, . Perianth bract-like with 4-7 lobes. Male flowers have stamens equal to, or twice in number to perianth parts (up to 40), filaments free. Female flowers in groups of 1 to 3, each group surrounded by basal involucre. Ovary inferior, styles 3 or 6.

Fruit: single-seeded nuts in groups of 1 or 3, surrounded or enclosed by an often hardened 'cupule'; seeds lack endosperm.

References

Hjelmquist, H. (1963). Some notes on Nothofagus from New Guinea and New Caledonia. Bot. Not. 116. Fasc. 2: 225 - 236.

Johns, R.J. (1984). Common Forest Trees of Papua New Guinea 3:

Langdon, L.M. (1949). The comparative morphology of the Fagaceae. I the genus Nothofagus. Bot. Gazette 108: 350.

Soepadmo, E. (1970). Flora Malesiana Praecursores XLIX. Malesian species of Lithocarpus Bl. Reinwardtia 8: 197 - 308.

Soepadmo, E. (1972). Flora Malesiana Series I. 7: 265 - 403.

van Steenis, C.G.G.J. (1952). Preliminary account of Papuan Nothofagus. Blumea 7: 146 - 147.

van Steenis, C.G.G.J. (1953). Results of the Archbold Expeditions. Papuan Nothofagus. Journ. Arn. Arb. 34: 301 - 374.

van Steenis, C.G.G.J. (1953). Results of the Archbold Expeditions. Additional Notes on Nothofagus. Journ. Arn. Arb. 35: 266.

van Steenis, C.G.G.J. (1955). Miscellaneous Notes on New Guinea Plants II. Nova Guinea 6: 279 -

van Steenis, C.G.G.J. (1971). Nothofagus, Key Genus of Plant Geography in time and space, living and fossil, ecology and phylogeny. Blumea 19: 65 - 98.

van Steenis, C.G.G.J. (1972). Nothofagus in Fagaceae. Flora Malesiana Series I. 7: 277 - 294.

van Steenis, C.G.G.J. (1975). Miscellaneous notes on New Guinea plants XI. Blumea 22: 170 -172.

van Steenis, C.G.G.J. (1978). Addenda, corrigenda et emendanda. Flora Malesiana Series I. 8: 549.

van Steenis, C.G.G.J. (1982). Addenda, corrigenda et emendanda. Flora Malesiana. Series I. 9: 563.

KEY TO THE GENERA IN PAPUASIA

(Source: modified from Soepadmo (1972)

1. Leaves glandular dotted beneath. Stipules peltately attached. .... 3. **Nothofagus**

1. Leaves not glandular dotted. Stipules not peltate............... 2

2. Cupule spiny; nuts rounded-angular in cross section, 1-15 in each cupule. Suckers usually present around the older trees. .... 1. **Castanopsis**

2. Cupule never lobed, saucer- or cup-shaped, variously squamose, lamellate or muricate. Nut always solitary in a cupule, circular in cross-section. Suckers very rare and only found on isolated trees. .... 2. **Lithocarpus**

Nuts of Fagaceous Genera
(A) Lithocarpus; (B) Castanopsis; (C) Nothofagus
(Source: R. J. Johns (1984)

1. CASTANOPSIS

● ● ● ● ● ○

In her monograph A. Camus (Chat. (1930) 243) recognised 120 species in this genus which ranges from India to Papuasia, with a single species in S W North America. In Malesia there are 34 species (Soepadmo 1972) and a single species in Papuasia.

Large trees to 35 m tall. Leaves spirally arranged, thin-coriaceous, to 17 by 6 cm, margin entire, tip sharply acuminate to 2.5 cm long. Inflorescences male, female or androgynous. Male rachis 5-10 cm, slender, bracts ovate-acute; male flowers densely arranged but solitary, perianth lobes 6, stamens 12-14, pistillode rudimentary. Female and androgynous rachis 5-10 cm, slender; androgynous rachis with a few male flowers in apical parts. Female flowers solitary, lobes 6, staminodes 12, styles 3, recurved, cylindrical. Young infructescence 10-15 cm, with 10-20 young cupules. Ripe cupules globose with the fruit partly emerging; outside with small flat-triangular spines. Fruit solitary, with dense rufous tomentum.

Key to Species in Papuasia

1 widespread species:

Castanopsis acuminatissima (BL.) A. DC. J. Bot. 1 (1863) 182.
Castanea acuminatissima Bl. Mus. Bot. 1 (1850) 283.

Notes:

1. This species is widespread from India, throughout S E Asia, to mainland New Guinea, the Papuan Islands, and New Britain. The altitudinal range is from 300-2500 m.

2. A common secondary species in New Guinea, reproducing from suckers. After gardening, or other disturbances, the stumps of the trees coppice and produce almost pure stands of this species. Consequently it is common on ridges in pure stands or often in disturbed sites with other species with light demanding seedlings ie. Lithocarpus, Intsia, Nothofagus or Anisoptera.

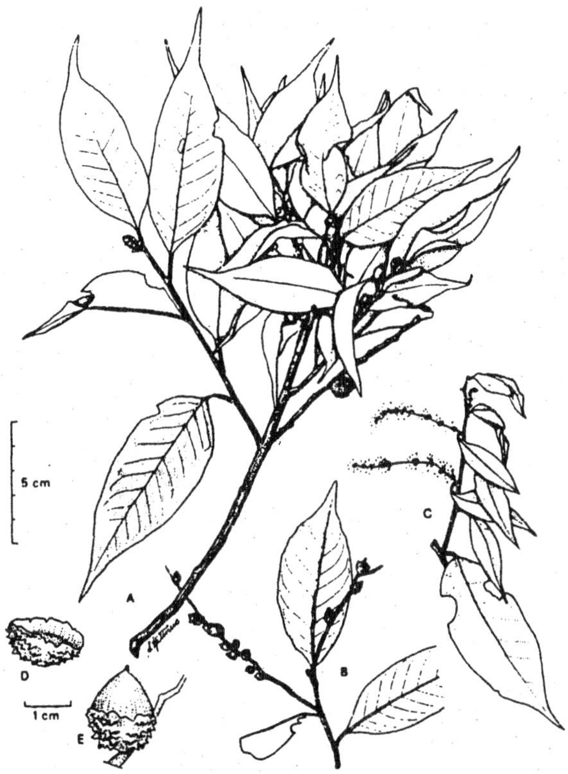

Castanopsis acuminatissima (A) twig with nuts; (B) twig with female flowers; (C) twig with male flowers. (Source: R. J. Johns (1984)

2. LITHOCARPUS

● ● ● ● o o

A genus of c. 300 species ranging from India to Papuasia with 1 species in S W United States. In Malesia c. 110 species with 10 species in Papuasia. In Papuasia the genus is common between sea level and 1500 m altitude, on a range of soil types including limestone. As evidenced by cave deposits in the Chimbu gorge, the fruits of Lithocarpus were collected and probably ground into a flour for food. This is one possible use for the mortars and pestles which are common in some parts of the Highlands of New Guinea.

Trees or shrubs. Leaves spirally arranged, entire, thin to thick-coriaceous, glabrous or variously hairy beneath. Petiole terete. Inflorescence male, female, androgynous or mixed, always erect, including bracts. Male rachis solitary in axil of a lower leaf or in dense paniculate cluster on lateral shoots, simple or much branched. Male flowers solitary or in dichasial clusters of 3-7(-30); perianth campanulate, (4-)6(-7)-lobed; pistillode always present. Female, androgynous or mixed rachis solitary in axil of a higher leaf or on upper part of the paniculate cluster. Female flowers solitary or in a dichasial cluster of 3-7(-15) along the rachis; perianth 6-lobed, staminodes 10-12, rudimentary or well developed and exceeding the perianth, sometimes polliniferous. Style 3-6(-15), conical or terete, densely tomentose at base; stigma terminal. Cupule solitary or in dichasial clusters of 3-7(-15), sessile or 0.5-2 cm stalked, enclosing a solitary fruit; cup-shaped, saucer shaped to almost globular, variously lamellate, squamose, tuberculate or muricate, never truly spiny. Pericarp of fruit woody or bony.

Key to Species in Papuasia
(Source: modified from Soepadmo (1972)

1. Cupule and mature fruit larger than 4 cm in diameter............... 2

1. Cupule and mature fruit smaller than 4 cm in diameter........... 5

2. Cupule obconical; covering more than 2/3 of the fruit; 3 to 3.5 cm high...... **L. schlechteri** Markgr. Bot. Jahrb. 59 (1924) 69. (34)*
L. perclusa Markgr. Bot. Jahrb. 59 (1924) 68

2. Cupule broadly saucer shaped; 1-2 cm high; 4-5 cm diameter; covering 1/4 to 1/2 of the fruit......... 3

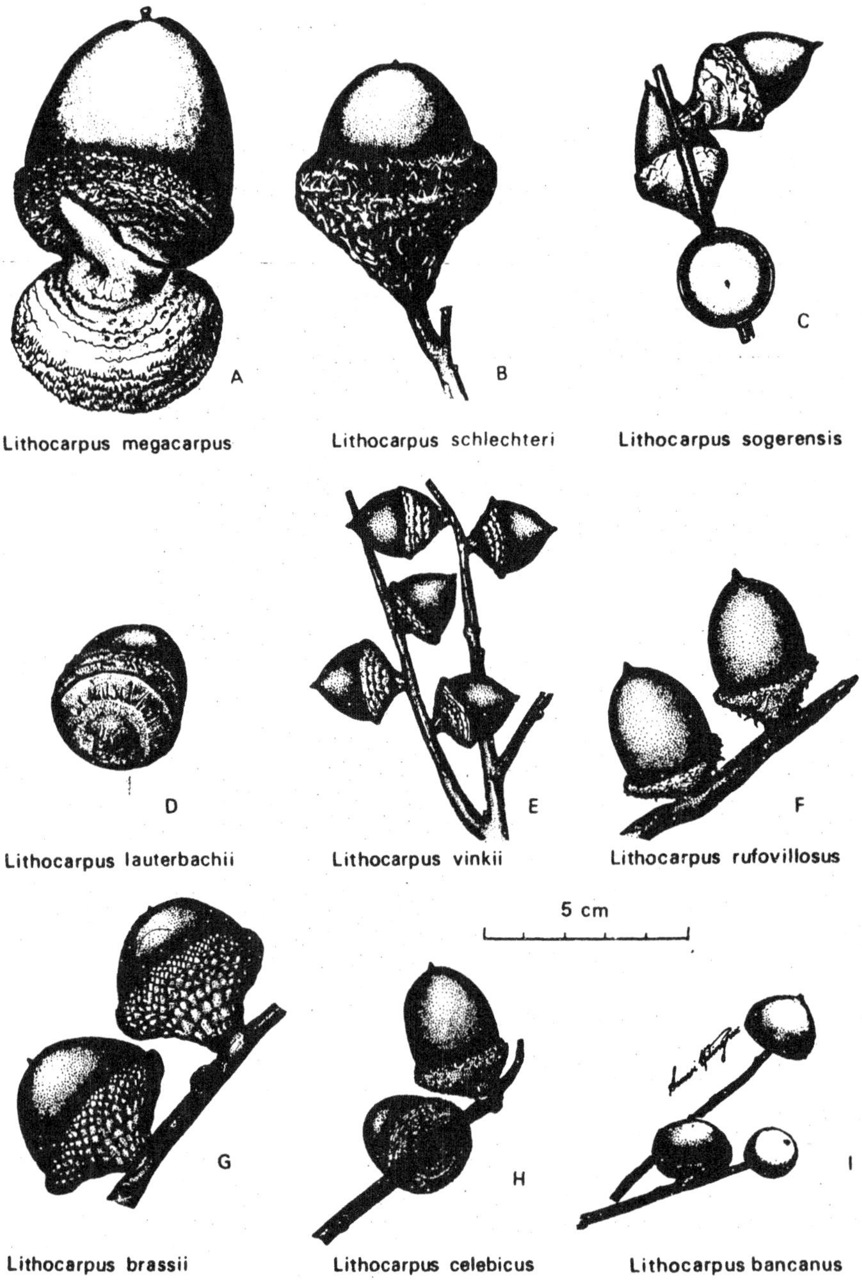

<u>Variation in the cupules and mature fruits of the species of Lithocarpus</u>

(Source: R. J. Johns (1984)

3\. Leaves thick, coriaceous, rigid; scar of fruit concave at centre.
.... **L. megacarpus**
Soepadmo Reinwardtia 8 (1970) 259. (35)

3\. Leaves thin, coriaceous, not rigid; scar of fruit strongly convex to conical........................ 4

4\. Cupule with thick rounded rim, scales adpressed, obscure; leaf tip acute, reticulations obscure.
.... **L. brassii**
Soepadmo Reinwardtia 8 (1970) 221. (37)

4\. Cupule with thin rim, woody scales thick, distinct; leaves acuminate, reticulations distinct beneath.
.... **L. lauterbachii**
(von Seeman) Markgr. Bot. Jahrb. 59 (1924) 69. (38)
Quercus lauterbachii von Seeman Bot. Jahrb. 57 (1897) 54
L. solanicarpa Markgr. Bot. Jahrb. 59 (1924) 67

5\. Cupule variously lamellate..... 6

5\. Cupule squamose to muricate.... 7

6\. Fruit tomentose.
.... **L. bancanus**
(63). Material described by Markgraf under Pasania bancana belongs to L. megacarpus

6\. Fruit glabrous.
.... **L. vinkii**
Soepadmo Reinwardtia 8 (1970) 286. (70)

7\. Cupule variously squamose...... 8

7\. Cupule variously muricate...... 9

8\. Scar of fruit conical; cupule with obscure lamellae on the lower half.
.... **L. sogorensis**
(S. Moore) Markgr. ex A. Camus Chenes 3 (1954) 795. (87)
Pasania sogerensis S. Moore J. Bot. 61 (1923) Suppl. 54.

8\. Scar of fruit flat to concave; cupule with distinct scales in concentric lines, or imbricate.
.... **L. celebicus**
(Miq.) Rehd. J. Arn. Arb. 1 (1919) 123. (88).
Quercus celebicus Miq. Ann. Mus. Bot. Lugd-Bat. 1 (1863) 110.
Quercus pseudomoluccana var. papuana O. Warb. Bot. Jahrb. 13 (1891) 286. Also under Pasania.

Lithocarpus celebicus (A) twig with nuts; (B) twig with female flowers; (C) twig with male flowers. (Source: R. J. Johns (1984)

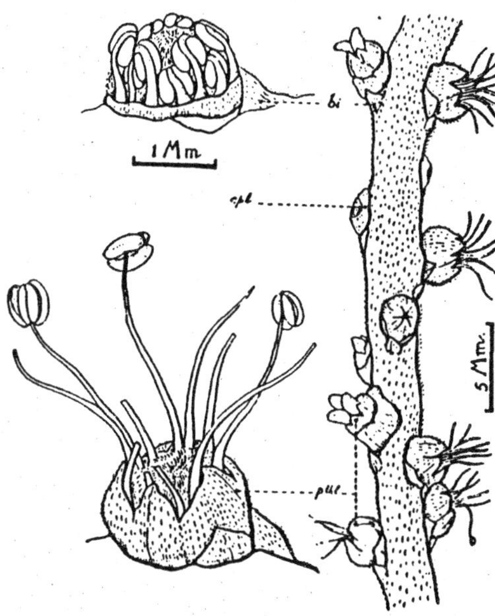

Lithocarpus Male catkins and male flowers. (Source: Soepadmo (1970)

9. Leaf 6-10 by 4-5 cm. Indumentum yellowish brown; cupule 1.5 cm diameter.. **L. aspericupula** (Markgr.) Rehd. J. Arn. Arb. 10 (1929) 123. (101).
Pasania aspericupula Markgr. Bot. Jahrb. 59 (1924) 78.

9. Leaf 10-20 by 3.5-7 cm. Indumentum reddish; cupule 2-3 cm diameter. .... **L. rufovillosus** (Markgr.) Rehd. J. Arn. Arb. 10 (1929) 133. (103).
Pasania rufovillosa Markgr. Bot. Jahrb. 59 (1924) 78.

* The numbers in brackets refer to the numbered descriptions of each species in the Flora Malesiana treatment by Soepadmo (1972).

Notes

1. The genus occurs over an altitudinal range from sea level to over 1600 m. On the ultrabasic belt, near Kui, plants of Lithocarpus occur at sea level, one tree was seen overhanging the sea. Most trees however occur between 700 and 1600 m altitude, in which zone species of the genus such as L. celebicus and L. rufovillosus can form almost pure stands following gardening and in secondary conditions following natural disturbances.

2. In several montane areas species of Lithocarpus commonly form forest margin communities or small almost pure patches of forest along streams in grassland.

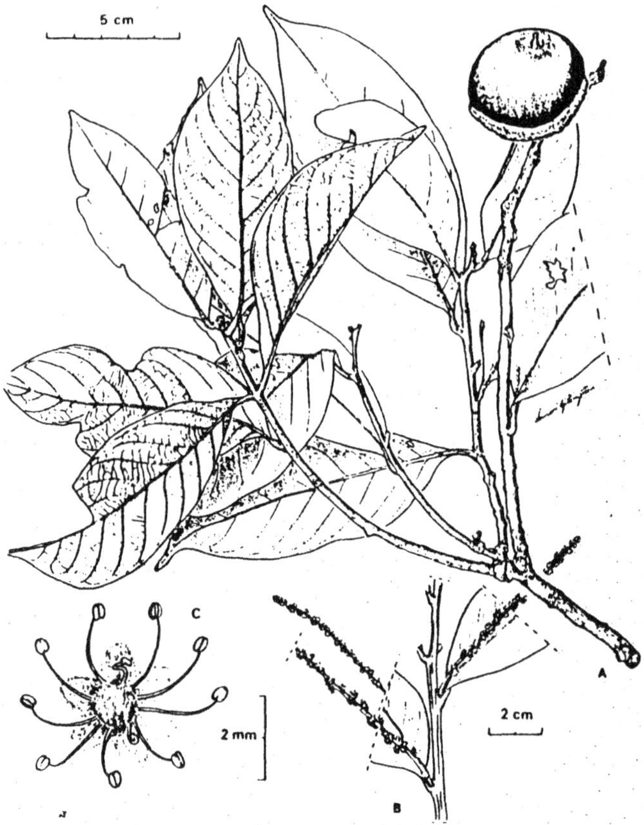

Lithocarpus megacarpus A. branch with leaves and mature cupules; B. male catkin; C. male flower.
(Source: National Herbarium, Lae)

3. NOTHOFAGUS

● ● ● ● ● ○

A genus of 37 species distributed in Papuasia (mainland New Guinea, Papuan Islands, New Britain), New Caledonia, E Australia, Tasmania, New Zealand and S America. The genus is strictly southern hemisphere in distribution (van Steenis 1972). All species in New Guinea and New Caledonia (5 species) belong to the Section Nothofagus, Subsection Bipartitae (van Steenis 1982).

Monoecious, evergreen, trees or shrubs. Leaves distichous, entire, rarely crenate, tip emarginate, beneath glandular dotted. Stipules peltate. Male flowers: in lower part of flush, solitary or in ebracteolate triads, stamens c. 12-18, filaments usually connate at base. Female flowers in upper part of flush, axillary, 1 or 3, usually in a cupule. Ovary sessile, flat; style short with 2 stigmatic arms. Cupule 2-lobed, usually with 1 or more lamellae, finally often woody and gaping; cupule rarely reduced to 2 minute free elamellar thin flaps or entirely absent. Nuts flat, 1-seeded.

Key to Species in Papuasia

(Source: van Steenis (1972))

1. Cupule 3-flowered............. 2

1. Female flowers solitary, cupule present or absent............. 5

2. Leaves upwards crenate, usually ovate to elliptic-oblong, c. 5-10 by 2.5-4.5 cm. Peduncle of cupule c. 1.5-3 cm.................. 3

2. Leaves entire, elliptic in outline, up to 5.5 by 2.5 cm........... 4

3. Cupule well-developed, lamellate, c. 1.5-2 cm through, the nuts not or hardly exserted. Leaves usually ovate-oblong, more or less pointed to the apex; glands underneath less than or at most up to 1 mm spaced. Stipules early caducous, at least not present on fruiting twigs. Peduncle of male triads 10-25 mm. .... **N. perryi**
Steenis Blumea 7 (1952) 146.

3. Cupule hardly developed, less than 0.5 cm long, consisting of 2 poorly developed valves each with an indistinct lamella, the nuts almost entirely protruding. Leaves elliptic-oblong, not or hardly pointed. Glands underneath more widely (1-2 mm) spaced. Stipules persistent, even on fruiting twigs. .... **N. nuda**
Steenis Flora Malesiana Series 1. 7: 285.

4. Midrib sulcate above, without trace of an elevated ridge over its entire length. Anthers c. 2.5 mm long. Peduncle of cupule 2-4 mm. .... **N. starkenborghii**
Steenis Blumea 7 (1952) 147.

4. Midrib at least in the basal half with an elevated ridge. Anthers c. 5-6 mm long. Peduncle of cupule 5-15 mm..... **N. brassii**
Steenis Blumea 7 (1952) 146.
N. recurva Steenis Blumea 7 (1952) 146.
N. pseudoresinosa var. microphylla Steenis Blumea 7 (1952) 147.

5. Cupule entirely absent......... 6

5. Cupule present, each valve consisting of at least an elamellate flap. 7

6. Leaves waxy beneath, elliptic, margin towards apex faintly crenate or minutely toothed, 4.5-10 by 2-5 cm; nerves c. 9 pairs, rather parallel and extending towards the margin; petiole 5-8 mm. Stipules 4-6 mm caducous. Male flowers 1, 6-7 by 3.5-4 mm; anthers 5-6 mm. Nut hairy, ovate-oblong. .... **N. resinosa**
Steenis Blumea 7 (1952) 147.

6. Leaves not waxy beneath, ± ovate - elliptic, entire, 2.5-4 by 1.5-2.25 cm; nerves c. 6 pairs, looped at apex; petiole 2-3 mm. Stipules 3-4 mm rather persistent. .... **N. pullei**
Steenis Blumea 7 (1952) 146.
(?) N. cornuta Steenis Blumea 7 (1952) 147.

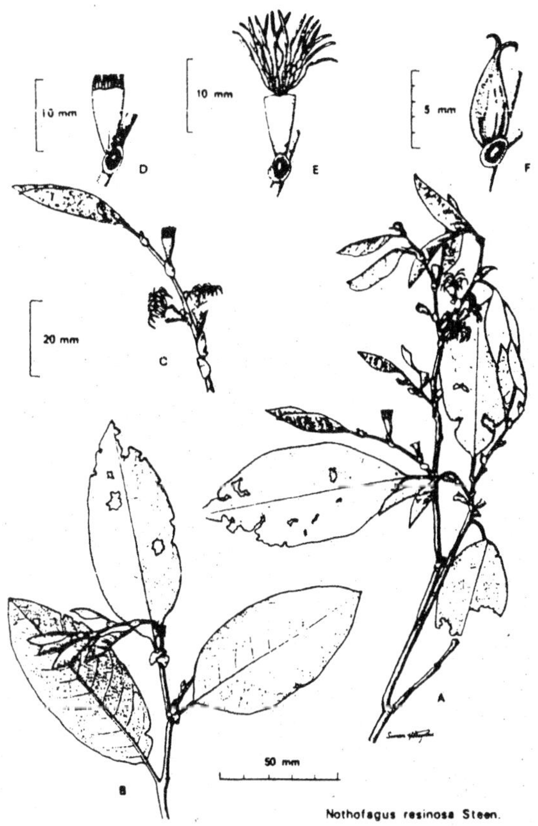

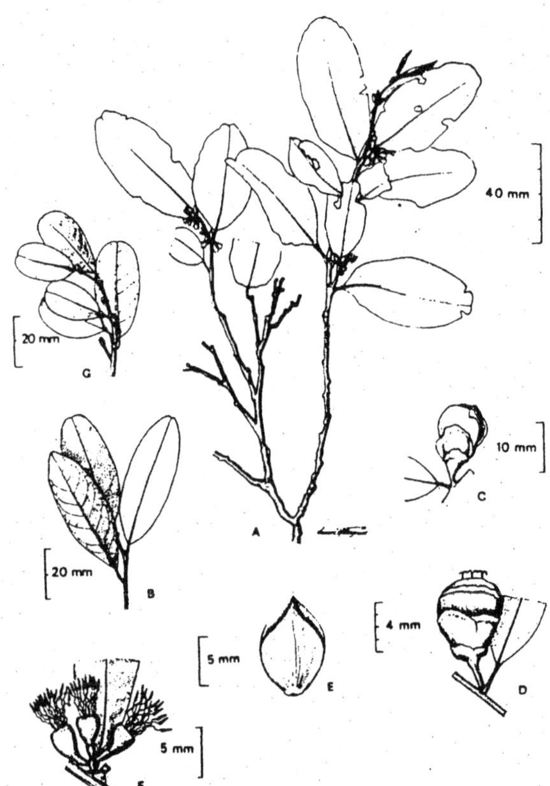

Nothofagus starkenborghii (A) twig with male and female flowers; (B) male flowers; (C) female flowers; (D) cupule.
(Source: National Herbarium, Lae)

7. Cupule only represented by 2 small free, elamellate flaps very much shorter than the nut.......... 8

7. Cupule valves at least provided with 1 lamella................ 9

8. Leaves ± obovate to obovate-oblong rarely also some ± elliptic, 2-4.5 by 1-2.5 cm; nerves 5-7 pairs; midrib above sulcate with ridge to halfway; petiole 2.5-5 mm. Ultimate twigs ± thin. Nuts c. 6-7 by 4-5 mm........ **N. carri**
Steenis Blumea 7 (1952) 147.

8. Leaves ovate-oblong, flat, acutish towards apex, or even ± acuminate (very rarely blunt and elliptic), 6.5-13 by 2.5-5.5 cm; nerves 8-10 pairs; midrib not sulcate, the ridge prominent to at least halfway, often to apex; petiole 7-10 mm. Nuts c. 8-10 by 5-7 mm. .... **N. flaviramea**
Steenis Nova Guinea 6 (1955) 281.
N. resinosa (non Steenis) Steenis J. Arn. Arb. 35 (1954) 267.
Shorea sp. Sloot. Reinwardtia 2 (1952) 61.

9. Cupule already in the young state distinctly stalked............ 10

9. Cupule sessile or almost so.... 14

10. Leaves ovate-oblong, distinctly acutish towards the apex....... 11

10. Leaves elliptic, not acutish towards the apex............... 13

11. Leaves smallish, c. 2.5-5 by 1.25-2 cm, crenate towards the apex. Cupule with c. 3 lamellae, the valves oval. Nut c. 5 mm, narrowly winged, about as large and of similar shape as the cupule halves. .... **N. crenata**
Steenis Blumea 7 (1952) 147.

11. Leaves much larger, c. 4.5-9 by 2.5-4 cm, entire............... 12

12. Cupule on a thin stalk, c.10-15 mm, with 1 rather apical lamella, much smaller and narrower than the nut, c. 6 mm long. Nut ovate-acute, c. 10 by 6 mm, winged towards apex. .... **N. womersleyi**
Steenis Flora Malesiana Series 1. 7: 294.

12. Cupule sessile to provided with a thick, obconical stalk-like extension c. 5 mm long, large, with 3-4 lamellate, finally woody, 12-13.5 by 7.5-10 mm. Nut rhomboidal, c. 7-10 by 9.5-10 mm, smaller than the cupular valves. Twigs coarse, greyish... **N. grandis**
Steenis Blumea 7 (1952) 147.

13. Leaves large, with distinct, somewhat prominent venation above, c. 4.5-7 by 2.5-4 cm. Twigs coarse, greyish. Mature cupules large, c. 12-20 by 7.5-10 mm. Nut large, rhomboid, c. 8-10 mm. .... **N. grandis**

13. Leaves much smaller, c. 1.75-2.75 by 7.5-10 mm. Nut rhomboid, c. 7-10 by 9.5-10 mm, smaller than the cupular valves. Twigs coarse, greyish... **N. rubra**
Steenis Blumea 7 (1952) 147. Steenis (1953) 368.
N. eymae Steenis Blumea 7 (1952) 147.
N. dura Steenis Blumea 7 (1952) 147.
N. bernhardii Steenis Blumea 7 (1952) 147.
N. decipiens Steenis Blumea 7 (1952) 147.

14. Leaves thin (minutely) crenate or minutely toothed towards apex, waxy underneath, c. 2.5-5 by 1.25-2.5 cm; nerves 7-8 pairs. Cupule c. 7-8 by 4-5 mm, with (1-)-2 lamellae. Male flowers solitary. .... **N. pseudoresinosa**
Steenis Blumea 7 (1952) 147.

14. Leaves quite entire, not waxy, not thin........................... 15

15. Rather coarse. Leaves longer than 3 cm. Cupules woody, not split to the base in mature state....... 16

15. Less coarse. Leaves less than 3 cm long.......................... 17

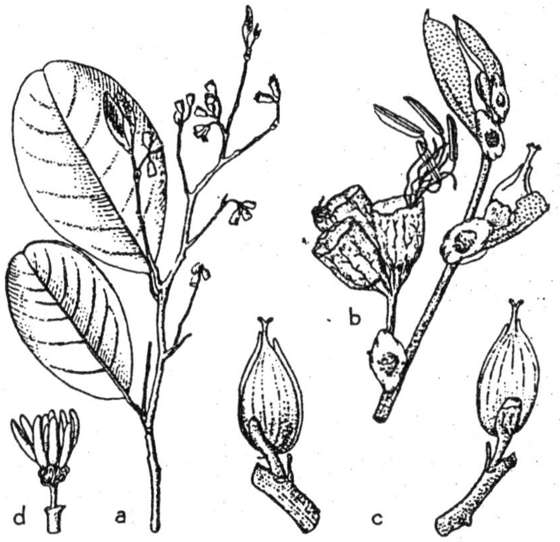

Nothofagus carri (A) twig with flowering flush; (B) triad of male flowers; (C) female flower.
(Source: van Steenis (1953)

Notes on Species in Papuasia

Details are given of the species of Nothofagus because of the economic importance of the genus and the scientific interest in the montane forests dominated by Nothofagus. The distributions of the species are mapped.

Nothofagus brassii

Generally a large tree but sometimes a dwarf shrub (Arfak). Often frequent or dominant in stands. Ranges from 1550-2700 m in altitude. The species is distributed throughout the Highlands of New Guinea. The species is variable in cupule characters. This has resulted in the inclusion of N. recurva into this species. The species is also variable in leaf texture.

Nothofagus carri

A gregarious species on ridges from 2000-3000 m altitude. At higher altitudes it is sometimes a dwarf shrub from 2-4 m tall. It is distributed from the Arfak to Mt Kaindi. Traditionally used as a source of timber, this species was planted around Mt Hagen and in the Chimbu along streams, and near gardens. The material of this species is not homogeneous (van Steenis 1972: 289). van Steenis includes N. cornuta as a synonym of N. carri. The type of cornuta lacks a cupule, but it is probably a freak specimen.

The specific name is spelt carri in the type description but carrii in the Flora Malesiana treatment.

Nothofagus crenata

A distinctive species known only from the type collection from near Lake Habbema at 2300 m altitude. It was described as being locally common. It has solitary male and female flowers.

van Steenis (1975) described a variety of this species from Lake Kutubu which is identical to N. crenata except for the entire leaf margins. It was collected by R. A. Hynes from a limestone ridge at 950 m.

var. sapeii Steenis. Blumea 22 (1975) 171.

Nothofagus flaviramea

A widespread tree of montane forests growing between 750 and 2450 m altitude. Often common or dominant, on Mt Cycloop co-dominant with Castanopsis and in the Star mountains co-dominant with Araucaria

Nothofagus grandis

One of the most common species in E New Guinea. Grows over an altitudinal range from 1350 to 2600 m. Often dominant and forming extensive forests from Wabag to Mt Dayman. Planted in Upper Chimbu. It is scarce in West Irian. In N. grandis the midrib is sulcate with a prominent ridge, in N. flaviramea the ridge is on the flat leaf surface.

Nothofagus nuda

Known from a single collection from the Tauri R. near Paina in the Gulf Province. Growing in mixed lower montane forest associated with Castanopsis. This species is allied to N. perryi from which it differs in the extremely small, non woody, 1-lamellate cupule, the leaf shape, the thin peduncle, prominent venation and the veinlet ending in the bay of the crenation instead of the tooth.

Nothofagus perryi

A widespread species from the Highlands of Papua New Guinea from Wabag to Lake Trist. Growing from 1600 - 2600 m altitude. Often associated with Lithocarpus, Castanopsis and Araucaria. Used locally for building. Considered by van Steenis as the most primitive of the species in Papuasia.

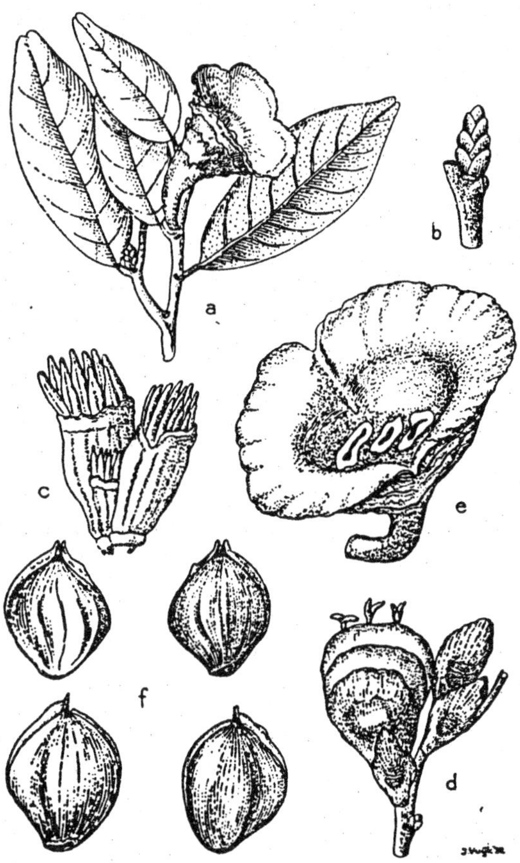

Nothofagus brassii (A) fruiting twig; (B) triad of male flowers; (C) nuts. (Source: van Steenis (1953)

16. Leaves large, c. 4.5-7 by 2.5-4 cm, rather thin-coriaceous (also on twigs with mature cupules), with distinct and ± prominent venation above. Twigs greyish, rather thick. Cupules c. 12-20 by 7.5-10 cm, with 3-4 lamellae. Nut c. 7-10 mm diameter.
.... N. grandis

16. Leaves mostly hard-coriaceous, some times convex or with recurved margin, smooth above, but the nerves (not the fine-tessellate veins) thickish and obtusely prominent, c. 2.5-5 by 1.25-3 cm. Cupules smaller, at most 12 by 6 (10-12) mm, with 2-3 lamellae. Nut c. 4-5 mm diameter.
.... N. rubra

17. Young twigs hairy to puberulous, at least minutely so on the innovations. Cupule with 1 lamella, not seldom irregularly incised-dentate, mostly split to the base, almost always narrower than the nut. Male flowers solitary.
.... N. pullei

17. Twigs glabrous. Leaves generally hard-coriaceous, smooth above, but the nerves thickish andd obtusely prominent, sometimes convex or with recurved margins. Cupule with 2-3 lamellae, woody, not split to base, as broad as or broader than the nut. Male flowers in triads.
.... N. rubra

Notes:

1. Leaf sizes cannot be taken from young trees or young branches. Cupule structure is that of mature cupules, it is difficult to ascertain and in some species unknown.

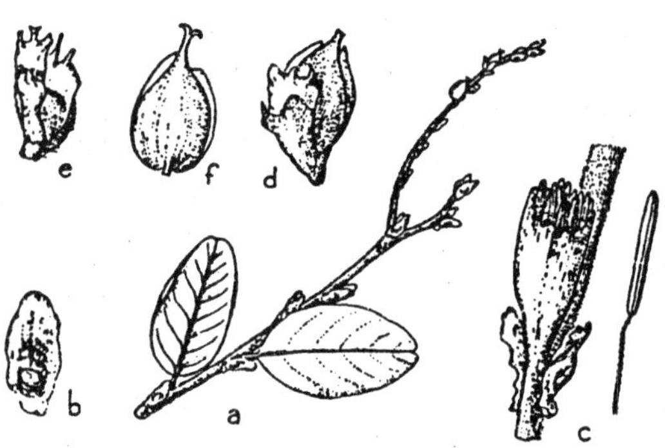

Nothofagus pullei (A) twig with flowering flush; (B) stipule; (C) male flower and stamen; (D) cupule with female flower. (Source: van Steenis (1953)

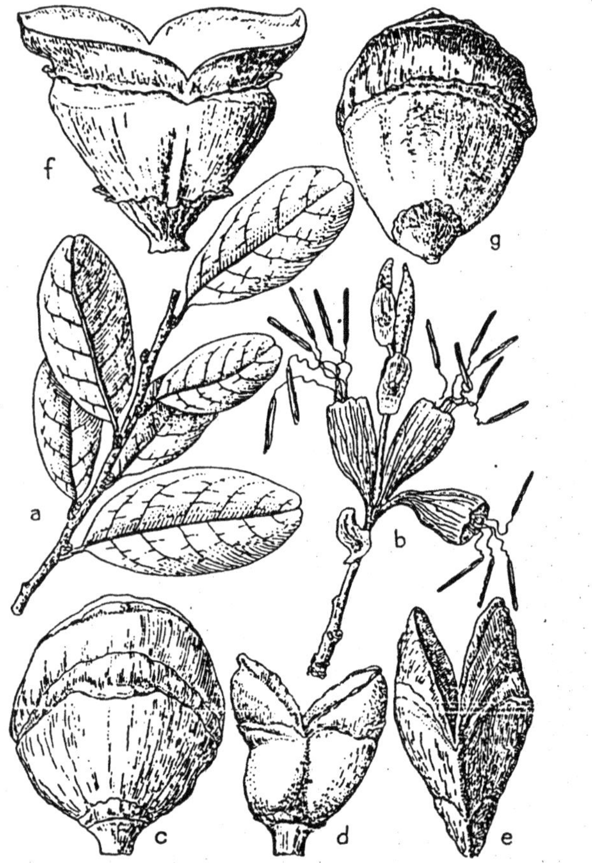

Nothofagus grandis (A) twig; (B) male flowers; (C) cupule.
(Source: van Steenis (1953)

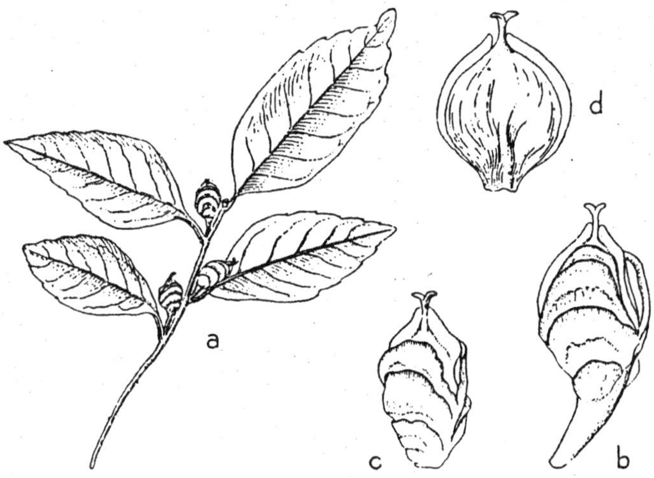

Nothofagus crenata (A) flowering twig; (B) cupule with female flowers.
(Source: van Steenis (1953)

Nothofagus pullei

A frequent species occuring on ridges above the zone dominated by N. grandis. Occurs over an altitudinal range from (1650-)2000-3000 m. The material of this species is not homogeneous. van Steenis (1972) tentatively reduced N. cornuta to this species, he suggested that the lack of a cupule was perhaps a freak condition.

Nothofagus pseudoresinosa

A tall tree to 45 m, from 2300-3100m altitude, very common often dominant on ridges and valleys. Distributed from L. Habbema, Wabag to the Finisterre Ranges. Similar to N. resinosa but differs by presence of a distinct lamellate cupule and the absence of minute hairs on the nut, flush or stipules.

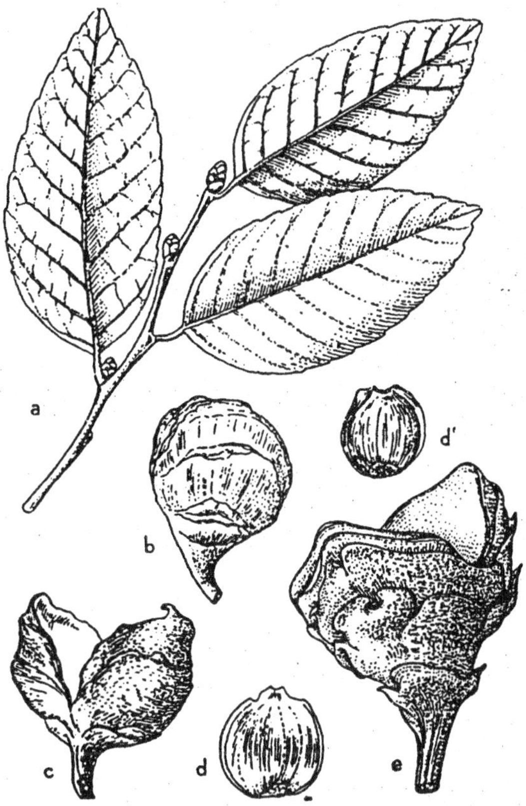

Nothofagus perryi (A) twig; (B) cupule; (C) nut. (Source: van Steenis (1953)

Nothofagus resinosa

A common species often reported as dominant between the altitudes of 2500 and 2800m. Young parts are studded with sulphur-yellowish waxy resin: a feature also of N. pseudoresinosa. It differs from that species in the complete absence of a cupule (found in all 5 fertile female specimens), the more acute leaf apex and the twig ends, petioles and stipules being finely white puberulous.

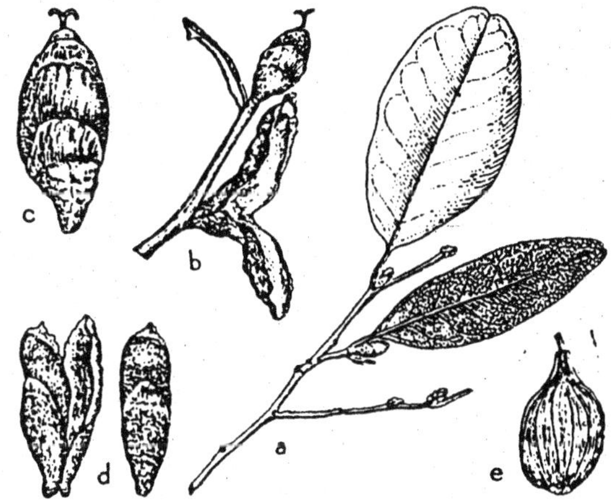

Nothofagus pseudoresinosa (A) twig; (B) twig with cupule; (C) flowering cupule; (D) nut. (Source: van Steenis (1953)

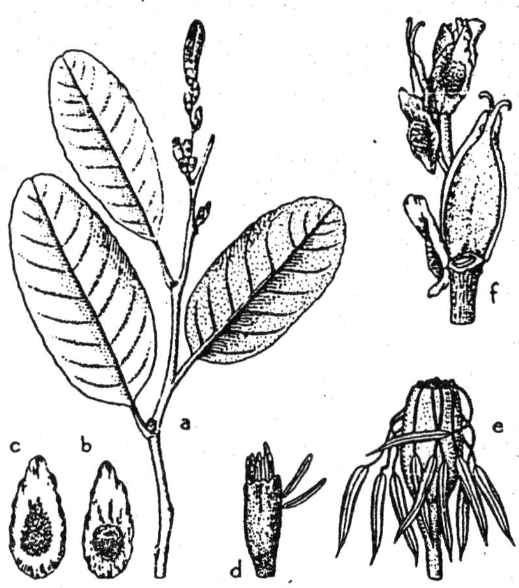

Nothofagus resinosa (A) twig with flowering flush; (B) stipules; (C) mal flower; (D) flush with 2 female flowers (Source: van Steenis (1953)

Nothofagus rubra

Throughout the range of the genus in Papuasia (except for New Britain); from Arfak to the d'Entrecasteaux Is. A common species, often abundant or even dominant. Often forms a shrub, to 4 m tall, in poor situations. van Steenis (1972) combined 5 previously described species into N. rubra due to his conclusions that the cupule stalk is of no taxonomic importance.

Nothofagus starkenborghii

Often a common large tree in mountain forest, 1200-1400 m, with low altitude stands at L Kutubu and on New Britain at 600m. In New Britain growing on limestone. Very common throughout the Highlands and is reported as dominant at Telefomin, Aiyura, Kainantu and Habbema.

Nothofagus womersleyi

Known from only a single collection by Versteegh from Kebar Valley in the Vogelkop Peninsula. Common in primary forest at 1200 m altitude. Resembles N. flaviramea but that species has an obovate nut and a much smaller, sessile, cupule consisting only of a tiny elamellate flap.

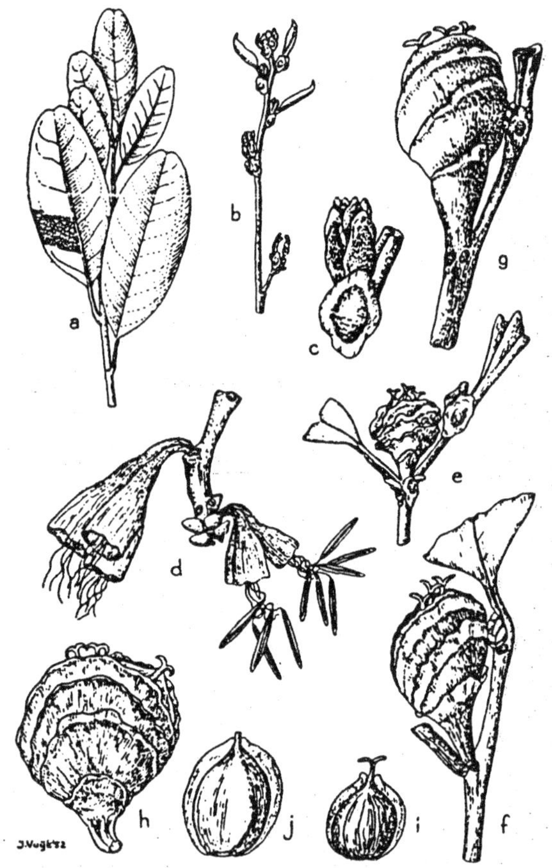

Nothofagus starkenborghii (A) twig; (B) flush; (C) male flower; (D) female flower; (E) cupule.
(Source: van Steenis (1953)

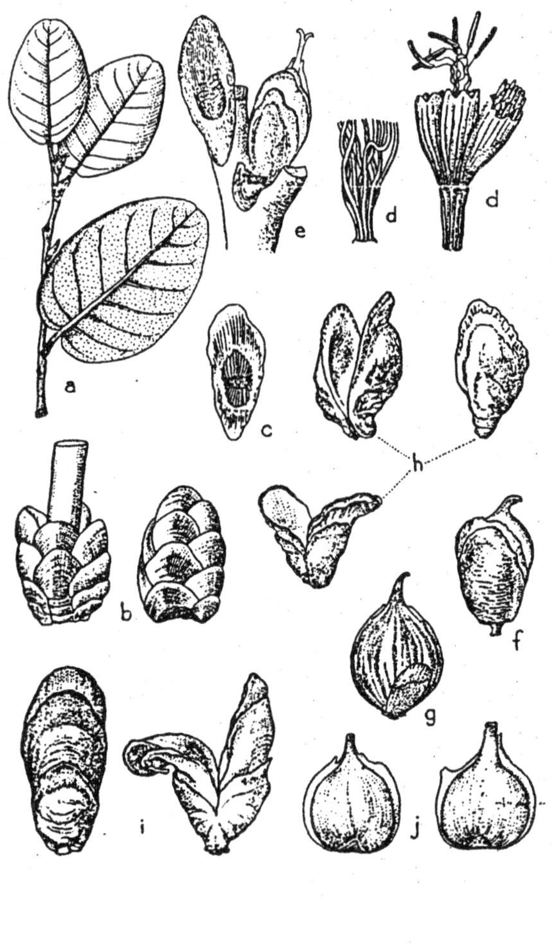

Nothofagus rubra (A) twig; (B) triad of male flowers; (C) cupules.
(Source: van Steenis (1953)

CASUARINALES

The order consists of one family treated traditionally as a single genus, but more recently it has been divided into 4 genera. In Papuasia 3 genera are represented in the flora.

The family includes 50 species, many of which are native to Australia, in particular in *Allocasuarina*, the largest genus. The other genera are spread Through Malesia and Papuasia to New Caledonia and Fiji. The coastal tree *Casuarina equisetifolia* is widespread on tropical coasts. The species *Casuarina oligodon* has long been used as a fallow tree in gardens in the Highlands of New Guinea (up to 900 years). Several other species are also cultivated.

CASUARINACEAE

	WORLD	PAPUASIA
Number of genera :	4	3
Number of species :	0.70	12-13

Distribution : S E Asia and S W Pacific. Throughout Papuasia.

Economic uses : Wood for furniture. Reforestation, N. fixation.

CASUARINACEAE

• • • • • •

The family Casuarinaceae comprises c. 70 species in 4 genera. The largest genus is *Allocasuarina* which is endemic to Australia where it is represented by some 44 species. *Casuarina* s. str. extends from Australia to S E Asia and Polynesia. It includes some 16 species, several undescribed. *Gymnostoma* extends from Sumatra and Luzon, throughout Papuasia to New Caledonia (Johnson and Wilson, 1981). *Ceuthostoma* includes 2 species: one endemic on Palawan, the other distributed from Sabah to Irian Jaya.

The genera can be distinguished by features of the fruits, infrutescences, branchlets, anatomy and chromosomes. Monoecious or predominantly dioecious trees or shrubs adapted to dry habitats or, in areas of rainforest, to secondary situations or soils of low fertility. *Gymnostoma papuana* is largely restricted, in the Solomon Islands , to areas of ultrabasic soils.

Fossil records, mostly from the Tertiary (Eocene to Miocene) of the most primitive genus, *Gymnostoma*, occur in S America and southern Australia and New Zealand (Barlow, 1983). There are less certain records from the Upper Cretaceous (Johnson and Wilson, 1981).

Diagnostic features

Leaves: reduced to scales, scale leaves are arranged in whorls of 4 - 15 or more, on needle-like, photosynthetic branches.

Flowers: simple structure; male flowers in spikes and female flowers in heads.

Fruits: female onflorescence develops into a woody cone composed of scales. The true fruit is a small samara with a terminal wing and a secons seed. Seeds often with more than one embryo.

References

Barlow, B. A. (1959). Chromosome numbers in the Casuarinaceae. *Aust. J. Bot.* 7: 230 - 237.

Barlow, B. A. (1983). Casuarinas- a taxonomic and biogeographic review. in Midglev, S. J., Turnvall, J. W. and Johnston, R. D. *Casuarina Ecology, Management and Utilisation*. C.S.I.R.O. Melbourne.

Johns, R. J. (1983). *Common Forest Trees of Papua New Guinea* 3: 80 - 83.

Johnson, L. A. S. (1972). New species and subspecies of *Casuarina* in Western Australia. *Nuytsia* 1: 261 - 265.

Johnson, L. A. S. (1980). Notes on Casuarinaceae. *Telopea* 2: 83 - 84.

Johnson, L. A. S. (1982). Notes on Casuarinaceae II. *J. Adelaide Bot. Gard.* 6: 73 - 87.

Johnson, L. A. S. (1988). Notes on Casuarinaceae III. *Telopea* 3: 133 - 137.

Johnson, L. A. S. and Wilson, K. L. (1981). Casuarinaceae: then and now. XIII International Botanical Congress Abstracts: 278

Johnson, L. A. S. and Wilson, K. L. (1983). Casuarinaceae in Morley, B. D. and Toelken, H. R. *Flowering Plants in Australia*. pp. 66 - 67.

van Steenis, C. G. G. J. (1931). On the occurence of *C. nodiflora* in Australia. *Journ. Arn. Arb.* 12: 197

van Steenis, C. G. G. J. (1939). New records of *Styrax* and *Casuarina* from the Solomon Islands. *Journ. Arn. Arb.* 20: 220 - 221.

Notes:

1. Barlow (9159) found that chromosome number support a suggested division of Casuarina into major species groups. Traditionally this isoltaed family, which is placed into a separate order, the Casuarinales, was treated as a single genus. The species groups are now treated as distinct genera by Johnson (1980, 1982, 1988). The chromosome numbers are:

x = 8	Gymnostoma
x = 9	Casuarina
x = 10, 11, 12	13, 14 Allocasuarina
x = unknown	Ceuthostoma

2. The largest genus, Allocasuarina described by Johnson (1982) is represented by 44 species confined to Australia.

3. The wood from several Australian species of Allocasuarina is extremely hard and is valued in furniture making. Casuarina equisetifolia is the most widespread cultivated species. It is used for stabilising coastal sand dunes.

4. In In Papua New Guinea Casuarina oligodon was widely cultivated before European contact. In the Simbu Province it played a significant role in the rehabilitation of garden areas after cultivation due its Nitrogen fixing association. It is also a valuable firewood and timber tree for building. In the last 30 years its use has spread widely throughout the Highlands of New Guinea.

5. Most species in New Guinea occupuy 'secondary ' habitats: landslide scars, river banks, or road sides etc., or occur as coastal forest. Several species are used as ornamental plants in tropical and subtropical dry regions throughout the world.

KEY TO GENERA IN PAPUASIA

(Source: Johnson (1988)

1. Furrows of branchlets shallow and open, exposing the stomates; teeth 4 per whorl; infrutescences on long or short branchlets that are similar to vegetative branchlets, and with a broad bract beneath each pair of bracteoles; samaras grey to pale brown; seed short lived. .... 3. Gymnostoma

1. Furrows of branchlets deep and narrow, the stomates not visible; infrutescences with more or less thin inconspicuous bracts....... 2

2. Infrutescences on long branchlets that are similar to vegetative branchlets, with relatively very long bracteoles protruding antrorsely; teeth 4; samaras pale brown; seed probably short-lived. .... 2. Ceuthostoma

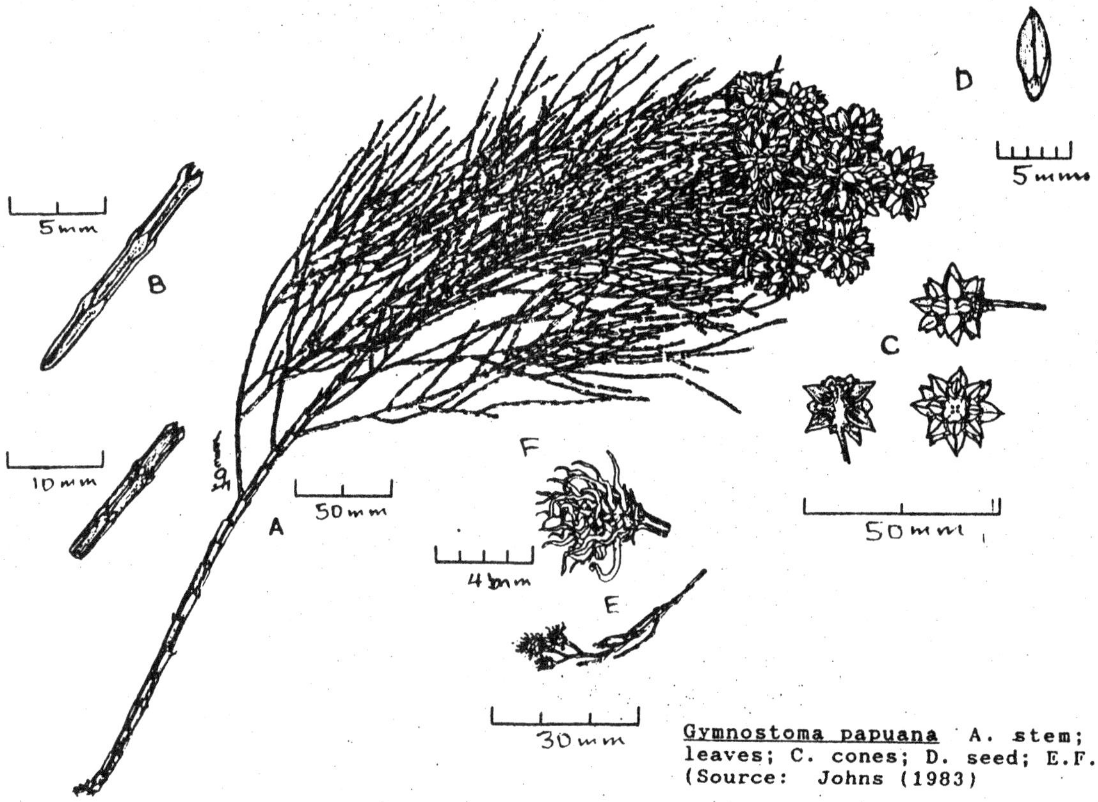

Gymnostoma papuana A. stem; B. scale leaves; C. cones; D. seed; E.F. flowers. (Source: Johns (1983)

2. Infrutescences on short modified woody branchlets ('peduncles') with very short to long bracteoles (thin and without any dorsal protuberances) protruding at c. 90 degrees to the vertical axis of the infrutescence; samaras grey or yellow brown, dull; teeth 5 to many; seed short lived.
.... 1. Casuarina

1. CASUARINA

• • • • • •

A genus of about 16 species from S E Asia to Australia and the Pacific.

Trees. Branchlets with 5-19 teeth per whorl. Cones with relatively thin flattish bracteoles without any dorsal protuberances (distinguishing it from Allocasuarina in Australia). Samaras grey or yellow-brown, rather dull.

List of Species in Papuasia

..... Casuarina angustifolia
K. Sch. & Laut. (1900: 254).
Also by Guppy from Solomon Is.

..... Casuarina equisetifolia
Forst. & Forst. f.
Common coastal species.
var. indica Gaud. Uranie Physicienne Bot. (1826) 56.
Mts of Rawak.

..... Casuarina grandis L. Johnson (1982: 83)
Local endemic at low altitudes from the Tufi-Biniguin area in Northern and Milne Bay Provinces.

..... Casuarina oligodon L. Johnson (1982: 79)
Common throughout the Highlands.
ssp. oligodon
ssp. abbreviata L. Johnson (1982: 83)

..... Casuarina sp. nov 1
(Casuarina mesostrobilium ms)
A species from rainforest at low altitudes. Dispersed over a wide range im Papuasia: Waigeo Is. to Sudest Is. Santa Isabel and San Cristobal.

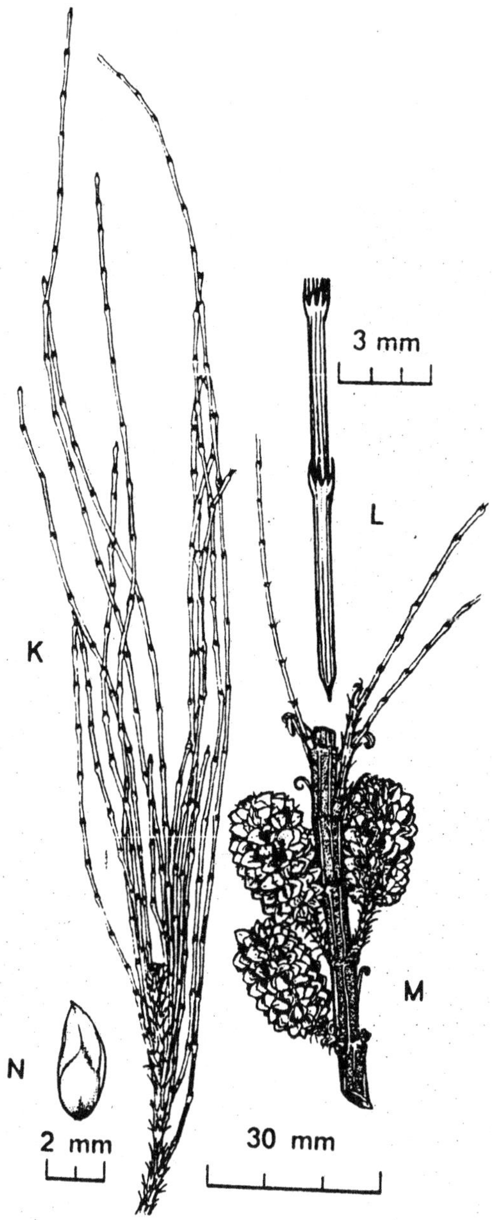

Casuarina equisetifolia K. habit; L. stem with scale leaves; M. stem with cones; N. seed. (Source: Johns (1983)

..... Casuarina sp. nov 2
(Casuarina spicigerum m.s.)
A local species around Vanimo in the West Sepik. Altitudinal range to 100 m. On sandy coral and river beds.

..... Casuarina sp. nov 3
A single collection by A.N. Vinus (UPNG 4996) cultivated around Minj.

..... Casuarina sp. nov 4
LAE 70902

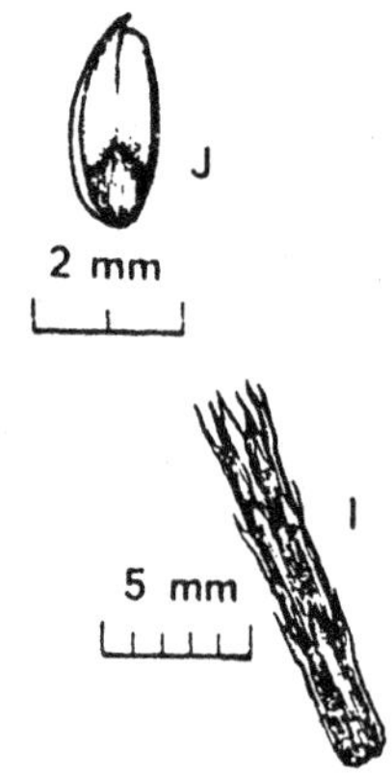

Casuarina oligodon I. stem with scale leaves; J. seeds.
(Source: Johns (1983)

2. CEUTHOSTOMA

● o o o o o

A genus of two species in Malesia. One species endemic in Palawan the other distributed from Borneo to Irian Jaya.

Apparently dioecious trees. All branchlets longitudinally 4-ribed, the furrows between the phyllichnia deep and almost closed, hiding the rows of stomata. Leaves 4 per whorl. Female inflorescences on at least moderately long branchlets that are similar to the vegetative branchlets. Cones usually found amongst the assimilatory branchlets; cone body longer than broad or subequal in dimensions; cone bracts somewhat expanded distally, having an obvious vertical abaxial face below the contracted acute apex; bracteoles not strongly woody in texture, convex dorsally and the dorsal excentric rib obvious, but without protuberances and not striate; smara pale brown, dullish.

Key to Species in Papuasia

1 species
.... Ceuthostoma terminale L. Johnson (1988: 135)

Notes:

1. It is known in isolated localities from Borneo, Halmahera, and, in Irian Jaya from near Djayapure (Hollandia). Johnson records 4 collections: Brass 8820; Versteegh 4704; and van Royen and Sleumer 6308 and 6497.

2. Near Djayapura it has been collected from sealevel to 200 m altitude in eroded places on lateritic soils near forest margins, or sometimes in the forest.

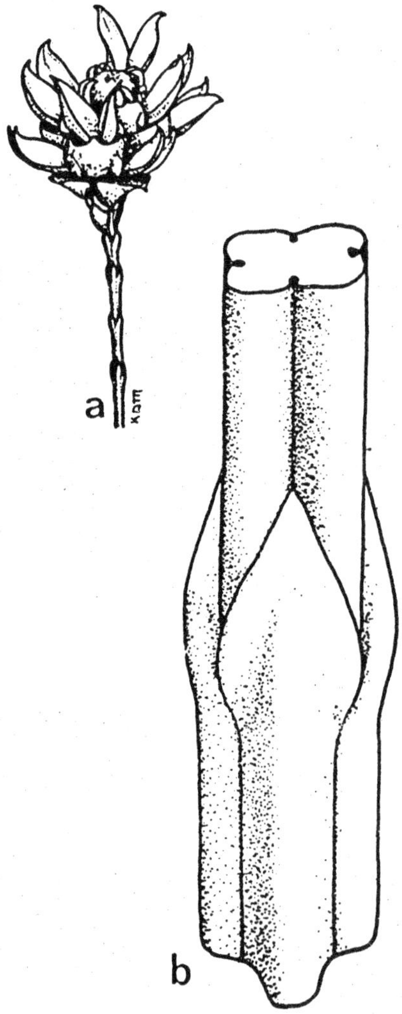

Ceuthostoma terminale A. cones; B. 4-ribed stem showing deep, closed furrows.
(Source: Johnson (1988)

3. GYMNOSTOMA

● ● ● ● ● ●

A genus of some 18 species distributed from Malesia to the Cape York Peninsula, Fiji and New Caledonia. Distributed throughout Papuasia. Johnson (1980) notes that in 1874 "Poisson recognised a 'Divisio Casuarinae Tetragonae seu Gymnostomae' but notes that this genus is not formally based on Poisson's taxon.

All vranchlets 4-ribed, the intercostal furrows shallow and open, exposing the stomata. Leaves in whorls of 4. Male inflorescence on branchlets similar to the vegetative ones; female inflorescences on branches that are short or elongated. Cones borne amongst the assimilatory brancjes; bracteoles protruding, convex on the back but neither split nor furnished with a protuberance. The type species, described from New Caledonia, is Gymnostoma nodiflorum (Thunb.) L. Johnson.

List of Species in Papuasia

..... Gymnostoma nodiflorum (Thunb.) L. Johnson (1982: 79).
Casuarina nodiflora Thunb. Nva Genera Plantarum (1782) 54.

..... Gymnostoma papuana (S. Moore) L. Johnson (1982: 84).
Casuarina papuana S. Moore J. Bot. 61. Suppl. 53 (1923).
Distributed throughout Papuasia, often in secondary situations; usually in the montane zone.

..... Gymnostoma rumphianum (Miq.) L. Johnson (1982:84)
Casuarina rumphianum Miq. Flora 48 (1865) 23, 38.
Collected on New Ireland and New Britain. Common in montane forest and on steep limestone. Also on Arfak Peninsula. At altitudes from 300 to 600 m.

..... Gymnostoma sumatrana (Jungh. ex de Vriese) L. Johnson (1982: 84)
Casuarina sumatranum Jungh. ex de Vriese Tijdschr. Natuur. Gesch. Physiol. 11.
Growing on limestone and along forest margins from 250 - 1200 m altitude. Frodin (NGF 26528) collected this, or a related species, at Omoi in New Britain near sealevel, adjacent to a sago swamp. van Steenis (1939: 221) recorded this species from the Solomon Islands.

..... Gymnostoma sp. nov. 1
(Gymnostoma gracillimum m.s.)
A species confined to the Solomon Islands. Collected on Bougainville amd Guadalcanal. Altitudes from 600 - 690 m.

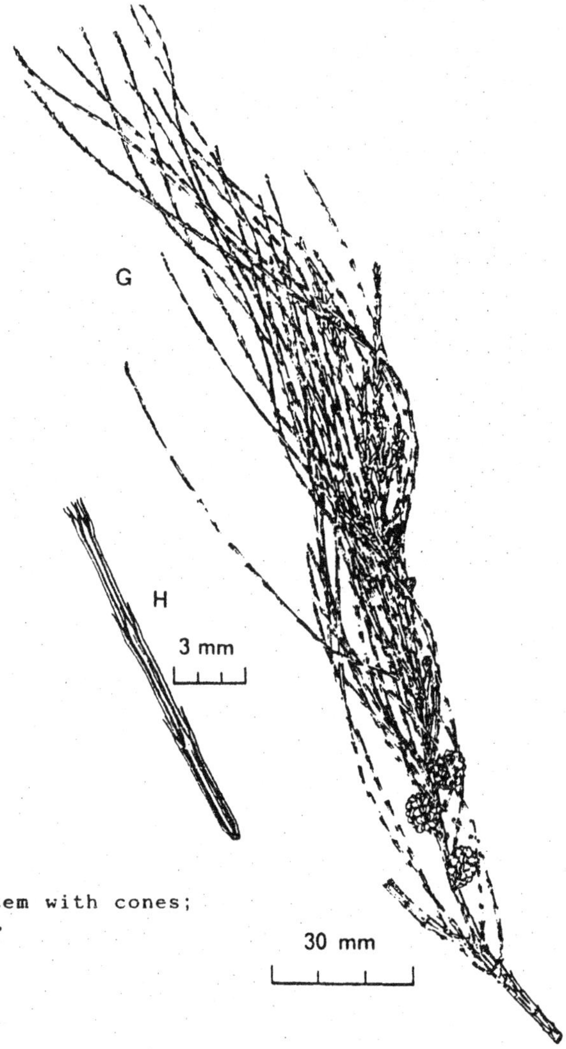

Casuarina oligodon G. stem with cones; H. stem with scale leaves.
(Source: Johns (1983)

APPENDIX 1.

Key to FICUS in New Guinea, based mainly on superficial characters
(Source: E. E. Henty, unpub.)

Notes on the key:

Only species which are represented in the Lae Herbarium are included.

Size measurements of figs refer to dry material.

Stipe/peduncle: many figs are stipitate, with the body at the base constricted and elongated to a stalk. A peduncle may also be present.

The position of the bracts is regarded as joining the division; at the base of the stipe, at the apex of the peduncle.

1. Stipules persistent after opening of leaf-bud.................... 2

1. Stipules falling as leaf-bud opens. 28

2. Climber, not becoming free-stand-standing...................... 3

2. Tree or shrub.................. 5

3. Figs ca. 2 cm across. Leaves leathery, not scabrid, and hairy beneath. ..... f. gymnorhygma

3. Figs to 1.5cm across, leaves chartaceous to leathery, not pitted beneath.......................... 4

4. Leaves scabrid and pubescent; ripe figs pink. ..... F. armiti

4. Leaves smooth; figs green. ..... F. ovacuta

5. Stipules briefly persistent, mostly at twig-tips only, lacking from older leaves.................. 6

5. Stipules long-persistent....... 18

6. Inflorescence cauliflorous or ramiflorous, often also axillary.... 7

6. Inflorescence only axillary; figs small on slender stalks........ 14

7. Figs scabrid, in clusters on small spurs from the trunk; also axillary ..... F wassA

7. Figs not scabrid, on spurs or branching rachises; also axillary. 8

8. Leaves glabrous, large, cordate or auriculate at the base, nerves prominent on the underside; figs ± globose, ca. 2 cm across. ..... F. sciaphila

8. Leaves ± pubescent............. 9

9. Figs on a long, slender, branching, ± pendulous, rachis, often near the base of the trunk, sometimes underground; leaves medium-sized, sometimes slightly scabrid, not very pubescent. ..... F arfakensis

9. Figs on spurs, or ± elongate rasch-ises, leaves usually large, mark edly scabrid, pubescent beneath. 10

10. Figs flattened-globose, with bracts on the body. ..... F. hispida

10. Figs without bracts on the body. 11

11. Rachis long and stout, sometimes ± drooping...................... 12

11. Rachis short.................. 13

12. Leaves elliptic-cordate; figs 2 cm across on peduncles to 1.5 cm; basal bracts conspicuous. ..... F. profusa

12. Leaves obovate, rounded at the base, oblique, figs ca.1 cm across, on peduncles less than 1 cm long; basal bracts very small. ..... F. d'albertisii

13. Leaves large, cordate, figs oval, ca. 2 cm long; many patent bracts at apex; basal bracts ca.5mm long. ..... F. pachyrachis var. porrecta

13. Leaves medium-sized, elliptic-rhombic; figs ± conical, ca 2 cm long, apical bracts small, not patent; basal bracts 2-3 mm long. ..... F. hispidioides

14. Twigs, petioles and undersides of leaves appressed-hairy......... 15

14. Twigs, petioles and leaves glabrous or nearly so................. 16

15. Fig 12 mm wide; internal bristles rather sparse; leaf-apex attenuate. Small tree. ..... F. servula

15. Fig 25-30 mm wide; internal bristles copious, leaf-apex accuminate. Rheophytic shrub.
..... *F. lapidaria*

16. Figs (often much) less than 1 cm across, on short slender stalks; leaves long-acuminate.
..... *F. tenella*

16. Figs 1-2 cm across, on rather stout stalks........................ 17

17. Figs with bracts on the body; leaves small, elliptic, acute to acuminate, base angular.
..... *F. arbuscula*

17. Figs without bracts on the body; leaves to 30 cm long, narrow-lanceolate, with small auricles at the base. *F. theophrastoides* var. *augustifolia*

18. Inflorescence cauliflorous..... 19

18. Inflorescence axillary, sometimes also cauliflorous (F. concephalifolia)........... 21

19. Figs very small, on long pendulous rachises.
..... *F. pungens*

19. Figs 2 cm or more across; rachis ± stiff......................... 20

20. Figs with large bracts on the body; figs gathered in dense clusters on stout branches from the trunk and exposed roots.
..... *F. praestans*

20. Figs with bracts at base and apex only, or a few small bracts on the body; figs dispersed on stout branching rachises near the base of the trunk.
..... *F. sublimbata*

21. Figs with bracts on the body... 22

22. Bracts clustered near the apex, long and narrow, twigs and stipules bearing long brown hairs.
..... *F. lancibracteata*

22. Bracts scattered, or near the base. 23

23. Bracts few, scattered; mature figs red, globose, 2-3 cm across, covered with stiff bristles; often cauliflorous clusters also, at the base of trunk.
..... *F. conocephalifolia*

23. Bracts numerous, covering figs. 24

24. Bracts broad and long; leaves large, elliptic, acute, smooth or scabridulous, usually glabrous, margin entire.
..... *F. longibracteata*

24. Bracts shorter and narrower, very numerous; leaves medium - sized, elliptic, caudate, scabrid, often densely hairy, margin serrate.
..... *F. complexa*

25. Leaves elliptic, long petiolate (to 15 cm) rounded at the base, stipules 6-8 cm long, attenuate, ± covering figs.
..... *F. cynaroides*

25. Leaves spathulate, truncate or auriculate at the base; petioles to 6 cm; stipules not exceeding 4cm. 26

26. Leaves sessile or nearly so, with stem-clasping auricles at the base, these often large, covering figs.
..... *F. theophrastoides*

26. Leaves distinctly petiolate; auricles, if present, not clasping stem or covering figs........ 27

27. Stipules triangular, acute, leaves tomentose beneath.
..... *F. megalophylla*

27. Stipules elliptic, obtuse; leaves glabrous.
..... *F. saccata*

28. Leaf-bud long, 10 cm or more; leaves large, plicate; figs in upper axils, large, covered by surface bracts.
..... *F. dammaropsis*

28. Leaf-bud long or short; leaves flat......................... 29

29. Figs cauliflorous............. 30

29. Figs axillary, or ramifiorous on twigs just below leaves........ 58

30. Bracts from base and body of figs, covering them when young, later deciduous; figs 2 cm across when mature ± sessile on an elongate rachis.
..... *F. tanypoda*

30. Bracts not covering figs at any stage......................... 31

31. Leaves glabrous or slightly pubescent; veins promminent on underside, leaf-margins recurved; figs scabrid or not, on slender stalks, solitrary or on small spurs on the trunk (also ramiflorous and axillary)......................... 32

31. Above characters not combined.. 34

32. Figs mostly to 1 cm diameter, red when ripe, usually scabrid; leaves scabrid or not, small to medium-sized, lanceolate with ± undulate or toothed margin, base usually angular; twigs less than 5mm thick; shrub or small tree.
..... *F. wassa*

32. Figs mostly 2 cm diameter or more, not scabrid; leaves medium to large, usually cordate or ± auriculate at the base.............. 33

33.
..... F. copiosa

33.
..... F. sciaphila

34. Leaves large, 50cm long or more, on a short (to 4 cm) stout petiole; twigs thick, pithy or hollow, sometimes inhabited by ants; small to medium tree, not much branched; figs in large clusters on the trunk
..... F. salomonensis

34. Not as above.................. 35

35. Figs covered with stiif bristles.
.... 36

35. Figs smooth, scabrid or soft-hairy.
.... 38

36. Twigs and leaves glabrous or nearly so; leaf-bases Very oblique, figs not more than 1cm across, on penduncles ca 5cm long; often scabrid.
..... F. erinobotrya

36. Twigs and leaves (at least on the underside densely pubescent.... 37

37. Figs scabrid and grey-bristly; about 1cm across, on slender stalks ca 1.5cm long (about 1/3 stipe, 2/3 pendunele; bracts ca 1cm above base of peduncle), Leaves and twigs greyish-pubescent; leaf-bases very oblique.
..... F. papuana

37. Figs covered with stiff red-brown irritant bristles; ca 2cm across, on stout peduncles from stout sparingly-branched spurs; leaves and twigs red-brown hairy. (The figs are also axillary).
..... F. pophyrochaete

38. Figs scabrid, usually 1cm or less in diameter.................... 39

38. Figs not scabrid.............. 40

39. Figs and petioles golden-hairy; leaf-base oblique; figs not pubescent, on slender penducles to 4 cm long (most often axillary).
..... F. chrysochaete

39. Twigs and petioles grey-hairy; leaf-base symmetrical, auriculate, figs pubescent, on peduncles not longer than 1cm (also axillary).
..... F imbricata

40. Figs small, to 1 cm across, on slender stalks from short, sometimes branching, spurs......... 41

40. Figs larger, or on stouter stalks, from spurs or elongate branching rachises..................... 44

41. Figs with a collar of basal bracts; peduncle less than 1cm long; spurs slender, branching, to 4 cm long; leaves narrow-elliptic to lanceolate.. F. pilulifera

41. Figs stipitate, the bracts at the base of the stipe; stalk (often much) longer than 1 cm......... 42

42. Leaves ovate, ± regularly cordate at the base, tip acuminate to caudate, margins serrate; sap serous; figs in small clusters.
..... F. griseifolia

42. Leaves rhombic-elliptic; sap milky
.... 43

43. Leaf-base ± cordate, often very oblique, margin serrate, surface scabrid and pubescent; figs usually in large masses low on the trunk.
..... F. bernaysii

43. Leaf-base angular or rounded, margin with wide teeth, or entire, surface glabrous, not or slightly scabrid; figs in small clusters or spurs.
..... F. arawaensis

44. Leaves regularly cordate, the widest part nearest the base...... 45

44. Leaves with rounded, angular, or even sub-cordate or cuneate-cordate base, but not regularly cordate
.... 47

45. Leaves large, scabrid, deeply cleft at the base; twigs thick, hollow; figs pubescent, flattened ovoid, to 6 cm across, on stout branching rachises.
...... F. robusta

45. Leaves not scabrid............ 46

46. Fig-walls thick, with inward processes bearing the flowers; basal bracts 2 mm long or more; leaf-apex blunt to acute; twigs stout; figs usually on a stout branching rachis.
..... F. nodosa

46. Fig walls not markedly thickened, without inward processes; basal bracts 1-2 mm long; leaf-apex often acuminate; twigs ± slender.
..... F. variegate

47. Figs on long, branching, ± pendulous rachises................. 48

47. Figs on short spurs, or the rachis, if elongated, stout and stiff.. 53

48. Leaves and twigs usually very pubescent..................... 49

48. Leaves and twigs not or sparsely pubescent (sometimes dense on veins on the underside).............. 50

49. Pubescence grey to yellowish-brown; figs flattened-globose.
..... F. baccaureoides

49. Pubescence dark reddish-brown (often paler on older leaves); figs globose-conical.
..... F. iodotrich

50. Figs globose, to ca 1 cm diameter, on stalks to 2 cm long, stipitate; stipe ca 5 mm long.
..... F. macrothyrsa

50. Figs flattened-globose, to ca 2 cm diameter, not stipitate; peduncle ca 1 cm long.................. 51

51. Leaves drying green; leaf profusely dotted with cystoliths on the upper side.
..... F. botryocarpa
var. subalbidoramea

51. Leaves drying brown; few or no cystoliths on the upper leaf-surface...................... 52

52. Fig orifice ± sunken, many apical bracts.
..... F. congesta

52. Fig orifice raised, 5-humped, 5 apical bracts.
..... F. subcongesta

53. Leaves not pubescent or scabrid, elliptic, chartaceous, angular at the base; figs ± conical, to 4 cm across, on rather long peduncles from short spurs.
..... F. itoana

53. Leaves pubescent on the under surface, usually markedly scabrid.
.... 53

54. Peduncles longer than the diameter of the fig-body; figs in usually small clusters on short spurs.. 55

54. Peduncles shorter than the diameter of the fig-body; rachis ± elongate, branching..................... 57

55. Figs ± conical; leaves elliptic-lanceolate with dentate margin.
..... F. primaria

55. Figs flattened-globose; leaf-margins serrate.................... 56

56. Leaves rhombic-ovate, base cordate fig-apex strongly raised; basa bracts patent, 5 mm long.
..... F. neobrittanica

56. Leaves rhombic - lanceolate, bas rounded, very oblique; fig-apex nc prominent, basal bracts to 3 m long. F. adelpha

57. Figs mostly flattened-globose, wit markedly hairy surface; basa bracts 5 mm long.
..... F. calopilina

57. Figs mostly conical, surface glat rous or nearly so, basal bracts 2m long. F. hispidioides

58. Climbers, never becoming fre standing......................

58. Trees or shrubs, or stangler finally becoming free standing.

59. Leaves scabrid on both surface figs ca 1 cm across, rose-pink.
..... F. armiti

59. Leaves not scabrid, or scabr beneath only..................

60. Figs sessile, hairy...........

60. Figs stalked, hairy or not.....

61. Hairs on figs ± soft, not irritar
..... F. amblisyce

61. Hairs on figs stiff, irritant..

62. Leaf-surface finely bullate.
..... F. insculpta

62. Leaf-surface coarsley bullate, flat. F. irritans

63. Figs with irritant hairs.......

63. Figs without irritant hairs....

64. Figs 3-4 cm across, red; leav chartaceous.
..... F. odoardi

64. Figs ca 1 cm across; leaves cori ceous........................

65. Lamina ± flat, or somewl bullate.
..... F. irritans

65. Lamina markedly convex.........

66. Leaf - underside with slight prominent primary veins, otherw not much patterned.
..... F. convexa

66. Leaf-underside with all veins pr inent, the depressions between

smallest reticulations filled with smooth, ± elliptic prominences.
..... F. ampulliformis

67. Figs ferruginous-tomentose beneath.
..... F. cinnamomea

67. Figs with grey, or grey-brown, hairs or glabrous............. 68

68. Leaves on fertile twigs small, not longer than 5 cm............... 69

68. Leaves on fertile twigs (usually much) longer than 5 cm......... 72

69. Figs not more than 5 mm across. 70

69. Figs ca 1 cm across, or more, sparsely hirsute, leaves ovate, with a blunt somewhat acute apex>
..... F. rhytidophloea

70. Basal nerves of leaf elongate, 1/3-1/2 as long as the lamina; leaf broad-elliptic to orbicular; fig ca 4 mm across.
..... F. calodictya

70. Basal nerves of leaf short..... 71

71. Leaves ± ovate, apex obtuse to emarginate; fig 3-4 mm across.
..... F. agapetoides

71. Leaves ± elliptic, attenuate near the apex but the tip blunt; figs 5 mm across.
..... F. phatnophylla

72. Figs stipulate, the stipe as long as the fig-body, or longer, often the peduncle lacking altogether, the stalk then consisting of stipe only, with bracts at its base.. 73

72. Figs not, or shortly stipitate, bracts at the top of the stalk, immediately under the fig body or a few millimeteres below......... 78

73. Leaves drying green; leaf buds and young petioles glabrous........ 74

73. Leaves drying brown; leaf-buds and young petioles ± hairy..... 75

74. Stalk of fig slender; figs rarely more than 1 cm across, never scabrid, orifice somewhat prominent with bracts not protruding; leaves chartaceous.
..... F. subulata

74. Stalk of fig ± stout; figs to 2 cm diameter, usually scabrid; orifice usually flush with the fig surface, bracts somewhat protruding at maturity, leaves coriaceous.
..... F. tinctoria

75. Leaves hairy on the veins beneath, buds and young petioles very hairy; figs 2-3cm across, red at maturity, with prominent apex, bracts not protruding, surface red, tomentose, rarely glabrescent.
..... F. baeuerlenii

75. Leaves glabrous or nearly so, buds and young petioles usually with scattered hairs only; figs to 1cm. across......................... 76

76. Figs brown-scurfy or short-tomentose; orifice prominent, the bracts hidden; stalk usually with bracts about halfway.
..... F. pantoniana

76. Figs glabrous or nearly so; stalk with bracts at base only....... 77

77. Fig-orifice prominent, with protruding bracts; leaves lanceolate, often long acuminate.
..... F. oxymitroides

77. Fig- orifice sunken; leaves cordate, acute.
..... F. camptandra

78. Leaves ovate or elliptic, the primary veins leaving the midrib at an angle of about 45 degrees... 79

78. Leaves lanceolate to oblong, the primary veins leaving the midrib at a wide angle.................. 82

79. Fig-orifice closed by a tuft of bristles; fig-surface ± tomentose; leaves ± cordate; figs ca 1 cm across.
..... F. hypophaea

79. Fig-orifice without bristles; fig-surface usually glabrous....... 80

80. Leaves elliptic-oblique, obtuse or emarginate; fig-orifice slighly prominent, with several rows of bracts; fig to 1 cm across, with patent bracts to 3 mm long at the base of a short stipe.
..... F. diotrichoidea

80. Leaves ovate or cordate........ 81

81. Figs to 2 cm across, orifice strongly prominent; patent bracts to 4 mm long at the base of a short stipe; leaves usually markedly cordate, acute.
..... F. nasuta

81. Fig ca 1 cm across, orifice slightly prominent with 5 small bracts; bracts to 2mm long against the base of the fig - body; leaves ovate, acute.
..... F. pleiadenia

82. Secondary venation not strongly prominent on the underside; lamina beneath dotted with cystoliths without pits; figs to 4 cm across, on a short stout stalk.
..... <u>F. homodroma</u>

82. Secondary venation not strongly prominent on the underside; lamina beneath with shallow pits between the tertiary veins, cystoliths not evident...................... 83

83. Figs to 3 cm across, on a short stout stalk (to 4 mm thick); lamina hairy on the underside, between the pits. <u>F. gymnorhygma</u>

83. Figs ca 1 cm across, on a longer, more slender stalk (4-10 x 1-2 mm); lamina, on the underside not hairy between the pits.
..... <u>F. scatchyleyana</u> var. <u>aurantiola</u>

84. Trees or shrubs, not beginning as stranglers.................. 106

84. Banyans, beginning as strangle developing lattice-stilt, or prop-roots...................... 85

85. Primary veins in 10-16 pairs, prominent, clearly differentiated from the intermediate venation, which is usually reticulate.... 86

85. Primary veins numerous, not prominent and not much differentiated from the, mostly parallel, intermediate veins between them ("Garcinia venation").......... 93

86. Figs on stalks 1 cm long or more.
.... 87

86. Figs sessile or on stalks to 5 mm long....................... 88

87. Figs ovoid, ca 1.5 cm across, with a rather hard rind when ripe.
..... <u>F. novoguineensis</u>

87. Figs globose with prominent apex, ca 0.5 cm across soft when ripe; often clustered on bare twigs just below leaves.
..... <u>F glaberrima</u>

88. Figs globose to ellipsoid, soft when mature; (usually 1.5 cm - 2 cm); twigs thick............. 89

88. Figs globose (usually much) less than 1 cm long; twigs thick or slender...................... 90

89. Figs globose to ellipsoid, soft when mature; basal bracts small, to 3 mm long.
..... <u>F. drupacea</u>

89. Figs globose, with a firm rind when mature; basal buds about half as long as the fig-body, forming a collar.
..... <u>F. crassiramea</u> var. <u>patellifera</u>

90. Figs seldom more than 5 mm across, clustered on bare twigs just below leaves; basal bracts persistent; twig thick; often a few unshed stipules about the bud.
..... <u>F. caulocarpa</u>

90. Figs 5-10mm across, mostly in leaf-axils; twigs ± slender..... 91

91. Bracts ca 1 mm long, persistent; figs sessile or shortly stipitate, light-coloured internally; terminal bud oblique to the twig axis.
..... <u>F. virgata</u>

91. Bracts 2-3mm long, caducous; figs sessile or nearly so, dark-coloured internally; terminal bud in line with the twig-axis............ 92

92. Fig with abundant internal chaffy scales; leaf-base usually ± angular
..... <u>F. virens</u>

92. Fig with a few internal bristles, or none; leaf-base obtusely rounded to sub-cordate.
..... <u>F. prasinocarpa</u>

93. Large sterile tree with many stilt-roots; stipules long, pink; not occuring wild.
..... <u>F. elastica</u>

94. Mature figs globose, not much more than 1cm across............... 95

94. Mature figs ± oval, 2 cm long or more......................... 99

95. Figs sessile, 1 cm across or more; basal bracts 3, conspicuous, free, persistent, ca 5 mm long.
..... <u>F. archboliana</u>

95. Figs less than 1 cm across; bracts caducous or very small......... 96

96. Leaves coriaceous, narrow-elliptic, obtuse to acute; venation faint, particularly on the underside; terminal bud 2-5cm long; peduncles 2-4mm long; basal bracts early-caducous.
..... <u>F. rhizophoriphylla</u>

96. Leaves characeous to coriaceous; venation clear on the underside.
.... 97

97. Basal bracts 2, acute, caducous, to 4 mm long; leaves coriaceous; twigs rather thick.
..... F. obliqua

97. Basal bracts 3, obtuse, usually shorter; leaves chartaceous to coriaceous; twigs ± slender... 98

98. Leaves drying green; basal veins usually rather short; figs sessile, mostly paired in the axils.
..... F. benjamina

98. Leaves drying brown; basal veins often long, to 1/3 as long as the lamina; figs sessile or stalked, sometimes single.
.....F. microcarpa

99. Figs with a rounded protuberance at the apex (umbonate).......... 100

99. Figs smoothly rounded at the apex, without a protuberance; sessile, ellipsoid to cylindrical, to 2 cm long........................ 105

100. Figs ellipsoid and cylindric, 1-2 cm across, peduncle relatively slender, to 3 cm thick, not dilated at the apex; fig-apex often ± rostrate; leaf-bud usually to 10cm long
..... F. xylosycia

100. Figs larger, on a stouter peduncle; leaf-bud often long and thick. 101

101. Peduncle 1 to several cm long, 3-5 mm thick.................... 102

101. Peduncle short, stout; figs ellipsoid to cylindric, hard-skimmed.
.... 102

102. Figs globose to ellipsoid or obovoid; skin rather thin, showing the fruits beneath when dried; peduncle glabrous or with a few hairs.
..... F. hesperidiiformis

102. Figs oval-cylindric; skin thick, firm wrinkled when dried but not showing structure beneath; peduncle often densely hairy.
..... F. sterrocarpa

103. Figs ellipsoid, pubescent.
..... F. heteromeka

103. Figs ± cylindrical, usually with 2 longitudinal ridges; not or slightly pubescent................. 104

104. Basal veins of leaf straight, conspicuous, to about 1/3 the length of the lamina.
..... F. augusta

104. Basal veins of leaf short, curved inconspicuous.
..... F. mafuluensis

105. Basal bracts fused to an adnate cupule about 1/3 the length of the fig body.
..... F. glandifera

105. Bracts small, hidden by the fig body. F. subcordata

106. Leaves or figs, or both scabrid.
.... 107

106. Neither leaves nor figs scabrid.
.... 128

107. Figs less than 1/2 cm across, covered with dark purplish-brown bristles; shrub or small tree.
..... F. phaeosyce

107. Figs with pale bristles or hairs, or glabrous.................. 108

108. Leaves, spiral or opposite.... 109

108. Leaves distichous............. 114

109. Figs covered with small spines, stipitate, no peduncle.
..... F. hystricocarpa

109. Figs at most roughly scabrid.. 110

110. Figs to 2 cm across, usually on a long stipe with no peduncle; apical bracts protruding.
..... F. copiosa

110. Figs mostly less than 2 cm across; apical bracts not prominent... 111

111. Leaves ± hispid beneath, sometimes with aculeate hairs........... 112

111. Leaves glabrous.............. 113

112. Leaves always alternate; figs rather numerous on twigs just below the leaves; usually a few bracts on fig body about the orifice.
..... F. gul

112. Leaves often opposite, often with aculeate hairs on the midribs beneath; figs mostly in the upper axil; sometimes bracts about the middle of the fig body.
..... F. opposita

113. Figs with (usually) a collar of basal bracts; stipules early-caducous; figs in upper axils.
.....F. cumingii

113. Figs without a collar of basal bracts; stipules often slow to fall; figs becoming ramiflorous.
..... F. wassa

114. Leaf-base ± deeply cordate, oblique, sometimes with overlapping auricles, or with a separate auricular lobe.................. 115

114. Leaf-base cuneate to subcordate, often oblique but not auriculate. 119

115. Leaves 30-50 cm long, oblong with cordate-auriculate base, the auricles often overlapping. ..... F. oleracea

115. Leaves smaller, leaf-base usually very oblique and auriculate on one side only.................... 116

116. Leaf-base usually rounded, with a single acute auricle 2-4 cm long, from one side, apex candate. ..... F. aurita

116. Leaf-base oblique-cordate, often with a rounded auricle extending across the petiole............ 117

117. Leaves ± reniform; figs to 1 cm across, on peduncles 5-10 mm long; fig-surface echinate (the figs may also be cauliflorous). ..... F. erinobotrya

117. Leaves oblong, caudate........ 118

118. Figs on slender stalks to 3 cm long, with bracts about halfway; stipules sometimes briefly persistent.. F. imbricata

118. Figs on stalks to 11/2 cm long; bracts near the top of the stalk; stipules caducous. ..... F. chrysochaete

119. Leaf only slightly scabrid; fig surface slightly scabrid but not hairy; fig-wall rather thick, no internal bristles. ..... F. tinctoria

119. Leaf often very rough; fig-wall thin, or if thick, then surface rough and hairy, often internal bristles..................... 120

120. Leaves mostly lanceolate, long-acuminate to caudate, internal bristles abundant, about as long as the seeds.................... 121

120. Leaves ± oblong, obtuse to acuminate; internal bristles not abundant, not more than half as long as the seeds, or absent.......... 122

121. Fig pedunculate, with a collar of basal bracts; leaf with a single gland at the base of the mid-rib on the underside; internal hairs white ..... F. schumanniana

121. Fig sessile or shortly stipitate; leaf with two glands, one on either side of the base of the midrib on the underside; internal hairs yellowish. ..... F. macrorrhyncha

122. Figs 1-1.5 cm across, wall ca 1 mm thick; stalk relatively stout, 1 mm thick, or more; internal parts dark (red when fresh)............. 123

122. Fig not more than 1 cm across, wall less than 1/2 mm thick; fig-stalk slender..................... 125

123. Conspicuous bracts, to 3 mm long, on stalk of fig just below body; figs and undersides of leaves often pubescent, short internal bristles. ..... F. melinocarpa

123. Bracts small, lower on stalk.. 124

124. Fig smooth or nearly so; bracts at base of stalk; no internal hairs; leaves glabrous, roughly scabrid. ..... F. umbonata

124. Fig roughly scabrid - echinate; bracts below fig, also often on body; short internal hairs; leaves pubescent beneath, roughly scabrid. ..... F. quercetorum

125. Leaves mostly ±cordate at the base; usually a collar of small bracts at the base of the fig-body; fig-stalk usually rather short, less than 1 cm long; sap serous, yellowish, internal parts of fig dark, with white hairs. ..... F. trachypison

125. Leaves mostly rounded to angular at the base; bracts usually on the stalk somewhat below the fig-body. 126

126. Figs to 5 mm across, the apex somewhat prominent, stalk short, to 5 mm; sap thin, whitish, internal parts of fig pale. ..... F. ampelas

126. Figs usually larger, on longer stalks, the apex markedly prominent, sap milky or clear........... 127

127. Stalk very slender, 1-3 cm long; internal parts of fig pale (figs also ramiflorous and cauliflorous. ..... F. pseudowassa

127. Stalk ca 1 cm long; internal parts of fig dark. ..... F. tonsa

128. Leaves long, to 50 cm, elliptic-lanceolate, narrowed to an auriculate base; small tree, with mostly apical leaves (stipules often persistent). ..... F. theophrastoides

128. Not as above................. 129

129. Leaves distinctly hairy beneath, at least on the primary veins, or, if the leaves almost glabrous, the

figs hairy.................... 130

129\. Leaves glabrous, or a few hairs on the underside on the midrib; figs glabrous...................... 146

130\. Leaves regularly cordate...... 131

130\. Leaf-base rounded or angular or cuneate and ± auriculate, rarely subcordate................... 132

131\. Figs to 3 cm across, with protruding apex; surface grey-tomentose; peduncle 2-3 cm long; leaves leathery, drying dark grey-brown; underside of leaf matted with short grey crimped hairs, the minor veins prominent.
..... *F. semivestita*

131\. Figs ca 1 cm across, apex inconspicuous; surface white-pilose; peduncle ± slender, ca 1 cm long; leaves chartaceous, drying green, underside of leaf ± white-pilose, the minor veins not prominent.
..... *F. albipila*

132\. Figs 2 cm or more across, on stout peduncles at least 2 mm thick, or ± sessile...................... 133

132\. Figs not more than 1.5cm across, on ± slender peduncles, or sessile
.... 138

133\. Leaves elliptic to obovate, acute from a broadly rounded apex, base rounded, oblique, drying green; figs symmetrically flattened-globose, on very short peduncles; fig-orifice inconspicuous; leaf-bud to 3 cm long.
..... *F. bougainvillei*

133\. Leaves elliptic, ovate or rhombic, symmetrical, drying brown or grey-brown. Figs ± globose or conical; leaf-bud to 2 cm long......... 134

134\. Primary veins leaving the midrib at a wide angle; leaves elliptic, base rounded or subcordate; pubescence on fig mainly at base and apex; fig-orifice prominent, with projecting bracts.
..... *F. edelfeltii*

134\. Primary veins leaving midrib at ± 45 degrees................... 135

135\. Figs without lateral bracts on the body; surface pubescent with straight hairs; leaves rhombic or elliptic, base narrowed but obtuse.
..... *F. subcuneata*

135\. Figs with lateral bracts on the body......................... 136

136\. Leaves elliptic to rhombic, narrowed to an acute base; figs pubescent with ± straight hairs, with no felty undercoat.
..... *F. lapidaria*

136\. Leaves rather broad, ovate or elliptic, the base rounded or ± obtuse or auriculate........... 137

137\. Figs sessile or nearly so, with conspicuous basal bracts to 5 m long; pubescence of figs mostly of straight hairs, with sometimes a felty undercoat; leaf-base cuneate-obtuse or cuneate-auriculate; twig slender.
..... *F. ochrochlora*

137\. Figs on stout peduncles to 2 cm long; bracts variously placed on the peduncle; indumentum of fig dense, felty, light-brown; leaf-base angular to round; twig thick.
..... *F. funiculosa*

138\. Leaves distchous, lanceolate, acuminate, pedulous, brown-hairy beneath; figs small, long-pedunculate.
..... *F. endochaete*

138\. Leaves spiral, mostly oval or oblong....................... 139

139\. Primary veins in about 15 pairs, leaving the midrib at a wide angle, prominent beneath D depressed on upper surface; figs densely brown-hairy....................... 140

139\. Primary veins in not more than 10 pairs; usually leaving the midrib at an angle of ca 45°......... 141

140\. Leaf underside brown-hairy; leaf-base often auriculate; leaf bud to 1cm long.
..... *F. pseudojaca*

140\. Leaves glabrous or nearly so; leaf base rounded; leaf-bud 2-5cm long.
..... *F. hadroneura*

141\. Cystoliths hypogenous......... 143

141\. Cystoliths amphigenous........ 144

142\. Basal pair of veins strong, reaching halfway along the lamina; peduncle rather stout, 0.5-2cm long, with bracts at the apex; fig with many internal bristles.
..... *F. glandulifera* var. *villosa*

142\. Basal pair of veins weak and short.
.... 143

143\. Figs with prominent bracts on the body, thick-walled, without internal bristles; pubescence on fig

persistent; leaves often densely hairy beneath.
..... *F. trichocerasa*

143. Figs without bracts on the body, thin-walled, with internal bristles; surface glabrescent; leaves usually sparsely pubescent.
..... *F. erythrosperma*

144. Leaves almost glabrous (cystoliths obscure on upper surface); figs ca 1cm across ± appressed-hairy, long pedunculate, with prominent apex.
..... *F. pedunculosa*

144. Leaves pubescent beneath...... 145

145. Pubescence on leaves usually dense; fig with dense straight hairs on the surface, and often prominent bracts; apex somewhat raised, with a few large bracts showing.
.....*F. mollior*

145. Pubescence on the major veins only; fig ribbed towards the apex, covered with very short pubescence, without noticeable bracts.
..... *F. indigofera*

146. Figs sessile, or nearly so, the stalk, if any, hidden by the basal bracts...................... 147

146. Figs distinctly stalked....... 148

147. Basal bracts conspicuous, 4-10mm long; riverside tree.
..... *F. pachystemon*

147. Basal bracts shorter, often hidden by the fig body.
..... *F. subtrinervia*

148. Peduncle lacking, stipe 0.5-2cm long........................ 149

148. Peduncle well-developed, with or without stipe................ 151

149. Leaf-bud short, not much exceeding 1 cm; leaves elliptic or obovate, rounded, oblique, at the base.
..... *F. tinctoria*

149. Leaf-bud 2-5 cm long; leaves ellipic, base cuncate............. 150

150. Leaves ± distichous; primary veins mostly in 6-8 pairs.
..... *F. subnervosa*

150. Leaves spiral; primary veins mostly in 10 or more pairs.
..... *F. hombroniana*

151. Fig diameter 1.5 cm or more... 152

151. Fig diameter 1cm or less...... 156

152. Primary veins in 10 or more pairs prominent on the underside an often contrasting in colour wit the lamina; leaf-bud ca 2 cm long curved, set at an angle to th twig-axis; leaves distichous.
..... *F. polyantha*

152. Primary veins usually in 7-8 pairs pairs...................... 153

153. Leaves coriaceous; figs orbicular.
.... 154

153. Leaves chartaceous; figs flattened globose..................... 15

154. Leaves elliptic-lanceolate, acute fig-stalk about 1 cm long (rati stipe/peduncle variable even on th same twig); fig-apex not prominent
..... *F. microdictya*

154. Leaves elliptic, obtuse; fig-stal to 4 mm long; fig-apex everted.
..... *F. mesotes*

155. Fig-body with usually 10 distinc longitudinal ridges; leaves dryin green.
..... *F. septica*

155. Fig-body not ridged; leaves dryin brown.
..... *F. austrina*

156. Figs with internal bristles... 15

156. Figs without internal bristles dark coloured internally...... 16

157. Cystoliths amphigenous; figs ligh -coloured internally......... 15

157. Cystoliths hypogenous; figs dark coloured internally.......... 15

158. Internal bristles few, much shorte than the seeds; leaves elliptic obtuse to acute, mostly rathe small, to 8cm long.
..... *F. adenosperma*

158. Internal bristles often abundant as long as the seeds; leaves elliptic, often acuminate, 1-20 cm long figs often ramiflorous just belo the leaves.
..... *F. comitis*

159. Internal bristles abundant; leave chartaceous, the primary vein leaving the midrib at ca 45°.
..... *F. erythrosperma*

159. Internal bristles few; leave coriaceous, the primary vein leaving the midrib at ca 30°.
..... *F. casearoides*

160. Cystoliths hypogenous; figs ca 5m across, on peduncles 5-10mm long

basal bracts patent, ca 2mm long; leaves lanceolate, 1-20cm long. ..... *F. smithii*

160. Cystoliths amphigenous........ 161

161. Leaves rhombic-ovate, 6 x 10 cm to 10 x 20 cm; figs ca 1 cm across, often puckered and ridged near the apex; stalk 5-10 mm long, 1-2 mm thick. ..... *F. indigofera*

161. Leaves elliptic, to 2.5x7cm; figs ca 5mm across, smooth-skinned; stalk 10-20mm long, less than 1mm thick. ..... *F. verticillaris*

INDEX TO HAMAMELIDAE

Antiaropsis	62
Antiaris	63
Aphananthe	98
Artocarpus	64
Boehmeria	10
Broussonetia	66
Calpidochlamys	66
Cannabis	58
Cannabidaceae	58
Castanopsis	110
Castilla	66
Casuarina	123
Casuarinaceae	121
Celtis	98
Ceuthostoma	124
Cudrania	67
Cypholophus	11
Dammaropsis	67
Daphniphyllaceae	107
Daphniphyllum	108
Debregeasia	15
Dendrocnide	16
Elatostema	21
Elatostemoides	28
Engelhardia	106
Fagaceae	109
Fatoua	67
Ficus	68
Fleurya	28
Gibbsia	29
Girardinia	29
Gironniera	101
Gymnostoma	125
Hamamelidaceae	4
Juglandaceae	106
Laportea	30
Leucosyke	32
Lithocarpus	111
Maclura	92
Malaisia	93
Maoutia	33
Memorialis	33
Missiessya	33
Moraceae	59
Morus	93
Myrica	6
Myricaceae	5
Nothocnide	34
Nothofagus	114
Oreocnide	35
Parartocarpus	94
Parasponia	103
Parietaria	35
Pellionia	36
Pilea	40
Pipturus	48
Poikilospermum	50
Pouzolzia	53
Prainea	94
Procris	54
Pseudopipturus	56
Robinnsoniodendron	56
Steblus	95
Sycopsis	5
Trema	104
Trophis	96
Ulmaceae	97
Urtica	56
Urticaceae	7
Villebrunea	57

Index to Families (Contd.)

Menispermaceae	**I**
Monimiaceae	**I**
Moraceae	IV
Moringaceae	IV
Myoporaceae	VIII
Myricaceae	**II**
Myristicaceae	**I**
Myrtaceae	V
Nepenthaceae	**I**
Nyctaginaceae	III
Nymphaeaceae	**I**
Ochnaceae	IV
Olacaceae	VI
Oleaceae	VIII
Opiliaceae	VI
Onagraceae	V
Orobanchaceae	VIII
Oxalidaceae	VII
Pandaceae	VI
Papaveraceae	**I**
Passifloraceae	VI
Pentaphragmataceae	IX
Peperomiaceae	**I**
Phytolaccaceae	III
Piperaceae	**I**
Pittosporaceae	V
Plantaginaceae	VIII
Plumbaginaceae	III
Podostemaceae	V
Polygalaceae	VII
Polygonaceae	III
Portulacaceae	III
Primulaceae	IV
Polygonaceae	III
Portulacaceae	III
Primulaceae	IV
Proteaceae	VI
Rafflesiaceae	VI
Ranunculaceae	**I**
Rhamnaceae	VI
Rhizophoraceae	V
Rosaceae	V
Rubiaceae	IX
Rutaceae	VII
Sabiaceae	VII
Santalaceae	VI
Sapindaceae	VII
Sapotaceae	IV
Saurauiaceae	IV
Saxifragaceae	V
Scropulariaceae	VIII
Simaroubaceae	VII
Solanaceae	VIII
Sonneratiaceae	V
Sphenostemonaceae	**I**
Stackhousiaceae	VI
Staphyleaceae?	VII
Sterculiaceae	IV
Stylidaceae	IX
Styracaceae	IV
Symplocaceae	IV
Theaceae	IV
Thymelaeaceae	V
Tiliaceae	IV
Trimeniaceae	**I**
Tropaeolaceae	V
Ulmaceae	IV
Umbelliferae	VII
Urticaceae	IV
Valerianaceae	IX
Verbenaceae	VIII
Violaceae	IV
Viscaceae	VI
Vitaceae	VI
Winteraceae	**I**
Zygophyllaceae	VII

INDEX TO FAMILIES

Acanthaceae	VIII
Aizoaceae	III
Alangiaceae	VI
Alseuomiaceae	IX
Amaranthaceae	III
Anacardiaceae	VII
Annonaceae	**I**
Apocynaceae	VIII
Aquifoliaceae	VI
Araliaceae	VII
Aristolochiaceae	**I**
Asclepiadaceae	VIII
Asteraceae (see Compositae)	IX
Avicenniaceae	VIII
Balanophoraceae	VI
Balsaminaceae	VIII
Basellaceae	III
Batidaceae	III
Begoniaceae	IV
Bignoniaceae	VIII
Bixaceae	IV
Bombacaceae	IV
Boraginaceae	VIII
Burseraceae	VIII
Byblidaceae	IX
Campanulaceae	IX
Cannabidaceae	**II**
Capparaceae	IV
Caprifoliaceae	IX
Cardiopteridaceae	VI
Caricaceae	IV
Caryophyllaceae	III
Casuarinanceae	**II**
Celastraceae	VI
Ceratophyllaceae	**I**
Chenopodiaceae	III
Chloranthaceae	**I**
Chrysobalanaceae	V
Clethraceae	IV
Cochlospermaceae	IV
Combretaceae	V
Compositae	IX
Connaraceae	VII
Convolvulaceae	VIII
Coriariaceae	VII
Cornaceae	VI
Corynocarpaceae	VI
Crassulaceae	V
Cruciferae	IV
Cucurbitaceae	IV
Cunoniaceae	V
Daphniphyllaceae	**II**
Datiscaceae	IV
Dichapetalaceae	VI
Dilleniaceae	IV
Dipterocarpaceae	IV
Droseraceae	V
Ebenaceae	IV
Elaeagnaceae	VI
Elaeocarpaceae	IV
Elatinaceae	IV
Epacridaceae	IV
Ericaceae	IV
Erythroxylaceae	VII
Euphorbiaceae	VI
Eupomatiaceae	**I**
Fabaceae (see Leguminosae	V
Fagaceae	**II**
Flacourtiaceae	IV
Gentianaceae	VIII
Geraniaceae	VII
Gesneriaceae	VIII
Gonostylaceae	VI
Goodeniaceae	IX
Guttiferae	IV
Haloragaceae	V
Hamamelidaceae	**II**
Hernandiaceae	**I**
Himantandraceae	**I**
Hypericaceae	IV
Icacinaceae	VI
Juglandaceae	**II**
Labiatae (see Lamiaceae)	VIII
Lamiaceae	VIII
Lauraceae	**I**
Lecythidaceae	IV
Leeaceae	VI
Leguminosae	V
Lentibulariaceae	VIII
Linaceae	VII
Lobeliaceae	IX
Loganiaceae	VIII
Lophopyxidaceae	VI
Loranthaceae	VI
Lythraceae	V
Magnoliaceae	**I**
Malphigiaceae	VII
Malvaceae	IV
Meliaceae	VII

THE FLOWERING PLANTS OF PAPUASIA

DICOTYLEDONS

PART 3
CARYOPHYLLIDAE

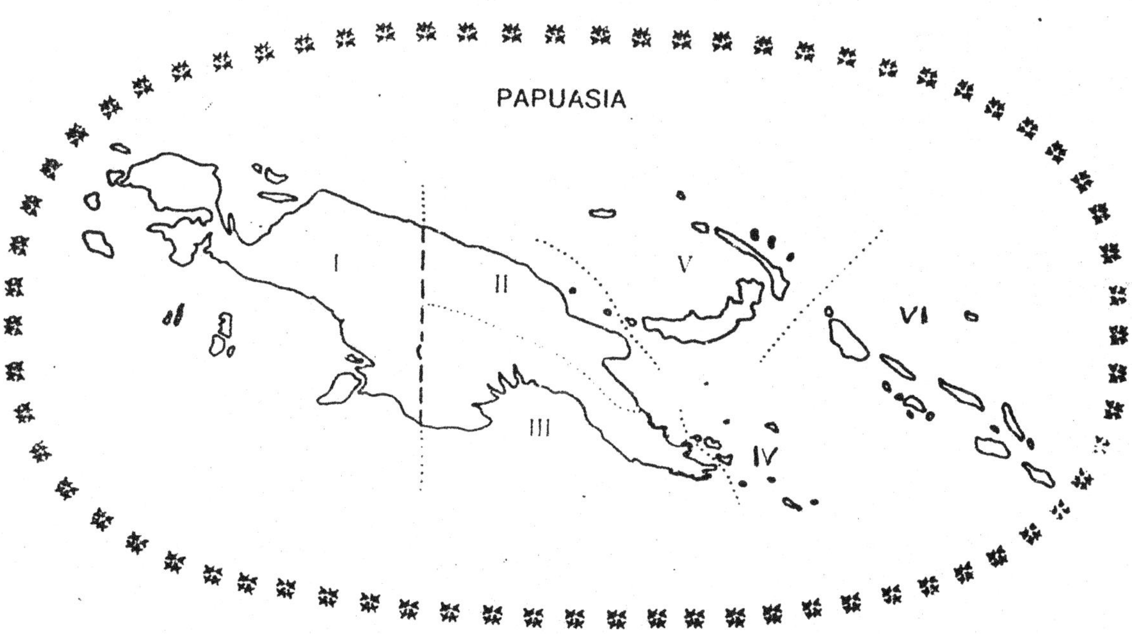

R. J. JOHNS 1989
FORESTRY DEPARTMENT

CHRISTENSEN RESEARCH INSTITUTE
Publication Number 2.

P.N.G. UNIVERSITY OF TECHNOLOGY
LAE
PAPUA NEW GUINEA

REGIONS OF PAPUASIA

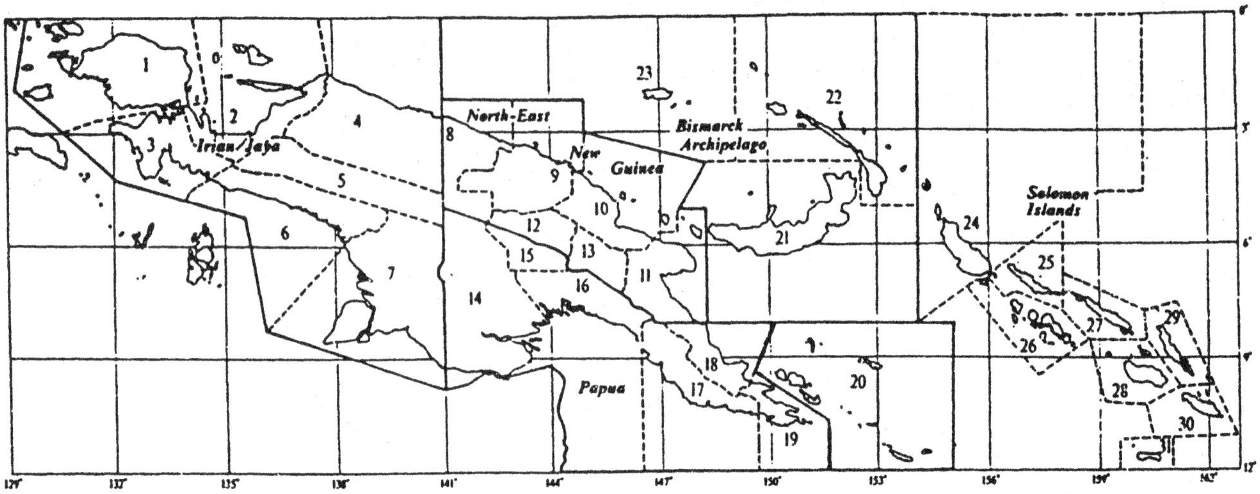

o I   o II   o III   o IV   o V   o VI

I. Irian Jaya

1. Vogelkop
2. Geelvink Bay
3. Fakfak
4. Jayapura
5. Snow Mountains
6. Mimika
7. Digul

II. North-east New Guinea

8. West Sepik
9. East Sepik
10. Madang
11. Morobe
12. Western Highlands
13. Eastern Highlands

III. Papua

14. Western
15. Southern Highlands
16. Gulf
17. Central
18. Northern
19. Milne Bay

IV. Papuan Islands

20. Papuan Islands

V. Bismarck Archipelago

21. New Britain
22. New Ireland
23. Manus

VI. Solomon Islands

24. Bougainville
25. Choiseul
26. New Georgia
27. Santa Isabel
28. Guadalcanal
29. Malaita
30. San Cristobal
31. Rennell Group

THE FLOWERING PLANTS OF

PAPUASIA

PART THREE: THE CARYOPHYLLIDAE

R. J. JOHNS

Johns, R. J. (Robert J.)

The flowering plants of Papuasia: Dicotyledons

1. Botany -- Papua New Guinea
2. Botany -- Indonesia - Irian Jaya
3. Botany -- Solomon Islands

I. Title. II. Papua New Guinea University of Technology.
Dept. of Forestry.

583.0995
ISBN 9980-56-010-X

October 1989

INTRODUCTION

The Caryophyllidae is the smallest of the volumes in the series on the Flowering Plants of Papuasia. Several of the families have been recently revised and published in the Handbooks to the Flora of Papua New Guinea. Few of the families are of economic importance although the Amaranthaceae includes several widely grown vegetables. The practise of giving dot distributions for all species is continued and this should prove of particular value to students working in Papuasia. It will allow them to develop regional floras as an aid to their own studies and research. It should again be noted that the flora of the region is poorly known, there are many undescribed species, a lot of which remain uncollected. If a genus or species is not shown for your region it probably means that it has not yet been collected, it does not mean that the species does not occur in your area.

ACKNOWLEDGEMENTS

I wish to express my sincere appreciation to the Research Committee of the Papua New Guinea University of Technology for a grant towards the cost of drawing the botanical plates and also the Christensen Research Institute which has generously given a grant towards the production of the plates and the printing costs for the third part of the Flowering Plants of Papuasia. Mr T. Iwagu has prepared many excellant illustrations for this volume. The National Herbarium, F.R.I., Lae, has provided illustrations from their extensive set of botanical plates. The following have carefully read the manuscript and provided useful notes and comments: Dr F. Arentz, Dr M. Jebb, Dr C. Sargent and Mrs R. Payet. Special thanks to Mr B. Vroone who assisted in converting the manuscript to Xerox Ventura Publisher.

The staff of the National Herbarium have been of considerable assistance and I am grateful for access to the collections and the Library facilities. I wish to express my sincere thanks to the staff of Herbarium Australiense in Canberra for access to the Card Index compiled by Dr R. Hoogland. This has made possible the inclusion of the list of species recorded from Papuasia. The check list of Solomon Island collections prepared by the Brisbane Herbarium has also been useful in compiling this volume.

This volume includes 11 families, 43 genera and 92 species. I would be grateful to receive any corrections, reprints of relevant articles, keys, or illustrations which could be included in revised editions of Parts 1, 2 or 3, or in the preparation of subsequent parts.

R. J. Johns
Department of Forestry
P.N.G. University of Technology
P.O. Private Bag
LAE
PAPUA NEW GUINEA

SUBCLASS: CARYOPHYLLIDAE

The Caryophyllidae consists, world wide, of three orders and fourteen families with some 11,000 species. Most species (c. 90%) belong to the largest order, the Caryophyllales. Both the Polygonales and the Plumbaginales each include only a single family. The other 12 families all belong to the order Caryophyllales which, in total, includes some 89 genera and 2070 species. The Bataceae was previously included in the Caryophyllales but Cronquist (1981) now places the family in the Dilleniidae (see Part IV).

All three orders in the subclass Caryophyllidae are represented in Papuasia. Locally the orders include 11 families, 43 genera and 92 species. In Papuasia all the families are small in size. The only families of economic importance in Papuasia are the Amaranthaceae, species of which are widely cultivated as vegetables, and the Nyctaginaceae which includes several introduced species of Bougainvillea that are cultivated as ornamentals.

The subclass probably originates from ancestors in the Ranunculaceae. Fossil plants are relatively recent, occurring in deposits only from 70 million years ago. This is not surprising as herbaceous plants played a 'negligible role" in the Cretaceous vegetation (Cronquist (1981). Orders and Families of Flowering Plants.)

Key to the Orders of the Caryophyllidae

Source: Cronquist (1981)

1. Ovule 1-many; seeds without true endosperm; most families producing betalains but not anthocyanins; perianth, ovary and placentation diverse, but the perianth not like that of typical members of the next two orders.
.... 1. CARYOPHYLLALES

1. Ovary solitary, seeds without perisperm; plants producing anthocyanins, but not betalains.. 2

2. Perianth not clearly differentiated into calyx and corolla; tepals 2-6 in a single whorl or similar sets of 3; stamens 2-9, seldom 5, generally in 2-3 whorls; leaves often with conspicuous, sheathing stipules..
.... 2. POLYGONALES

2. Perianth differentiated into 5-merous, sepal-like calyx and a 5-merous petal-like corolla; stamens 5, opposite the corolla-lobes; leaves exstipulate.
.... 3. PLUMBAGINALES

LIST OF ORDERS AND FAMILIES IN THE CARYOPHYLLIDAE

CARYOPHYLLALES

Phytolaccaceae
Achatocarpaceae
Nyctaginaceae
Aizoaceae
Didiereaceae
Cactaceae
Chenopodiaceae
Amaranthaceae
Portulacaceae
Basellaceae
Molluginaceae
Caryophyllaceae

POLYGONALES

Polygonaceae

PLUMBAGINALES

Plumbaginaceae

List of Families

Aizoaceae
Amaranthaceae
Basellaceae
Caryophyllaceae
Chenopodiaceae
Molluginaceae
Nyctaginaceae
Phytolaccaceae
Plumbaginaceae
Polygonaceae
Portulacaceae

The orders and families that are underlined occur in Papuasia.

Putative Relationships among the Families and Orders of the Caryophyllidae.
Source: Cronquist (1981)

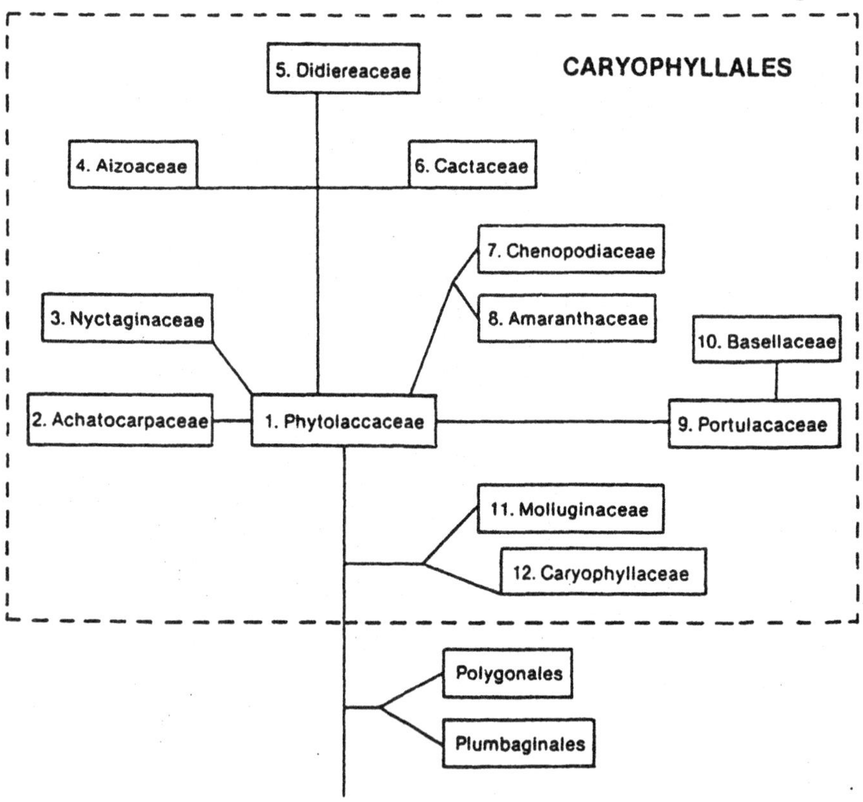

CARYOPHYLLALES

The Caryophyllales consists, world wide, of 12 families and c. 10,000 species. The species are in the three families: Aizoaceae, Cactaceae and Caryophyllaceae which account for some two thirds of the species in the order. The Phytolaccaceae are considered the most primitive family: possibly ancestral to all others. Cronquist (1981) lists a series of characters which often occur in the order: all members (examined) of the order have characteristic sieve tube plastids, unknown elsewhere in the angiosperms; most produce betalains and lack anthocyanins; many species have a characteristic pollen grain which is comparable in appearance to a golf-ball; several species show C4 photosynthesis, an adaptation to high temperatures and/or aridity; many species have a succulent habit, and associated crassulacean acid metabolism; anomalous cambial activity is common.

AIZOACEAE

o o o o o o

A family of 115 genera (or 12 genera if Mesembryanthemum is not divided into its c. 100 segregate genera) and 2410 species distributed mainly in the tropics and subtropics. Africa is the main center of diversity, with a secondary centre of speciation in Australia. Most plants are leafy succulents adapted to extremely dry climates (xeromorphic), often with C4 or crassulacean acid metabolism.

Succulent herbs, or more rarely shrubs or subshrubs. Leaves opposite to spiral, usually ± succulent, rarely stipulate. Flowers solitary or in small cymes, bisexual, rarely monoecious. Calyx: (3-)5(-8), mostly succulent. Staminodes often many in 1-6 cycles, rarely 0. Androecium: 4-5 or 8-10 or numerous, distinct or basally connate; nectaries commonly in a ring at the inner base of the androecium. Gynoecium inferior or superior (2-5 -many). Fruit a loculicidal capsule often enclosed in a persistent calyx. Seeds often arillate.

Diagnostic features:

Leaves: succulent, alternate or opposite, mostly simple, entire, with or without stipules.

Flowers: solitary or in cymes, regular, with their parts in whorls, usually bisexual. Sepals 4-8, usually 5, imbricate or rarely valvate. Petals numerous and of staminodal origin, occasionally absent. Stamens perigynous, numerous, sometimes filaments connate. Ovary superior or inferior; 1-20, usually 5 stigmas; ovules usually numerous.

Fruit: a dry capsule, rarely a berry or nut. Seeds with large curved embryo.

References

Backer, C. A. (1951). Aizoaceae (including Molluginaceae). Flora Malesiana I: 4: 267 - 275. Fig 1 - 2.

Melville, (1952). Some critical species allied to Trianthema pentandra. Kew Bull. 261 - 269.

Merrill, E. D. & Perry, L. M. (1942). Plantae Archboldianae. X. Journ. Arn. Arb. 23: 386. (see Molluginaceae)

Pax, F. and Hoffmann, K. (1934). Aizoaceae. Pflanzenfamilien ed. 2, 16C: 179 - 233. Fig. 83 - 103.

AIZOACEAE

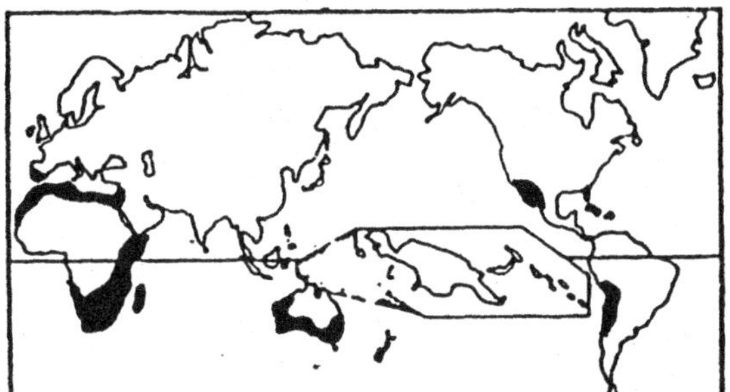

	WORLD	PAPUASIA
Number of genera :	115	3
Number of species :	2,410	3

Distribution : Pantropical but mainly in S Africa.

Economic uses : Many ornamental plants.

Prescott, A. & Venning, J. (1984). Aizoaceae. Flora of Australia 4: 19 - 62.

Pulle, A. (1910). Aizoaceae. Nova Guinea 8: 355.

KEY TO GENERA IN PAPUASIA

Source: Prescott and Venning (1984)

1. Leaves spirally arranged. Ovary semi-inferior. Fruit indeniscent.
...... 3. TETRAGONIA

1. Leaves opposite or pseudo-whorled. Ovary superior. Fruit dehiscing by a lid on the capsule... 2

2. Styles 1. Capsule partly exserted from perianth.
...... 2. TRIANTHEMA

2. Styles 3 - 4. Capsule included by perianth.
...... 1. SESUVIUM

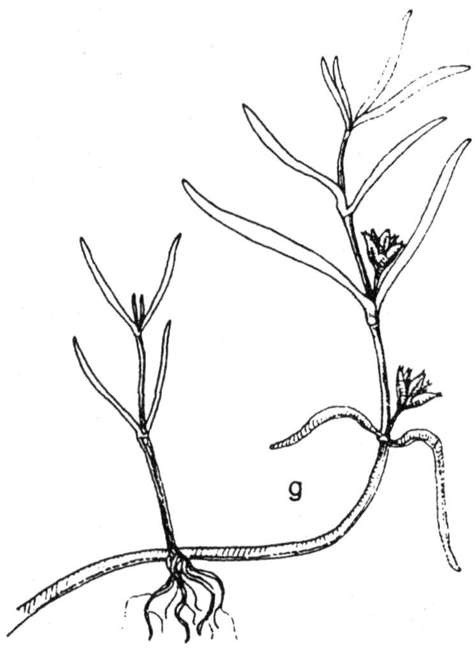

Sesuvium portulacastrum A. flowering branch. (Source: Morley & Thompson. Fl. Pl. Aust. 1983).

1. SESUVIUM

o o o o o o

A genus of 6 species, one of which is pantropical. The other species are local endemics in Angola (4 spp.) and Galapagos (1 sp.). Sesuvium portulacastrum is a coastal halophyte of tropical and subtropical regions; on mudflats or in sand, often associated with mangroves.

Prostrate, succulent, perennial herbs; stems rooting at nodes. Leaves opposite, fleshy, connate at base. Flowers solitary, axillary, pedicellate; lacing bracts. Perianth segments 5. Stamens 5-numerous, staminodes absent. Ovary superior, styles (and locules) 2- 5; axile placentation. Fruit a capsule. Seeds several, pea- to comma-shaped, black, smooth.

Key to Species in Papuasia

1 cosmopolitan species:
..... Sesuvium portulacastrum

List of Species recorded from Papuasia

Sesuvium portulacastrum (L.) L. Syst. Nat. (1759) 1058

Portulaca portulacastrum L. Sp. Pl. 1 (1753) 446; Backer (1951) 272

o o o o o o

Notes

1. Widespread species along coasts and islands, on sand and mudflats, often growing in association with mangroves. A cosmopolitan tropical species. The leaves, if continually cooked (to remove the salt), yield what Backer (1951) considers an inferior vegetable.

2. TRIANTHEMA

o o o o o o

A genus of 50-60 S Hemisphere species. Most species are in S Africa and Chile; a few occur in Australia and New Zealand. Some species are used for medicines and soap. A source of alkaloids. In Papuasia a single cultivated species.

Prostrate or diffuse herbs, sometimes woody at base. Leaves opposite, entire, unequal, terete, linear or flat. Flowers axillary, solitary or clustered, sessile or pedicellate, each subtended by one or more bracts. Perianth of 5 shortly- fused segments; lobes unequal, margins scarious, coloured inside. Stamens 5 to many, adnate to upper edge of perianth, staminodes absent. Ovary superior. Fruit a capsule, woody or membranous, perianth usually persistent. Seeds 1 to many, black.

Key to Species in Papuasia

A single pantropical weedy species:
..... Trianthema portulacastrum

List of Species recorded from Papuasia

Trianthema portulacastrum L. Sp. Pl. (1753) 223; Backer (1951) 274; Jeffrey Fl. Trop. E. Africa. (1961) 23; Pax and Hoffm. (1933) 230; Pulle (1910) 355; Tard. Fl. Cambodge, Laos, Vietnam (1967) 85. Fig. 1 - 5; Pulle (1910) 355

o o o o o o

Notes:

1. A native of Africa and tropical Asia. Weed of cultivation, possibly introduced. Often associated with disturbed areas. Mostly growing in mountain areas from 1000 - 1700 m. Commonly called New Zealand spinach.

Trianthema portulacestrum A. habit; B. flower; C. seed; D. lower leaf surface. UPNG 8576. (Drawn T. Iwagu)

3. TETRAGONIA

o o o o o o

A genus of 50 - 60 species mainly in the S Hemisphere, especially in Chile and S Africa. In Malaysia only a single cultivated species.

Erect, ascending, prostrate or climbing herbs or undershrubs, covered with small white papillae. Leaves spirally arranged, flat or slightly undulate, fleshy, exstipulate. Flowers axillary, solitary or fascicled, sessile or stalked, greenish or yellowish. Perianth tube short, with 3-5 often unequal segments. Stamens 1+ at mouth of perianth tube, alternating with the segments, solitary or in groups. Styles 2-9, linear, stigmatose along the inner side. Fruit turbinate, endocarp woody, indehiscent.

Key to Species in Papuasia

A single introduced species:
..... Tetragonia tetragonioides

List of Species recorded from Papuasia

Tetragonia tetragonioides (Pallas) O.K. Rev. Gen. (1891) 264

Demidovia tetragonioides Pallas En. Pl. Demidorf (1781) 150. t. 1

o o o o o o

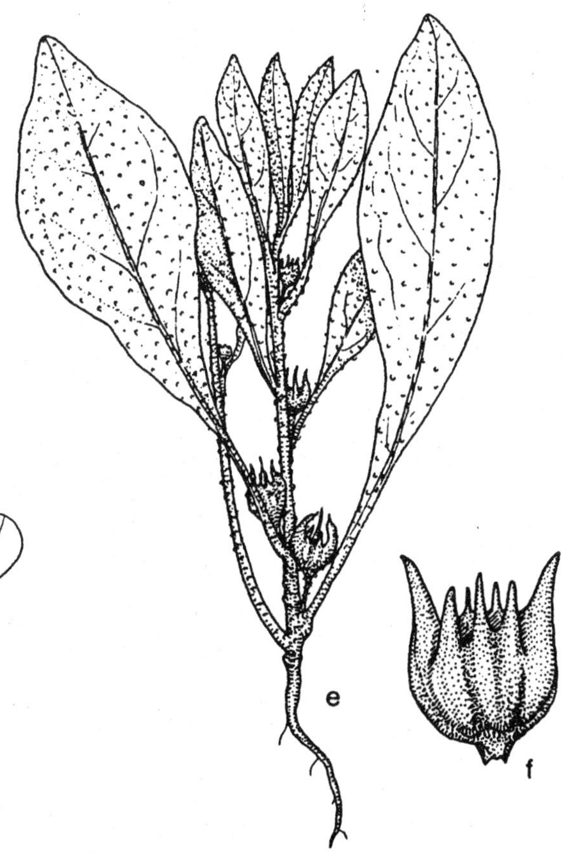

Tetragonia tetragonoides A. habit; B. fruit. (Source: Morley & Thompson. Fl. Pl. Aust. 1983).

Notes:

1. Cultivated as a pot herb at altitudes between 1000 and 1700 m: native to New Zealand. Adventitious buds occasionally develop from the persistent calyx and produce new flowers which are sometimes fertile (Prescott & Venning 1984).

AMARANTHACEAE

o o o o o o

A family of 65-71 genera and 800- 900 tropical, warm and a few temperate species. Some species are cultivated as ornamentals; others have edible seeds (Amaranthus). Some are grown as pot herbs. The genera are divided between two subfamilies: Amaranthoideae (A 4- loc, 1-many ovules) and the Gomphrenoideae (A 2-loc, 1 ovule). Pupalia was listed erroneously for Papuasia; there are no authentic records (Kanis 1972: 16).

Mainly herbs or climbers, shrubs or rarely trees. Leaves spiral or opposite, simple usually entire. Flowers small, solitary or in cymose or thyrsoid inflorescences often with bristly bracts and bracteoles. Wind or insect pollinated, rarely unisexual. Calyx: (1-2) 3-5, connate at base. Corolla absent. Stamens usually same number as the calyx then usually connate at base into a tube, often toothed or lobed; nectary often as a ring within the tube. Gynoecium 2-3 rarely 4, 1 loculate with a single style. Fruit an achene or small nut or a 1-seeded capsule, rarely a berry.

AMARANTHACEAE

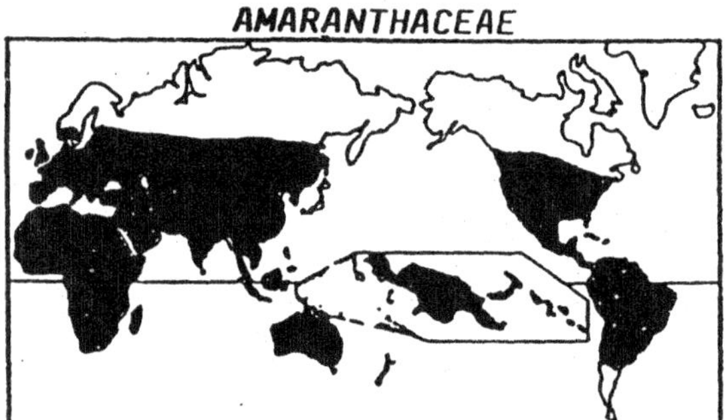

	WORLD	PAPUASIA
Number of genera :	71	10
Number of species :	800	24

Distribution : *Cosmopolitan, many in Africa and America.*

Economic uses : *Garden plants, herbs and vegetables or grain crops.*

Diagnostic features:

Leaves: entire, opposite or alternate, without stipules.

Flowers: solitary or in axillary cymes arranged in spike- or head-like inflorescences, usually bracteate. Bisexual, rarely unisexual, regular. Perianth 4-5, sometimes united; stamens 1-5, free or often united at base into a tube. In some genera petaloid appendages arise between the stamens. Ovary superior, free or united to perianth. Sometimes lateral flowers are sterile and develop spines, wings or hairs which function as dispersal agents. Flowers subtended by well- developed dry, scales which are often showy.

Fruit: a berry, pyxidium or nut; seeds usually with shiny testa.

References

Backer, C. A. (1949). Amaranthaceae. Flora Malesiana Ser. I. 4: 69 - 98, 593 - 594. Fig. 8.

Kanis, A. (1972). A review of the Amaranthaceae in Papuasia. Contr. Herb. Aust. 1: 3 - 18.

Kanis, A. (1978). Amaranthaceae. Handbooksof the Flora of Papua New Guinea. 1: 12 - 42.

Kanis, A. (1974). Further notes on the Amaranthaceae in Papuasia. Contr. Herb. Aust. 7: 7 - 13.

Pulle, A. (1910). Amaranthaceae. Nova Guinea 8: 351 - 352.

Schinz, H. (1934). Amaranthaceae. Pflanzenfamilien. ed. 2. 16C: 7 - 85. Fig. 1 - 46.

Schumann, K. M. & Lauterbach, C. A. G. (1900). Die Flora der deutschen Schutze gebiete in der Sudsee. pp. 303 - 306.

Schumann, K. M. & Lauterbach, C. A. G. (1905). Nachtrage zur Flora der deutschen Schutz gebiete in der Sudsee. p. 261.

Steenis, C. G. G. J. van (1965). Miscellaneous notes on New Guinea Plants VIII. Nova Guinea 23: 495 - 496.

Steenis, C. G. G. J. van (1972). Addenda, corrigenda et emendanda. Flora Malesiana Ser. I. 6: 915 - 917.

Veldkamp, J. F. (1971). Alternanthera paronychioides in Indo- Malesia. Blumea 17: 167 - 169.

KEY TO GENERA IN PAPUASIA

Source: Kanis (1978)

1. Leaves alternate.. 2

1. Leaves opposite.. 5

2. Fruits always 1-seeded and dry.................... 3

2. Fruits 2-many-seeded, if partly 1- seeded then usually berries... 4

3. Bracts, bracteoles and tepals distinctly hairy outside; flowers bisexual; stigmas 2, capitate, on branched style; erect perennial herbs.
...... 2. AERVA

3. Bracts, bracteoles and tepals glabrous; flowers unisexual; stigma 2-3(-4), elongate, ± filamentous, sessile; erect, ascending or prostrate herbs.
...... 4. AMARANTHUS

4. Fruit a membranous circumsessile capsule; stigma one, capitate, on filamentous style; annual herbs.
...... 5. CELOSIA

4. Fruit a berry; stigmas 2-4, usually recurved, ± sessile; perennial herbs, shrubs or woody climbers.
...... 7. DEERINGIA

5. Inflorescence paniculate; flowers minute c. 1.5 mm long, unisexual (in Papuasia female only); leaves broadly ovate with acuminate apex to obcordate, sometimes deeply bilobed.
...... 9. IRESINE

5. Inflorescences apparently unbranched; flowers 2.5 mm long or more, bisexual; leaves narrower, never obcordate.................................. 6

6. Inflorescences slender, rather lax towards base, 4 cm long; older flowers or flower clusters distinctly reflexed. 7

6. Inflorescences compact, 0.5-4 cm long; flowers or flower clusters never distinctly reflexed....... 8

7. Flowers in shortly stalked clusters of 1-3, older ones accompanied by bundles of pungent c. 2 mm long hooks; bracteoles not spiniform though sometimes rather long-acuminate.
...... 6. CYATHULA

7. Flowers single, sessile, without pungent hooks; bracteoles firmly spiniform with short wings at base. .1. ACHYRANTHES

8. Bracteoles keeled, with a dorsal crest at least in upper part, 6 mm or more; staminal filaments almost completely joined into tube of same length as tepals; stigmas 2, filamentous.
...... 8. GOMPHRENA

8. Bracteoles neither keeled nor crested, 4 mm long or less; staminal filaments joined only at base into cup or short tube distinctly shorter than tepals; stigma 1, capitate..................................... 9

9. Outer tepals with strong midrib, often accompanied by 2 convergent strong nerves in lower part, 2-2.5 mm long; flower clusters 1-3 per leaf axil, ± globose at least when young.
...... 3. ALTERNANTHERA

9. Outer tepals with 3 - 5 strong straight parallel nerves over almost full length, 2-2.5 mm long; flower clusters 1(-2) per leaf axil conical when young.
...... 10. PSILOTRICHUM

1. ACHYRANTHES

o o o o o o

A palaeotropical genus with some 6 poorly defined species. In Papuasia there are 2 widely distributed and polymorphic species. The distribution of the weedy species is aided by the spiny bracteoles to which the nuts remain attached.

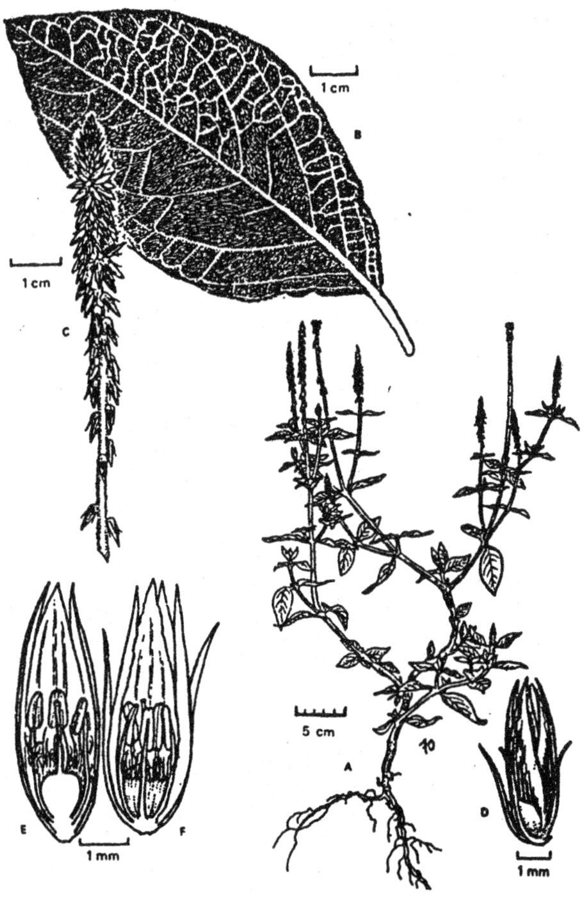

Achyranthes aspera A. habit; B. leaf; C. inflorescence; D.E.F. flowers. (Source: Kanis 1978)

Perennial herbs, both stem and branches strongly ribbed, ± quadrangular, swollen above nodes. Soft white hairs especially on fertile parts. Leaves opposite, ± hairy. Inflorescences terminal, sometimes lateral, unbranched; bracts acuminate, membranous, glabrous; bracteoles with short membranous wings at the base. Flowers bisexual, sessile, reflexed after flowering; tepals 5, glabrous; stamens 5, joined at base, alternating with truncate staminodes, anthers 2-celled. Fruit a thin walled nut, falling with perianth and bracteoles.

Key to Species in Papuasia

Source: Kanis (1978)

1. Inflorescences terminal only; membranous wings of bracteoles attached lengthwise to spiny part, sometimes breaking away laterally in old flowers; tepals 3 or more with conspicuous nerves, green with pale margin.
..... Achyranthes aspera

1. Inflorescences both terminal and in upper leaf axils; membranous wings of bracteoles laterally free from spiny part; tepals with only 1 distinct nerve, pale, often tinged violet during flowering.
..... Achyranthes bidentata

List of Species recorded from Papuasia

Achyranthes aspera L. Sp. Pl. (1753) 204.
A. grandifolia Miq. (1849)

o o o o o o

Achyranthes bidentata Bl. Bijdr. Fl. Ned. Ind. (1826) 545

o o o o o o

Achyranthes grandifolia Miq. (1849)
= Achyranthes aspera L.

Notes:

1. A. aspera is a weedy species of gardens and roadsides from sea level to 200 m altitude, probably introduced into coastal settlements.

2. A. bidentata is a forest species but also a weed in shady areas between 250-2100 m. Probably indigenous.

2. AERVA

o o o o o o

A tropical and temperate old-world genus of c. 10 species. One widely distributed species found in Papuasia.

Perennial herbs; stems and branches terete, with white woolly hairs. Leaves alternate, ± densely hairy. Inflorescences lateral and sessile, compact, terminal parts form a compound inflorescence; bracts and bracteoles membranous, hairy outside. Flowers bisexual, sessile; tepals 5, hairy outside; stamens 5, filaments joined at base, alternating with staminodes. Ovary broadly ovoid, narrowing into short style;

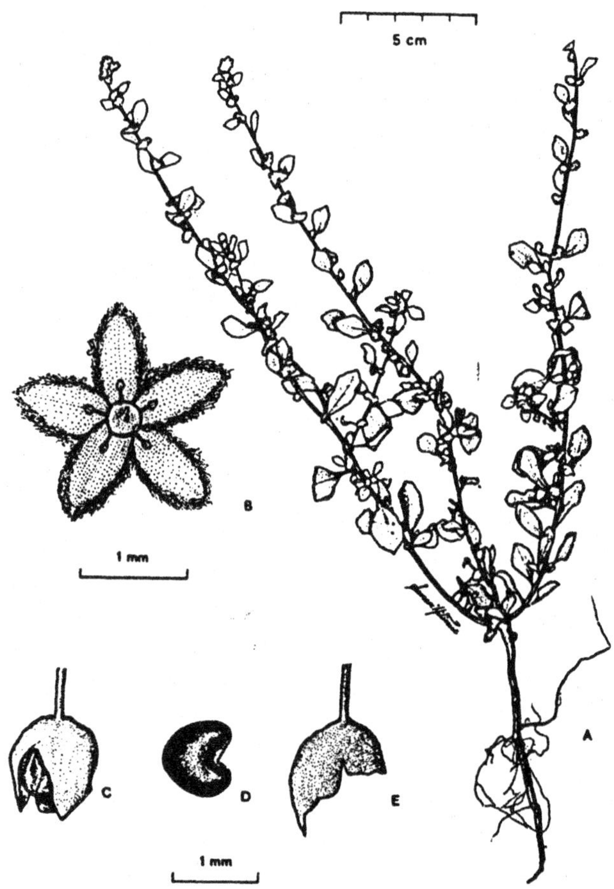

Aerva lanata A. habit; B. flower; C.E. fruit; D. seeds. (Source: Kanis 1978)

stigmas minute. Fruit a thin walled nut, 1-seeded, bursting irregularly, falling with perianth.

Key to Species in Papuasia

One widely distributed species:
..... Aerva lanata

List of Species recorded from Papuasia

Aerva lanata (L.) Schultes Syst. Veg. 16.5. (1819) 564

o o o o o o

Notes:

1. A widely distributed (Africa to Malesia) weed of dry open places. In Papuasia only collected in S New Guinea. Possibly restricted to areas with a pronounced dry season (Kanis 1978).

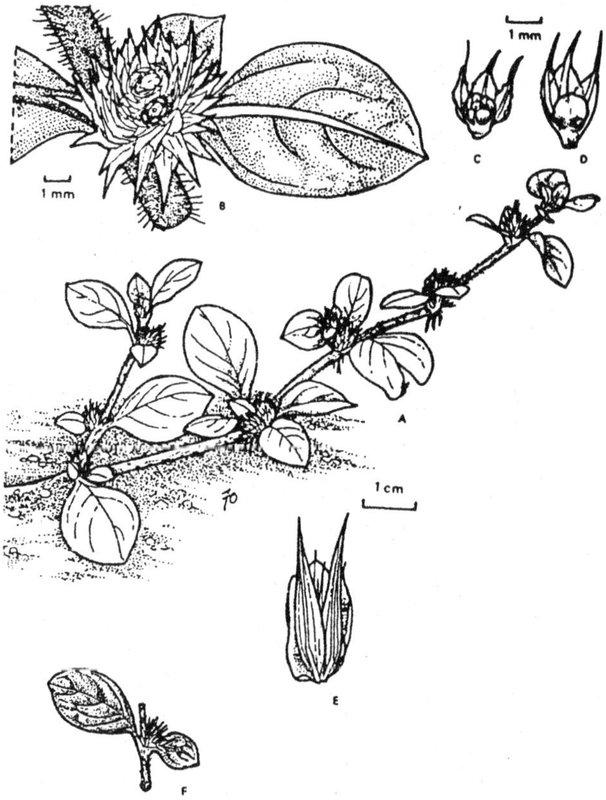

Alternanthera pungens A. habit; B.F. inflorescence; C.E. flower; D. fruit. (Source: Kanis 1978)

3. ALTERNANTHERA

o o o o o o

A pantropical and subtropical genus with c. 170 species, mainly American in distribution. Three species in Papuasia, two introduced from America, 1 indigenous.

Perennial herbs; stems many and much branched, terete or quadrangular, ± hairy. Leaves opposite. Inflorescences lateral, usually 1-3 per axil, sessile, short, compact; bracts and bracteoles dry-membranous, glabrous or hairy outside. Flowers bisexual, sessile. Tepals 5, outer 3 with distinct midrib. Stamens 5, rarely 2 with reduced anthers, filaments joined at base forming a small tube, alternating with staminodes. Ovary subglobose, styles short, stigma capitate. Fruit a thin- walled nut, falling with perianth. Seeds lens shaped.

Key to Species in Papuasia

Source: Kanis (1978)

1. Tepals ± equal, shortly acuminate, with only 1 distinct nerve, 2.5-3.3 mm long; bracts and bracteoles shortly acuminate, c. 1 mm long; stems with only 2 rows of hairs lengthwise and single rows across the nodes; hairs woolly, smooth.
..... Alternanthera sessilis

1. Tepals distinctly unequal, acuminate or awned, 3 outer ones with 3(-5) distinct nerves in lower half, 3 mm long or more, 2 inner ones much smaller; bracts and bracteoles acuminate or awned, c. 2 mm long or more; stems usually evenly hairy at least on younger parts; hairs rather straight, minutely denticulate under high magnification.. 2

2. Bracts, bracteoles and tepals not pungent, with long white hairs dispersed on outside, those on 2 outermost tepals projecting slantingly upwards from the midrib; leaves densely hairy when young, glabrescent, green or variegated.
..... Alternanthera bettzickiana

2. Bracts and especially 2 outermost tepals pungent; bracts and bracteoles glabrous, 3 outer tepals with hair tufts restricted to base, projecting perpendicularly sideways, 2 inner ones with a central tuft on outside; leaves glabrous or nearly so, never variegated.
..... Alternanthera pungens

List of Species recorded from Papuasia

Alternanthera bettzickiana (Regel) Nichols Ill. Dict. Gard. 1 (1885) 59
A. manillensis auct. non. (Walp.) Kanis (1972); Kanis (1978) 18

o o o o o o

Alternanthera denticulata auct. non R.Br. S Moore (1923); Kanis (1978) 20
= Alternanthera sessilis

Alternanthera manillensis auct. non. (Walp.) Kanis (1972); Kanis (1978) 18
= Alternanthera bettzickiana

Alternanthera nodiflora auct. non R.Br. Sch. & Laut. (1900)
= Alternanthera sessilis

Alternanthera pungens Kunth Nov. Gen. Sp. 2 (1818) 206
A. repens (L.) Link (1821) non Gmel. (1791); Kanis (1978) 18

o o o o o o

Alternanthera repens (L.) Link (1821) non Gmel. (1791); Kanis (1978) 18
= Alternanthera pungens

Alternanthera sessilis (L.) DC. Cat. Hort. Monspel. 77 (1813)
Alternanthera tenuissima Suess (1939)
A. sessilis var. tenuissima (Suess.) Backer (1949) A. nodiflora auct. non R.Br. Sch. & Laut. (1900).
A. denticulata auct. non R.Br. S. Moore (1923); Kanis (1978) 20

o o o o o o

A. sessilis var. tenuissima (Suess.) Backer (1949)
= Alternanthera sessilis

Alternanthera tenuissima Suess (1939)
= Alternanthera sessilis

Notes:

1. Alternanthera sessilis is a pantropical weed growing from sea level to 2000 m altitude. Throughout Papuasia, in open, moist habitats, in plantations, gardens and along roadsides. By report smoked as a medicine in the Baliem Valley.

2. The ornamental form of Alternanthera bettzickiana probably does not produce seed (Kanis 1978). A fertile form introduced in 1960's is spreading rapidly from major centres. The other species introduced from S America is a weed of dry open habitats up to 700 m in altitude.

4. AMARANTHUS

o o o o o o

A genus of c. 50 species, mostly tropical and subtropical in distribution, but extending into the temperate zone. In Papuasia 9 species, possibly 1 indigenous. The genus is important in Papuasia with 4 species commonly grown as green vegetables. No species are cultivated as grain crops in Papuasia.

Annual herbs; stems and branches angular. Leaves alternate, glabrous or ± hairy on nerves. Inflorescence of lateral cymes, inflorescences compound. Bracts and bracteoles dry-membranous, glabrous, sometimes awned, rarely spiny. Flowers unisexual, sessile; tepals 3-(4- 5), membranous, glabrous. Stamens as many as tepals, filaments free, anthers 2-celled; staminodes absent; ovary with 1 sessile ovule; often present in reduced form in male flowers. Stigma (2-)3(-4). Fruit 1-seeded circumcisile capsule or a nut, falling with perianth and bursting irregularly. Seed shiny dark brown or rarely yellowish-white.

Key to Species in Papuasia

Source: Kanis (1978)

1. Mature fruits not opening by circumcision, usually falling entirely with perianth.................. 2

1. Mature fruits opening by circumcision, upper part falling with seed, lower part persisting with perianth.. 4

2. Tepals 5 in most female flowers; fruit usually with smooth light-coloured conical beak, distinct from wrinkled darker main body.
..... Amaranthus interruptus

2. Tepals 3 in most female flowers; fruits without distinct top part, entirely wrinkled or smooth..... 3

3. Ripe fruits rather smooth; leaf-blades ovate to obovate, 1-3 cm long.
..... Amaranthus lividus

3. Ripe fruits distinctly wrinkled; leaf-blades ± trullate, 3-7 (-10) cm long.
..... Amaranthus viridus

4. Tepals 3 per flower, distinctly awned and strongly recurved like bracts, 3-5 mm long in female flowers.
..... Amaranthus tricolor

4. Tepals 5 per flower (occasionally 3 or 4 in some male flowers), ± mucronate, 1.5 - 2.5 mm long in female flowers.. 5

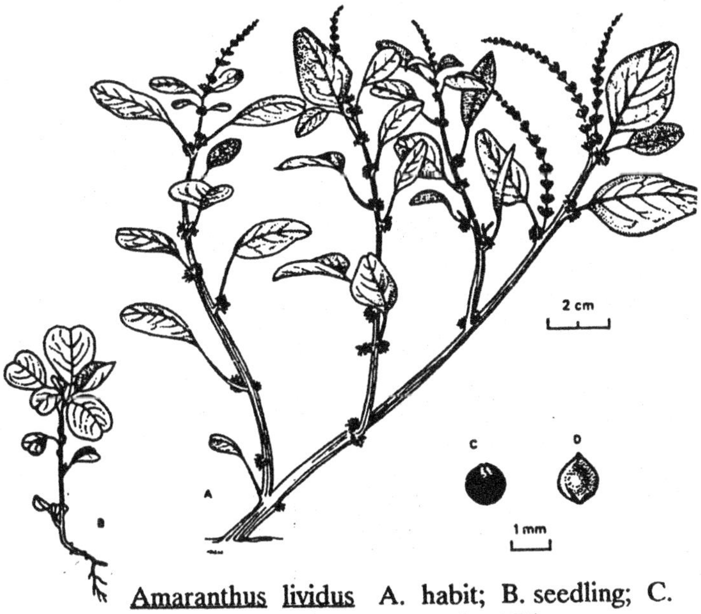

Amaranthus lividus A. habit; B. seedling; C. seed; D. fruit. (Source: Kanis 1978)

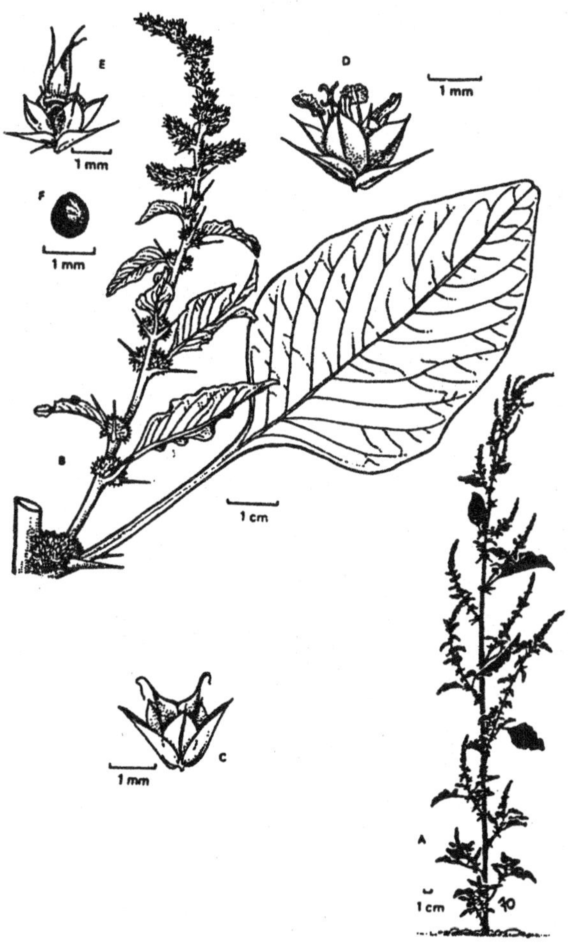

Amaraanthus spinosus A. habit; B. inflorescence; D. flowers. F. seed. (Source: Kanis 1978)

5. Tepals about equal in length, perianth ± incurved against ripe fruit.. 6

5. Tepals distinctly unequal in length, perianth ± spreading away from ripe fruit............................ 7

6. All flower clusters unarmed, awns of all bracts mm long; flowers mostly female, only initial one of each cluster male.
..... Amaranthus dubius

6. Flower clusters in leaf axils with 1 or 2 spines (metamorphosed bracts) 5-20 mm long; flowers all female in lower clusters all male in upper clusters.
..... Amaranthus spinosus

7. Bracts mucronate, shorter than perianth, about as long as wrinkled dark fruit.
..... Amaranthus leptostachyus

7. Bracts long-awned, overall equal to or slightly longer than perianth, shorter than smooth pale fruit.. 8

8. Tepals ± recurved in fruit, with distinctly overlapping margins, inner ones obtusely spathulate; stigmas divergent, ± recurved, widened at base and together forming 3-sided saddle.
..... Amaranthus caudatus

8. Tepals straight in fruit, margins not or hardly overlapping, inner ones acutely oblanceolate, stigmas erect, straight, slender at base and attached to short cylindrical beak.
..... Amaranthus cruentus

List of Species recorded from Papuasia

Amaranthus caudatus L. Sp. Pl. (1753) 990
o o o o o o

Amaranthus cruentus L. Syst. Nat. (1759) 1279
A. paniculatus L. (1763)
A. hybridus var. paniculatus (L.) Uline & Bray (1894)
A. hybridus auct. non L. Kanis (1972); Kanis (1978) 23
o o o o o o

Amaranthus dubius Thell. Fl. Adv. Montpell. (1912) 203
A. hybridus auct. non L. Kanis (1972); Kanis (1978) 24
o o o o o o

Amaranthus hybidus auct. non L. Kanis (1972); Kanis (1978) 24
= Amaranthus dubius

Amaranthus interruptus R. Br. Prodr. (1810) 414
Euxolus interruptus (R. Br.) Miq. (1849); Kanis (1978) 24

o o o o o o

Amaranthus leptostachyus Benth. Fl. Aust. 5 (1870) 214 ; Kanis (1978) 25

Amaranthus lividus L. Sp. Pl. (1753) 990. Kanis (1978) 25.

o o o o o o

Amaranthus spinosus L. Sp. Pl. (1753) 991.

o o o o o o

Amaranthus tricolor L. Sp. Pl. (1753) 989
A. melancholicus L. (1753)
A. gangeticus L. (1759)
A. melancholicus var. tricolor (L.) Miq. (1849)
A. blitum 'forma monstrosa phyllomanica' K. Sch. (1887)
A. melancholicus forma phyllomanica K. Sch. & Laut. (1900)
A. oleraceus auct. non L. Warb. (1891)
A. caudatus auct. non L. (1907) Kanis (1978) 27.

o o o o o o

Amaranthus viridus L. Sp. Pl. (1763) 1405. Kanis (1978) 29
A. gracilis Desf. (1804)
A. spinosus forma inermis Laut. & K. Sch. (1900)
A. caudatus auct. non L. H. J. Lam (1927)

o o o o o o

Notes:

1. The following list includes the species eaten as vegetables (where a question mark is placed after the name the record is doubtful).

A. caudatus
A. cruentus
A. dubius
A. lividus (?)
A. tricolor (?)
A. viridus

5. CELOSIA

o o o o o o

A genus of some 60 species mainly distributed in Africa and America. A single cosmopolitan species is found in Papuasia.

Annual herbs; stems and branches strongly ribbed, glabrous. Leaves alternate, glabrous. Inflorescences terminal or lateral, simple or branched, compact or interrupted; bracts and bracteoles membranous, 1-nerved, glabrous. Flowers bisexual, sessile; tepals 5, glabrous, spreading during flowering. Stamens 5, filaments joined at base, alternating with stamens. Style filamentous, stigma capitate, slightly lobed. Fruit a circumcisile capsule. Seeds reniform, lenticular.

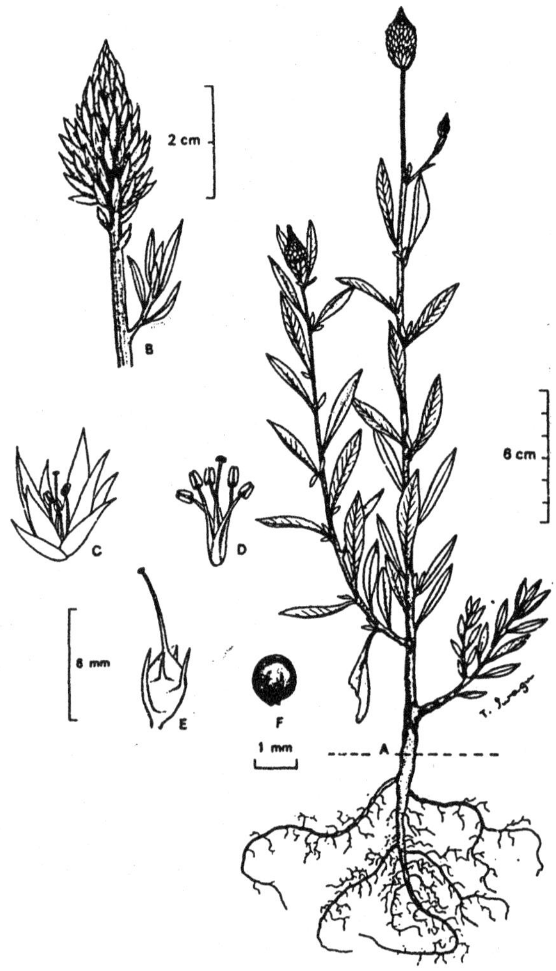

Celosia argentea A. habit; B. inflorescence; C.D. flowers. E. fruit with stamens; F. seed. (Source: Kanis 1978)

Key to Species in Papuasia

A single pantropical species:
..... Celosia argentea

List of Species recorded from Papuasia

Celosia argentea L. Sp. Pl. (1753) 205
Celosia crista L. (1753) Kanis (1978) 30
o o o o o o

Celosia crista L. (1753) Kanis (1978) 30.
= Celosia argentea

Notes:

1. Possibly of Indian origin, growing as a weed in dry open areas, roadsides, wastelands and stream beds from sea level to 1800 m altitude. A common ornamental, occasionally used as a vegetable (E Sepik) and for dyeing nets (Madang).

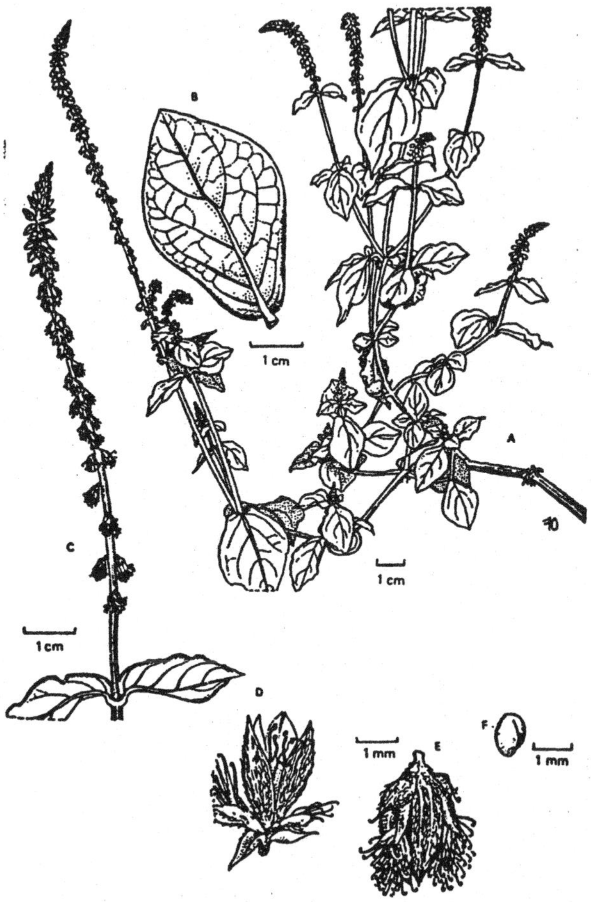

Cyathula prostrata A. habit; B. leaf; C.E. inflorescence; D. flowers; F. seed. (Source: Kanis 1978)

6. CYATHULA

o o o o o o

A genus of mainly African distribution with some 25 species. A single species in Papuasia. The flower clusters, which contain the ripe fruits, are covered with spiny hooks which facilitate dispersal.

Perennial herbs with angular stems and branches, ± densely hairy. Leaves opposite, ± hairy. Inflorescences terminal, sometimes lateral, compact. Flowers shortly stalked, the clusters comprising bundles of hooks (deformed sterile flowers); bracts and bracteoles membranous, hairy outside; clusters and persistent subtending bracts reflexed after flowering. Flowers bisexual. Tepals 5, 3- to 5-nerved, with long hairs outside; filaments joined at base, alternating with staminodes. Style filamentous; stigma capitate. Fruit 1- seed, indehiscent, each cluster falling as a whole together with tepals, bracteoles and spiny hooks.

Key to Species in Papuasia

A single widely distributed species:
..... Cyathula prostrata

List of Species recorded from Papuasia

Cyathula prostrata (L.) Bl. Bijdr. Fl. Ned. Ind. (1826) 549
C. geniculata auct. non Lour. Miq (1858)
Pupalia atropurpurea auct. non (Lamk.) Moq. F. Muell. (1886); Kanis (1978) 32
o o o o o o

Cyathula geniculata auct. non Lour. Miq. (1858)
= Cyathula prostrata

Notes:

1. This species grows in primary and secondary forest, and also as a weed in moist shady areas from sea level to 1200(-2000) m altitude.

7. DEERINGIA

o o o o o o

A tropical genus of c. 7 species, distributed from Madagas-car, India and S China to N E Australia. Three species are indigenous in Papuasia.

Perennial herbs, shrubs or woody climbers; stems often angular when young. Leaves alternate, ± glabrous. Inflorescences lateral, sometimes terminal, simple or branched; bracts and bracteoles membranous, glabrous. Flowers bisexual or functionally unisexual, sessile or shortly stalked; tepals 4(-5), glabrous, persistent. Stamens 4(-5), filaments ± joined at base, staminodes absent or only in female flowers. Stigmas 2-3(-4), ± sessile. Fruit a berry with reniform seeds.

Key to Species in Papuasia

Source: Kanis (1978)

1. Herb or soft shrub, usually erect; inflorescences lateral only; berr ies white, with many seeds; free parts of staminal filaments about as long as joined basal parts.
..... Deeringia polysperma

1. Climber or scandent shrub; inflor- escences terminal and lateral; berries red with 1-5(-9) seeds; free parts of staminal filaments several times longer than joined basal parts.. 2

2. Lateral inflorescences racemose, 5- 40 cm long; sometimes with 1 diver- gent branch at base; flowers bisexual; tepals ± reflexed in fruit; fruit 4-7 mm across, with 1- 5(-9) seeds; scandent shrub or low climber.
..... Deeringia amaranthoides

2. Lateral inflorescences paniculate, 5-20 cm long; flowers functionally unisexual; tepals appressed in fruit; fruit 3 - 4 mm across, with 1-2 seeds; tall climber.
..... Deeringia arborescens

List of Species recorded from Papuasia

Deeringia altissima (Moq.) F. Muell (1860); Kanis (1978) 34.
= Deeringia arborescens

Deeringia amaranthoides (Lamk.) Merr. Interpr. Rumph. Herb. Amb. (1917) 211
D. celosioides R. Br. (1810) nom. illeg.
D. indica Retz. ex. Bl. (1826)
D. baccata (Retz.) Moq. (1849), Kanis (1978) 34

o o o o o o

Deeringia arborescens (R. Br.) Druce. Rep. Bot. Exch. Club Br. Isles (1917) 619
D. altissima (Moq.) F. Muell (1860); Kanis (1978) 34

o o o o o o

Deeringia baccata (Retz.) Moq. (1849) Kanis (1978) 34.
= Deeringia amaranthoides

Deeringia celosioides R. Br. (1810) nom. illeg.
= Deeringia amaranthoides
Deeringia indica Retz. ex. Bl. (1826)
= Deeringia amaranthoides

Deeringia indica Zoll. ex. Moq. (1849) non Retz. ex. Bl. (1826); Kanis (1978) 35
= Deeringia polysperma

Deeringia polysperma (Roxb.) Moq. in C. DC Prodr. 13 (1849) 236.
D. indica Zoll. ex. Moq. (1849) non Retz. ex. Bl. (1826); Kanis (1978) 35

o o o o o o

Notes:

1. Most species occur in primary and secondary rainforest between sea level and 1200 m altitude, often in moist places but also on ridges.

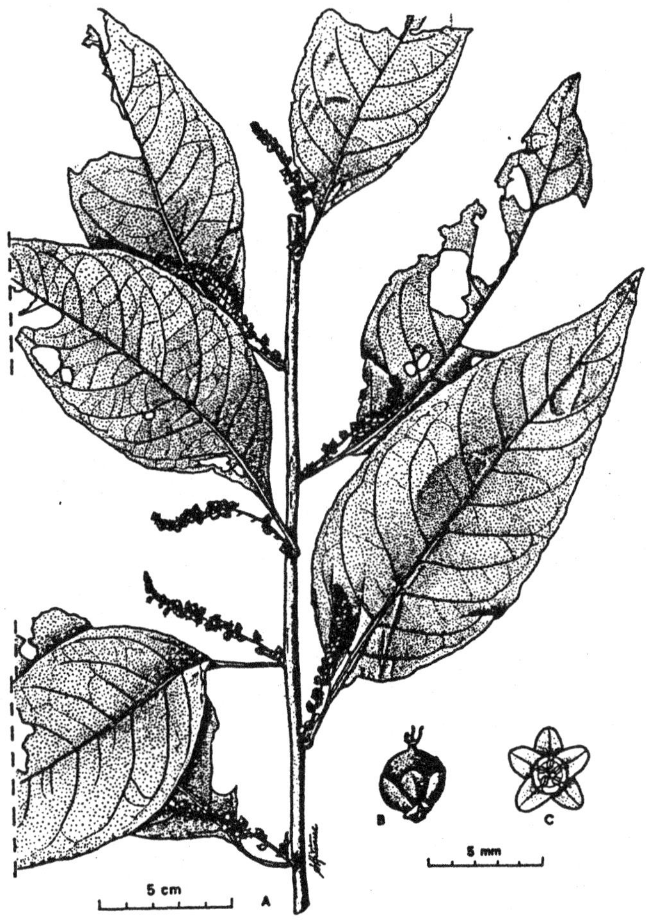

Deeringia polysperma A. habit; B. fruit; C. flower. (Source: Kanis 1978)

8. GOMPHRENA

o o o o o o

A neotropical genus of around 90 species. In Papuasia two introduced species.

Annual herbs; stems ± quadrangular, sometimes hairy. Leaves opposite. Inflorescences terminal, sessile, subglobose at first, ± lengthening during flowering; bracts and bracteoles dry-membranous, glabrous, bracteoles keeled with dorsal crest, longer than bracts. Flowers bisexual, sessile; tepals 5, hairy outside; stamens 5, filaments almost completely joined into a tube, alternating with staminal lobes of almost equal length. Style and 2 stigmas enclosed by staminal tube. Fruit 1-seeded, falling unopened with perianth and bracteoles.

Key to Species in Papuasia

Source: Kanis (1978)

1. Inflorescences c. 1 cm across initially ± globose, becoming cylindrical and to 5 cm long; bracteoles with narrow dorsal crest in upper part only, ending ± abruptly below apex, white, sometimes yellow towards apex.
..... Gomphrena celosioides

1. Inflorescences c. 2 cm across, initially depressed-globose, becoming ovoid and to 3 cm long; bracteoles with distinct dorsal crest from base, exceeding apex, deep purple, pink or white.
..... Gomphrena globosa

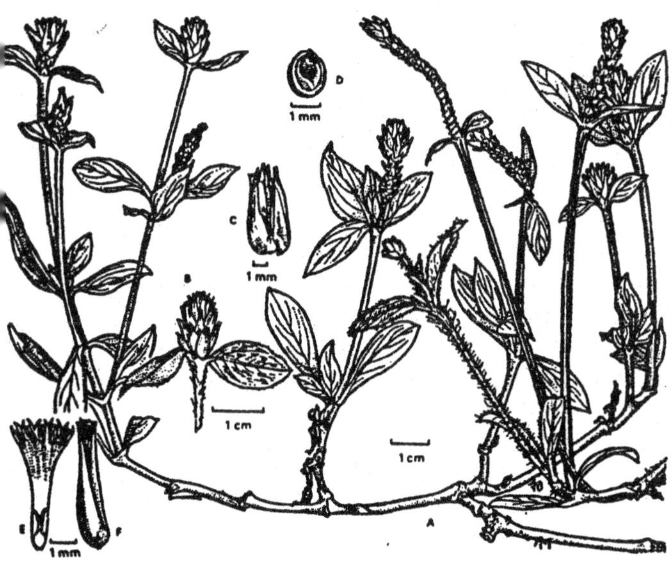

Gomphrena celosioides A. habit; B. inflorescence; C.D.E.F. flowers. (Source: Kanis 1978)

List of Species recorded from Papuasia

Gomphrena celosioides Mart. Beitr. Amaranth. (1925) 93; Kanis (1978) 37

o o o o o o

Gomphrena globosa L. Sp. Pl. (1753) 224; Kanis (1978) 37

o o o o o o

Notes:

1. Both species occur as weeds in dry open areas and wastelands at low altitudes. Mainly growing near coastal settlements, and along roads. They are sometimes used as ornamentals.

9. IRESINE

o o o o o o

This genus of c. 70 species is American in origin. Several are introduced elsewhere. A single species occurs in Papuasia.

Perennial herbs with angular stems, ± hairy. Leaves opposite, sparsely hairy. Inflorescences terminal, sometimes also lateral, paniculate, many-flowered. Bracts and bracteoles membranous, glabrous. Flowers unisexual, ± sessile. Tepals 5, ± hairy outside. Stamens 5, in female flowers minute and staminodial, in male flowers filaments joined at base, usually alternating with staminodes. Ovary absent in male

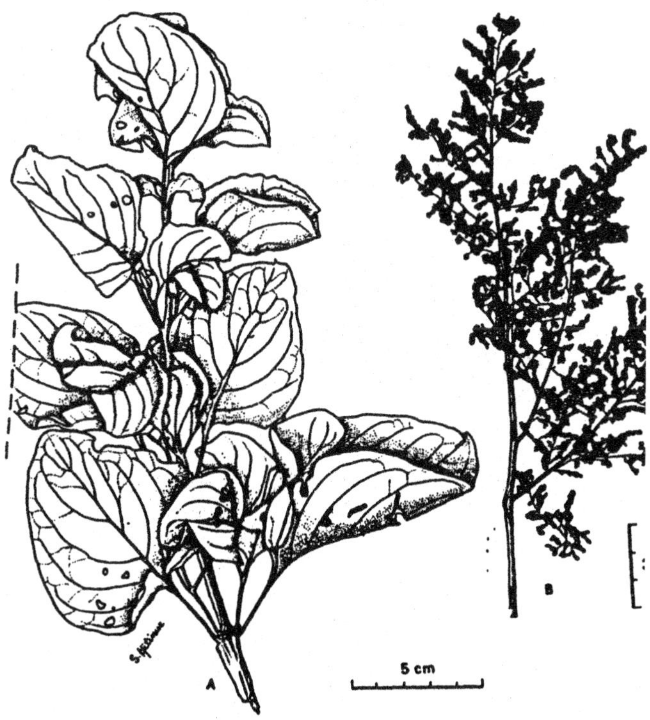

Iresine herbstii A. habit; B. inflorescence. (Source: Kanis 1978)

flowers; style very short; stigmas 2. Fruit 1-seeded, not opening.

Key to Species in Papuasia

A single species:
..... Iresine herbstii

List of Species recorded from Papuasia

Iresine herbstii Hook. f. Gard. Chron. 654 (1864) 1206 ; Kanis (1978) 39

o o o o o o

Notes:

1. This S American species is now widely dispersed. In Papuasia, where it is commonly used as an ornamental in mountain areas, it is propogated by vegetative cuttings only. It is well established in moist, shady places from 500 - 2500 m altitude. In the Simbai (Madang) it was reported as a vegetable (Kanis 1978).

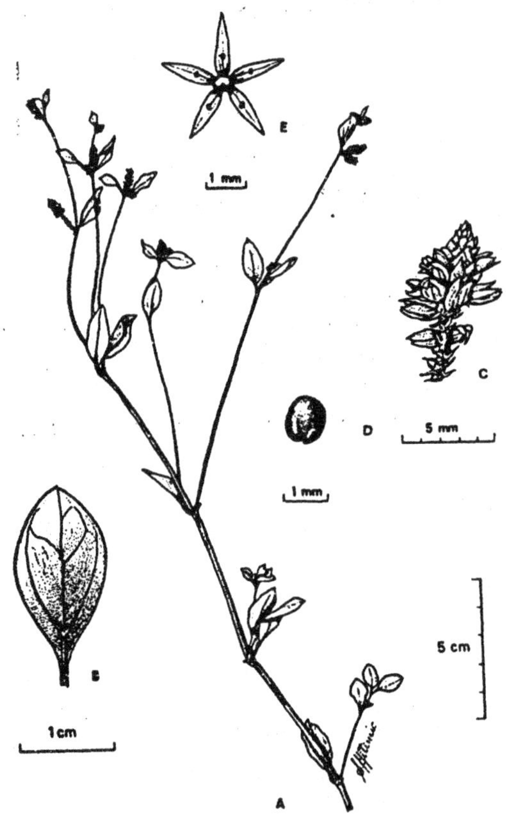

Psilotrichum ferrugineum A. habit; B. leaf; C. inflorescence; D. seed; E. flowers. (Source: Kanis 1978)

10. PSILOTRICHUM

o o o o o o

A paleotropical genus with 14 species mostly distributed in Africa. The single Asian species occurs in Papuasia.

Annual herbs; stems angular, ± hairy at nodes. Leaves opposite, glabrous. Inflorescences terminal, sometimes also lateral, solitary or in pairs, conical, lengthening during flowering; bracts and bracteoles membranous, glabrous. Flowers bisexual, sessile; tepals 5, with conspicuous parallel nerves, glabrous. Stamens 5, filaments joined at base. Staminodes absent. Style short, stigma capitate. Fruit 1-seeded, falling unopened with perianth and bracteoles.

Key to Species in Papuasia

A single species;
..... Psilotrichum ferrugineum

List of Species recorded from Papuasia

Psilotrichum ferrugineum (Roxb.) Moq. in C. DC. Prodr. 13 (1849) 279
P. trichotomum Bl. (1826); Kanis (1978) 41.

o o o o o o

Notes:

1. Found in moist areas and open grassland, often along tracks and in ditches. From sea level to 1200 m altitude. Sometimes a weed in other sites, rare.

BASELLACEAE

o o o o o o

A small family of 4 genera and 15 to 22 tropical and warm temperate species of vines found particularly in America. One pantropical species, Basella alba, is cultivated as a vegetable. It is introduced into Malesia. The rootstock is often tuberous. Andredera baselloides is cultivated as a decorative climber.

Diagnostic features:

Leaves: simple, broadly-ovate and fleshy, no stipules, arranged alternately on the long climbing stem of the vine.

Flowers: small, regular, bisexual or unisexual, supported by a pair of bracts, arranged in spikes

BASELLACEAE

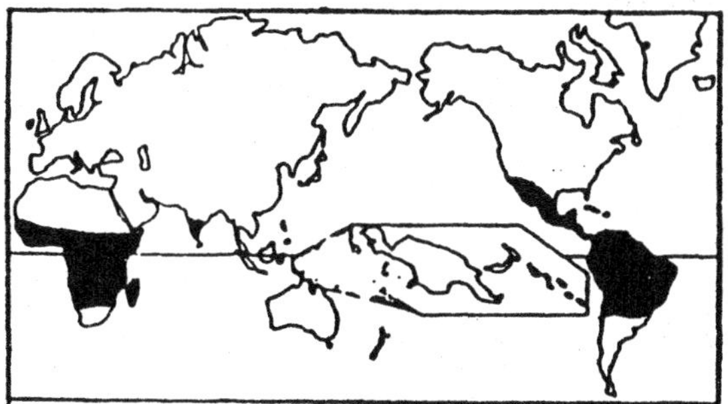

	WORLD	PAPUASIA
Number of genera :	4	2
Number of species :	15	2
Distribution : Mainly T America		

Economic uses : Limited use as vegetables and ornamentals.

Andredera cordifolia A. habit; B. cuplet; C. bud; D. flower. (Source: van Steenis 1957)

or racemes in leaf axils. Perianth has 5-segments, sometimes fused at base. Ovary superior. A single style and 3 stigmas.

Fruit: a drupe, enclosed within the persistent fleshy perianth.

References

Franz, E. (1908). Beitrage zur Kenntnis der Portulacaceen und Basellaceen. Bot. Jahrb. Beilblatt. No. 97. 42: 1 - 46. Figs. 1 - 43.

Mansfield, J. (1959). Kulturpflanze Beih. 2: 62.

Steenis, C. G. G. J. van (1957). Basellaceae. Flora Malesiana Series I. 5: 300 - 304.

Ulbrich, E. (1934). Basellaceae. Nat. Pfl. fam. 2 Aufl. 16C: 263 - 271. Fig. 117 - 118.

KEY TO GENERA IN PAPUASIA

Source: van Steenis (1957)

1. Flowers sessile, fleshy, hardly open at anthesis. Perianth very fleshy, enclosing the fruit. Filaments erect in bud. Embryo spirally twisted.
..... 2. BASELLA

1. Flowers with pedicel, not fleshy, patent at anthesis. Perianth not berry-like. Filaments reflexed in bud. Embryo semi-circular or horseshoe-shaped.
..... 1. ANDREDERA

1. ANREDERA

o o o o o o

An American genus of 5-10 species. Introduced as an ornamental plant in many tropical countries.

Branched vines with annual shoots from a fleshy rhizome. Leaves slightly fleshy, sessile or petioled. Inflorescences axillary, racemes or spikes, simple or branched. Bracts caducous or persistent; pedicels if present persistent, articulate below the perianth. Flowers bisexual rarely unisexual. Tepals connate at base, concealing the ovary, segments thin, later thickened and enclosing the fruit. Filaments broadened towards the base. Style 1; stigma 3-lobed. Fruit globose, enclosed by the perianth; pericarp fleshy. Seeds lens-shaped.

Key to Species in Papuasia

A single cultivated species:
..... Anredera cordifolia

List of Species recorded from Papuasia

Anredera cordifolia (Tenore) Steenis. Flora Malesiana Series 1. 5: 303.
Boussingaultiacordifolia Tenore Ann. Sc. Nat. III 19 (1853) 355.

o o o o o o

Anredera cordifolis A. flowering branch. (Source: Morley & Thompson. Fl. Pl. Aust. 1983).

2. BASELLA

o o o o o o

A genus of 5 species, natives of T Africa and Asia. Three species described as endemic to Madagascar. The single species in Papuasia can be used as a spinach, a pot herb and in leafy vegetable stews. The leaves are very nutritive being rich in several vitamins, calcium and iron. Pulped leaves are used as a poultice for sores. Also used to treat acne. The berries are dried and used for colouring edibles, jellies, pastry etc.

Succulent rhizomatous herbs. Flowers bisexual, in simple or branched spikes. Bracts minute, caducous. Bracteoles and perianth connate, fleshy, opening very little in anthesis, after flowering very fleshy and enclosing the fruit, forming a pseudo-berry. Stamens included, filaments very short, inserted near the top of the corolla-tube, erect in bud. Perianth segments short, blunt. Styles 3, stigmas linear. Embryo spiral.

Key to Species in Papuasia

A single cultivated species:
..... Basella alba

List of Species recorded from Papuasia

Basella alba L. Sp. Pl. 1753: 272.

o o o o o o

Notes:

1. Several cultigens, previously described as distinct species, are known. Burkill states that this species is probably native in the Old World, both in Africa and T S E Asia. It is often planted up to an altitude of 500 m.

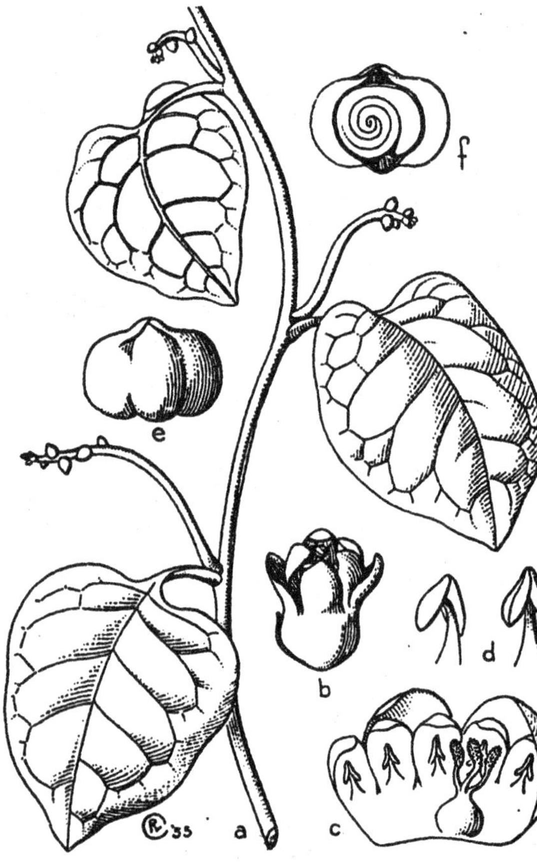

Basella alba A. habit; B.C. flower; D. anthers; E.F. pseudo berry. (Source: van Steenis 1957)

CARYOPHYLLACEAE

o o o o o o

An almost cosmopolitan family, throughout the temperate areas of the world and in the tropics common on mountains. It includes about 80 genera and 2200 species. In Papuasia some 5 genera with about 15 species. Many species are commonly used as garden ornamentals (carnations and pinks). The family is relatively uniform and easily recognised. Plants are usually herbaceous.

CARYOPHYLLACEAE

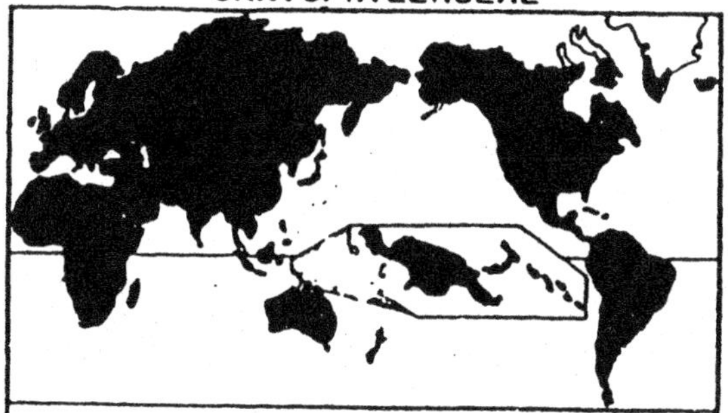

	WORLD	PAPUASIA
Number of genera :	89	7
Number of species :	2,070	13

Distribution : *Temperate. Mainly Mediterranean.*

Economic uses : *Garden plants, particularly carnations.*

Diagnostic features:

Leaves: simple, opposite, with or without stipules.

Flowers: bisexual, either singly or in a cymose cluster. Sepals 3-6. Petals 3- 6, each petal with a short or long claw. Stamens 2-12 (usually 8-10).

Fruit: usually a capsule opening at the top.

References

Bakker, C.A. (1956). Revision of the genus Polycarpaea in Malaysia. Acta. Bot. Neerlandica 6: 48 - 53.

Baker, C.A. (1963). Caryophyllaceae. Flora of Java 1: 206 - 214.

Duke, J. A. (1961). Preliminary revision of the genus Drymaria. Ann. Miss. Bot. Gard. 48(3): 173 - 268.

Gartner (1939). Fedde Rep. Beih. 113: 1 - 96. Figs. 1 - 101. T. 1 - 19.

Mattfeld, J. (1929). Caryophyllaceae in Diels, L. Beitrage zur Flora des Saruwaged Gebirges. Bot. Jahrb. 62: 475 - 479. (452 - 501).

Mattfeld, J. (1938). Key to Cerastium and Sagina. Bot. Jahrb. 69: 267 - 273.

Mattfeld, J. (1939). Biologische und morphologische Blutenformen bei den Caryophyllaceen. Notizbl. Berl. - Dahlem 14: 470 - 482. Figs. 1 - 15.

Merrill, E. D. (1923). An enumeration of the Philippine Flowering Plants. Manilla. Bureau of Science. (4 volumes).

Merrill, E. D. & Perry, L. M. (1942). Plantae Papuanae Archboldianae. X. Journ. Arn. Arb. 23: 386 - 388.

Merrill, E. D. & Perry, L. M. (1949). Plantae Papuanae Archboldianae XVIII. Journ. Arn. Arb. 30: 39 - 40.

Mizushima, M. (1957). A revision of Drymaria caudata Willd. Journ. Jap. Bot. 32: 69 - 81.

Pax, F. and Hoffmann, K. (1934). Caryophyllaceae. Pflanzen familien ed. 2. 16C: 275 - 367. Figs. 120 - 130.

Royen, P. van (1982). The Alpine Flora of New Guinea. 3: 2345 - 2371.

Steenis, C. G. G. J. van (1934). Caryophyllaceae in On the origin of the Malaysian mountain flora. Part 1. Bull. Jard. Bot. Btzg. III 13: 180 - 181.

Warburg, O. (1893). Bergpflanzen aus Kaiser Wilhelms-Land. Caryophyllaceae. Bot. Jahrb. 16: 21.

Williams (1898). Bull. Herb. Bois. 6: 893 - 904.

KEY TO GENERA IN PAPUASIA

1. Styles 3, or 1 style with 3 branches. 2

1. Style 4 - 5.......... 4

2. Styles 3, free. Leaves without stipules. Capsule 3- or 6-valved, if 3-valved then 1-seeded. ...... 7. STELLARIA

2. Styles 3, united towards the base. Leaves with stipules. capsule 3-valved, more than 1-seeded. 3

3. Annual herbs with conspicuous scarious silvery stipules and floral bracts. Basal leaves obovate, the rest linear and often clustered.
...... 3. POLYCARPAEA

3. Perennial herb with slender or subulate stipules.
...... 2. DRYMARIA

3. Styles 2. Stamens 1. Ovules usually 13.
...... 5. SCLERANTHUS

4. Petals entire. Fruit dehiscent by 4-5 valves, down to base. Leaves slender - linear, less than 3 mm wide.. 5

4. Petals usually with a shallow or deep notch at apex. Fruit dehiscent by 8-10 teeth. Leaves wider than 3 mm.
..... 1. CERASTIUM

5. Leaves of each pair joined at base. Stipules none.
...... 4. SAGINA

5. Leaves free. Membranous stipules present.
...... 6. SPERGULA

1. CERASTIUM

o o o o o o

A genus of some 60 to 70 species, almost cosmopolitan in distribution. Many weeds such as the chickweeds. Moschl (1978) made a detailed study of Cerastium papuanum. In Papuasia 3 species, one introduced species at low altitudes, the other 2 are mountain plants.

Hairy annuals or perennials. Leaves exstipulate, larger ones over 0.5 cm wide. Flowers in terminal, forked cymes, 4-5-merous. Sepals free. Petals oblong- obovate, emarginate or 2-lobed, white, rarely absent. Stamens 4-10. Styles 3- 5. Capsule cylindric, straight or slightly curved, with 8-10 teeth or valves. Seeds compressed, brown verrucose.

Key to Species in Papuasia

Source: van Royen (1982)

1. Pedicel both in flower and fruit, shorter or almost as long as calyx. No glands at base of filaments of stamen.
...... Cerastium glomeratum

1. Pedicel, both in flower and fruit, considerably longer than calyx. Stamens with a gland at base of filament.
..... Cerastium papuanum

Infraspecific taxa of Cerastium papuanum

Source: Moschl (1978)

ssp. geminiflorum Tattfield (1938) 269.

ssp. phaenops var. eciliatum Mattfield (1938) 269.

List of Species recorded from Papuasia

Cerastium glomeratum Thuill.

o o o o o o

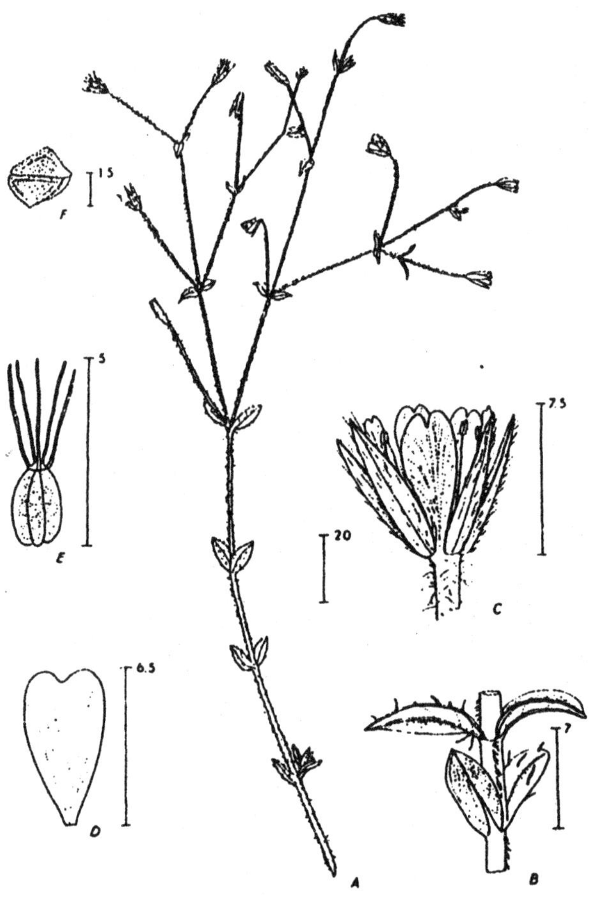

Cerastium papuanum A. habit; B. leaf; C. capsule and calyx; D. petal; E. seed. (Source: van Royen 1982)

Cerastium keysseri Schlechter ex. Diels (1929) 478. Mattfeld in Diels (1938) 270. van Royen (1982) 2358; Hoogland Blumea Suppl. 4 (1958) 228
= Cerastium papuanum

Cerastium papuanum Schlechter ex Mattfeld (1929) 477 emend. Moschl.
Cerastium keysseri Schlechter ex. Diels (1929) 478; Mattfeld in Diels (1938) 270; van Royen (1982) 2358; Hoogland Blumea Suppl. 4 (1958) 228

o o o o o o

Notes:

1. C. glomeratum is an introduced plant growing at low altitudes. A creeping herb with white flowers.

2. Cerastium holosteoides (previously C. vulgatum L. Sp. Pl. ed.2 (1762) 627 (Sargent pers com.) is a widespread weed of N Temperate Zone, extending into the tropics. Merrill reported it from the Philippines in 1927 and probably it occurs in Papuasia.

2. DRYMARIA

o o o o o o

A largely subtropical genus (Duke 1961), all but 2 of the 48 species being strictly American. Mizushima (1937) made a detailed study of Drymaria cordata and treated ssp. diandra, which includes most material collected in Papuasia, as a distinct species.

Annual or perennial, glabrous or pubescent herbs; prostrate, scrambling, creeping, spreading or erect. Leaves usually opposite, glabrous to villose or glandular. Sessile to long pedunculate, usually with persistent or fugaceous stipules.

Flowers few to many, in cymes or solitary or clustered in leaf axils. Sepals 5, not connate. Pedicels and sepals covered with minute, sticky, glandular hairs. Petals (0-)3-5, white or yellow, usually 2-cleft, often auriculate. Stamens 2-5, flattened filaments slightly connate at base, rarely alternating with prominent staminodia. Ovary superior, styles 3, ± united below. Capsule ovoid to sphaeroid, usually dehiscing into 3 entire valves. Seeds 1 to many.

Key to Species in Papuasia

A single species:
..... Drymaria cordata

Key to Subspecies of Drymaria cordata

Source: Duke (1961)

1. Flowers broadest at or below the middle; the sepals not inflexed; seeds 0.8 - 1.5 mm broad.
...... Drymaria cordata ssp. cordata

1. Flowers broadest above middle, the sepals usually incurved; seeds 1.5 - 2.0 mm broad.
...... Drymaria cordata ssp. diandra

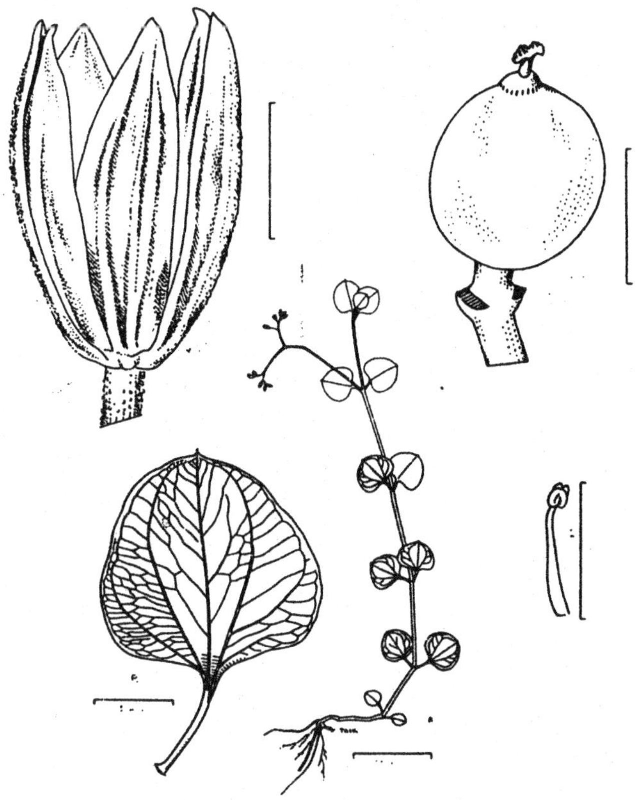

Drymaria cordata A. habit; B. leaf; C. flower; D. immature fruit; E. stamen. Hartley 10349. (Drawn T. Iwagu)

List of Species recorded from Papuasia

Drymaria cordata (L.) Willd ex. Roem. & Schult. Syst. Veg. 5 (1819) 406.

Holosteum cordatum L. Sp. Pl. (1753) 88. Hook. f. Fl. Brit. India 1 (1871) 244.

o o o o o o

Drymaria cordata ssp. diandra (Bl.) J. Duke (1961).

Drymaria diandra Bl. Bijdr. Fl. Nederl. Ind. (1825) 62.

Drymaria gerontogea F. Muell. Desc. Notes Pap. Pl. 1 (1887) 87. Sch. &Laut. (1901) 309.

o o o o o o

Drymaria diandra Bl. Bijdr. Fl. Nederl. Ind. (1825) 62.

= Drymaria cordata ssp. diandra

Drymaria gerontogea F. Muell. Desc. Notes Pap. Pl. 1 (1887) 87. Sch. & Laut. (1901) 309.

= Drymaria cordata ssp. diandra

Notes:

1. This species is morphologically very variable. It is common along ditches, in garden fallows and along the margins of drained swampland. A common weed of cultivation throughout the Highlands. Cooked and rubbed as an ointment on kaskas (scabies). Pantropical, probably a native of tropical America.

2. Common in Papuasia at altitudes from 440 m (Koitaki) to 1700 m (Kaindi and Kasanombe).

3. POLYCARPAEA

o o o o o o

A genus of 50 tropical and warm temperate species from the Old World. The genus is poorly represented in the region. Most specimens belong to Polycarpaea corymbosa which is of limited distribution. Polycarpaea spirostylis from Queensland occurs on the islands of the Torres Staits and 'probably' occurs in the Western Province. This species is well known as a copper indicator and is frequently used as a guide to deposits (Bailey Qld. Flora 1 (1899) 91). Bakker (1957) notes that the taxon from the Torres St. could be undescribed.

Annual or perennial, dichotomously branched herbs, sometimes woody at base. Leaves opposite, often with developed axillary buds, stipules scarious. Inflorescences poor or rich flowered cymes, provided with scarious bracts. Sepals scarious, not keeled. Petals, which Mattfeld (1938) interprets as staminodes, are structurally similar to the sepals, but frequently toothed at apex. In section Planchonea they are connate in the lower half into a tube or cup (P. spirostylis). Stamens 5 or less, sometimes with alternating minute staminodes (possibly in P. corymbosa (Bakker 1957). Style simple, stigma punctate to globular, or lobed. Capsule 3-valved.

Key to Species in Papuasia

Source: Bakker (1957)

1. Petals free, at the base annularly adnate to the stamens (and staminodes if present).
   ..... Polycarpaea corymbosa

1. Petals connate in their lower half into a tube or cup.
   ..... Polycarpaea spirostylis

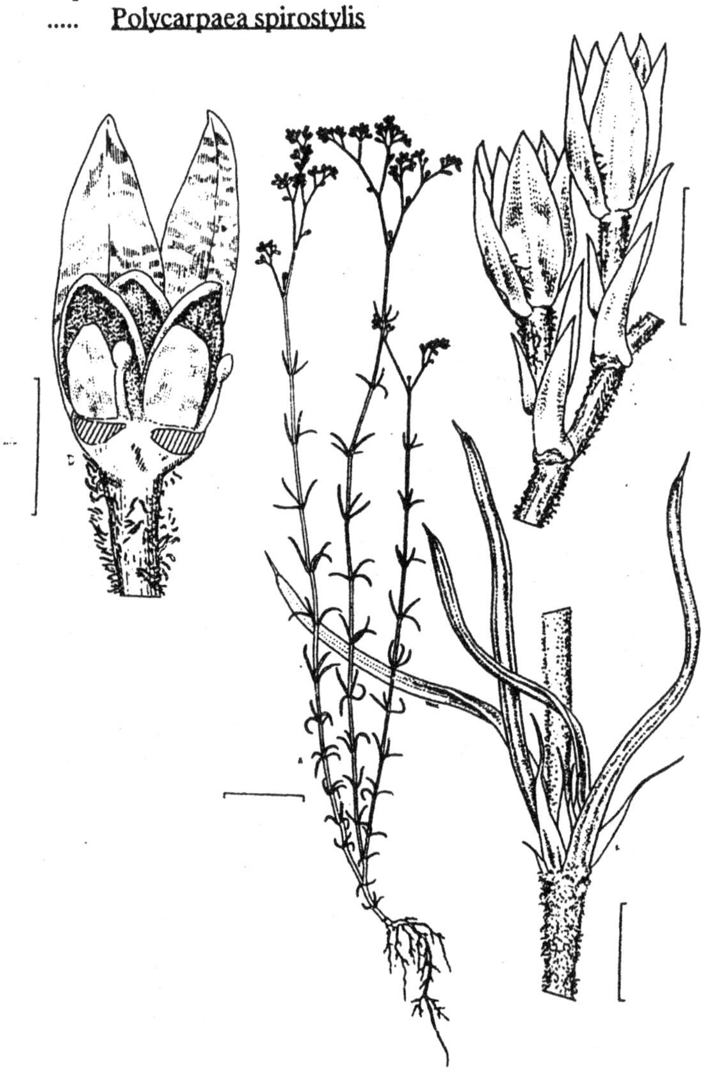

Polycarpaea corymbosa A. habit; B. leaves; C.D. flower. Pullen 6952. (Drawn T. Iwagu)

List of Species recorded from Papuasia

Polycarpaea corymbosa (L.) Lamk. Tabl. Encycl. 2 (1797) 129; Hook. f. Fl. Brit. India. 1 (1874) 245; Merrill in Philip. J. Sci. 5 (1910) Bot. 347

Achyranthes corymbosa L. Sp. Pl. (1753) 205.

o o o o o o

Polycarpaea spirostylis F. Muell. Rep. Bab. Exp. (1858) 8; Fl. Aust. 1 (1863) 165; Domin, Bibl. Bot. Heft. 69 (1925); F. Muell. Desc. Notes Pap. Pl. 2 (1886) 43; Bailey Qld. Flora 1 (1899) 91

o o o o o o

Notes:

1. P. corymbosa was collected at Kosipe Swamp in C. P. at 1935 m (ANU 28037). It is pantropical in distribution in open, dry, sandy places from sealevel to 1300 m altitude in the Philippines (Merrill 1927).

4. SAGINA

o o o o o o

A genus of circa 25 species from N temperate regions (13 in Europe) and tropical mountains. One species is a bad weed in temperate regions, others are cultivated as rock garden plants.

Small branched herbs. Leaves decussate, exstipulate, filiform or narrowly linear, minutely mucronate, basal portion forming a sheath. Sheaths of each pair of leaves connate at base. Flowers axillary, solitary or in few flowered cymes, 4- to 5-merous. Sepals free. Petals entire or subemarginate, white, rarely absent; tamens as many as sepals or twice their number. Styles 4- 5. capsule 4- to 5-valved, splitting to base. Seeds minutely verrucose, wingless.

Key to Species in Papuasia

Source: van Royen (1982)

1. Plants growing in cushions, sometimes in mats. Bracteoles absent. Flowers sessile or subsessile. Pedicels glabrous or scattered pilose with glandular hairs in between.. 2

1. Plants with long decumbent, glabrous runners, often forming intertwined mats. Bracteoles present or not. Flowers on long stalks. Pedicels scattered to densely pilose.
.... 3

2. Stamens very much shorter than petals. Leaves 2-7 by 1-2 mm, along elongate stems sometimes up to 12 by 1.5 mm.
...... Sagina donatoides

2. Stamens about as long as petals or slightly shorter. Leaves 2 - 7 by 0.5 - 1.0 mm.
...... Sagina belanophylla

3. Bracteoles 2. Stem erect, finely hairy in upper nodes. Leaves conferted at base of stem. Leaves 25 - 40 mm long, about 1mm wide.
...... Sagina monticola

3. Bracteoles none. Stems prostrate, glabrous. Leaves scattered along entire stem. Leaves 4-18 by 0.3 - 1.0 mm.
...... Sagina papuana

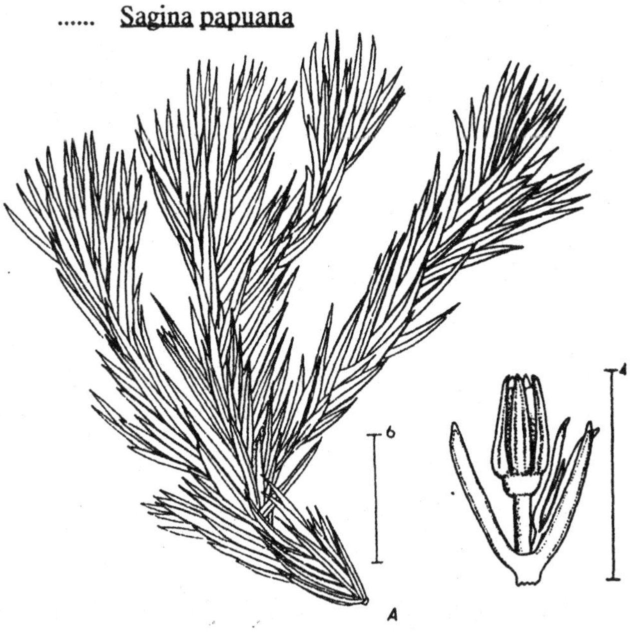

Sagina belanophylla A. habit; B. flower. (Source: van Royen 1982)

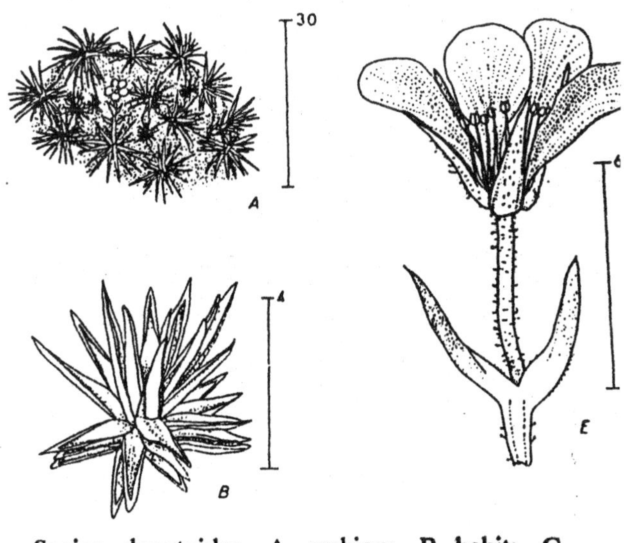

Sagina donatoides A. cushion; B. habit; C. flower. (Source: van Royen 1982)

List of Species recorded from Papuasia

Sagina belanophylla Mattfeld (1940) 468; Merrill & Perry (1942) 387; van Royen (1982) 2349.

o o o o o o

Sagina donatoides F. Muell. Trans. Roy. Soc. Vict. (1889) 3; Mattfeld (1938) 271; van Royen (1982) 2347.

o o o o o o

Sagina echinoperma Hayata sensu. Merrill and Perry (1942).
= Sagina monticola

Sagina monticola Merrill & Perry (1949)
Sagina echinoperma Hayata sensu. Merrill and Perry (1942).

o o o o o o

Sagina papuana Warb. (1892) 14; Mattfeld (1938) 270; Merrill & Perry (1942) 388; van Royen (1982) 2351

o o o o o o

Notes:

1. Sagina papuana is a variable species. The type was collected by Hellwig (No. 350) from the Finisterre ranges at c. 1000 m. Brass collected this species at high altitudes from 2800 to 3400 m.

2. Sagina belanophylla was described based on a collection made in 1937 from Mt Saawaket.

3. Sagina monticola was collected by Brass (Brass 11569) from vegetated limestone pinnacles.

5. SCLERANTHUS

o o o o o o

A world wide genus of 10 to 15, often weedy, species distributed from Europe to Asia, Africa and Australia. A single species in Papuasia restricted to alpine regions.

Prostrate, tufted perennial herbs forming dense mats. Stems terete, glabrous or papillate. Leaves linear, base slightly sheath-like, connate to opposite leaf, 1-nerved. Flowers solitary, sessile or subsessile, axillary. Bracts 2, tips apiculate. Pedicels glabrous. Calyx-tube cylindric, lobes erect or convergent. Petals absent. Stamens 1-3, filamenst connate at base. Styles 2. Capsule similar to ovary but larger, indehiscent.

Key to Species in Papuasia

A single high altitude species:
...... Scleranthus singuliflorus

List of Species recorded from Papuasia

Scleranthus singuliflorus (F. Mueller) Mattfeld (1938) 272
Mniarium singuliflorum F. Muel. Trans. Phil. Soc. Vict. 1 (1855) 13
Scleranthus mniaroides F. Muell. nom. illeg.

o o o o o o

Notes:

1. In boggy grassland from 2640 to 4350 m altitude. Growing amongst short Poa in Deschampsia tussock grassland at 3750 m altitude (Craven 2665). The species is distributed in Australia and Papuasia.

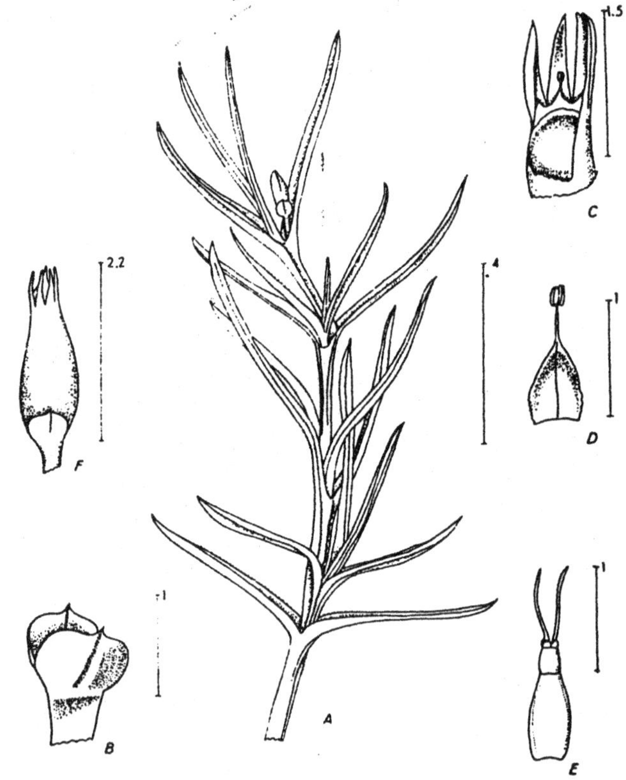

Scleranthus singuliflorus A. habit; B. bracts; C. corolla; D. stamens; E. fruit. (Source: van Royen 1982)

6. SPERGULA

o o o o o o

Represented by a single species in Papuasia. A cosmopolitan weed at higher altitudes. Backer (19) stated that it runs wild in Java at altiudes from 1450 to 2500 m altitude in gardens, along roadsides and in forest margins particularly in loose soils.

Erect herbs. Stem erect or ascending, fetid. Leaves with scarious- margined stipules; leaves linear filiform, bearing a bundle of leaves in axils (seemingly verticillate). Flowers in lax, terminal cymes, 5-merous; pedicels recurved after flowering. Sepals free, ovate, with membranaceous margins. Petals entire, white. Stamens 5 to 10. Styles 3(-5), short. Fruit dehiscent by 5-valves down to the base; valves alternating with sepals, entire. Seeds lenticular, winged.

Key to Species in Papuasia

1 introduced species:
...... Spergula arvensis

List of Species recorded from Papuasia

Spergula arvensis L. Sp. Pl. 1 (1753) 440; Hook. f. Fl. Brit. India 1 (1874) 243

o o o o o o o

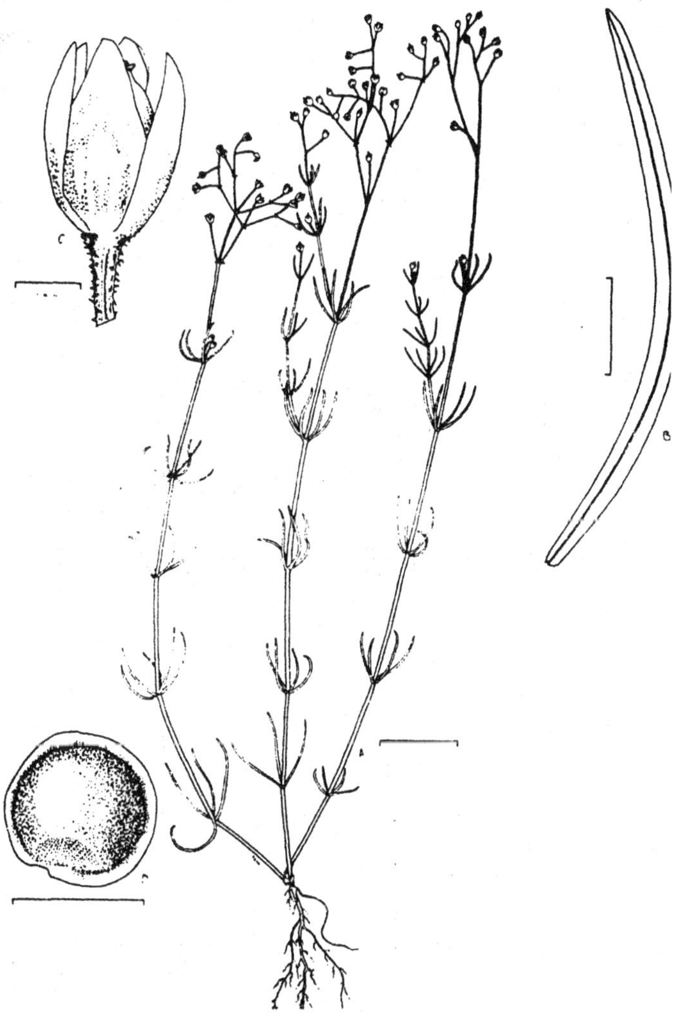

Spergula aruensis A. habit; B. leaf; C. flower; D. seed. Brass 29662. (Drawn T. Iwagu)

Notes:

1. A native of Europe which has become a cosmopolitan weed.

7. STELLARIA

o o o o o o

A cosmopolitan genus of some 120 species. Some are cultivated as ornamentals, others as weeds (ie. chickweed). In Papuasia 2 species occur at mid altitudes, one native and one introduced.

Annual or perennial herbs. Leaves exstipulate, distinctly longer than wide. Flowers solitary or cymose, 4- to 5-merous, chorisepalous. Petals bifid or bilobed, white, rarely absent. Stamens 2-10; styles 2-5. Capsule mostly ellipsoid, 3- to 6-valved. Seeds 1 to many, compressed; smooth, rugose or verrucose.

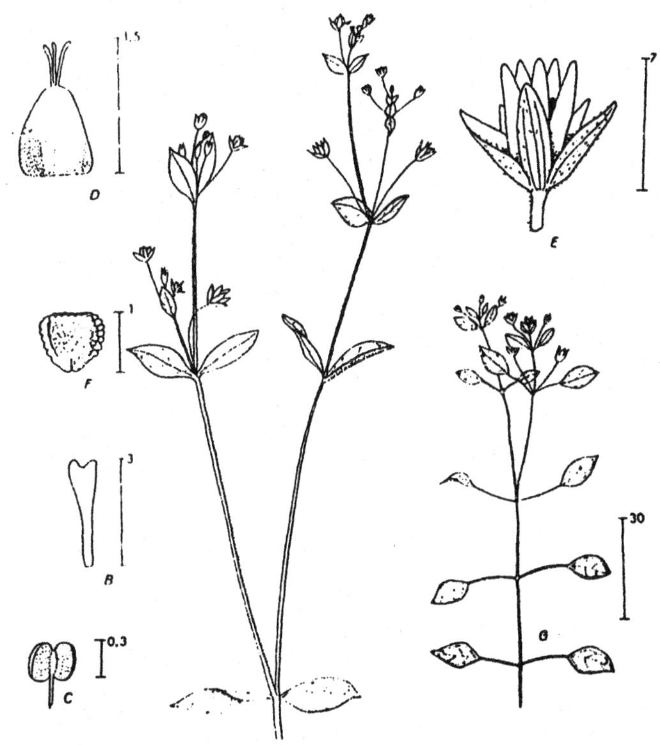

Stellaria media A. habit; B. petal; C. anther; D. ovary; E. fruit and calyx; F. seed; G. flowering stem. (Source: van Royen 1982)

Key to Species in Papuasia

1. Leaves with stellate hairs.
   ...... Stellaria stellato-pilosa

1. Leaves glabrous.
   ...... Stellaria media

The following species has been reported from Irian Jaya. It is not mentioned in van Royen (1982) but is a third species from Papuasia.
...... Stellariasaxatilis

List of Species recorded from Papuasia

Stellaria media (L.) Cyr. Char. Comm. (1784) 36; Hook. f. Fl. Brit. India 1 (1874) 230
Alsine media L. Sp. Pl. (1753) 272.

o o o o o o

Stellaria saxatilis Buch-Ham in Don Prodr. Fl. Nepal (1825) 215; Hook. f. Fl. Brit. India 1 (1874) 232

o o o o o o

Stellaria stellato-pilosa Hayata

o o o o o o

Notes:

1. Sprague in Kew Bull. (1920: 308 - 318) includes a detailed discussion of Stellaria v. Alsine.

2. Species have been collected as a weed in kaukau gardens at 1850 and along wet road margins up to 2720 m in altitude. Stellaria saxtilis was first reported by Merrill & Perry (1942) from Irian Jaya (Brass 10728 & Brass 11582). Brass reported them as being very abundant near L. Habbema from 2200 to 2800 m altitude, scrambling over fences, fernery and shrubs.

CHENOPODIACEAE

o o o o o o

A large family of over 100 genera and some 1500 species. It is cosmopolitan in distribution, but most species occur in the tropical and temperate zones of both hemispheres. Most are common in deserts, semi-arid and saline environments, or areas a pronounced dry season. Several have C4 photosynthesis. In Malesia 7 genera are represented, with only 3 in Papuasia (Backer 1949). Many species are weeds of cultivation.

Suaeda australis A. habit; B. flowering branch; C. flower. (Source: Fl. Aust. 4: 311).

CHENOPODIACEAE

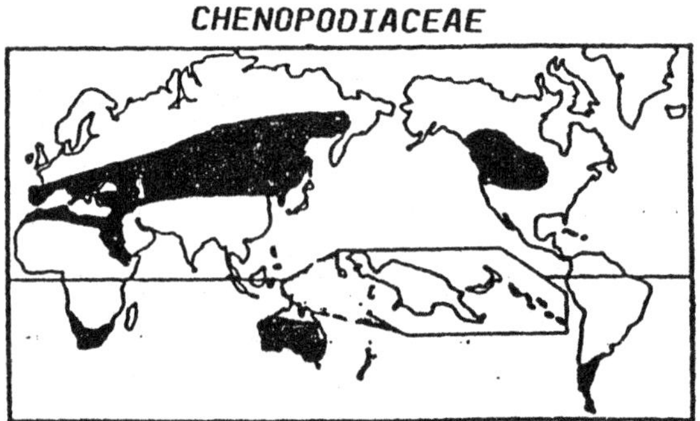

	WORLD	PAPUASIA
Number of genera :	120	5
Number of species :	1,300	9

Distribution : Temperate and sub tropical. Saline or desert.

Economic uses : Sugar beet, beetroot and spinach

Only the two genera Salsola and Tecticornia appear to be indigenous to Papuasia. Kanis includes two genera Arthrocnemum and Suaeda in his revision of the family because they both occur in adjacent areas and it is possibly they occur here. Beta vulgaris, sugar beet, is the only important agricultural crop.

Diagnostic features:

Leaves: male-textured or hairy, without stipules. Some genera lacking leaves; stems fleshy jointed.

Flowers: small, regular, unisexual or bisexual, wind-pollinated. Tepals 5, 3 or 2; brown or greenish lobes. Anthers equal in nu,ber to perianth lobes. Style terminal with 2 (rarely 1 or 3) stigmas.

Fruit: a small round nut or achene.

References

Backer, C. A. (1949). Chenopodiaceae. Flora Malesiana Ser. I. 4: 99 - 106.

Backer, C. A. (1954). Chenopodicaeae. Flora Malesiana Ser. I. 4: 594 - 595.

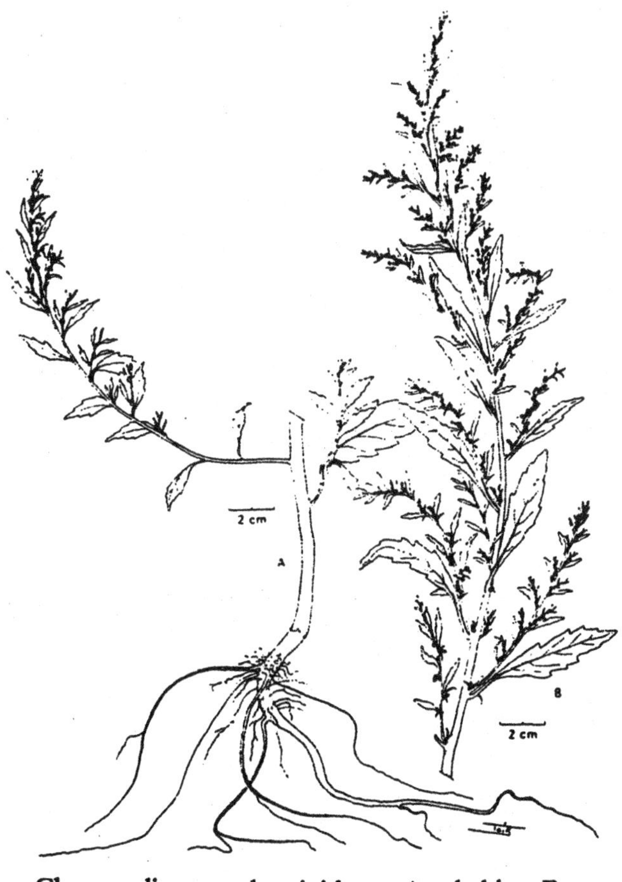

Chenopodium ambrosioides A. habit; B. flowering stem. (Source: Kanis 1982).

Kanis, A. (1984). A review of the Chenopodiaceae of Papuasia. pp. 1 - 6.

Merrill, E. D. & Perry, L. M. (1948). Journ. Arn. Arb. 29: 194.

Pulle, A. (1910). Chenopodiaceae. Nova Guinea 8: 349.

Scott, A. J. (1977). Reistatement and revision of Salicorniaceae. Bot. Journ. Linn. Soc. 75: 357 - 374.

Scott, A. J. (1978). A revision of Camphorosmioideae. Feddes Repert. 89: 101 - 119.

Steenis, C. G. G. J. van (1972). Addenda, corrigenda et emendanda. Flora Malesiana I. 6: 932.

Steenis, C. G. G. J. van (1967). Miscellaneous Botanical Notes XVIII. Blumea 15: 145 - 155.

Ulbrich, E. (1934). Chenopodiaceae. Nat. Pflanzenfam. 16C: 379 - 584. figs. 155 - 224.

Valeton, (1907). Bull. Dep. Agric. Ind. Neerl. 10: 9.

Wilson, P. G. (1972). A taxonomic revision of the genus Tecticornia. Nuytsia 1: 277 - 288, figs. 1 - 7.

Wilson, P. G. (1980). Nuytsia 3: 28 - 69.

Wilson, P. G. (1984). Chenopodiaceae. Flora of Australia. 4: 81 - 317.

KEY TO GENERA IN PAPUASIA

Source: Kanis (1984)

1. Branches apparently jointed, ± swollen, with opposite very reduced leaves. Inflorescences of sessile cymes of 3-5 flowers ± concealed between bracts, combined into compound, compactly ovoid or cylindrical structures............................ 2

1. Branches not jointed, slender, with alternate, clearly developed leaves (lowest leaves sometimes opposite); inflorescences of single flowers or flower clusters, sometimes combined into compound but rather lax structures......................... 3

2. Inflorescences terminal, sometimes with 1 or 2 sessile lateral ones at base; bracts ± free, half circular, with thick outer rim.
...... 5. TECTICORNIA

2. Inflorescences terminal and lateral, not sessile (terminal on short side branches); bracts of opposite pairs jointed into faintly 2-lobed cup-shaped structures with thin outer rim.
...... 2. HALOSARCIA

3. Plants shortly hairy (elsewhere often mealy), often with glandular hairs; leaves ovate to narrowly lanceolate, flat, mostly with distinct petioles; flowers without branches.
...... 1. CHENOPODIUM

3. Plants glabrous or almost so, without glandular hairs; leaves linear, ± succulent, sessile; flowers with bracteoles.. 4

4. Leaves pungent; bracteoles similar to bracts, longer than perianth; tepals thickened across middle, in fruit usually with horizontally extending wing.

...... 3. SALSOLA

4. Leaves not pungent; bracteoles transparent, shorter than perianth; tepals not thickened, though in fruit ± evenly succulent, without wing.

..... 4. SUAEDA

BETA

Beta vulgaris L. is cultivated in the Highlands of Papuasia. it is commonly referred to as beetroot. Another variety of this species is used for the production of sugarbeet in temperate areas, particularly in Europe.

1. CHENOPODIUM

o o o o o o

A genus of 70 (to 200) species, with 23 species recorded from Australia.

Annual or perennial herbs, occasionally weak shrubs, with minute sessile hairs which may collapse to form a scaly covering, or with glandular hairs, or glabrous. Leaves alternate, flat, entire or dissected. Flowers small, usually clustered, axillary or in panicles, bisexual or unisexual, the terminal often male or bisexual, the lateral female. Perianth 3- to 5-lobed, sometimes hardened in fruit. Stamens 1- 5, free or united at base into a disc. Stigmas 2 or 3. Pericarp membranous, rarely succulent. Seed lenticular.

Key to Species in Papuasia

1. Flower clusters combining into eafy thyrses as well as partly into terminal panicles; leaves subglabrous, covered with sessile yellow glands below, malodorous; stamens in bisexual flowers (1-)4 - 5; fruit at maturity concealed in tepals, with yellow glandular hairs in upper half and with (2-) 3 - 5 stigmas.

...... Chenopodium ambrosioides

1. Flower clusters not combining into distinct compound inflorescences; leaves with short hairs, at least some tipped with yellow glands, aromatic; stamens in bisexual flowers 1(-2); fruit at maturity not concealed in tepals, glandless and with 2 stigmas.. 2

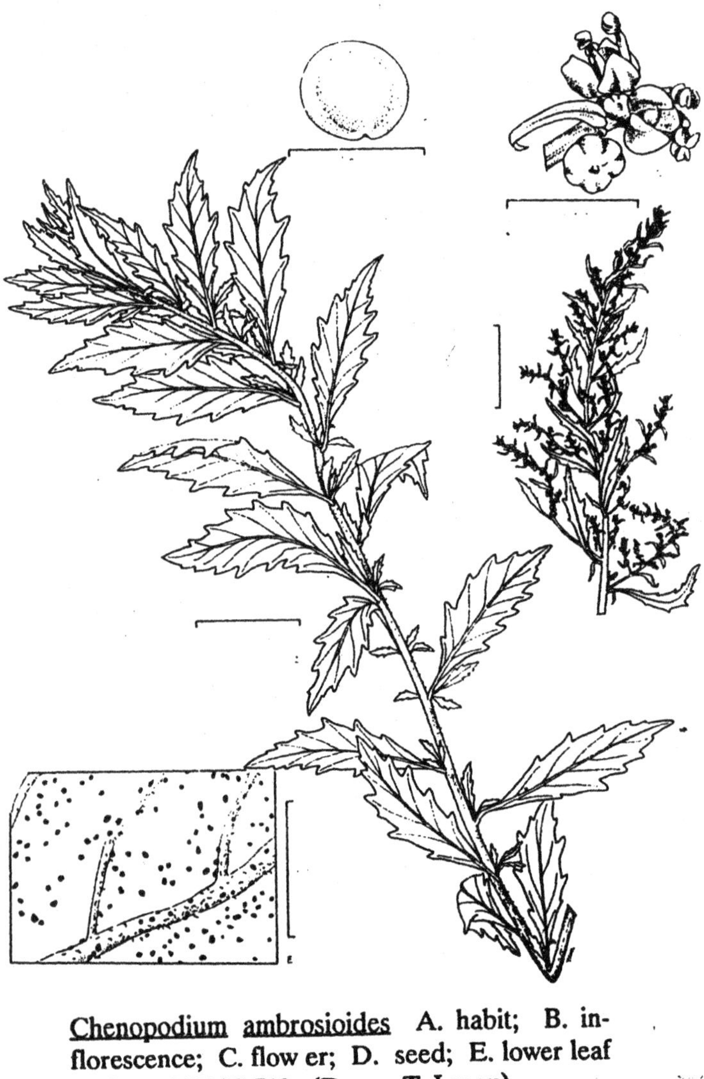

Chenopodium ambrosioides A. habit; B. inflorescence; C. flow er; D. seed; E. lower leaf surface. UPNG 712. (Drawn T. Iwagu)

2. Tepals distinctly keeled with truncated triangular crests, giving perianth a star-like appearance; seed distinctly keeled along most of circumference.

...... Chenopodium carinatum

2. Tepals concave but not keeled; seed keeled along half the circumference only.

...... Chenopodium pumilio

List of Species recorded from Papuasia

Chenopodium ambrosioides L. Sp. Pl. (1753) 219; Hartley Lloydia 36 (1973) 246; Kanis (1984) 3

o o o o o o

Chenopodium carinatum R. Br. Prodr. Fl. Nov. Holl. (1810) 407; Merrill & Perry (1948) 154; Steenis (1954) 595; Kanis (1984) 3

o o o o o o

Chenopodium pumilio R. Br. Prodr. Fl. Nov. Holl. (1810) 407; Steenis (1967) 154, (1972) 932; Kanis (1984) 3

o o o o o o

Notes:

1. Kanis (1984: 2) refers to notes from E. E. Henty regarding C. ?album. A collection of this was made by Henty but the condition of the specimen is so bad that positive identification is not possible.

2. C. carinatum is presumably recently introduced from Australia where it is widely distributed (van Steenis 1967). Clemens first collected the species at Kajabit Mission in November 1939 (Clemens 10814). It is reportedly a weed of tobacco cultivation in the Markham Valley. C. carinatum and C. pumilio intergrade in Australia (Kanis 1984) but the two collections from Morobe Province are quite distinct.

3. Hartley (1973) first recorded C. ambrosioides as collected in 1962. They are spread in Central and Morobe Provinces: village tracks and river banks from sea-level to 2000 m altitude.

HALOCNEMIUM

Halocnemum cinereum F. Muell. is the basionym for Tecticornia cinerea (F. Muell.) F. M. Bailey (1901).

2. HALOSARCIA

o o o o o o

Wilson (1980, 1984) lists the species previously described under Arthrocnemum in the genus Halosarcia. The genus includes 23 species all endemic except for Halosarcia indica (Willd.) P. G. Wilson which occurs around the Indian Ocean, Malaysia and N Australia.

The species is not yet collected from Papuasia but occurs in neighbouring areas so will probably occur on tidal mudflats in this region. Backer (1949) records Halosarcia (Anthrocnemum) indicum from Java to Timor. It or a closely related species probably occurs in the southern parts of the New Guinea mainland which have a pronounced dry season.

Glabrous shrubs or herbs. Young branches with cylindrical succulent, segments, the apex representing the reduced opposite leaves. Inflorescence a spike-like thyrse, usually terminal to a branchlet, consisting or triads of flowers in the axils of each bract. Flowers sessile, usually bisexual but the stamens or ovary sometimes not developed, some plants only female. Perianth

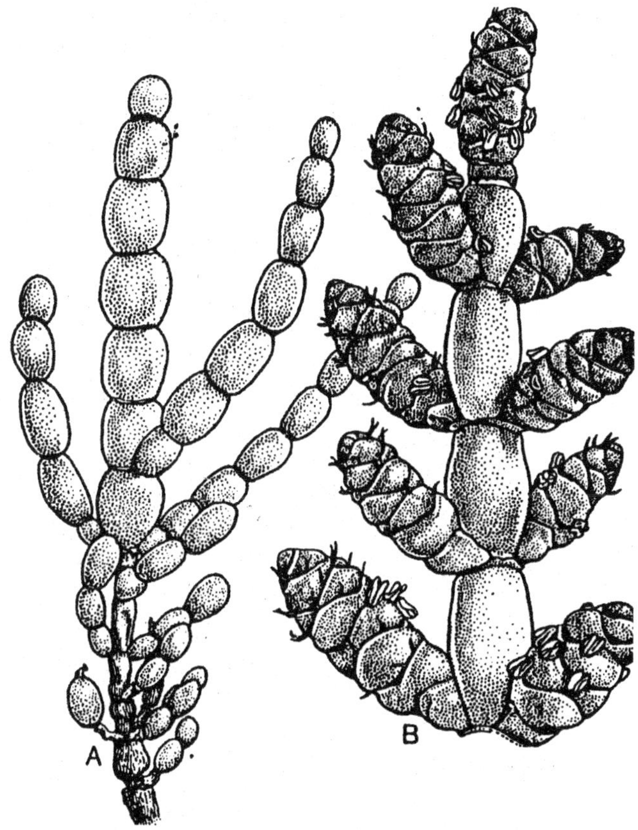

Halosarcia bulbosa A. vegetative branch; B. flowering branch. (Source: Fl. Aust. 4: 295).

gamopetalous, initially membranous or succulent; lobes 3 (2 lateral 1 abaxial). Stamens 1, abaxial. Ovary vertical, membranous or succulent, style 2(-3)-lobed, slender. Fruiting perianth membranous, succuoent, spongy.

Key to Species in Papuasia

A single species;
..... Halosarcia indica

List of Species recorded from Papuasia

Halosarcia indica (Willd.) P.G. Wilson Nuytsia 3 (1980) 63
Anthrocnemum indicum (Willd.) Moq. Chenop. Monogr. Enum. (1840) 113
Salicornis indica Willd. Ges. Naturf. Freunde Berlin Neue Schriften 2 (1799) 111, t.4, f.2; Wilson (1984) 298

subsp. indica. Wilson (1980) notes that this plant appears to occur only as female plants in Africa, India and Australia.

3. SALSOLA

o o o o o o

The genus Salsola is represented by over 100 species in Europe, Asia and Africa, introduced into Australia and America. It is represented by a single species in Papuasia which probably is naturally occuring.

Key to Species in Papuasia

A single species in Papuasia:
...... Salsola kali

List of Species recorded from Papuasia

Salsola kali L. Sp. Pl. (1753) 222; F. Muell. (1886) 28; Warb. (1891) 303; Pulle (1910) 349; Booberg (1933) 5; Backer (1949) 106; Steenis (1954) 595; Kanis (1984) 4; Wilson (1984) 314

o o o o o o

Notes:

1. It is restricted in distribution to the southern Coasts of New Guinea, particularly to areas with a pronounced dry season. An annual herb with yellow flowers, growing on salt flats at Cape Suckling, also on dry sandy beach material and on sandy coral soils. Warburg (1891) collected it from the Aru Islands.

2. The complex of S. kali is recognised in Papuasia. Elsewhere taxonomists recognise several species and infraspecific taxa. The complex needs comparative studies with materials from outside the region to understand its taxonomic affinities.

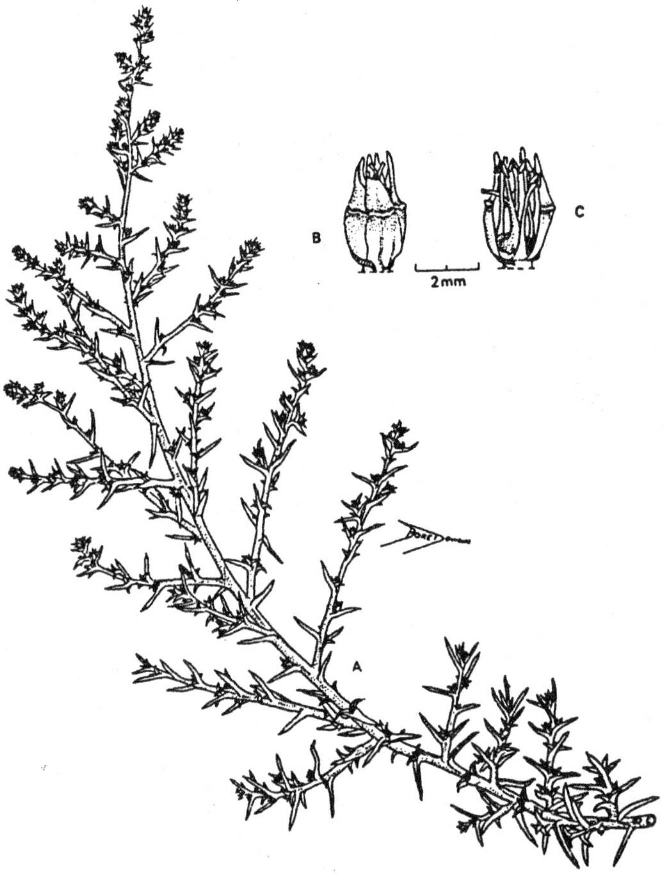

Salsola kali A. habit; B.C. flowers. (Source: Kanis 1981)

4. SUAEDA

o o o o o o

Records relating to this genus in Backer (1949) refer to Batis argillicola (Batacaeae) which was described by van Royen in 1956 (Nova Guinea n.s.)

Suaeda maritima (L.) Dumort, reported from New Guinea was retained in the treatment by Kanis as he felt that 'a species of Suaeda may well be discovered in Papuasia.

5. TECTICORNIA

o o o o o o

A genus of 3 species with two endemic in Australia, the third species also occuring in Java and Papuasia.

Annual or short-lived perennial, herbaceous to suffruticose, apparently leafless plants. Branches apparently jointed and succulent due to the opposite leaves enveloping the stem in a cylindrical sheath; leaf blade minute or absent. Inflorescence of terminal and lateral spike-like cymes consisting of 3(5) flowers in the axils of opposite and decussate rows of free bracts. Bracts closely imbricate, standing vertical to the spike axis, broadly deltoid to semicircular, the outer margin fleshy with an upturned rim.

Flowers hermaphrodite, concealed within bracts and standing vertical to the spike axis, free or early united to the upper bract and each other. Perianth succulent consisting of two free tepals, enclosing the ovary; stamen solitary and abaxial. Style slender, weak with a pair of filiform stigmas. Fruiting perianth with tepals completely free. Seed erect with endosperm.

Key to Species in Papuasia

A single species in Papuasia:
...... Tecticornia australasica

List of Species recorded from Papuasia

Tecticornia australasica (Moq.) P. G. Wilson Nuytsia (1972) 280
..... Halocnemum australasicum Moq. Chenop. Enuk. (1840) 110
..... Halocnemum cinereum F. Muell. Fragm. Phytogr. Austral. 1 (1859) 140
.... Salicornia cinerea (F. Muell.) F. Muell. (1885); Kanis (1984) 4; Pulle (1910) 349; Backer (1949) 104
..... Tecticornia cinervia (F. Muell.) F. M. Bailey (1887) 185

o o o o o o

Tecticornia cinervia (F. Muell.) F. M. Bailey (1887) 185
= Tecticornia australasica

Notes:

1. On salt flats and coastal mudflats at Kapakapa in the Central Province and also at Merauke in Irian Jaya. Purple and green forms of the plants grow intermingled. van Royen (1956) provides a detailed discussion of Tecticornia from this region. Most plant are less than 40 cm tall.

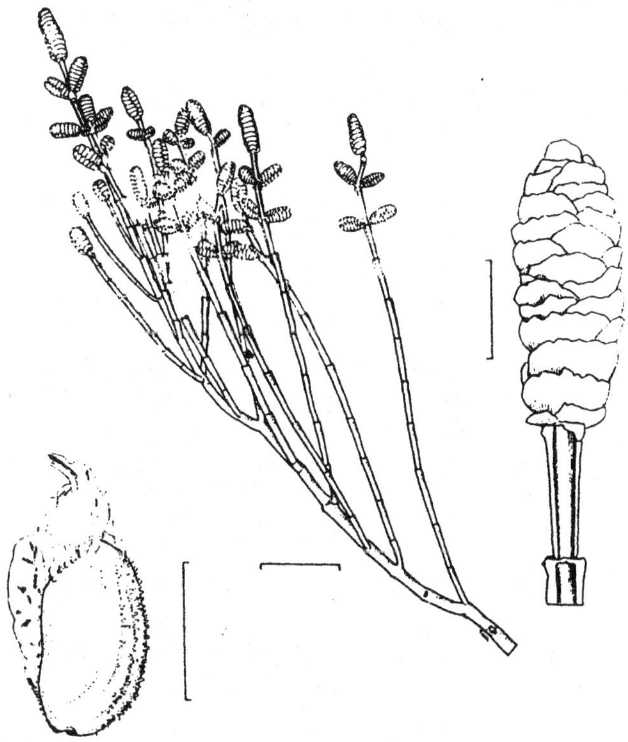

Tecticornia australasica A. habit; B. inflorescence; C. seed. McKee 1915. (Drawn T. Iwagu)

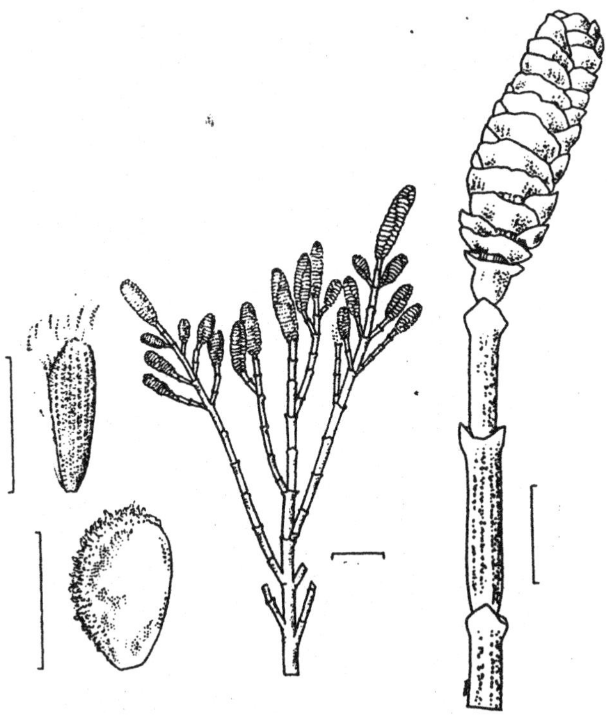

Tecticornea cinerea A. habit; B. flowering head; C. D. seeds. Rau 250. (Drawn T. Iwagu)

MOLLUGINACEAE

o o o o o o

A family of 9 genera and some 90 species distributed in tropical and warm areas, centered in Africa. Often with unusual patterns of secondary thickening: concentric rings of vascular bundles or alternate rings of phloem and xylem. The family was previously included in the Aizoaceae but is scarcely succulent and has free sepals. The major distinguishing character is the presence of anthocyanins.

Herbs, rarely shrubs. Leaves simple scarcely succulent, opposite to spirally arranged or whorled. Stipules small and deciduous or absent. Flowers usually bisexual, regular, often small; solitary or in cymes. Calyx 5, rarely basely connate. Staminodes 7, small or absent, rarely forming a tube. Stamens (2-)5- 10 (-more) sometimes basely connate; nectary as a ring around or on lower part of gynoecium. Styles distinct. Fruits usually dehiscent, often with a persistent calyx.

MOLLUGINACEAE

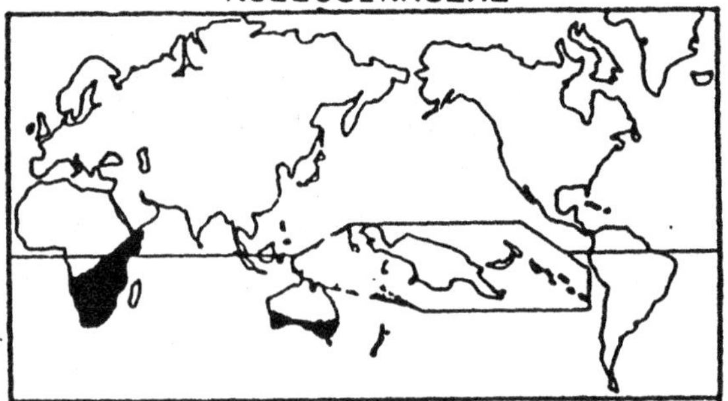

	WORLD	PAPUASIA
Number of genera :	9	2
Number of species :	90	3

Distribution : Tropical and warm temperate. esp. Africa

Economic uses :

Diagnostic features:

Leaves: simple, stipules small, deciduous or absent.

Flowers: bisexual, regular. Tepals 5, free. Staminodes present. Stamens usually 5-10. Styles distinct.

Fruit: dehiscent, calyx persistent.

References

Backer, C. A. (1951). Aizoaceae (including Molluginaceae). Flora Malesiana I. 4: 267 - 275.

Merrill, E.D. & Perry, L.M. (1942). Plantae Papuanae Archboldianae X. J. Arn. Arb. 23: 386.

Pulle, A. (1910).

KEY TO GENERA IN PAPUASIA

1. Seed with an appendage. Flowers in axillary clusters. Leaves not very narrow, often hairy.
   ...... 1. GLINUS

1. Seeds without an appendage. Flowers in terminal cymes or stalked umbels. Leaves narrow, glabrous.
   ...... 2. MOLLUGO

1. GLINUS

o o o o o o

A genus of some 10 species from the tropics and subtropics of both hemispheres. Two species in Malaysia, sometimes used for medicinal purposes.

Diffuse, glabrous or hairy, often much branched annuals. Leaves opposite or spuriously whorled, entire or obscurely dentate. Flowers axillary in few-flowered clusters or fascicles, pedicelled or subsessile. Tepals with scarious, frequently white margins, often unequal. Stamens 3-20, free. Styles erect, persistent. Capsule 3- to 5-valved. Seeds numerous, distinctly strophiolate. Testa finely granulate.

Key to Species in Papuasia

A single species:
..... Glinus oppositifolius

List of Species recorded from Papuasia

Glinus oppositifolius (L.) C. DC. Bull. Herb. Bois. 2 (1901) 552; Pulle (1910) 355; Backer (1951) 270; Merrill & Perry (1936) 386.

o o o o o o

Notes:

1. A second species is doubtfully recorded: Glinus spergula (L.) Steud. Grows in dried-up pools and ditches in dry regions.

2. M. cerviana occurs in S E Asia and Australia and will probably be collected in Papuasia.

Glinus oppositifolius A. habit; B. leaf; C. inflorescence; D. seed. Brass 8446. (Drawn T. Iwagu)

2. MOLLUGO

o o o o o o

A genus of some 15 species, tropical and temperate in distribution. Two species in Papuasia or 1 species if M. stricta is accepted as a synonym of M. pentaphylla. Used as a pot herb and for medicinal purposes.

Erect or diffuse, slender, glabrous herbs, mostly annual. Leaves partly subopposite, entire. Flowers small, in terminal cymes, single flowers pedicelled. Tepals herbaceous, with scarious, often white margins. Stamens 3-5, alternating with tepals, less often 6-10, free; filaments filiform. Styles 3, filiform. Fruit leaving a persistent pedicel.

Key to Species in Papuasia

1. Ripe seeds very distinctly granulate. Leaves often more than 3 mm wide.
...... Mollugo pentaphylla

1. Ripe seeds not at all granulate, but very faintly reticulately ribbed. Leaves less than 3 mm wide.
...... Mollugo cerviana

List of Species recorded from Papuasia

Mollugo cerviana (L.) Ser. in C.DC. Prod. 1 (1824) 392
Pharnaceum cerviana L. Sp. Pl. (1753) 272; Backer (1951) 268.

o o o o o o

Mollugo pentaphylla L. Sp. Pl. (1753) 89; F. Muell. (1888) 43.
Mollugo stricta L. Sp. Pl. 2 (1762) 131; Backer (1951) 268.

o o o o o o

Mollugo stricta L. Sp. Pl. 2 (1762) 131
= Mollugo pentaphylla

Notes:

1. M. pentaphylla grows throughout the tropics and subtropics of Old World, throughout Malaysia to New Britain.

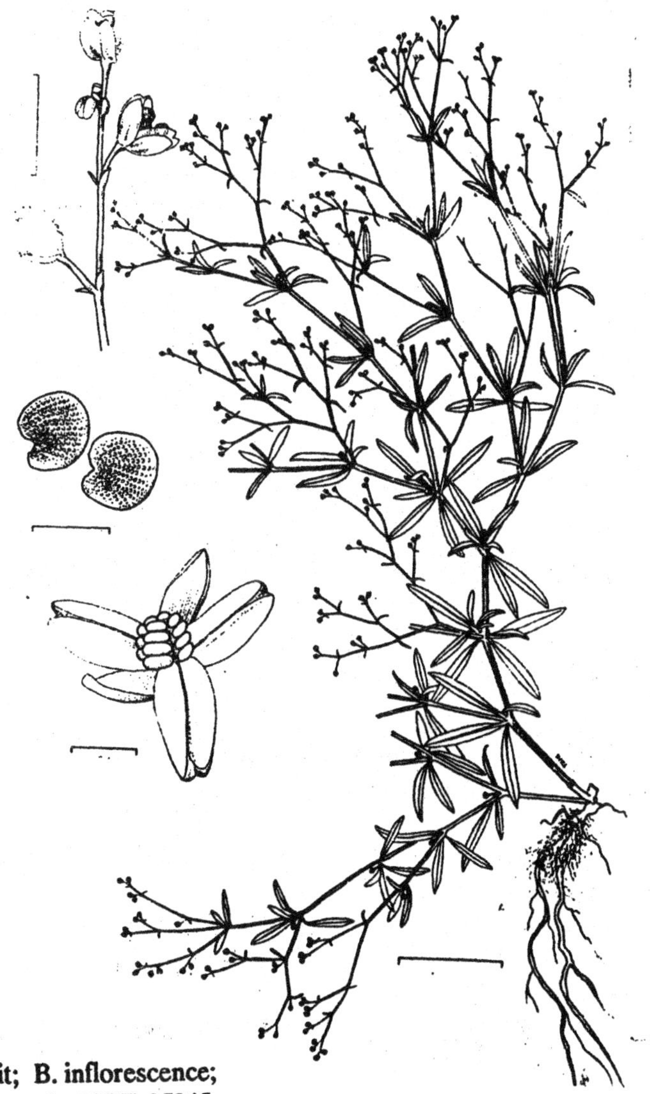

Mollugo pentaphylla A. habit; B. inflorescence; C. capsule with seeds; D. seed. NGF 35245. (Drawn T. Iwagu)

NYCTAGINACEAE

o o o o o o

A family of tropical trees, shrubs and herbs. Pantropical but particularly American in distribution. The family consists of some 30 genera and 300 species distributed in tropical and subtropical regions; abundant in the New World but also common in the Old World. Fosberg (1978) noted that the major problems within the Nyctaginaceae are associated with generic limits. The strand and coastal species are polymorphic, particularly in the genus Boerhavia. Members of the genus Bougainvillea are often grown as decorative hedges. Many cultivars have been produced. Many species of Mirabilis are grown as garden plants. The leaves of Pisonia are edible (P. grandis and P. alba). Extracts of other species of Pisonia (P. aculeata and P. capitata) are used as medicines.

The family is closely related and probably derived from the Phytolacc aceae.

NYCTAGINACEAE

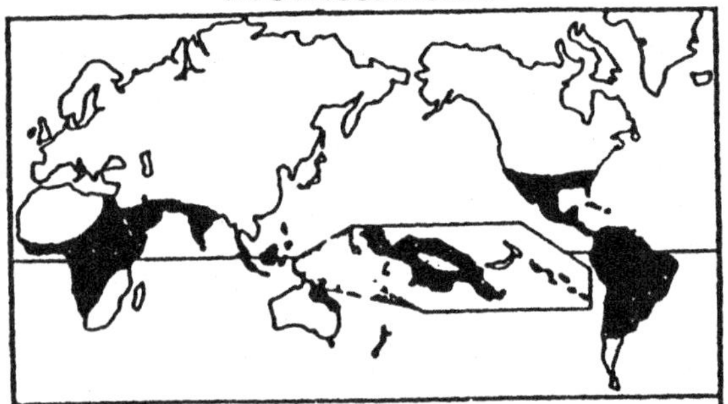

	WORLD	PAPUASIA
Number of genera :	*34*	*4*
Number of species :	*350*	*14*

Distribution : *Pantropical, mainly from America.*

Economic uses : *Ornamentals ie. Bougainvillea and vegetables - Pisonia. also medicinals.*

Diagnostic features:

Leaves: alternate or simple, without stipules.

Flowers: inflorescences cymose, flowers bisexual or unisexual, sometimes surrounded by coloured bracts, resembling the calyx. Perianth petal- like. Petals lacking, 1 to many stamens, usually 5, alternating with the 5 perianth lobes. Filaments free or fused at the base. Ovary superior, surmounted by a long style.

Fruit: an indehiscent achene, sometimes enclosed by persistent calyx.

References

Airy Shaw, (1952). Kew Bull. 87.

Bargagali-Petrucci, G. (1901). Le specie di Pisonia della Regione dei Monsoni. Nuov. Giorn. Bot. Ital. II, 8: 603 - 645.

Fosberg, F. R. (1978). Studies in the genus Boerhavia 1 - 5. Smithsonian Contributions to Botany 39: 1 - 20.

Gillis, W. T. (1976). Bougainvilleas of Cultivation. Baileya 20: 34 - 41.

Heimerl, A. von (1913). Die Nyctaginaceen-Gattungen Calpidia und Rockia. Osterre Bot. Zeit. 63: 279 - 290.

Heimerl, A. von (1934). Nyctaginaceae. Nat. Pfl. fam. 2 Aufl. 16C: 86 - 134.

Heimerl, A. von (1937). Nyctaginaceae of S E Polynesia and other Pacific Is. Occ. Papers of Bernice P. Bishop Museum. 13(4): 27 - 47.

Holttum, R. E. (1955). The Cultivated Bougainvilleas I. MAHA Magazine 12(2): 2 - 9.

Holttum, R. E. (1955). The Cultivated Bougainvilleas II. B. x buttiana, its varieties and hybrids. MAHA Magazine 12(3): 2 - 11.

Holttum, R. E. (1955). The Cultivated Bougainvilleas III. The varieties of B. glabra. MAHA Magazine 12(4): 25 - 36.

Holttum, R. E. (1956). The Cultivated Bougainvilleas IV. B. spectabilis and its varieties. MAHA Magazine 13(1): 13 - 22.

Holttum, R. E. (1956). The Cultivated Bougainvilleas V. The recorded hybrids. MAHA Magazine 13(2): 66 - 74.

Lauterbach, (1929). Bot. Jahrb. 62: 287 - 556.

Lauterbach, (1930). Bot. Jahrb. 63: 449.

Merrill, E. D. and Perry, L. M. (1939). J. Arn. Arb. 20: 327.

Mueller, F. von (1876). Descr. Notes Pap. Pl. 1: 58.

Pulle (1910). Nova Guinea 8: 353.

Pulle (1912). Nova Guinea 8: 629.

Schumann, K. (1895). Notizbl. Bot. Gart. Berlin 1: 49.

Schumann, K. (1896). Notizbl. Bot. Gart. Berlin 1: 208.

Schumann, K. (1898). Notizbl. Bot. Gard. Berlin. 2: 118.

Schumann, K. and Hollrung, K. (1889). Fl. Kais. Wilh. 42.

Schumann, K. and Lauterbach, C. (1900). Die Flora der deutschen Schutzgebiete in der Sudsee 306 - 308.

Skottsberg, (1926). Act. Hort. Gotheb. 2: 231.
Skottsberg, (1941). Svensk Bot. Tidskr. 35: 364.
Sprague (1928). Bull. Misc. Inf. Kew 348.
Stemmerik, J. F. (1964). Notes on Pisonia in the Old World. Blumea 12: 275 - 284.
Stemmerik, J. F. (1964). Nyctag- inaceae. Flora Malesiana I. 6: 450 - 468.
Valeton, (1907). Bull. Dept. Agric. Ind. Neerl. 10: 10.
Valeton, (1914). Bot. Jahrb. 52: 101 - 103.
Warburg, O. (1891). Bot. Jahrb. 13: 303 - 304.
White, C. T. (1929). J. Arn. Arb. 10: 211

KEY TO GENERA IN PAPUASIA

Source: Stemmerik (1964)

1. Herbs without thorns....................................... 2

1. Woody plants, sometimes with thorns........ 3

2. Leaves equal. Inflorescence involucrate. Perianth not articulated with the pedicel, 4.5 - 6.25 cm long. Anthocarp with faint ribs, not viscid.
...... 3. MIRABILIS

2. Leaves sub-equal. Inflorescence without an involucrum. Perianth articulated with the pedicel, 1.5 -12 mm long. Weeds.
...... 1. BOERHAVIA

3. Each pedicel adnate to a subsessile coloured bract 3 - 6 cm long. 3-4 cultivated species as well as hyb rids and cultivars.
...... 2. BOUGAINVILLEA

3. Each pedicel bearing 1-3 small, not coloured, bracteoles.
...... 4. PISONIA

1. BOERHAVIA

o o o o o o

A genus of some 40 species, with some 200 published names (Fosberg, 1978). Used as local medicines, vege tables and often growing as weeds. Many species are tropical or subtropical, often with a wide geographical range. Represented in Papuasia by 2 species. The genus was revised by Stemmerik in 1964.

Herbs, rarely vine-like, creeping or scrambling, sometimes woody at the base. Leaves opposite, simple, entire to sinuate or lobed; usually petiolate. Inflorescence various, bracteate, invol ucral bracts not connate. Perianth of a united series of parts: a perigone; coralloid, tubular or funneliform, cadu cous, persistent (at maturity closely investing the fruit). Stamens basally attached, filaments free; ovary usually exserted; pistil 1, basally attached, style filiform, stigma capitate. Antho carp 3-5 or 10, sulcate, indehiscent, containing 1 erect seed.

Key to Species in Papuasia

Source: Stemmerik (1964)

1. Anthocarp club-shaped, roundish in section, with short hairs and small sessile to stalked glands, top rounded. Stamens 2-3.
..... Boerhavia diffusa

1. Anthocarp obconical, star-shaped in section, without glands or hairs, top truncate. Stamens 1-2.
..... Boerhavia erecta

Boerhavia diffusa A. habit; B. flower; E. upper leaf surface. NGF 1062 (Drawn T. Iwagu)

List of Species recorded from Papuasia

Boerhavia diffusa L. Sp. Pl. (1753) 3
Boerhavia erecta L. Sp. Pl. (1753) 3

o o o o o o

Boerhavia diffusa f. repens Pulle (1910) 353; Pulle (1912) 629; Valeton (1907) 10
Boerhaviarepens L. Pulle (1910)
Boerhavia mutabilis R.Br. forma repens L. Valeton (1914) 101

o o o o o o

Boerhavia erecta L. Sp. Pl. (1753) 3; Stemmerik (1964) 454

o o o o o o

Notes:

1. B. erecta is a pantropical weed, common in open sandy places up to 700 m altitude.

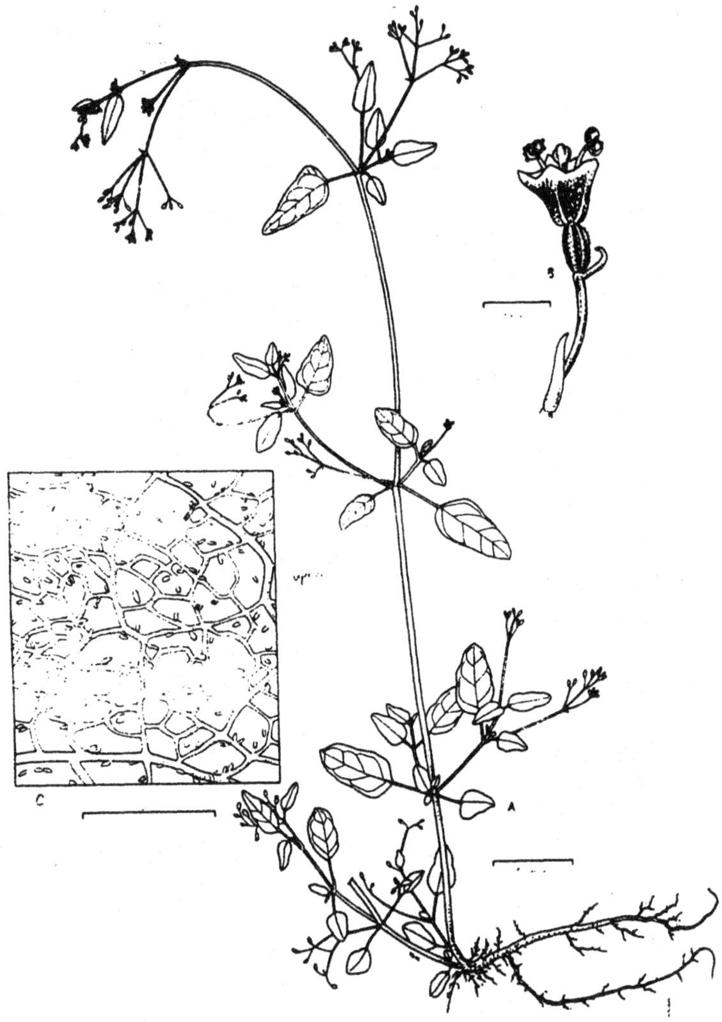

Boerhavia erecta A. habit; B. inflorescence; C. flower; D. fruit. UPNG 8596. (Drawn T. Iwagu)

2. BOUGAINVILLEA

o o o o o o

A genus of 14 central and tropical S American species, with flowers arising from persistent coloured bracts. Culti vated spiny ornamentals: most cultivars are hybrids of B. glabra X B. peruviana. Originally published as Buginvillaea (Juss. 1789), Spach in 1841 recorded it as Bougainvillea (nom. cons. Taxon 8 (1959) 268). Stemmerik (1964) noted the occurrence of 4 species in Malesia, all of which are introduced for cultivation.

Coarse climbers to 25 m, with supra-axillary spines (abortive inflor escences). Leaves opposite, ovate to elliptic -oblong. Inflorescences supra axillary above the bud, the peduncle bearing a single apical triad (3) of flowers, or the triad in dichasis branched 1-2(-3) times. Perianth tubul ar, 5(-4)-lobed, tube with 5(-4) ribs. Stamens (4-)5-8(-10), unequal, not ex serted. Gynoecium shorter than tube; stigma fimbriate. Anthocarp spindle- shaped, coriaceous, 5-ribbed, not viscid.

Key to Species in Papuasia

Source: Gillis (1976)

1. Flower tube less than 2.2 mm in diameter, glabrous.
...... Bougainvillea peruviana

Bougainvillea glabra A. habit; B. inflorescence; C. flower; D. LS flower. Live material UNITECH. (Drawn T. Iwagu)

1. Flower tube more than 2.2 mm in diameter, swollen at base, more or less pubescent........... 2

2. Flower tube distinctly angled, bearing abundant, spreading, straight hairs up to 1 mm long; stem and leaves villous.
...... Bougainvillea spectabilis

2. Flower tube not distinctly angled, with upward curving hairs 0.5 mm long or less; stem and leaves glabrous or pubescent.. 3

3. Leaves broadly ovate; bracts crimson or orange when fresh, apices obtuse, not crinkled; thorns straight; leaves becoming reduced in size below inflorescences, flowers closely grouped.
...... Bougainvillea x buttiana

3. Leaves narrowly ovate to elliptic; bracts magenta or purple when fresh, apices acute, often crinkled; plant thornless or thorn curved; leaves not reduced in size below inflorescences; flowers scattered.
...... Bougainvillea glabra

Notes:

1. Bougainvillea peruviana occurs in Colombia, Ecuador and Peru. The bracts are magenta to pink.

2. Bougainvillea spectabilis has purple coloured bracts.

3. Bougainvillea X buttiana is a pre sumed hybrid between B. peruviana and B. glabra. Some 'double' forms are known from cultivars of this hybrid.

4. Bougainvillea glabra, unlike other species, flowers throughout the year. It flowers best after heavy pruning. The hybrids with B. peruviana is the widely cultivated B. x buttiana. Holttum suggests B. x spectoglabra for hybrids with B. spectabilis.

5. Most species of Bougainvillea flower in response to dry weather.

CALPIDIA

Calpidia was reduced to Pisonia by Choisy in 1849. This is generally accepted but Skottsberg (1936) more recently suggested that it is a synonym of Ceodes (see the discussion under Pisonia by Stemmerik (1964) 276). Heimerl (1913) revised both Calpidia and Rockia.

List of Synonomy for Species in Papuasia

Calpidia cauliflora (Scheffer.) Heimerl. (1913) 283 Icones Bogoriensis Fasc. I Tafel Tafel XXII. Pisonia caulifera Scheff. Ceram and Amboinea to Waigeu.
= Pisonia cauliflora

Calpidia corniculata Heimerl (1913) 283
Ceodes corniculata Merr. & Perry (1939)
= Pisonia corniculata

Calpidia cuspidata Heimerl. (1913) 283
T Beccari Rami Exsicc. Nr. 340
= Pisonia muelleriana

Calpidia excelsa (Bl.) Heimerl. (1913) 283 - 284 New Guinea and Bismark Archipelago.
= Pisonia umbelliferax

Calpidia grandifolia (Warburg) Heimerl. (1913) 286
Pisonia grandifolia Warburg (1891) 303 Insel Aru.
= Pisonia longirostris

Calpidia lauterbachii (Warburg) Heimerl. (1913)286
Pisonia lauterbachii Warburg 1891); Schuman & Laut. (1901) 307
= Pisonia longirostris

Calpidia longirostris (Teysmann et Binnendijk.) Heimerl. (1913) 286; (1934) 125 (Natuurkundig Tidjschrift v. Nederl. Indie. XXV
Pisonia longirostris Teysm. Binn. (1863) 40; Valeton Icon. Bogor. Fasc. 1. Tafel XXI. Boeroe, New Guinea.
= Pisonia longirostris

Calpidia mulleriana (Warburg) Heimerl (1913)287
Pisonia mulleriana (1891) 304 New Guinea
= Pisonia mulleriana

Calpidia rostrata (Warburg) Heimerl. (1913) 288; (1934) 125
Pisonia rostrata (1891) 304
= Pisonia longirostris

Calpidia spathiphylla (K. Sch.) Heimerl (1913) 288
Pisonia spathiphylla K. Schumann in Schumann & Laut. (1900)
= Pisonia longirostris

Calpidia triandra (Bargagali-Petrucci) Heimerl (1913) 289
Pisonia triandra Bargagali-Petrucci (1901) 610 New Guinea

Not listed in Stemmerik (1964)
= Pisonia longirostris

CEODES

The genus Ceodes was described by Forst et f. (Char. Gen. Pl. 1776· 141. t.71). The type species was Ceodes umbellifera. Seemann in 1862 placed Ceodes as a synonym of Pisonia. Its position was discussed in detail by Skottsberg (1936: 722 - 743). Stemmerik (1964) also listed Ceodes as a synonym of Pisonia. Skottsberg (1936: 722 - 743) includes extensive notes on this genus.

List of synonyms

Ceodes corniculata (Barg.-Petr.) Merr. & Perry (1939) 327
Pisonia corniculata Barg.-Petr. (1901) 615
Note: Brass 6729 listed as this species is P. mulleriana
= Pisonia corniculata

Ceodes excelsa (Bl.) Skottsb. (1926) 231
Pisonia excelsa Blume.
= Pisonia umbellifera

Ceodes longirostris (Teysm. & Binn.) Merr. & Perry (1939) 328
Pisonia longirostris Teysm. & Binn. (Nat. Tijdschr. Ned. Ind. 25 (1863) 401
= Pisonia longirostris

Ceodes umbellifera Forst. et. f. (Char.Gen. Pl. (1776) 142; Merr. & Perry (1939) 327; Skottsberg (1936) 723, 730
= Pisonia umbellifera

Ceodes urocarpa Merr. & Perry (1939) 329
T Brass 2972
= Pisonia longirostris

HEIMERLIA

Heimerlia was described by Skotts berg in 1936. It was based on Heimerlia brunoniana (Endl.) Skottsb. the type and only species. In 1941 Skottsberg described Heimerliodendron as the generic name Heimerlia had already been used by Hohnel. Stemmark (1964) discus sed the position of Heimerlia (latter Heimerliodendron) and concluded there were no characters to distinguish Heimerlia brunoniana (Endl.) Skottsb. from Pisonia umbelliferae (see Stemmerik 1964).

List of Species recorded from Papuasia

Heimerlia brunoniana (Endl.) Skottsb (1936) 738
Heimerliodendron brunoniana (Endl.) Skottsb. (1941) 364
Basionym Pisonia brunoniana Endler (1833)
= Pisonia umbellifera

HEIMERLIODENDRON

This genus, described by Skottsberg in 1941, was based on Heimerlia Skotts berg (1936) non v. Hohnel (1903). In 1964 both genera were listed as a synonym of Pisonia. Heimerliodendron brunonianum (Endl.) Skottsb. is Pisonia umbellifera. See discussion under Heimerlia.

3. PISONIA

o o o o o o

A genus of 35 warm temperate trees and shrubs. Some species have edible leaves. The genus was revised by Stemm erik in 1964. Pisonia is the only genus which has a considerable number of species outside the Old World. Some species are widespread. Stemmerik (1964) treats Calpidia, Ceodes and Rockia as part of Pisonia. He recog nises only 13 species.

Erect trees or shrubs to 30 m tall, usually unarmed. Leaves opposite or alternate, entire. Inflorescence axill ary or terminal, small or large thyrse. Cauliflorous or ramiflorous in one species. Pedicel with 1-3 small, caduc ous bracts. Dioecious or bisexual; male and female flowers sometimes of differ ent shape. Perianth fleshy, valvate in bud, tubular, 5- or rarely 10-lobed, basal part persistent. Stamens 2-40, in 1-2 whorls, mostly exserted, sterile in female flowers. Ovary on a small disc, elongate; style longer than ovary; stigma capitate. Anthocarp coriaceous, smooth or initially with 5(-6) ribs; sometimes a rostrum is produced. Seeds oblong with a deep longitudinal furrow.

Key to Species in Papuasia

Source: Stemmerik (1964)

Based on Flowering material

1. Plants unarmed, erect.................................... 2

1. Plants spinose, climbing. Perianth limb with 5 large lobes, alternating with 5 smaller ones. Anthocarp with 5 rows of biserial prickles.
...... Pisonia aculeata

2. Leaves distinctly petioles.............................. 3

2. Leaves subsessile, elliptic to obovate, tapering towards the base. Flowers densely hairy with truncate lobes. Stamens 5. Anthocarp smooth when mature, rostrum 10-40 cm.
...... Pisonia longirostris

3. Inflorescence cauliflorous, sometimes ramiflorous. Perianth lobes keeled on thc inside. Stamens 13-15. Anthocarp 7 cm long, not viscid.
...... Pisonia cauliflora

3. Inflorescences terminal, at least not cauliflorous or ramiflorous. Perianth lobes not keeled on the inside.. 4

4. Leaves with distinct dark veins contrasting with lighter coloured intervenium. Nerves and veins hairy beneath. Perianth glandular. Anthocarp with 5 rows of monoserial prickles.
...... Pisonia grandis

4. Leaves without distinctly contrasting dark veins, glabrous on lower surface. Perianth never glandular. Anthocarp without prickles........... 5

5. Perianth lobes truncate. Stamens 3-6. Anthocarp c. 5 cm, rostrum c.3 cm.
..... Pisonia mulleriana

5. Perianth lobes not truncate. Perianth dark-brown hairy, glabrescent. Stamens less than 15.
.... 6

6. Perianth 5- or 10-lobed, the lobes short and widely separated by shallow sinuses, the margin as a whole nearly sinuate rather than lobed. Stamens 2 or 4; in male flower longer than the vestigal gynoecium; staminodes in female flower shorter than the gynoecium. Anthocarp with a long rostrum.
...... Pisonia diandra

6. Perianth distinctly 5 - lobed. Stamens 5 - 14.
.... 7

7. Stamens 6-14. Anthocarp 2-4 cm, with 5 viscid ribs, without rostrum.
...... Pisonia umbellifera

7. Stamens 5-6. Leaves 18-24 by 7-11 cm. Anthocarp with a rostrum.
...... Pisonia corniculata

Based on Fruiting Specimens

1. Overhanging woody climber with mostly recurved axillary spines. Male and female flowers of different shape. Perianth lobes 10, unequal. Anthocarp provided with 5 biserial rows of viscid prickles.
...... Pisonia aculeata

1. Unarmed shrubs or trees. In dioecious species male and female flowers of similar shape. Perianth lobes 5 (but see P. diandra).............. 2

2. Perianth with 5 rows of black glands slightly concelaed by a ather dense indument. Anthocarp c. 1.25 cm long, with 5 ribs, each soon provided with a row of viscid stiff prickles. Leaves 10-20 by 6-10 cm, with dark or dark-red veins contrasting with a paler intervenium, hairy on the nerves beneath.
...... Pisonia grandis

2. Perianth without 5 rows of black glands, sometimes with 5 faint ribs. Anthocarp without prickles. Leaves glabrous beneath............................ 3

3. Anthocarp not produced into a rostrum
.... 4

3. Rostrum of anthocarp 3-40 cm long 5

4. Inflorescence cauliflorous or ramiflorous. Perianth lobes keeled inside. Ribs of the anthocarp not viscid. Leaves 24-60 cm long.
...... Pisonis cauliflora

4. Inflorescence terminal. Perianth lobes not keeled inside. Anthocarp with viscid ribs. Leaves shorter than 23 cm. Stamens 6-14. Ribs of anthocarp not sulcate. Perianth sparsely short brown hairy or puberulous.
...... Pisonia umbellifera

5. Perianth sparsely hairy to puberulous, in bud constricted halfway, in the single specimen the lower portion thicker than the upper. Stamens 5-6. Immature rostrum thin, at least 7 cm long.
...... Pisonia corniculata

5. Perianth densely hairy.................................. 6

6. Buds almost cylindric or at least hardly constricted, not club-shaped. Inflorescence branches very thin. Perianth campanulate, c. 5 mm long, lobes not truncate. Stamens 2 or 4. Immature rostrum at least 40 cm long, thin.
...... Pisonia diandra

6. Buds club-shaped. Inflorescence branches not particularly thin. Perianth lobes truncate......... 7

7. Leaves c. 15-24 cm long, rather elliptic; petiole (1-)2-8 cm. Flowers 4.5 mm long. Stamens 3-6. Rostrum c. 3-4 cm long, thick.
...... Pisonia mulleriana

7. Leaves large, c. 25-50 cm long, rather obovate-oblong, sessile to subsessile; petiole 0-2 cm, coarse. Flower c. 6 mm long. Stamens 5. Rostrum thin, 10-40 cm long.
...... Pisonia longirostris

List of Species recorded from Papuasia

Pisonia aculeata L. Sp. Pl. (1753) 1026; Stemmerik (1964) 284; (1964) 467. Port Moresby - Gerehu.

o o o o o o

Pisonia alba Span Linnaea XV (1841) 342; Warburg (1891) 303; K. Sch. & Laut. (1900) 307; Merril noted (Interpr. Rumph. Herb. Amb. (1917) 216 considered this a cultivated form of Pisonia sylvestria. It is cultivated in the Mal. Arch. and India.
= Pisonia grandis

Pisonia aruensis Bargagali-Petrucci (1901) 618
= Pisonia umbellifera

Pisonia beccariana Bargagli-Petrucci, G. (1901) 612
= Pisonia longirostris

Pisonia brunoniana Endl. Prod. Fl. Norfolk (1883) 43; F. Muell. (1877) 98; K. Schumann (1898) 115
Valeton (1914) = Pisonia spathiphylla
= Pisonia umbellifera

Pisonia buxifolia see Ebenaceae (Part IV)

Pisonia cauliflora Scheff. Ann. Jard. Bot. Btzg. 1 (1876) 45; F. Muell (1877) 98; Valeton Ic. Bog. 1: 67 t. 21- Ceram; Stemmerik (1964) 281. Collected by Teysmann on Rawak.

o o o o o o

Pisonia corniculata Bargagali-Petrucci (1901) 615; Stemmerik (1964) 464; Beccari 650. Andai. Vogelkop.

o o o o o o

Pisonia diandra Pulle (1912) 629
Stemmerik (1964) 283; (1964) 463 Sepik River.

o o o o o o

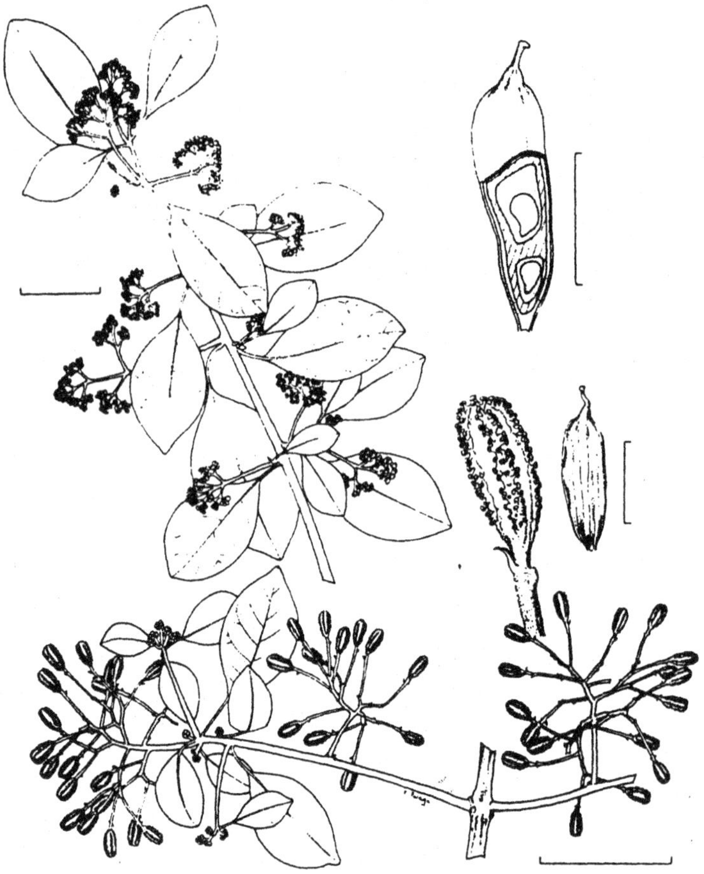

Pisonia aculeata A. habit; B. inflorescence; C. D. fruit;
E. dissection of flower. Specht 1008. (Drawn T. Iwagu)

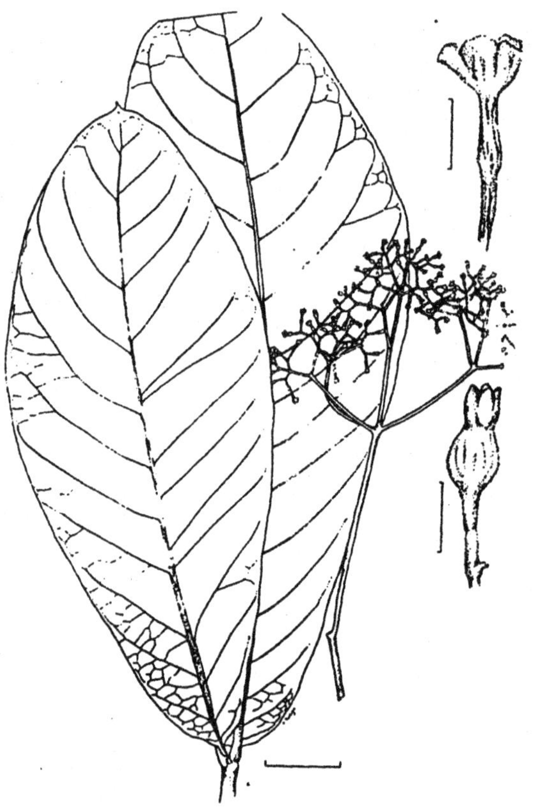

Pisonia longirostris A. leaf; B. inflorescence; C. female flower; D. male flower. NGF 14847. (Drawn T. Iwagu)

Pisonia grandifolia Warburg (1913) 303
Valeton (1914) 102 - Aru Is.
= Pisonia longirostris

Pisonia grandis R. Br. Prod. 1 (1810) 412; Heimerl (1934) 127; Skottsberg (1936) 736; Airy Shaw (1952) 87; Stemmerik (1964) 283; (1964) 464 - Laing Is. nr Awar. Aru.

o o o o o o

Pisonia lauterbachii Warburg ex. K. Sch. & Laut. (1900) 307; Laut. (1930) 454; Valeton (1914)
= Pisonia spathiphyllum

Pisonia lineatipilum C.DC ex. Ind. Kew.
= ? Piper

Pisonia longirostris Teysmann et Binnedyk (1863) 401; Stemmerik (1964) 281; (1964) 463; Icon. Bog. 1: 69 t XXI.

o o o o o o

Pisonia major Baillon Adansonia X (1872) 185 Rawak
= Pisonia cauliflora

Pisonia membranacea K. Sch. in K. Sch.& Hollr. (1889) 43; Warburg (1889) 43; Warburg (1891) 303; K.Sch. & Laut. (1900)

Pisonia micrantha Valeton (1914) 102
= Pisonia diandra

Pisonia micrantha var. augustifolia Valeton (1914) 103
= Pisonia diandra

Pisonia mulleriana Warburg (1891) 304; Laut. (1929) 556; K. Sch. & Laut. (1900) 307; Stemmerik (1964) 281

o o o o o o

Pisonia rostrata Warburg (1891) 304; Valeton (1914) 102; Heimerl (1913); Stemmerik (1964)
= Pisonia longirostris

Pisonia spathiphylla K.Sch. in Sch. & Laut. (1901) 308; Laut. (1930) 449; Valeton (1914) 103; Stemmerik (1964)
= Pisonia longirostris

Pisonia triandra Bargagali-Petrucci (1901) 610; Stemmerik (1964)
= Pisonia longirostris

Pisonia umbellifera (Forst. et f.) Seem. Blonplandia 10 (1862)
Ceodes umbellifera Forst. et f. (1776); White (1929) 211; K. Sch. (1895) 49; K. Sch. & Laut. (1900) 307; Laut. (1929) 14; (1930) 424, 439, 446, 455; Stemmerik (1964) 280, (1964) 460; Valeton (1910) 102

o o o o o o

Notes:

1. The distribution of Pisonia grandis R.Br. was mapped by H. St John (Webbia 8 (1951) 225 - 228). The fruits are unique in Pisonia and the species is widely distributed by far ranging sea birds.

2. Pisonia lineatipilum C. DC. in Lorentz (Nova Guinea 8 (6) 1914: 1009 was incorrectly recorded. It is a species of Piper.

PHYTOLACCACEAE

o o o o o o

A tropical and subtropical family of some 22 genera and c. 125 species. Many plants have medicinal uses, yield red dyes, used as ornamentals and herbs. Most members are native to America and the West Indies. The family is divided into four subfamilies: Phytolaccoideae; Stegnospermaroideae; Microteoideae and Archato-carpoideae. Recently several of the genera have been split off into monotypic families. In Malesia three genera, one of which occurs in Papuasia. A single species of Phytolacca, native

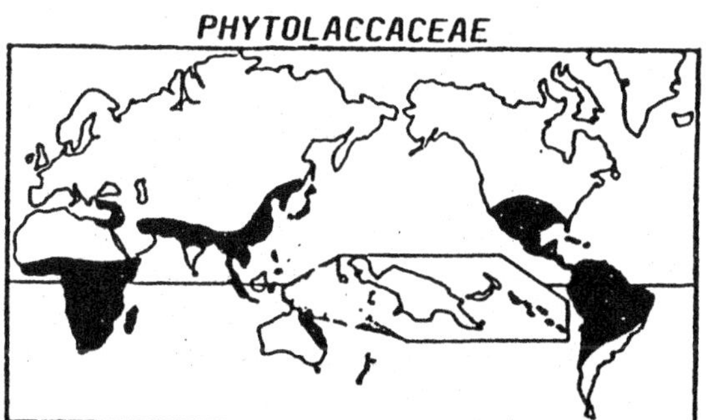

	WORLD	PAPUASIA
Number of genera :	18	1
Number of species :	65	1

Distribution : Tropical and sub tropical. America

Economic uses : Medicinal plants, herbs and ornamentals.

to tropical America, is naturalised in Papuasia. Several species of Phytolacca have edible leaves. Many genera have medical uses.

Diagnostic features:

Leaves: alternate, simple and entire, typically lacking stipules or with min ute stipules.

Flowers: bisexual, rarely unisexual, monoecious or dioecious. Perianth one whorl of 4 or 5 free, persistent seg ments; petals sometimes present. Stamens hypogynous, sometimes united at base, equal in number to perianth lobes or more numerous by branching. Ovary superior, rarely inferior.

Fruit: a fleshy berry, dry nut or rare ly a capsule.

References

Backer, C. A. (1951). Phytolacc aceae. Flora Malesiana Series I. 4: 229 - 232.

Heimerl, A. (1934). Phytolaccaceae. Pflanzenfamilien ed. 2. 16C: 135 - 164.

Walter, H. (1909). Phytolaccaceae. Pflanzenrich. Heft 39: 1 - 154. Figs. 1 - 42.

KEY TO GENERA IN PAPUASIA

A single naturalised genus.
...... 1. PHYTOLACCA

1. PHYTOLACCA

o o o o o o

A genus of 25 species, mostly native in tropical America, some weeds of cultivation. A single species natur alised in Papuasia.

Erect herbs, monoecious, stems angular. Inflorescence an erect raceme, sometimes spike-like. Flowers mostly bisexual, bracteate. Perianth 5-merous. Stamens 6-33, 1- or 2-seriate, irregul arly inserted on or below the disc. Ovary or 5-16 carpels. Fruit a berry.

Key to Species in Papuasia

A single introduced species:
..... Phytolacca octandra

List of Species recorded from Papuasia

Phytolacca octandra L. Sp. Pl. (1762) 631

o o o o o o

Notes:

1. Locally a common large weed along roadsides, particularly in exposed areas. Collected at 600 m near the Sirinumu dam. The young shoots and leaves can be used as a vegetable.

PORTULACACEAE

o o o o o o

A cosmopolitan family of 19 genera and c. 500 species, centered in S Africa and America. Annual or perennial succu lent herbs or sub-shrubs. Several species are grown as ornamentals and a single pot herb, Portulaca oleracea is grown as a salad and pot herb. P. grandiflora (possibly of hybrid origin) is widely grown as a bedding annual.

Diagnostic features:

Leaves: fleshy, spiral or opposite (in Montia), entire with stipules. More or less fleshy.

Flowers: small, except in a few cultivated species of Portulaca. Regular, bisexual typically with 2 green sepals, (4-)5(-6) free petals and stamens.

PORTULACACEAE

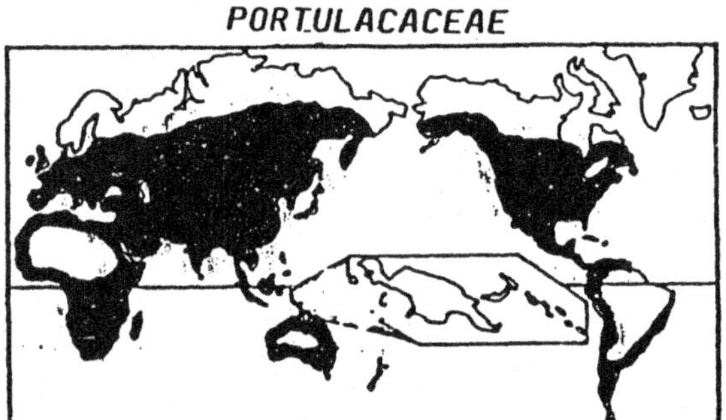

	WORLD	PAPUASIA
Number of genera :	22	3
Number of species :	400	7

Distribution : Cosmopolitan, mainly S Africa and America.

Economic uses: Ornamentals and a few edible.

Ovary ± superior, style usually divided. Nectar present, flowers insect pollinated.

Fruit: a capsule dehiscing by 2 or 3 valves or a lid.

References

Geesink, R. (1968). An account of the genus Portulaca in Indo-Australia and the Pacific. Blumea 17: 275 - 301.

Geesink, R. (1971). Portulacaceae. Flora Malesiana I. 7: 121 - 133.

Lam, H. J. (1927) Nat. Tijdschr. Ned. Ind. 87: 157. Translated Sargentia 5 (1945) 28

Mansfeld, . (1959). Kulturpflanze 2: 60.

Merrill and Perry, L. M. (1942). J. Arn. Arb. 23: 386.

Pax. & K. Hoffm. (1934). Nat. Pfl. fam. 2 Aufl. 16C 234 - 262. Fig. 104 - 116.

Poellnitz, van (1934). Monographie der Gattung Talinum. Repert. Sp. Novarum Regni Vegetabilis. No. 911/917. Fedde. Rep. 35: 1 - 34.

Pulle, A. (1910). Portulacaceae. Nova Guinea 8: 357.

Schumann, K. (1898). Notizbl. Bot. Gart. Berlin 2: 115.

Schumann, K. & Lauterbach (1900). Fl. Deut. Schutzgeb. Sudsee. 809.

Warburg, O. (1891). Bot. Jahrb. 13: 305

KEY TO GENERA IN PAPUASIA

Source: Geesink (1971)

1. Ovary half-inferior. Fruit with a caducous lid.
...... 2. PORTULACA

1. Ovary superior. Fruit valved or irregularly dehiscent.. 2

2. Leaves ca. 4 mm long, opposite. Aquatic plants.
...... 1. MONTIA

2. Leaves more than 3 mm long, spirally arranged, the lowermost ones sometimes opposite.
...... 3. TALINUM

1. MONTIA

o o o o o o

A genus of some 50 species, most American. The only representative in Papuasia is a high mountain species of world wide distribution. This genus is not well defined in relation to closely related genera such as Claytonia.

Small, perennial or annual, much branched herbs. Lacking stipules. Leaves fleshy, opposite. Flowers small, in few-flowered racemes; inflorescences axillary or terminal, sometimes solit ary. Sepals 2 (rarely 3), elliptic or ovate, persistent. Petals 5, equal or unequal, connate in lower half, later separating. Stamens 3 rarely 4 to 5, inserted at base of smaller petals. Capsule 3-valved, the valves twisting after opening, membranous. Seeds 1 - 5.

Key to Species in Papuasia

A single species:
...... Montia fontana

List of Species recorded from Papuasia

Montia fontana L. Sp. Pl. (1753) 87; Geesink (1971) 126

Montia lamprosperma Cham. Linnaea 6 (1831) 565; Merr. & Perry (1942) 386; van Royen (1978) 2341

o o o o o o

Notes:

1. Found in stagnant parts of streams between 2660 and 3660 m in altitude.

Montia fontana A.B. habit; C. flowers; D. fruit. (Source: National Herbarium, Lae).

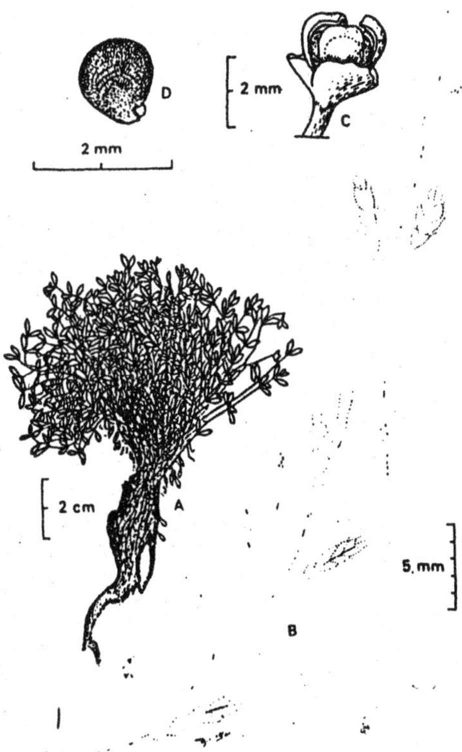

2. PORTULACA

o o o o o o

A genus of c. 40 species, in trop ics and subtropics, several are quite polymorphous, others are widespread weeds.

Succulent, copiously branched herbs. Leaves linear to orbicular, most with axillary hairs. Flowers in (1-)2- 30 flowered terminal capituli, surround ed by a whorl of 3-30 involucral leaves. Sepals occasionally keeled or hooded, persistent or caducous with the petals, stamens and styles. Petals 4-6(-8), obovate, occasionally emarginate or mucronate. Stamens 4 to numerous in 1 whorl. Style with 2-18 arms. Capsule with a caducous circumcisile operculum. Seeds numerous.

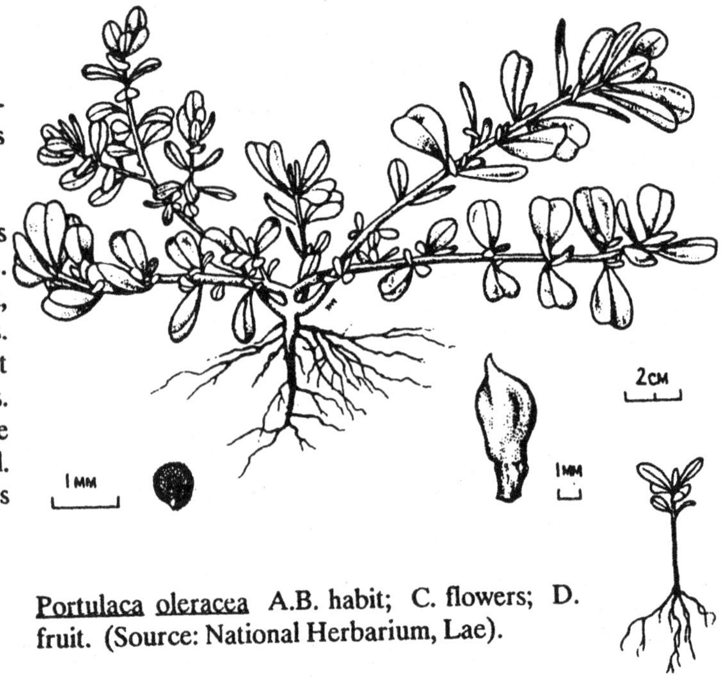

Portulaca oleracea A.B. habit; C. flowers; D. fruit. (Source: National Herbarium, Lae).

Key to Species in Papuasia

Source: Geesink (1971)

1. All leaves opposite. Hairs intra- and interpetiolar.
 ..... Portulaca quadrifida

1. At least the middle-cauline leaves spirally arranged. Hairs if present only axillary........... 2

2. The largest leaves obovate to spathulate. Axillary hairs inconspicuous. Sepals distinctly carinate (20X mag.)... 3

2. The largest leaves linear to elliptic. Hairs mostly conspicuous. Sepals not carinate (20X mag.), occasionally with an apical, dorsal, dome-shaped hood.
 ...... Portulaca pilosa

3. Mostly 1-3 involucral leaves with an axillary axis. Capituli (2-)3-30 flowered. Stamens 7-15. Fruit c. 4 mm long.
 ...... Portulaca oleracea

3. Involucral leaves mostly without axillary axes. Capituli (1-)2-3(-6) -flowered. Stamens 18-50. Fruit c. 7 mm long.
 ...... Portulaca lutca

List of Species recorded from Papuasia

Portulaca lutca Forster Pl. Esc. (1786) 72; Geesink (1969) 291; Geesink (1971) 129

o o o o o o

Portulaca oleracea L. Sp. Pl. (1756) 715; Warburg (1891) 305; K. Sch. (1898) 115; K. Sch. & Laut. (1900) 309; Lam (1927)157; Laut.

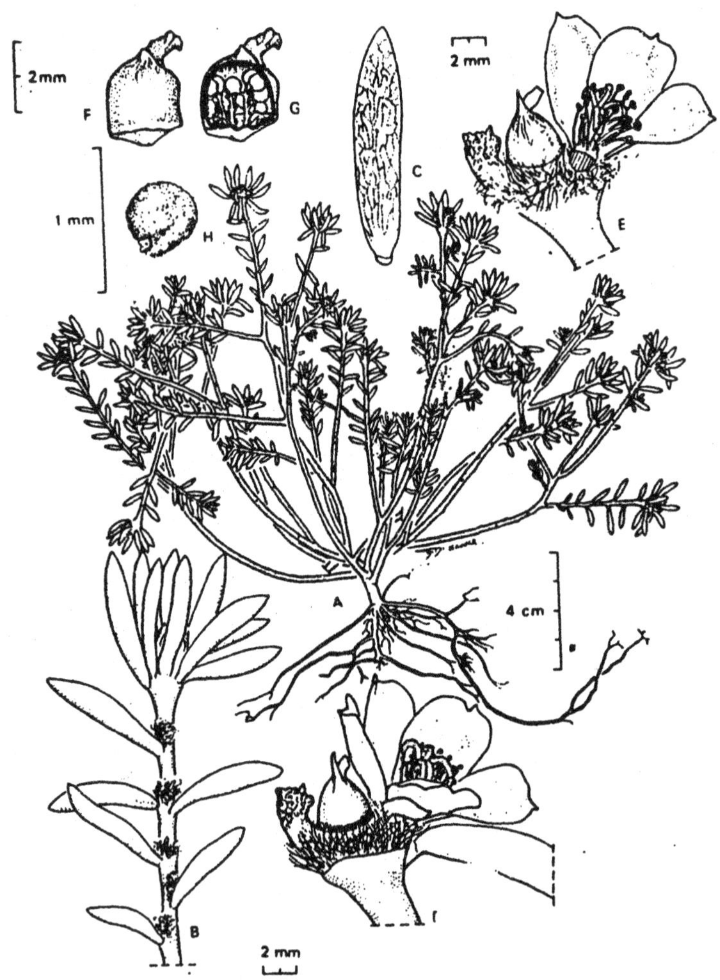

Portulaca pilosa A.B. habit; C. leaf; D.E. flowers; F.G.H. fruits. (Source: National Herbarium, Lae).

(1928)301; Laut. (1929)18; Laut. (1930) 453; Dukleus (1930) 240; Valeton (1907) 10; Pulle (1910)357; Pax & K. Hoffm. (1934) 247

o o o o o o

Portulaca pilosa L. Sp. Pl. (1753) 445 ; Geesink (1969) 294; Geesink (1971) 131

ssp. pilosa Geesink (1969) 295; Geesink (1971) 131

o o o o o o

Portulaca quadrifida L. Mant. Pl. (1767) 73; Gibbs (1917) 211

o o o o o o

Notes:

1. For Portulaca portulacastrum L. see Aizoaceae

2. Portulaca quadrifida is a wide spread tropical weed common on beach sands. Plants are very fleshy with yellow flowers.

3. Portulaca samoensis Poelln. has been reported from the Solomon Islands.

3. TALINUM

o o o o o o

About 50 species, native to C and S America and S Africa. Two species naturalised pantropically.

Herbs or semi-shrubs, glabrous. Leaves spirally arranged, linear to obovate. Flowers in terminal thyrsi, or cymosely arranged, seldom axillary or solitary. Sepals mostly caducous. Petals mostly 5. Stamens 5-numerous. Ovary superior, style with 3 arms. Fruit globular; 3-valved or irregularly caducous.

Key to Species in Papuasia

Source: Geesink (1971)

1. Flowering axes sharply triangular in cross-section. Branches with 2 lateral, basal buds. Stamens 20-40.

...... Talinum triangulare

1. Flowering axes terete. Branches without basal buds. Stamens up to 15

..... Talinum paniculatum

List of Species recorded from Papuasia

Talinum paniculatum (Jacq.) Gaert. Fruct. 2 (1791) 219; Backer and Bakh. (1963) 217; Geesink (1971) 124

o o o o o o

Talinum triangulare (Jacq.) Willd. Sp. Pl. 2 (1799) 862; Geesink (1971) 24

o o o o o o

Notes:

1. Pantropical weeds. Native of tropical America but still extending their range. Both species with edible leaves.

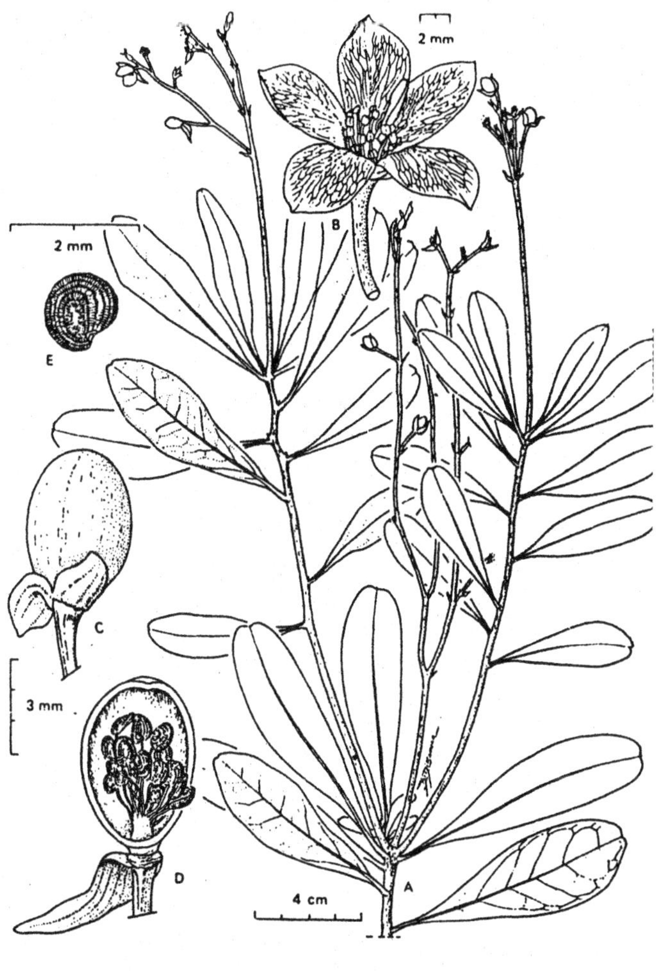

Talinum triangulare A. habit; B.C.D. flowers; E. fruit. (Source: National Herbarium, Lae).

POLYGONALES

A monotypic order, represented by a single family the Polygonaceae. Most species occur in north temperate regions. In Papuasia there are 4 genera, one represented by an important naturalised ornamental plant. The family includes some vegetables such as Rheum (rhubarb) and several serious agricultural weeds such as Rumex (docks) and Polygonum (knotweeds). Several genera are cultivated as ornamentals. These include Antigonon, Eriogonum, Homalocladium, Muehlenbeckia, Polygonum and Rheum (Mabberley 1987).

POLYGONACEAE

o o o o o o

A family with some 40 genera and c. 800 species. Mabberley (1987) states that the family includes 51 genera and c. 1050 species. It is distributed world-wide but most species occur in north-temperate regions. Four genera occur in Papuasia. One genus, Antigonon, is represented by a single introduced ornamental species which has escaped from cultivation. The other three genera are represented by some 18 species. Rheum rhaponticum L. (rhubarb) is cultivated as a vegetable in the montane zone. Many species have unusual vascular structure.

Trees, shrubs, lianes or herbs often with unusual vascular structure. Stems often with swollen nodes. Leaves simple, alternate. Flowers mostly bisexual, if unisexual then either monoecious or dioecious. Flowers actinomorphic; perianth usually biseriate, tepals 3-6, often petaloid, imbricate; corolla absent. Stamens 6-9 in 2 series. Filaments free or basally connate. Ovary superior, single syncarpus pistil. Fruits angled, round or a winged achene. Seeds 1 per fruit. Tepals often enlarged and membranous in the fruit.

Diagnostic features

Leaves: usually alternate and simple, with a characteristic membranous sheath (an ochrea) uniting the stipules.

Flowers: small, white greenish or pinkish, solitary or in racemose inflorescences, usually bisexual, rarely unisexual; sepals 3-6, often becoming large and membranous in the fruit; petals absent; stamens 6-9; ovary superior, composed of 2-4 carpels united into one locule; styles 2-4 usually free.

Fruits: a triangular nut; seeds with abundant endosperm.

POLYGONACEAE

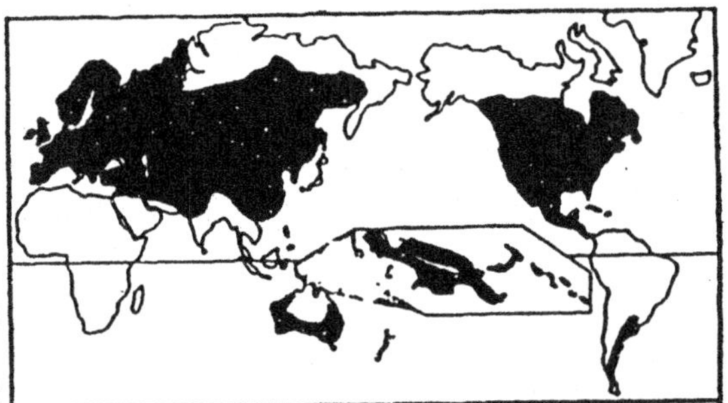

	WORLD	PAPUASIA
Number of genera:	51	4
Number of species:	1,050	19
Distribution:	Cosmopolitan, mainly N temperate.	
Economic uses:	Ornamentals, rhubarb and weeds.	

References

Backer, C. A. and Bakhuizen van den Brink, R. C. (1963). Polygonaceae. Flora of Java 1: 219 - 226 (rev. B. H. Danser).

Brass, L. J. (1941). J. Arn. Arb. 22: 311.

Dammer, (1892). Bot. Jahrb. 15: 260 - 285. Fig. 1 - 2.

Danser, B. H. (1927). Die Polygonaceen Niederlandisch- Ostindiens. Bull. Jard. Bot. Btzg. III 8: 117 - 261.

Danser, B. H. (1927). Polygonaceae. Nova Guinea 14: 333 - 336.

Danser, B. H. (1935). Notes on a number of New Guinea Polygonaceae mainly collected by L. J. Brass in Papua on the 1933 New Guinea Expedition of Natural History. Bull. Jard. Bot. Btzg. III 13: 429 - 431.

Danser, B. H. (1940). Notes on the Muehlenbeckias of New Guinea and Queensland. Bull. Jard. Bot. Btzg. III 16: 324 - 328.

Gross, (1913). Bot. Jahrb. 49: 234 - 339. Fig. 1 - 36.

Henty, E. E. (1978). Polygonaceae. Handbooks of the Flora of Papua New Guinea. Vol. 1: 222 - 248.

Merrill, E. D. and Perry, L. M. (1942). J. Arn. Arb. 23: 384 - 385.

Merrill, E. D. and Perry, L. M. (1948). J. Arn. Arb. 29: 154.

Mueller, F. von (1858).

Pulle, A. (1910). Polygonaceae. Nova Guinea 8: 359.

Pulle, A. (1911). Polygonaceae. Nova Guinea 8: 625 - 626.

Rechinger, (1932). Beih. Bot. Centralbl. 49: 1 - 132.

Royen, P. van (1982). Polygonaceae. The Alpine Flora of New Guinea. 2373 - 2392.

Steenis, C. G. G. J. van (1934). Bull. Jard. Bot. Btzg. III 13: 236.

Steward, A. N. (1930). The Polygonaceae of E Asia. Cont. from Grey Herbarium 88: 1 - 129.

KEY TO GENERA IN PAPUASIA

Source: Henty (1978)

1. Climbing plant with tendrils.
...... 1. ANTIGONON

1. Tendrils absent... 2

2. Perianth 6-lobed; in fruit 3 outer lobes much smaller than inner ones.
...... 4. RUMEX

2. Perianth 5-lobed, lobes subequal or outer ones larger..3

3. Flowers separately male and female: styles 3 with ± fringed stigmas; shrubs or ± woody climbers.
...... 2. MUEHLENBECKIA

3. Flowers bisexual; styles 2 or 3, free or connate at base, with capitate stigmas; herbs, rarely somewhat woody at base.
...... 3. POLYGONUM

1. ANTIGONON

o o o o o o

A genus of 8 species, native to tropical America. In Papuasia 1 introduced species, widely cultivated, known by the common name corallita.

Climbing perennial herbs and sub-shrubs. Ochreas almost lacking. Flowers bisexual, in fascicles combined into axillary racemes; tepals 5 in two whorls, outer tepals (3) larger than the inner pair; after anthesis accrescent and enclosing the fruit. Stamens 7-9, filaments united at base to form a short tube. Styles 3 with stigmas capitate. Fruit trigonous.

Key to Species in Papuasia

A single introduced species:
..... Antigonon leptopus

List of Species recorded from Papuasia

Antigonon leptopus Hook. & Arn. Bot. Beech. (1841) 308; Danser (1927) 16; Henty (1978) 223

o o o o o o

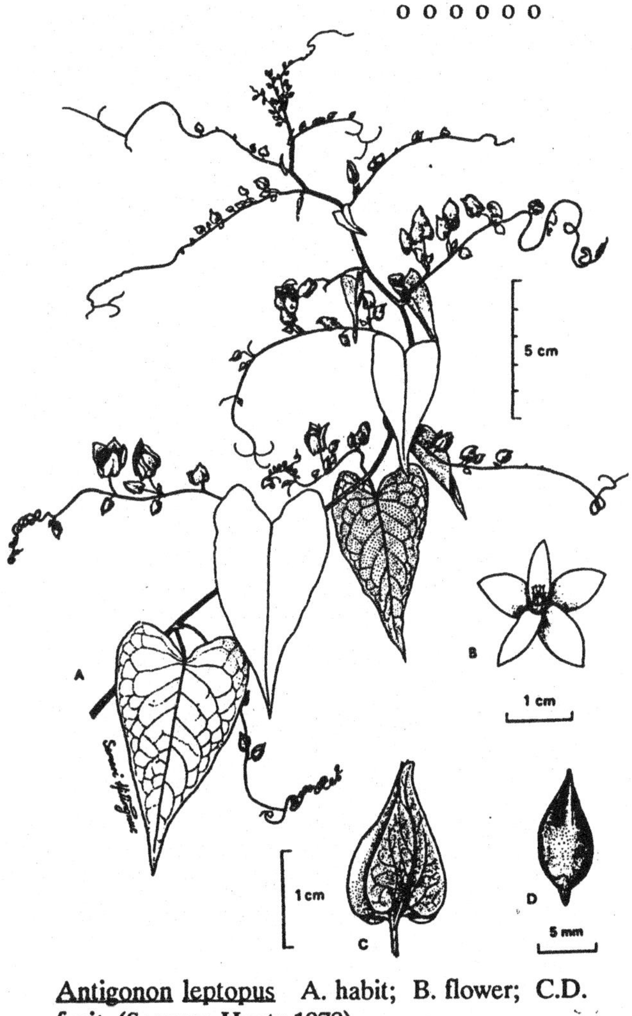

Antigonon leptopus A. habit; B. flower; C.D. fruit. (Source: Henty 1978)

Notes:

1. This species is native to Mexico. It is introduced throughout the tropics as an ornamental plant, grown for its bright pink flowers. Widely cultivated in lowland regions; sometimes a garden escapee found growing on raodside, particularly on banks and road-cuttings. Domestic poultry eat the fruits (Henty 1978). The edible tubers have a nut-like flavour.

CALACINUM

A single species:
..... Calacinum platycladum

List of Species recorded from Papuasia

Calacinum platycladum (F. Muell.) Macbride.
= Polygonum platycladum F. Muell. (1858)

COCCOLOBA

A genus of some 150 T and warm Temperate American species. Some species are cultivated ie. C. uvifera the seaside grape. Used for jelly. Most are littoral, pachycaul trees. A single species was described from Papuasia: Coccoloba platyclada (F. Muell.) F. Muell. (1863: 5382). Muehlenbeckia platyclada (F. Muell.) Meissn. (1865). Howard (1961: 108). It was transferred to the genus Homalocladium by Bailey in 1929. At present it is treated under the genus Muehlenbeckia (Henty 1978).

2. MUEHLENBECKIA

o o o o o o

A genus of c. 15 species distributed in Australia, New Zealand, South America and in Papuasia by 3-4 species. Some species are cultivated as ornamentals.

Erect or scandent shrubs or sub-shrubs; monoecious or dioecious. Leaves spiral, petiolate or sessile. Ocrea present. Flowers small, clustered in axils of stem-sheathing bracts. clusters in lax, axillary or terminal pseudospikes, simple or in panicles. Perianth deeply 5-lobed, lobes subequal or outer 3 larger. Male flowers 8 stamens, rarely less; gynoecium absent or rudimentary. Female flowers; stamens small, imperfect or abortive; ovary 3- angled; styles 3, short with ± fringed or lobed stigmas. Nut trigonous or ± globose, enclosed in a persistent perianth.

Key to Species in Papuasia

Source: Henty (1978)

1. Stem ± compressed, ultimate branches completely flattened to phylloclades; leaves fugaceous.
..... Muehlenbeckia platyclada

1. Stems terete with normally persistent leaves.. 2

2. Scandent shrub; leaves ± oval with cuneate base.
..... Muehlenbeckia monticola

2. Half-woody climber; leaves ovate-triangular with a ± truncate base.
..... Muehlenbeckia zippelii

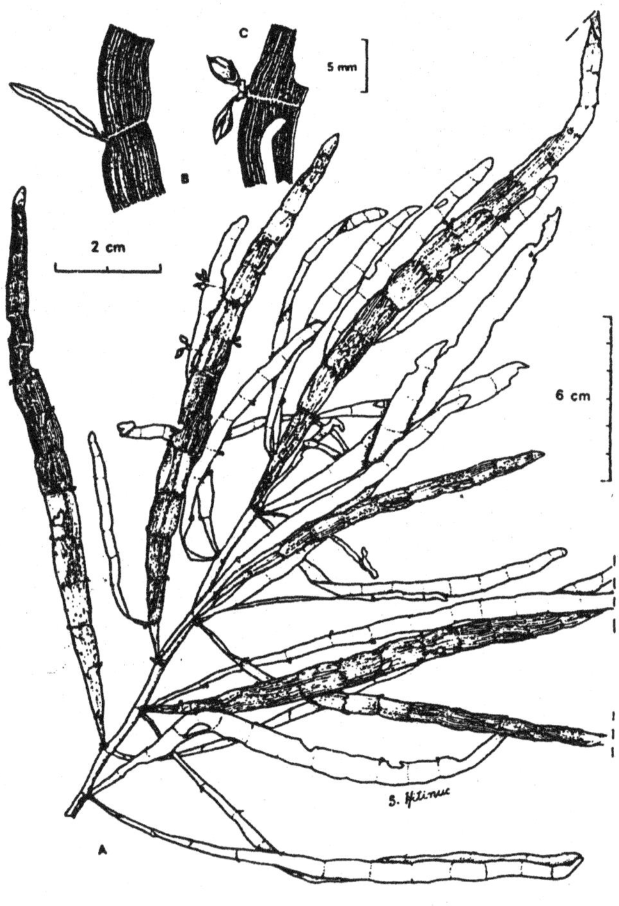

Muehlenbeckia platyclada A. habit; B. branch node with leaf; C. fruit. (Source: Henty 1978)

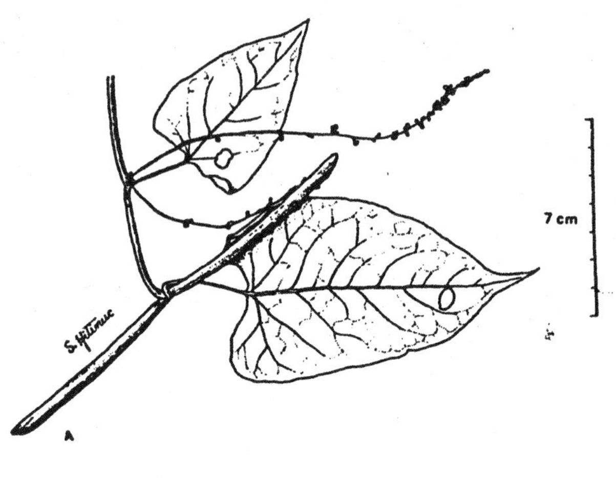

Muehlenbeckia zippelii A. habit. (Source: Henty 1978)

List of Species recorded from Papuasia

Muehlenbeckia monticola Pulle (1912) 625; Danser (1927) 335; van Royen (1982) 2375

o o o o o o

Muehlenbeckia platyclada (F. Muell.) Meissn. Bot. Zeit. 23 (1865) 313

..... Coccoloba platyclada F. Muell. (1863).

..... Calacinum platycladum Macbride (1927).

Notes

1. M. monticola and M. platyclada are endemic to Papuasia. The former occurs throughout the mainland but does not grow on the islands. It is common in forest, particularly disturbed forest, from 1500 - 3000 m altitude. M. platyclada is common in grassland and abandoned garden areas between 1500 and 2500 m altitude. It has also been reported as an epiphyte. It is widely cultivated as an ornamental plant.

2. M. zippelii grows in Australia and Papuasia. In Papuasia it only occurs in Fakfak and Western Province. Locally common in forest margins and garden regrowth. It is limited to areas with a marked seasonal rainfall and to low altitudes.

3. Mueller (1875: 60) reported M. gracillima Meissn. in DC. as collected by d'Albertis on the Fly River. Danser (1940: 326) states that this resembles M. zippelii, though smaller.

3. POLYGONUM

o o o o o o

A cosmopolitan genus mainly centered in the north temperate regions. Considerable confusion exists in the delimitation of genera; some authors treat Polygonum as a large genus, others split the genus based on minor differences. This has resulted in recognition of many small genera such as Persicaria and Cephalophilum. In consequence the number of species also varies from c. 50-300. Vegetative characters often vary and this has led to the description of many subspecies and varieties, many of which are difficult to define. In New Guinea 13 species are recognised by Henty (1978). Several species are important weeds of cultivation and some are dangerous to livestock.

Perennial and annual herbs or rarely sub-shrubs; creeping, erect or scandent. Leaves spiral, petiole long to almost absent; sheaths ± developed; longitudinal nerves often extend beyond the leaf apex as bristles. Leaves often glandular. Flowers bisexual or unisexual by abortion; stalks short; axillary in axils of bracts; bract fascicles arranged in pseudo-spikes, or solitary to form compound inflorescences. Perianth of 5 tepals, deeply divided, tepals greenish, white or pink, subequal; after anthesis sometimes succulent. Stamens 4-9; ovary trigonous with 3 styles or compresseed with 2 styles; stigma capitate. Fruit trigonous or lenticular, sometimes subglobose, often apiculate.

Polygonum barbatum A. habit; B. ocrea; C. fruit. (Source: Henty 1978)

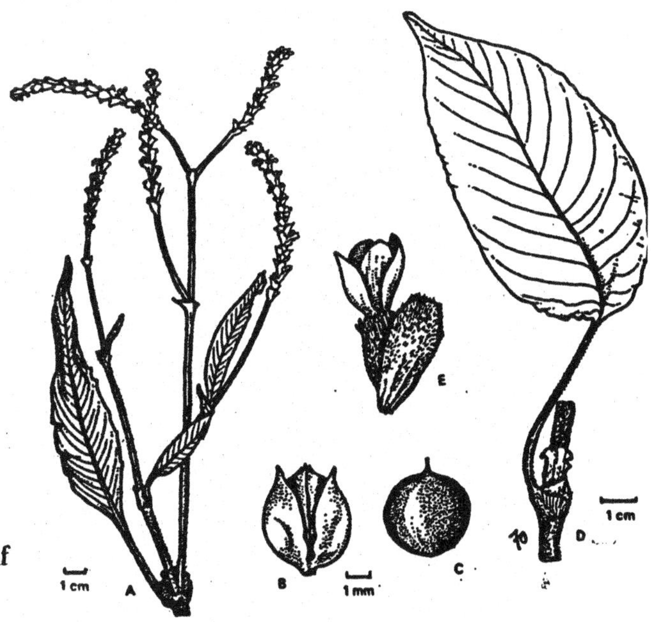

Polygonum orientale A. habit; B.C. fruit; D. leaf and ocrea; E. flowers. (Source: Henty 1978)

Key to Species in Papuasia

Source: Henty (1978)

1. Ocrea with distinct green border...... 2

1. Ocrea not different at apex...................... 3

2. Border expanded to a broad salviform leaf-like collar; leaves deltoid, slightly peltate; climber with prickly stems; fruits globose.
..... Polygonum perfoliatum

2. Border not much expanded; leaves ovate to lanceolate, rounded to cordate at base; decumbent to erect with soft pubescent stem; fruit flattened flask-shaped.
..... Polygonum orientale

3. Pseudospikes elongate, lax or dense, cylindrical... 4

3. Pseudospikes short, compact, globose to ovate.
.... 9

4. Fruit flattened flask-shaped, with 2 flat or slightly convex faces and rounded edges; ocrea without bristles at apex, or with a few, very short and fine... 5

4. Fruit trigonous or lenticular, edges sharp, at least near the apex; ocrea usually with well-developed bristles at apex.............................. 6

5. Fruit at least 3 mm long; lower stems 5 mm thick or more; leaves, young stems and inflorescence axes uniformly and usually densely hairy.
..... Polygonum attenuatum

5. Fruit c. 2 mm long; stems slender; leaves rarely pubescent; stems and axes with a few scattered hairs.
..... Polygonum lapathifolium

6. Fruit all, or mostly lenticular; styles 2.
..... Polygonum minus

6. Fruits all, or mostly trigonous; styles 3......... 7

7. Stems moderately stout (3-4 m); ocreas usually with dense appressed hairs, with stout, long (1-2 cm) apical bristles; pseudospikes fairly dense and stout.
..... Polygonum barbatum

7. Stems slender; ocreas with scattered surface hairs and fine apical bristles not 1 cm long; pseudospikes slender, lax...................................... 8

8. Ocrea bristles 5-10 mm long; bracts densely ciliate; stems c. 3 mm thick.
..... Polygonum longisetum

8. Ocrea bristles to 3 mm long, very fine; bracts sparsely ciliate; stems not 2 mm thick.
..... Polygonum decipiens

9. Leaf margins bearing short, stiff bristles; usually down- ward-pointing bristles on the midribs beneath; ring of downward- pointing appressed stiff bristles on base of each; often a few strong downward-pointing bristles on stem.........
....10

9. Leaf margins smooth; midribs, sheaths and stems glabrous or with soft, patent hairs........... 11

10. Leaves 4 cm long or more; ocreas 1.5 cm long or more; fairly robust decumbent plants.
..... Polygonum strigosum

10. Leaves not 1 cm; ocreas not 7 mm long; very small creeping plants.
..... Polygonum benguetense

11. Petiole not winged; fruiting perianth ± succulent; ocrea very obliquely truncate, 3-4 cm long.
..... Polygonum chinense

11. Petiole ± winged; fruiting perianth scarious or herbaceous.. 12

12. Leaf-blade entire, apex obtuse to acute, base ± cuneate; petiole winged the whole length to a uniform width; ocrea obliquely truncate, 3-10 mm long.
..... Polygonum nepalense

12. Leaf-blade often lobed at the base or, if not lobed, base truncate or very obtuse-angled; apex acute to acuminate; petal often ± winged near the top only, or winged whole length and margin lobed; ocrea transversely truncate, 1 cm long or longer.
..... Polygonum runcinatum

List of Species recorded from Papuasia

Polygonum alatum Buch.-Ham. ex Spreng (1827); van Royen (1982) 2384
= Polygonum nepalense

Polygonum attenuatum R. Br. Prodr. (1810) 420; Danser (1927) 162

o o o o o o

Polygonum barbatum L. Sp. Pl. (1753) 362; Danser (1927) 145

o o o o o o

Polygonum benguetense Merr. Phil. J. Sc. Bot. 10 (1915) 301; van Royen (1982) 2382

o o o o o o

Polygonum chinense L. Sp. Pl. (1753) 363; van Royen (1982) 2380

o o o o o o

Polygonum decipiens R. Br. Prodr. (1810) 420
..... P. minus Benth (1870)
... P. minus Huds. ssp. decipiens (R.Br.) Danser (1927)

o o o o o o

Polygonum lapathifolium L. Sp. Pl. (1753) 360

o o o o o o

Polygonum longisetum De Br in Miq. Pl. Jungh. (1854) 307; Danser (1927) 170

o o o o o o

Polygonum microcephalum var. papuanum Warb. (1893)
= Polygonum nepalense

Polygonum minus Huds. Fl. Angl. 1 (1762) 148

o o o o o o

Polygonum minus Benth (1870) non Huds.
= Polygonum decipiens

Polygonum minus Huds. ssp. decipiens (R.Br.) Danser (1927)
= Polygonum decipiens

Polygonum nepalense Meissn. Mon. Gen. Polygon. 84, t.7 (1826) Ic. 2
..... Polygonum microcephalum var. papuanum Warb. (1893)
..... Polygonum alatum Buch.-Ham. ex Spreng (1827); van Royen (1982) 2384

o o o o o o

Polygonum orientale L. Sp. Pl. (1753) 362

o o o o o o

Polygonum perfoliatum L. Sp. Pl. (1762) 521

o o o o o o

Polygonum runcinatum D. Don Prodr. Fl. Nepal (1825) 73; Danser (1927) 205

van Royen (1982) includes the following infraspecific taxa:

ssp. papuanum Danser (1935) 13; Merrill & Perry (1942)

var. papuanum
= Polygonum runcinatum D. Don ssp. papuanum Danser

var. alpinum van Royen (1982) 2389.

Polygonum strigosum R. Br. Prodr. (1810) 420; Danser (1927) 227

o o o o o o

Notes

1. Species show various distribution patterns:

a). local endemics :-

P. benguetense restricted to L. Habbema (c. 3000 m.) but also in Philippines;

b). widespread on New Guinea mainland and rare or absent on Islands :-

P. attenuatum (c. sea-level to c. 1000 m.); P. decipiens L. Habbema, W Highlands and C. Province (2500 - 3000 m +); P. lapathifolium Baliem R. and Mt Wilhelm (+ 1500 m); P. longisetum (c. 2000 m.); P. nepalense (1200 to above 3000 m.); P. orientale S New Guinea (low altitudes); P. perfoliatum only India to Irian Jaya (1000-1750 m.); P. runcinatum Asia to mainland New Guinea (2000 to 3500 m.); P. strigosum (c.1000 to 3000 m.)

c). widespread throughout Papuasia :-

P. barbatum (sea-level to 2000 m.); P. chinense mainland and Bougainville (1000-2500 m.); P. minus throughout Papuasia (sea level to 2500 m.).

Altitudinal ranges in metres are given in brackets after the species names.

2. The following species are suspected of causing disease or death in cattle: P. lapathifolium, P. minus, P. orientale. Several are reported as weeds of cultivation: P. chinense, P. minus, P. nepalense, P. runcinatum and P. strigosum. Several species have minor uses: P. attenuatum - as a substitute for tobacco; P. barbatum - - a source of an indigo-like dye; P. minus is recorded as a fish poison in Australia (Henty 1978).

3. The following species have been reported from Papuasia but are not accounted for in Henty (1978):

P. cumingianum Gandoger: Danser (1927) reports this species from the N coast of Irian Jaya. It is very closely related to P. attenuatum.

P. paniculatum Bl. and P. pubescens Bl. are both reported, presumably from near Port Moresby, by Mueller (1876: 59).

4. RUMEX

o o o o o o

A cosmopolitan genus with 170 - 200 species (depending on interpretation), found especially in N Hemisphere. It includes weedy species, culinary herbs and tannins. Seeds are long lived.

Perennial or annual herbs. Leaves spirally arranged,often with lower leaves in a basal rosette; sheaths short, ocreas well developed. Inflorescence racemose or paniculate; flowers bisexual, monoecious or dioecious, stipitate, articulate on arched pedicels, fasicles bracteate. Tepals 6, in 2 whorls of 3; tepals in outer whorl smaller; inner whorl much enlarged after anthesis, enclosing the fruit. Stamens 6, ovary trigynous; 3 short styles with large dissected stigmas. Fruit trigynous.

Key to Species in Papuasia

Source: Henty (1978)

1. At least one of the larger, inner tepals (in fruit) with a prominent basal boss; margins of tepals slightly toothed, some of the teeth mucronate.
..... Rumex crispus

1. None of tepals with basal boss; margins of tepals armed with strong hooked spines.
..... Rumex brownii

List of Species recorded from Papuasia

Rumex brownii Campd. Monogr. Rumex (1819) 64: 81; Danser (1927) 131
..... Rumex fimbriotus R. Br. (1810) non Poir

o o o o o o

Rumex crispus L. Sp. Pl. (1753)335; Danser (1927) 125, f. 1

o o o o o o

Rumex fimbriotus R. Br. (1810) non Poir
= Rumex brownii

Notes

1. R. brownii is widespread in Papuasia, occurring throughout mainland New Guinea and extending to Australia and Java. It occurs in disturbed, open habitats at 2000 to 3000 m altitude. Common on creek-banks, roadsides and gardens.

2. Rumex crispus, or curled dock, is native to Old World temperate regions, now almost a cosmopolitan weed. In Papuasia at altitudes around 1700 m. In Papuasia restricted to the E Highlands. It is harmful to stock causing dermatitis and gastric disturbances.

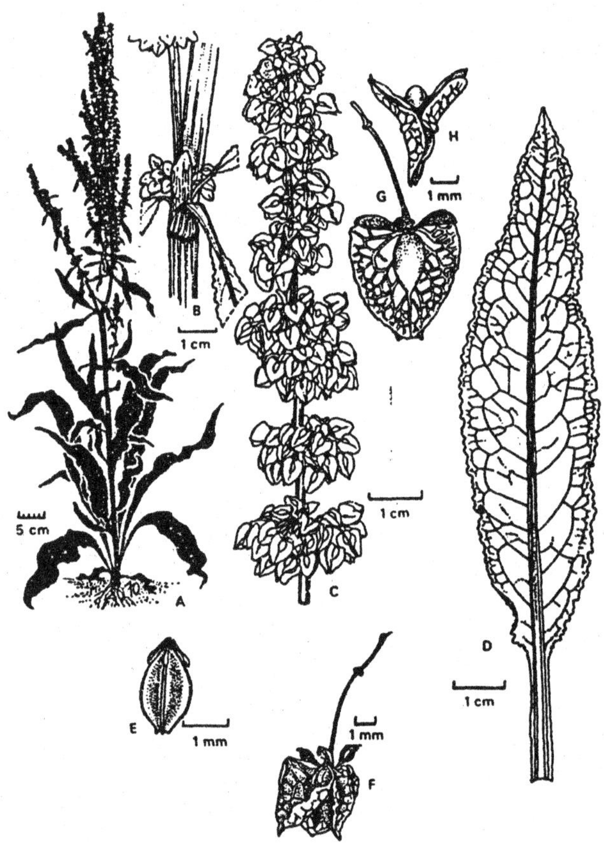

Rumex crispus A. habit; B. leaf base and ocrea; C. inflores cence; D. leaf; E.-H. fruit. (Source: Henty 1978)

PLUMBAGINALES

This order consists of a single family the Plumbaginaceae. There are some doubts as to the position of this family. A relationship to the Primulales (Dillenidae) had been suggested but serological studies by Froline and John in 1978 show that no relationship exists (Cronquist 1981).

PLUMBAGINACEAE

o o o o o o

A cosmopolitan family of some 10 to 12 genera and 400 - 560 species (the number depending on the taxonomic view of the larger genera). The family is best developed in the Mediterranean region, W and C Asia. In Papuasia only two genera are recorded, each represented by a single species. The family occurs particularly in saline and dry habitats, most species being xerophytic or maritime. The family was first recorded from the Middle Miocene.

Several species are used as medicinal plants: for example extracts from Plumbago europaea and P. scandens are used to treat dental ailments; skin diseases are treated with extract of P. zeylanica, and bronchial haemorrhages are treated with extracts of the roots of the European species, Limonium vulgare. Many plants are cultivated as garden ornamentals.

Diagnostic features

Leaves: either in a basal rosette or leaves alternate on branches and stems. Simple, glandular, without stipules.

Flowers: in a cymose or racemose inflorescence. Flowers are bisexual; regular, with parts in fives. Sepals 5, persistent, fused to form a 5-lobed tube which is often membranous, ribbed and coloured. Petals 5, free or connate at base, or fused into a long tube. Stamens 5, opposite the petals, free or inserted at base of corolla. Ovary superior, with 5 styles and 5 sessile stigmas.

Fruit: enclosed by calyx and normally indehiscent. Seeds with mealy endosperm.

PLUMBAGINACEAE

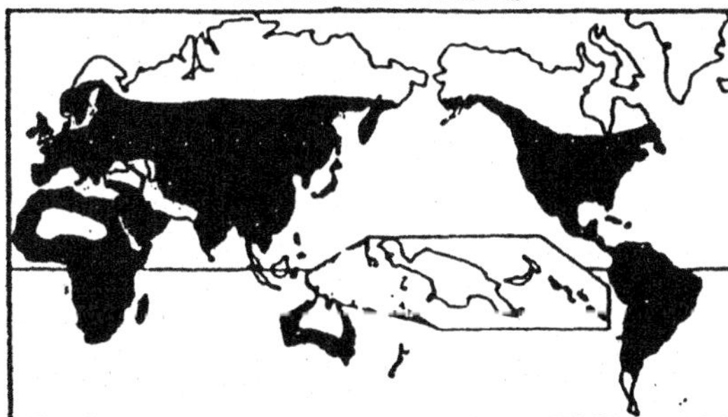

	WORLD	PAPUASIA
Number of genera :	22	2
Number of species :	44	2

Distribution : *Cosmopolitan, mainly dry or saline.*

Economic uses : *Ornamentals and medicinal plants.*

References

Mueller, F. von (1876). Descriptive Notes on Papuan Plants. 1: 58.

Pulle, A. (1910). Plumbaginaceae. Nova Guinea 8: 397.

Steenis, C. G. G. J. van (1937). Loodkruid. Trop. Natuur. 26: 12 - 13.

Steenis, C. G. G. J. van (1949). Plumbaginaceae. Flora Malesiana I. 4: 107 - 112.

Tomlinson, P. B. (1986). The Botany of Mangroves. Plumbaginaceae. pp. 310 - 312.

Warburg, O. (1891). Beitrage zur Kenntnis der papuanischen Flora. Bot. Jahrb. 13: 230 - 455 (p.400).

White, C. T. (1929) Ligneous Plants of Papua. Journ. Arn. Arb. 10: 197 - 274 (p. 258).

KEY TO GENERA IN PAPUASIA

1. Petiole alate throughout, clasping the branch. Blade orbicular. Style 5.
.... 1. AEGIALITIS

1. Petiole not winged. Blade not orbicular. Style 1.
.... 2. PLUMBAGO

1. AEGIALITIS

o o o o o o

A genus of 2 species from S. E. Asia to Australia. Associated with open mangrove communities. van Steenis (1949) stated that the genus consisted of 2 widely separated populations with differences based on unimportant flower characters. As no intermediates are known van Steenis accepted the populations as distinct species. Further studies are required. Note that there are ecological differences between the two populations. The genus is sometimes segregated into a distinct family because of its numerous distinctive features. This includes anomalous secondary thickening, abundant sclereids and viviparous seeds. As discussed by Tomlinson (1986) the greatest affinity is with Plumbago.

Simple-stemmed shrub to 3 m tall, branches with pith. Leaves alternate, orbicular, glabrous, entire, gland-dotted; nerves and veins parallel; petiole long, thick, winged, clasping the stem, leaving scars. Flowers solitary, in the axil of a bract. Bracts sheathing, enclosing 2 similar , persistent, boat-shaped sheathing bracts. Calyx persistent, with 5 prominent ribs. Corolla white, longer than calyx, lobes elongate-spathulate, connate at the base with the stamens. Stamens inserted at the apex of the tube; anthers basifix, c. as long as the corolla. Styles free, articulate at the base, stigma small, reaching just above the stamens. Capsule linear, long-exserted, dehiscing.

Key to Species in Papuasia

A single species:
..... Aegialitis annulata

List of Species recorded from Papuasia

Aegialitis annulata R. Br. Prod. 1 (1810) 426; Warb. (1891) 400; Pulle (1910) 397; White (1929) 258

o o o o o o

Notes

1. This species grows as isolated individuals or in small isolated colonies, usually in sandy and rocky places often on back edges of mangroves. It does not occur in closed mangrove communities; but it can withstand wave and tidal action. It occurs from N W Australia to E Malesia: S New Guinea, Aru Is., the Sunda Is., and Flores. Within Papuasia it occurs only along the south coast of New Guinea, having been collected at Idlers Bay, Daru, Merauke, Dauan Island and on the Aru Islands.

2. Aegialitis rotundifolia, which grows from India to the Malay Peninsula (?) and possibly in N Sumatra, occurs in low muddy mangrove communities. This species can be distinguished from A. annulata (characters in backets) by calyx 13 mm long (7-8 mm) and corolla with 12 lobes (9-10 lobes).

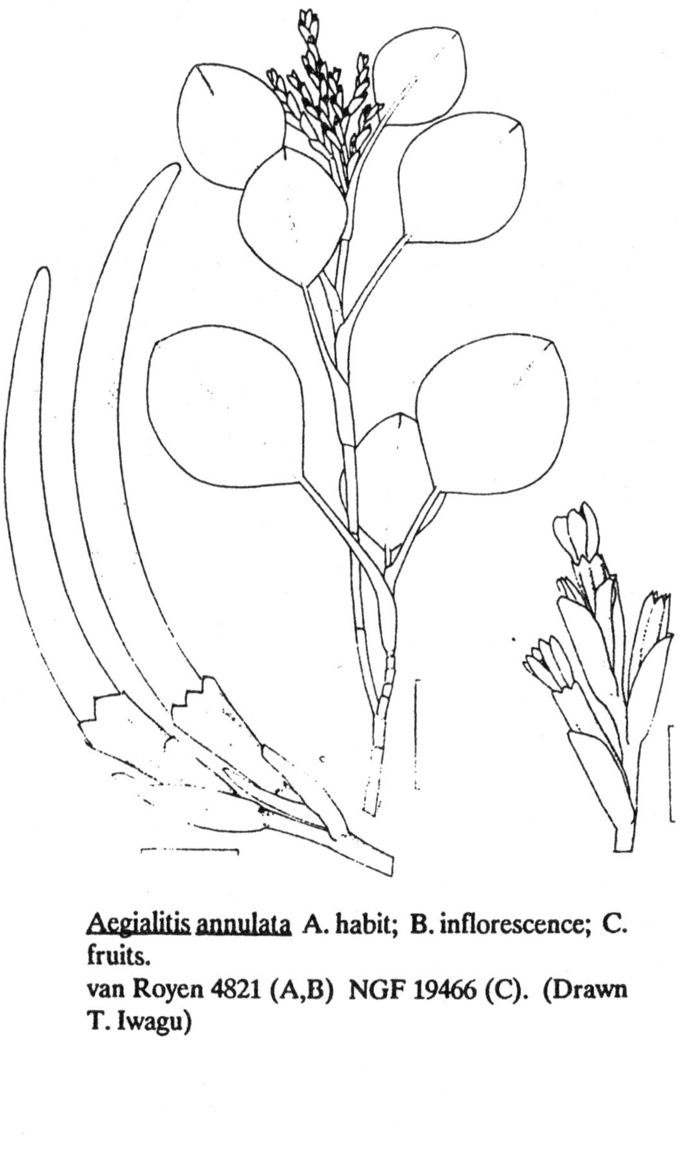

Aegialitis annulata A. habit; B. inflorescence; C. fruits.
van Royen 4821 (A,B) NGF 19466 (C). (Drawn T. Iwagu)

2. PLUMBAGO

o o o o o o

A pantropical genus of some 10 species, often used as ornamentals. Most grow in semi-arid conditions, not limited to saline localities. Propogates vegetatively. Strictly none are native in Papuasia, although P. zeylanica might be accepted as such.

Perennial herbs or undershrubs, often straggling. Leaves spread, entire, older leaves often pale-lepidote below (from excreted carbonates), or reduced on the flowering stems. Petiole often auriculate at the base. Flowers in terminal racemes or spikes, often united into a leafy panicle. Calyx tubular, outside often with glands, sessile or stalked; teeth erect, not enlarged in fruit. Corolla funnel-shaped, lobes spreading. Stamens free, broadened at base, Style short with 5 branches. Capsule included in the persistent calyx and corolla (often twisted); pericarp thin, often splitting towards the apex with 5 valves.

Key to Species in Papuasia

A single species:
..... Plumbago zeylanica

List of Species recorded from Papuasia

Plumbago zeylanica L. Sp. Pl. (1753) 151

o o o o o o

Notes

1. Distributed in the tropics of the Old World to Hawaii, possibly only indigenous in S E Asia. Cultivated. Always in anthropogenic localities to 1000 m altitude. In Papuasia collected at Rouna Falls in secondary forest; as a scrambling shrub in coconut plantations aroung Sialum, and in pockets of monsoon forest amongst eucalypt forest near Port Moresby.

Plumbago zeylanica A. habit; B. inflorescence; C. leaf sur face. NGF 49767. (Drawn T. Iwagu)

INDEX

Achyranthus 8
Aegialitis 55
Aerva 9
Aizoaceae 4
Alternanthera 10
Amaranthaceae 7
Amaranthus 11
Andredera 18
Antigonon 48
Basella 19
Basellaceae 17
Boerhavia 36
Bougainvillea 37
Calpidia 38
Caryophyllaceae 20
Celosia 13
Ceodes 39
Cerastium 21
Chenopodiaceae 27
Chenopodium 29
Cyathula 14
Drymaria 22
Glinus 33
Gomphrena 16
Halocnemium 30
Halosarcia 30
Heimerlia 39
Heimerliodendron 39
Molluginaceae 33
Mollugo 34
Montia 44
Muehlenbeckia 49
Nyctaginaceae 35
Phytolacca 43
Phytolaccaceae 42
Pisonia 39
Plumbaginaceae 54
Plumbago 56
Polycarpaea 23
Polygonaceae 47
Polygonum 50
Portulaca 45
Portulacaceae 43
Psilotrichum 17
Rumex 53
Sagina 24
Salsola 31
Scleranthus 25
Sesuvium 5
Spergula 26
Stellaria 26
Suaeda 31
Talinum 46
Tecticornia 32
Tetragonia 6
Trianthema 5

Index to Families (Contd.)

Menispermaceae	I
Mollugnaceae	III
Monimiaceae	I
Moraceae	IV
Moringaceae	IV
Myoporaceae	VIII
Myricaceae	II
Myristicaceae	I
Myrtaceae	V
Nepenthaceae	I
Nyctaginaceae	III
Nymphaeaceae	I
Ochnaceae	IV
Olacaceae	VI
Oleaceae	VIII
Opiliaceae	VI
Onagraceae	V
Orobanchaceae	VIII
Oxalidaceae	VII
Pandaceae	VI
Papaveraceae	I
Passifloraceae	VI
Pentaphragmataceae	IX
Peperomiaceae	I
Phytolaccaceae	III
Piperaceae	I
Pittosporaceae	V
Plantaginaceae	VIII
Plumbaginaceae	III
Podostemaceae	V
Polygalaceae	VII
Polygonaceae	III
Portulacaceae	III
Primulaceae	IV
Proteaceae	VI
Rafflesiaceae	VI
Ranunculaceae	I
Rhamnaceae	VI
Rhizophoraceae	V
Rosaceae	V
Rubiaceae	IX
Rutaceae	VII
Sabiaceae	VII
Santalaceae	VI
Sapindaceae	VII
Sapotaceae	IV
Saurauiaceae	IV
Saxifragaceae	V
Scropulariaceae	VIII
Simaroubaceae	VII
Solanaceae	VIII
Sonneratiaceae	V
Sphenostemonaceae	I
Stackhousiaceae	VI
Staphyleaceae	VII
Sterculiaceae	IV
Stylidaceae	IX
Styracaceae	IV
Symplocaceae	IV
Theaceae	IV
Thymelaeaceae	V
Tiliaceae	IV
Trimeniaceae	I
Tropaeolaceae	V
Ulmaceae	IV
Umbelliferae	VII
Urticaceae	IV
Valerianaceae	IX
Verbenaceae	VIII
Violaceae	IV
Viscaceae	VI
Vitaceae	VI
Winteraceae	I
Zygophyllaceae	VII

INDEX TO FAMILIES

Acanthaceae	VIII
Aizoaceae	**III**
Alangiaceae	VI
Alseuosmiaceae	IX
Amaranthaceae	**III**
Anacardiaceae	VII
Annonaceae	**I**
Apocynaceae	VIII
Aquifoliaceae	VI
Araliaceae	VII
Aristolochiaceae	**I**
Asclepiadaceae	VIII
Asteraceae (see Compositae)	IX
Avicenniaceae	VIII
Balanophoraceae	VI
Balsaminaceae	VIII
Basellaceae	III
Batidaceae	IV
Begoniaceae	IV
Bignoniaceae	VIII
Bixaceae	IV
Bombacaceae	IV
Boraginaceae	VIII
Burseraceae	VIII
Byblidaceae	IX
Campanulaceae	IX
Cannabidaceae	**II**
Capparaceae	IV
Caprifoliaceae	IX
Cardiopteridaceae	VI
Caricaceae	IV
Caryophyllaceae	III
Casuarinanceae	**II**
Celastraceae	VI
Ceratophyllaceae	**I**
Chenopodiaceae	III
Chloranthaceae	**I**
Chrysobalanaceae	V
Clethraceae	IV
Cochlospermaceae	IV
Combretaceae	V
Compositae	IX
Connaraceae	VII
Convolvulaceae	VIII
Coriariaceae	VII
Cornaceae	VI
Corynocarpaceae	VI
Crassulaceae	V
Cruciferae	IV
Cucurbitaceae	IV
Cunoniaceae	V
Daphniphyllaceae	**II**
Datiscaceae	IV
Dichapetalaceae	VI
Dilleniaceae	IV
Dipterocarpaceae	IV
Droseraceae	V
Ebenaceae	IV
Elaeagnaceae	VI
Elaeocarpaceae	IV
Elatinaceae	IV
Epacridaceae	IV
Ericaceae	IV
Erythroxylaceae	VII
Euphorbiaceae	VI
Eupomatiaceae	**I**
Fabaceae (see Leguminosae	V
Fagaceae	**II**
Flacourtiaceae	IV
Gentianaceae	VIII
Geraniaceae	VII
Gesneriaceae	VIII
Gonostylaceae	VI
Goodeniaceae	IX
Guttiferae	IV
Haloragaceae	V
Hamamelidaceae	**II**
Hernandiaceae	**I**
Himantandraceae	**I**
Hypericaceae	IV
Icacinaceae	VI
Juglandaceae	**II**
Labiatae (see Lamiaceae)	VIII
Lamiaceae	VIII
Lamiaceae	**I**
Lecythidaceae	IV
Leeaceae	VI
Leguminosae	V
Lentibulariaceae	VIII
Linaceae	VII
Lobeliaceae	IX
Loganiaceae	VIII
Lophopyxidaceae	VI
Loranthaceae	VI
Lythraceae	V
Magnoliaceae	**I**
Malphigiaceae	VII
Malvaceae	IV
Meliaceae	VII

CPSIA information can be obtained at www.ICGtesting.com
Printed in the USA
BVOW10s0315081215

429541BV00007B/4/P

9 789980 945853